Progress in Inflammation Research

More information about this series at http://www.springer.com/series/4983

Maria Chiara Maiuri • Daniela De Stefano
Editors

Autophagy Networks in Inflammation

Springer

Editors
Maria Chiara Maiuri
Centre de Recherche des Cordeliers
UMRS 1138 – Team 11
Paris
France

Daniela De Stefano
Multi Medica Fondazione Onlus
Milan
Italy

Progress in Inflammation Research
ISBN 978-3-319-30077-1   ISBN 978-3-319-30079-5 (eBook)
DOI 10.1007/978-3-319-30079-5

Library of Congress Control Number: 2016943729

Printed on acid-free paper

This Springer imprint is published by Springer Nature
The registered company is Springer International Publishing AG Switzerland

Somewhere, something incredible is waiting to be known.

Carl Sagan

Preface

Autophagy, a highly conserved process from yeast to mammals, is a cellular catabolitic process in which cellular components, including organelles and macromolecules, are delivered to the lysosomes for degradation. It can be classified in three principal types: microautophagy, chaperone-mediated autophagy (CMA), and macroautophagy. Microautophagy leads to degradation of sequestered portions of cytosol by direct invagination of lysosome membrane. CMA delivers cytosolic protein containing a KFERQ-like motif to the lysosomal lumen. Macroautophagy (hereafter referred to as autophagy) is the well-characterised form of autophagy that leads to the formation of a double-membraned vacuole, the autophagosome, which engulfs cytoplasmic contents (macromolecules and organelles) and fuses with lysosomes (autophagolysosomes) for cargo delivery and degradation. Materials degraded within autophagolysosomes are recruited to anabolic reactions in order to maintain energy levels and to provide macromolecules for the synthesis of more complex structures (nucleic acids, proteins, or organelles), thereby promoting cell metabolism, homeostasis, and survival. Autophagy represents for cells the crucial protective mechanism to both escape multiple stressors and aid organisms to defend against inflammatory-based, infectious, and neoplastic diseases. Intriguingly, autophagy's vital role is also strictly balanced with cell death. In fact, excessive or inefficient cell death frequently occurs during development of tissues as well as organs, or even in pathological conditions.

The aim of this book is to describe the state of the art about the *in fieri* comprehension of the interface between autophagy, immunity, and inflammation. We discuss how emerging concepts about the functions of the autophagy pathway and the autophagy proteins may reshape our understanding of immunity and diseases.

Chapter Organization/Brief Content of the Chapters

This book is composed of two parts. The first part is dedicated to the molecular mechanism underlying autophagy in inflammation, whereas the second part is consecrated to describe how targeting autophagy can represent a novel therapeutic strategy in inflammation-based pathologies. More specifically, the book aims to cover all the aspects of both basic autophagy and its role in the development/resolution of several diseases. Relevant scientists in this field have preciously contributed to create a comprehensive excursus, starting from the basics of autophagy and going in depth in highlighting complex pathways constituting a link between autophagy and inflammation and, even, the intricate role of autophagy in the immune response.

Part II addresses relevant pathological conditions, underpinning the importance of a balanced autophagy process. In fact, a *leitmotiv* is represented by the fact that autophagy may result both in excess and/or be defective, when not dysregulated.

The complexity of the autophagy process makes this book a guide for a wide audience representing a valuable support for undergraduate and graduate students who approach to the basic concepts and a precious tool for more experienced scientists who are moving toward the study of autophagy process in inflammation-based diseases.

Paris, France Maria Chiara Maiuri
Milan, Italy Daniela De Stefano

Contents

List of Contributors

Kelly Airiau INSERM U1218 ACTION, Institut Bergonié, Bordeaux, France

University of Bordeaux, Bordeaux, France

Johan Arnold CNRS UPR3572- Immunopathologie et Chimie Thérapeutique, Institut de Biologie Moléculaire et Cellulaire, Université de Strasbourg, Strasbourg, France

Izela Bernal-Sore Advanced Center for Chronic Diseases (ACCDiS), Facultad Ciencias Químicas y Farmacéuticas & Facultad Medicina, Universidad de Chile, Santiago, Chile

Mario Bustamante Advanced Center for Chronic Diseases (ACCDiS), Facultad Ciencias Químicas y Farmacéuticas & Facultad Medicina, Universidad de Chile, Santiago, Chile

Michela Campolo Department of Chemical, Biological, Pharmaceutical and Environmental Sciences, University of Messina, Messina, Italy

Ignacio Casado-Naranjo Centro de Investigación Biomédica en Red sobre Enfermedades Neurodegenerativas (CIBERNED), Cáceres, Spain

Departmento de Neurologia, Hospital San Pedro de Alcántara, Cáceres, Spain

Giovanna Casili Department of Chemical, Biological, Pharmaceutical and Environmental Sciences, University of Messina, Messina, Italy

Pablo F. Castro Advanced Center for Chronic Diseases, Facultad Medicina, Pontificia Universidad Catolica de Chile, Santiago, Chile

Virve Cavallucci Department of Experimental Neurosciences, IRCCS S. Lucia Foundation, Rome, Italy

Min Chen Department of Pathology and Immunology, Baylor College of Medicine, Houston, TX, USA

Alexander J. Clarke Kennedy Institute of Rheumatology, University of Oxford, Headington, Oxford, UK

Marika Cordaro Department of Chemical, Biological, Pharmaceutical and Environmental Sciences, University of Messina, Messina, Italy

Alfredo Criollo Laboratory of Cellular and Molecular Biology, Research Institute of Dental Science and Advanced Center for Chronic Diseases (ACCDiS), Faculty of Dentistry, University of Chile, Santiago, Chile

Salvatore Cuzzocrea Department of Chemical, Biological, Pharmaceutical and Environmental Sciences, University of Messina, Messina, Italy

Marcello D'Amelio Department of Experimental Neurosciences, IRCCS S. Lucia Foundation, Rome, Italy

Department of Medicine, University Campus Bio-Medico, Rome, Italy

Mojgan Djavaheri-Mergny INSERM U1218 ACTION, Institut Bergonié, Bordeaux, France

University of Bordeaux, Bordeaux, France

Emanuela Esposito Department of Chemical, Biological, Pharmaceutical and Environmental Sciences, University of Messina, Messina, Italy

Mathias Faure CIRI, International Center for Infectiology Research, Inserm, U1111, CNRS UMR5308, Université Lyon 1, Ecole Normale Supérieure de Lyon, Lyon, France

José M. Fuentes Centro de Investigación Biomédica en Red sobre Enfermedades Neurodegenerativas (CIBERNED), Cáceres, Spain

Departamento de Bioquímica y Biología Molecular y Genética, F. Enfermería y Terapia Ocupacional, Universidad de Extremadura, Cáceres, Spain

Rubén Gómez-Sánchez Centro de Investigación Biomédica en Red sobre Enfermedades Neurodegenerativas (CIBERNED), Cáceres, Spain

Departamento de Bioquímica y Biología Molecular y Genética, F. Enfermería y Terapia Ocupacional, Universidad de Extremadura, Cáceres, Spain

Lorena García Advanced Center for Chronic Diseases (ACCDiS), Facultad Ciencias Químicas y Farmacéuticas & Facultad Medicina, Universidad de Chile, Santiago, Chile

Jianning Ge Department of Molecular Microbiology and Immunology, Keck School of Medicine, University of Southern California, Los Angeles, CA, USA

Rosa A. González-Polo Centro de Investigación Biomédica en Red sobre Enfermedades Neurodegenerativas (CIBERNED), Cáceres, Spain

Departamento de Bioquímica y Biología Molecular y Genética, F. Enfermería y Terapia Ocupacional, Universidad de Extremadura, Cáceres, Spain

Frédéric Gros CNRS UPR3572- Immunopathologie et Chimie Thérapeutique, Institut de Biologie Moléculaire et Cellulaire, Université de Strasbourg, Strasbourg, France

Frédéric GROS, Institut de Biologie Moléculaire et Cellulaire, UPR3572 CNRS, Immunopathologie et Chimie Thérapeutique, Strasbourg, France

Seungmin Hwang Department of Pathology, The University of Chicago, Chicago, IL, USA

Valentina Izzo "Apoptosis, Cancer & Immunity", Team 11, INSERM U1138 Cordeliers Research Center, Paris, France

Cell Biology and Metabolomics Platform, Gustave Roussy Comprehensive Cancer Center, Villejuif, France

Xi Jin Department of Internal Medicine, University of Michigan, Ann Arbor, MI, USA

Eun-Kyeong Jo Department of Microbiology, Infection Signaling Network Research Center, College of Medicine, Chungnam National University, Jungku, Daejeon, South Korea

Rui Kang Department of Surgery, University of Pittsburgh Cancer Institute, University of Pittsburgh, Pittsburgh, PA, USA

Jin Kyung Kim Department of Microbiology, Infection Signaling Network Research Center, College of Medicine, Chungnam National University, Jungku, Daejeon, South Korea

Sergio Lavandero Advanced Center for Chronic Diseases (ACCDiS), Facultad Ciencias Químicas y Farmacéuticas & Facultad Medicina, Universidad de Chile, Santiago, Chile

Cardiology Division, Department of Internal Medicine, University of Texas Southwestern Medical Center, Dallas, TX, USA

Qiming Liang Department of Molecular Microbiology and Immunology, Keck School of Medicine, University of Southern California, Los Angeles, CA, USA

Györgyi Műzes 2nd Department of Internal Medicine, Semmelweis University, Budapest, Hungary

Christophe Macri Department of Biochemistry and Molecular Biology, Bio21 Molecular Science and Biotechnology Institute, The University of Melbourne, Parkville, VIC, Australia

Justine D. Mintern Department of Biochemistry and Molecular Biology, Bio21 Molecular Science and Biotechnology Institute, The University of Melbourne, Parkville, VIC, Australia

Eugenia Morselli Department of Physiology, Faculty of Biological Sciences, Pontificia Universidad Católica de Chile, Santiago, Chile

Sylviane Muller CNRS, Immunopathology and Therapeutic Chemistry/ Laboratory of Excellence MEDALIS, Institut de Biologie Moléculaire et Cellulaire, University of Strasbourg, Strasbourg, France

UPR3572 CNRS, Immunopathologie et Chimie Thérapeutique, Institut de Biologie Moléculaire et Cellulaire, Strasbourg, France

Diane Murera CNRS UPR3572- Immunopathologie et Chimie Thérapeutique, Institut de Biologie Moléculaire et Cellulaire, Université de Strasbourg, Strasbourg, France

Annalisa Nobili Department of Experimental Neurosciences, IRCCS S. Lucia Foundation, Rome, Italy

Department of Medicine, University Campus Bio-Medico, Rome, Italy

José M. Bravo-San Pedro Centro de Investigación Biomédica en Red sobre Enfermedades Neurodegenerativas (CIBERNED), Cáceres, Spain

Departamento de Bioquímica y Biología Molecular y Genética, F. Enfermería y Terapia Ocupacional, Universidad de Extremadura, Cáceres, Spain

Marcela Pizarro Advanced Center for Chronic Diseases (ACCDiS), Facultad Ciencias Químicas y Farmacéuticas & Facultad Medicina, Universidad de Chile, Santiago, Chile

Elisa Pizarro-Estrella Centro de Investigación Biomédica en Red sobre Enfermedades Neurodegenerativas (CIBERNED), Cáceres, Spain

Departamento de Bioquímica y Biología Molecular y Genética, F. Enfermería y Terapia Ocupacional, Universidad de Extremadura, Cáceres, Spain

Clara Quiroga Advanced Center for Chronic Diseases, Facultad Medicina, Pontificia Universidad Catolica de Chile, Santiago, Chile

Mario Rodríguez-Arribas Centro de Investigación Biomédica en Red sobre Enfermedades Neurodegenerativas (CIBERNED), Cáceres, Spain

Departamento de Bioquímica y Biología Molecular y Genética, F. Enfermería y Terapia Ocupacional, Universidad de Extremadura, Cáceres, Spain

Andrea E. Rodriguez Advanced Center for Chronic Diseases (ACCDiS), Facultad Ciencias Químicas y Farmacéuticas & Facultad Medicina, Universidad de Chile, Santiago, Chile

Valentina Sica Ecole doctorale de Cancerologie, University of Paris SUD, Paris, France

"Apoptosis, Cancer & Immunity", Team 11, INSERM U1138 Cordeliers Research Center, Paris, France

Cell Biology and Metabolomics Platform, Gustave Roussy Comprehensive Cancer Center, Villejuif, France

A. Katharina Simon Kennedy Institute of Rheumatology, University of Oxford, Headington, Oxford, UK

Ferenc Sipos 2nd Department of Internal Medicine, Semmelweis University, Budapest, Hungary

Daolin Tang Department of Surgery, University of Pittsburgh Cancer Institute, University of Pittsburgh, Pittsburgh, PA, USA

Rodrigo Troncoso Advanced Center for Chronic Diseases (ACCDiS), Facultad Ciencias Químicas y Farmacéuticas & Facultad Medicina, Universidad de Chile, Santiago, Chile

Instituto de Nutrición y Tecnología de los Alimentos (INTA), Universidad de Chile, Santiago, Chile

Christophe Viret CIRI, International Center for Infectiology Research, Inserm, U1111, CNRS UMR5308, Université Lyon 1, Ecole Normale Supérieure de Lyon, Lyon, France

Jin Wang Department of Pathology and Immunology, Baylor College of Medicine, Houston, TX, USA

Francisco Westermeier Advanced Center for Chronic Diseases (ACCDiS), Facultad Ciencias Químicas y Farmacéuticas & Facultad Medicina, Universidad de Chile, Santiago, Chile

Maud Wilhelm CNRS, Immunopathology and Therapeutic Chemistry/Laboratory of Excellence MEDALIS, Institut de Biologie Moléculaire et Cellulaire, University of Strasbourg, Strasbourg, France

Sokhna M.S. Yakhine-Diop Centro de Investigación Biomédica en Red sobre Enfermedades Neurodegenerativas (CIBERNED), Cáceres, Spain

Departamento de Bioquímica y Biología Molecular y Genética, F. Enfermería y Terapia Ocupacional, Universidad de Extremadura, Cáceres, Spain

Part I
Molecular Mechanisms Underlying Autophagy in Inflammation

The Basics of Autophagy

Rosa A. González-Polo, Elisa Pizarro-Estrella, Sokhna M.S. Yakhine-Diop, Mario Rodríguez-Arribas, Rubén Gómez-Sánchez, Ignacio Casado-Naranjo, José M. Bravo-San Pedro, and José M. Fuentes

Abstract Autophagy can be defined as a catabolic process that maintains cellular homeostasis by the degradation of damaged or excess cellular organelles and protein aggregates from the cytoplasm, thereby enabling cell survival. Cell culture and in vivo studies have revealed the importance of autophagy in numerous diseases, including cancer, aging, neurodegenerative, infectious and inflammatory diseases. Therefore, understanding the molecular basis of the formation and composition of the different structures involved in autophagy, as well as the regulation of this pathway, is an important goal for converting autophagy into a potential therapeutic target in a plethora of diseases.

Abbreviations

AD	Adenylate cyclase
Akt	v-akt murine thymoma viral oncogene homolog 1
Ambra1	Activating molecule in Beclin-1-regulated autophagy
AMPK	AMP-activated protein kinase
Atg	Autophagy-related genes
Bcl	B-cell lymphoma
Bnip3	Bcl-2/adenovirus E1B 19 kDa-interacting protein 3
CMA	Chaperone-mediated autophagy
DEPTOR	DEP-domain containing mTOR interacting protein

R.A. González-Polo (✉) • E. Pizarro-Estrella • S.M.S. Yakhine-Diop • M. Rodríguez-Arribas
R. Gómez-Sánchez • J.M. Bravo-San Pedro • J.M. Fuentes (✉)
Centro de Investigación Biomédica en Red sobre Enfermedades Neurodegenerativas (CIBERNED), Cáceres, Spain

Departamento de Bioquímica y Biología Molecular y Genética, F. Enfermería y Terapia Ocupacional, Universidad de Extremadura, Avda Universidad, s/n, Cáceres 10003, Spain
e-mail: rosapolo@unex.es; jfuentes@unex.es

I. Casado-Naranjo
Centro de Investigación Biomédica en Red sobre Enfermedades Neurodegenerativas (CIBERNED), Cáceres, Spain

Departmento de Neurologia, Hospital San Pedro de Alcántara, Cáceres 10002, Spain

M.C. Maiuri, D. De Stefano (eds.), *Autophagy Networks in Inflammation*,
Progress in Inflammation Research, DOI 10.1007/978-3-319-30079-5_1

4E-BP1	Translation initiation factor 4E-binding protein-1
Epac	Exchange protein directly activated by cAMP
ER	Endoplasmic reticulum
ERK1/2	Extracellular-signal-regulated kinase 1/2
ESCRT	Endosomal sorting complex required for transport
FIP200	Focal adhesion kinase family-interacting protein of 200 kDa
FoxO3	Forkhead box O3
GAP	GTPase-activating protein
GPCRs	G-protein-coupled receptors
HOPS	Homotypic fusion and protein sorting vacuoles
hVps	Mammalian homologue of vacuolar protein sorting
IKK	Inhibitor of nuclear factor κB kinase
IMPase	inositol monophosphatase
IP3R	Inositol 1,4,5-triphosphate receptor
I1R	Imidazoline-1 receptor
JNK1	c-Jun N-terminal kinase 1
LC3	Microtubule-associated protein light chain 3
LKB1	Liver Kinase B1
mLST8	Lethal mammalian protein SEC13 With 8
mSIN1	mammalian stress-activated protein kinase mitogen activated-interacting protein 1
mTOR	Mammalian target of rapamycin
mTORC	mTOR complex
p70S6K	Ribosomal protein S6 kinase-1
PDK1	Phosphoinositide-dependent kinase 1
PE	Phosphatidylethanolamine
PI3K	Phosphoinositide 3-kinase
PI3KC1a	Class Ia PI3K
PI3KC3	Class III PI3K
PI3KK	PI3K-related protein kinase
PRAS40	Proline-rich Akt substrate of 40 kDa
PTEN	Phosphatase and tensin homologue deleted from chromosome 10
Raptor	Regulatory-associated protein of mTOR
Rheb	Ras homologue enriched in brain
Rictor	Rapamycin-insensitive companion of mTOR
SLC	Solute carrier
SNARE	*N*-ethylmaleimide-sensitive factor-attachment protein receptor
TFEB	Transcription factor EB
TOR	Target of rapamycin
TSC	Tuberous sclerosis complex
UPS	Ubiquitin-proteasome system
ULK1	UNC-51-like kinase 1
UVRAG	UV irradiation resistance-associated gene
v-ATPase	Vacuolar H^+-ATPase
Vps	Vacuolar protein sorting

1 Introduction

The cell is in a continuous process of replacing organelles and proteins, and therefore, it is necessary to discard material that has been synthesized but is no longer beneficial to the cell. The correct maintenance of a balance between synthesis and degradation of cellular content is vital for cell survival. The cell has two mechanisms of degradation: the ubiquitin-proteasome system (UPS) and autophagy [12, 37, 41].

The term autophagy is derived from two Greek words: "*auto*", which means self, and "*phagia*", which indicates the action of eating (autophagy literally means "to eat oneself"). Autophagy is an intracellular catabolic mechanism, highly conserved through evolution, whereby the cell recycles or degrades damaged proteins or cytoplasmic organelles [42]. It was first described by Christian de Duve in the 1960s [4]; however, it was not until the 1990s when the genes involved in this process were

Table 1 Nomenclature in yeast and mammalian cells and function of each Atg protein

Atg proteins		
Yeast	Mammals	Functions
Atg1	ULK1,ULK2	Kinases that form the complex: Atg1-Atg13-Atg17-Atg29-Atg31 (phagophore initiation and organization)
Atg2	Atg2	Atg9/Atg2-Atg18 complex (phagophore formation)
Atg3	Atg3	Enzyme required for Atg8 lipidation
Atg4	Atg4 A, B, C, D	Cysteine protease: Atg8s activation and delipidation
Atg5	Atg5	Member of the Atg12-Atg5 complex, necessary for Atg8 lipidation
Atg6	Beclin-1	Subunit of PI3K-Vps34
Atg7	Atg7	Enzyme which is conjugated with Atg12 and LC3B
Atg8	LC3B, GATE-16, GABARAP	Conjugation with PE in the autophagosome
Atg9	Atg9 L1, L2	Interaction with Atg2-Atg8 complex binding to membrane
Atg10	Atg10	Conjugation with Atg12
Atg11	Atg11	Interaction with phospho-Atg29. Important for phagophore elongation
Atg12	Atg12	Atg5 complex
Atg13	Atg13	mTOR signaling from: Complex- Atg1-Atg13-Atg17-Atg29-Atg31 (phagophore initiation and organization)
Atg14	Barkor	Subunit of Vps34 PI3K complex, participates in reticulophagy
Atg15	¿?	Lipase located in the ER and is required for ER-derived vesicles fusion
Atg16	Atg16L	Complex Atg12-Atg5
Atg17	FIP200	Complex Atg1-Atg13-Atg17-Atg29-Atg31
Atg18	WIPI-1, 2, 3, 4	Complex Atg9/Atg2-Atg18
Atg29	¿?	Complex Atg17-Atg31-Atg29 (phagophore initiation and organization)
Atg31	¿?	Complex Atg17-Atg31-Atg29 (phagophore initiation and organization)

identified in yeast. These genes were thereafter named autophagy-related genes (*Atg* genes; Table 1, [20, 61]), and the discovery of these genes was a breakthrough in understanding the autophagic mechanism [57, 78]. At present, the number of works related to autophagy grows exponentially because such studies are revealing the importance of this mechanism in the development of many diseases, including cancer [38], cardiomyopathies [55], problems with skeletal muscle [49] or adipose tissue [75] and neurodegeneration [25, 48, 65, 86].

Paradoxically, although autophagy is initially a protective process for the cell, for example, in the recycling of protein aggregates that might be toxic [6], it also plays a key role in triggering cell death, the execution of a program of irreversible self-destruct [9].

There are also studies linking autophagy to aging processes [17, 73]. Thus, a high-calorie diet has been shown to accelerate the aging process compared to a low-calorie diet (without reaching malnutrition) [46]. Individuals with a low-calorie diet have a reduced chance of developing cancer, cardiovascular disease, diabetes and early mortality [14]. Autophagy is also interrelated with apoptosis [10, 47], as in the case of neurodegenerative disorders.

2 Basic Autophagy Machinery

During autophagy, the elimination of certain proteins or cellular organelles takes place when they are damaged or are no longer needed for the cell. For this purpose, the content is enclosed in membranous structures, which have different structural stages and subsequently fuse with lysosomes to degrade their content through the action of lysosomal enzymes. The origin of these membranes can be varied; there are tomographic studies relating ER (endoplasmic reticulum) with autophagosomes [88], although several studies point to the Golgi apparatus, the nucleus or even mitochondria as membrane sources.

To gain a better understanding of the autophagic process, we differentiated autophagy in several phases:

(a) *Initiation.* A start signal results in the implementation of autophagy. This initiation can occur by nutrient deprivation, lack of amino acids or fatty acids, or energy requirement [54, 78].
(b) *Nucleation.* A recruitment of the necessary Atg proteins to the membrane leads to the phagophore isolation. The origin of this pre-phagophore membrane seems to be related with ER by a structure called "omegasome" that links the ER with the inner membrane of phagophore [83].
(c) *Elongation.* Phagophore expansion occurs to engulf internalized material, resulting in a structure called an autophagosome. The autophagosome is closed in a double-membrane structure.
(d) *Fusion.* Autophagosomes merge with endosomes and lysosomes to proceed to eliminate material that has been captured by autophagosomes.
(e) *Maturation.* The degradation and release of the material and gradient occurs.

This process is very complex and is highly regulated by many proteins.

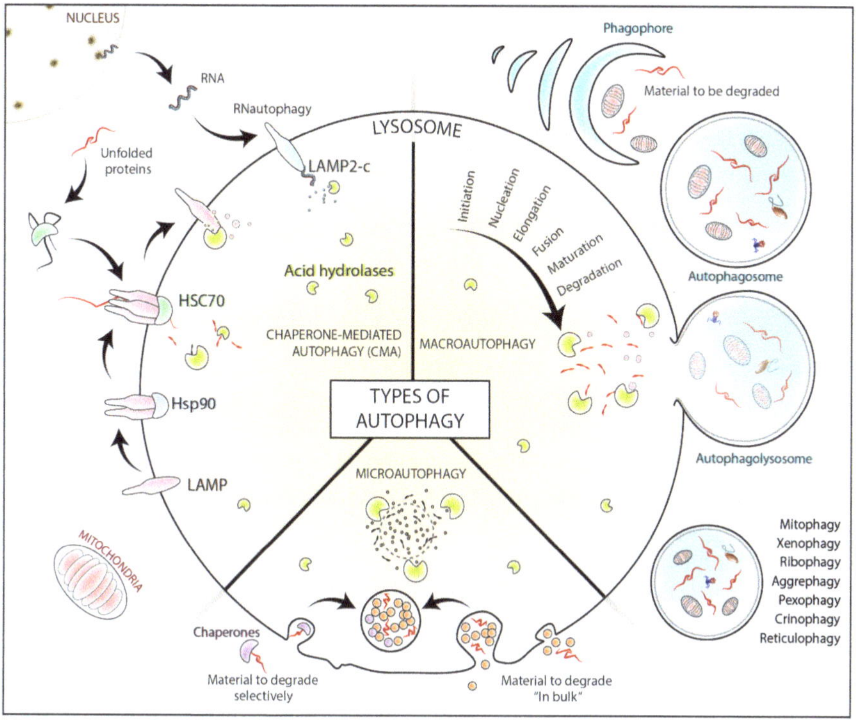

Fig. 1 Scheme of three different types of autophagy presents in mammalian cells. Chaperone-mediated autophagy (CMA) Macroautophagy and Microautophagy

3 Classification of Autophagy Types

We created a classification of the different types of autophagy, depending on the functional mechanism used to degrade the substrate into the lysosomal lumen [36]. Based on this approach, we were able to distinguish three types of autophagy (Fig. 1).

3.1 *Macroautophagy*

Often referred to as autophagy. In this process, the material to be degraded is sequestered into double-membrane vesicles called autophagosomes [5, 21]. A subset of classification can be developed depending on the type of material to be degraded within the autophagosomes:

(a) Mitophagy: kidnapping and selective degradation of mitochondria.
(b) Xenophagy: selective degradation of microbes (bacteria, fungi, parasites or virus).

(c) Ribophagy: kidnapping and selective degradation of ribosomes.
(d) Aggrephagy: selective degradation of protein aggregates.
(e) Pexophagy: kidnapping and selective degradation of peroxisomes (this may also occur in microautophagy).
(f) Crinophagy: direct fusion of secretory vesicles with lysosomes.
(g) Reticulophagy: kidnapping and selective degradation of ER.

3.2 *Microautophagy*

In this process, lysosomes do not fuse with autophagic vesicles but directly engulf cytosoplasmic cargos (selectively, using chaperones or "in bulk") via invaginations of the lysosomal membrane (Fig. 1). Next, there is vesicle scission into the lysosomal lumen, and degradation of the content occurs inside the lysosomes [52]. In yeast, microautophagy is constitutive, but it can also be induced by starvation or treatment with rapamycin (an mTOR inhibitor drug) [19]. Similar to macroautophagy, microautophagy functions also as a "housekeeping" mechanism for the degradation of cytosolic materials. Microautophagy can occur simultaneously with macroautophagy to regulate lysosomal membrane size. As macroautophagy can result in a large flow of membranes to lysosomes, microautophagy can regulate this flow and reduce lysosomal size by consuming lysosomal membranes [82]. Additionally, microautophagy was shown to play an important role in early mammalian development. Indeed, microautophagy was used to deliver endosomes to lysosomes in the visceral endoderm of mouse embryos [84]. This process was found to be important for the proper delivery of maternal nutrients as well as signaling molecules to the embryo.

3.3 *Chaperone-Mediated Autophagy*

Chaperone-mediated autophagy (CMA) differs from the macroautophagy pathway by the absence of vesicular traffic. Instead, only single proteins are delivered to lysosomes for degradation [2]. Macroautophagy and CMA pathway were shown to be coordinated processes. Indeed, the inhibition of macroautophagy induced up-regulation of CMA, even under basal conditions [34]. It has also been studied a link between CMA and aging, with a decrease in the activity of this type of autophagy in old organisms [72]. CMA degrades only unfolded protein substrates with a KFERQ or a KFERQ-like motif, which are present in approximately 30 % of all cytosolic proteins [16, 68]. CMA was first described in human fibroblasts cultured under starvation conditions [16]. CMA can also be induced *in vitro* in the presence of isolated lysosomes using a protein substrate with the specific pentapeptide sequence and a molecular chaperone complex [81]. CMA occurs in mammalian cells but not in yeast [44]. In mammalian cells, chaperone heat shock cognate protein 70 (HSC70) and other cooperating chaperones recognize cytosolic proteins with a KFERQ motif, unfold these proteins, and target them to lysosomal membranes. The interaction between the chaperone complex and the lysosomal- associated membrane protein 2 (LAMP-2) channel present at the lysosomal membrane facilitates the translocation of unfolded proteins to the lysosomal lumen, where they are rapidly degraded by lysosomal proteases [2]. The translocation of unfolded proteins through LAMP-2 requires another chaperone protein, HSC70 (lys- HSC70), a protein present in the lysosomal lumen [3]. Once proteins are captured by the lysosome, the HSC70-cochaperone complex is released from the lysosomal membrane and is now available to bind other cytosolic proteins with a KFERQ motif [2, 33].

As a variation to the theme, a novel type of autophagy called RNautophagy was recently described [23]. RNautophagy is a RNA degradation process by the lysosome. Unlike CMA, RNautophagy is independent of chaperone complexes and uses LAMP2-c.

4 Regulation of Autophagy Process

Autophagy is a mechanism with very complex regulation, and there are still many gaps in the understanding of the mechanism behind autophagy. This is because there is an extensive network of routes with a potential regulation of cellular recycling process.

4.1 Formation of Autophagosomes

Autophagy is a highly conserved mechanism from yeast to mammals where Atg proteins are orchestrating the initiation, elongation, maturation and melting of autophagosomes [35]. There are two systems of ubiquitin conjugation involved in the formation of autophagosomes.

In a first step, the C-terminal domain of Atg12 is activated by Atg7 (E1-like), forming the intermediate Atg7-Atg12. Subsequently, the protein Atg12 is transferred to Atg10 to form the Atg12-Atg10 complex. The C-terminal domain of Atg12 covalently binds a lysine [11] in Atg5. This new (Atg12-Atg5) complex binds noncovalently to the homologous protein of yeast Atg16L [53]. The Atg12/Atg5/Atg16L complex resides mostly in the cytosol, but a small fraction is at the autophagosome membrane. During the process of distribution and elongation, the Atg12/Atg5/Atg16L associating asymmetric complex is largely localized to the outer membrane. Finally, this complex dissociates from the membrane upon completion of autophagosome formation [83]. Some of the LC3 protein (microtubule-associated proteins 1A/1B light chain 3A) in mammals, or yeast Atg8, undergoes post-translational modifications before joining the autophagosomal membrane. Immediately after synthesis, 22 and 5 amino acids are removed from the C-terminal domain in rats and in humans, respectively. This processing is catalyzed by the protein Atg4. The processed form, LC3-I, resides mainly in the cytosol and retains a glycine residue at the C-terminus. Upon activation by Atg7 protein, which also functions as an activating enzyme for Atg12, the LC3-I is transferred to an E2 enzyme ubiquitination system, homologous to Atg3, and conjugated with phosphatidylethanolamine (PE). The final product, called LC3-II, is associated with the precursor membrane, and unlike Atg12/Atg5/Atg16L complex, remains bound to the inner membrane after complete formation of the autophagosome and is degraded by fusion with the lysosome. LC3-II protein found in the cytosolic side of the autolysosome membrane is delipidated by Atg4 [32, 39, 79]. The sequestosome 1 protein (SQSTM1, also known as p62) is a ubiquitin-binding protein involved in oxidative stress and autophagy. The p62 protein forms aggregates that may be degraded through its autophagosomal engulfment; mediated by its interaction with autophagosome membrane bound LC3. Lysosomal degradation of autophagosomes leads to decreased levels of LC3-II and p62 as part of autophagic flow [8, 58, 63].

In mammalian cells, there are different classes of PI3K (phosphatidylinositide 3-kinase) that regulate autophagy. PI3K class III (PI3KC3) activity is essential for the biogenesis of autophagosomes, whereas the activity of Class I PI3K (PI3KC1) can be stimulated through the mTOR pathway and inhibit autophagy [62]. hVps34 kinase phosphorylates phosphatidylinositol to form phosphatidylinositol-3-phosphate (PtdIns3P), which enables the synthesis of autophagosomes. hVps34 kinase is part of the macro-initiator complex that includes autophagy BECN1/Atg6 and Atg14L. hVps34 activity increases when it interacts with BECN1, which has several binding partners that direct autophagosome formation [50]. When BECN1 interacts with Atg14L, Ambra1 (activating molecule in Beclin 1-regulated autophagy), UVRAG (UV irradiation resistance-associated gene) and Bif-1 (endophilin B1), it positively regulates autophagy. While interacting with Rubicon, anti-apoptotic proteins Bcl-2-BCL-XL and pro-apoptotic Bim protein, it negatively regulates autophagy [27]. The GTPase Rab5, also binds to this complex, macro-activating hVps34 to induce autophagy [64]. The ULK1-ATG13-FIP200 macro-complex is also involved in autophagosome biogenesis, and the mechanisms involved in this regulation are described below. The Atg9 protein localized in the trans-Golgi network is responsible for the elongation of the phagophore for autophagosome formation [89]. Autophagosomes first join late endosomes to form amphisomes, which are merged with lysosomes forming autolysosomes. These fusion events are mediated by a complex containing Rab7/HOPS (Homotypic fusion and protein sorting vacuoles), ESCRTs (endosomal sorting complexes required for transport), SNARE (N-ethylmaleimide-sensitive factor-attachment protein receptors) and class C vps proteins. In forming the amphisome, the bond between the vesicle membranes is directed by the GTPase Rab7 complex and its effector/activator HOPS. The SNARE and ESCRT complex regulates the formation of amphisomes and the maturation of autophagosomes [66, 77]. Although the mechanisms involved in the formation and regulation of autophagosomes are relatively well-characterized, much less is known about the general organization of the route and to what extent the various functional elements communicate with each other. A proteomic analysis of human cells at baseline autophagy has revealed the complexity of the interactions, revealing a network of 751 interactions among 409 candidates, with extensive connectivity among subnetworks. Many new components of the interaction network are involved in vesicle trafficking, lipid and protein phosphorylation and ubiquitination [7].

4.2 *Regulation of Autophagy by mTOR-Dependent Pathway*

mTOR is one of the most important kinases in regulating cellular response to a decrease or absence of nutrient sensors. mTOR is a serine/threonine kinase that is part of two protein complexes: mTORC1 and mTORC2 [74]. The mTORC1 complex comprises Raptor (Regulatory-associated protein of mTOR), PRAS40 (proline-rich Akt substrate of 40 kDa), mLST8 (lethal mammalian protein SEC13 With 8) and DEPTOR (DEP-domain containing mTOR-interacting protein). The mTORC2 complex comprises Rictor (rapamycin-insensitive companion of mTOR), Protor (Observed With rictor protein), mSIN1 (mammalian stress-activated protein kinase mitogen activated-interacting protein 1), DEPTOR and mLST8. In these complexes, mTOR plays an important role in the regulation of autophagy as its kinase activity is very sensitive to the depletion of nutrients, nitrogen, ATP or rapamycin. Unlike the mTORC1 complex, mTOR activity in mTORC2 is not inhibited by rapamycin [43, 92]. During periods of starvation, mTORC1 activity is inhibited and induces autophagy to recycle cellular components, acting as a power source. mTOR inhibition and induction of autophagy is associated with reduced phosphorylation of two downstream effectors of mTOR, p70S6K (ribosomal protein S6 kinase-1) and 4E-BP1 (translation initiation factor 4E-binding protein-1) at Thr389/Thr421/Ser424 and Thr37/Thr46, respectively [31]. We will briefly describe the mechanisms controlling autophagy that are linked to mTOR.

4.2.1 Regulation of Autophagy by Nutrient Deficiency

Amino acids are necessary for the activation of mTOR. However, the mechanism by which mTOR detects the intracellular levels of amino acids has not been fully clarified. Recent studies have demonstrated that the cellular uptake of L-glutamine and its subsequent discharge in the presence of the essential amino acids is the limiting step that activates mTOR. The absorption of L-glutamine is regulated by the SLC1A5 amino acid transporter. SLC1A5 inactivation inhibits cell growth and activates autophagy. The bidirectional transporter SLC7A5/SLC3A2 regulates the transport of L-leucine and amino acids into the cell simultaneously and secretes L-glutamine from cells. Intracellular levels of L-glutamine modulate the activity of SLC1A5 and SLC7A5/SLC3A2 transporter, thus providing a switch for absorbing nonessential amino acids and regulating mTORC1 signaling [30, 51, 56, 69].

The GTPase Rag (Ras-related GTP-binding protein) exists as a heterodimer between RagA or RagB joined to RagC or RagD. The Rag GTPases with Ragulator and v-ATPase form a complex on the surface of lysosomes that conducts mTOR signaling when the amino acid concentration varies. When the amino acid concentration is low, the Rag GTPases remain in an inactive state, where the RagA/B complex is bound to GDP and the RagC/D complex to GTP. In the presence of amino acids, Rag GTPases undergo an active conformation change where the RagA/B complex is bound to GTP and the RagC/D complex to GDP. The active Rag heterodimer physically interacts with Raptor, kidnapping and activating mTORC1 on the surface of lysosomes and thereby negatively regulating autophagy [56]. The recruitment of mTORC1 to the lysosomal surface promotes the interaction with TFEB (transcription factor EB) and its phosphorylation and inactivation. When there are nutritional deficiencies, mTORC1 frees TFEB, which is targeted to the nucleus, thus activating the expression of related pathways and autophagy lysosomal degradation genes [69].

Low glucose levels also promote mTORC1 translocation to the cytoplasm. When glucose levels are normalized, mTORC1 is relocated to lysosomal surface. However, in cells expressing constitutively active Rag, the mTORC1 lysosome is located at the surface regardless of glucose concentrations. The v-ATPase interaction regulates lysosomal Ragulator lysosome surface in response to the availability of amino acids and glucose. Rag GTPases are multi-input sensors of nutrients, converging amino acids and glucose, in a v-ATPase dependent manner, upstream of mTORC1.

4.2.2 Regulation of Autophagy by ULK1

Under conditions of nutrient abundance, mTOR inhibits autophagy through direct binding to the ULK1-ATG13-FIP200 complex, thereby inactivating phosphorylation activity of Atg13 and ULK1 kinase. Under conditions of starvation or treatment with rapamycin, mTOR dissociates from this complex, leading to dephosphorylation and activation of the kinase ULK1.

ULK1 phosphorylates Atg13, FIP200 and itself, triggering the start of autophagy. Various signals, such as the incorporation of amino acids, growth factors, a high rate of ATP/AMP turnover, activate mTORC1 and inhibit autophagy [69].

4.2.3 Regulation of Autophagy by Growth Factors

The most common signaling pathway responsible for the regulation of the mTORC1 is the PI3KC1a pathway [62]. The binding of growth factors, insulin, integrins, proteins GPCRs (G-protein-coupled receptors) or Ras oncogenes to their respective receptors activate the PI3KC1a membrane complex. The PI3KC1a complex catalyzes the conversion of PtdIns (4,5) P2 to PtdIns (3,4,5) P3. The PtdIns (3,4,5) P3 binds to proteins with PH domain (pleckstrin homology), the serine/threonine kinase PDK1 (phosphoinositide-dependent kinase 1) and Akt (v-akt murine thymoma viral oncogene homolog 1), thus activating and translocating proteins to the plasma membrane. This activation inhibits autophagy. The tumor suppressor PTEN (phosphatase and tensin homologue deleted from chromosome 10) antagonizes dephosphorylating

PI3KC1a PtdIns (4,5) P2 and PtdIns (3,4,5) P3, thereby preventing the activation of Akt and PDK1 and ultimately activating autophagy [18, 45]. Activation of Akt also activates the mTORC1 complex through the TSC/Rheb pathway. The TSC1/2 complex (tuberous sclerosis complex) is a GAP (GTPase-activating protein) that inhibits Rheb function by binding GDP. Akt activation induces phosphorylation of TSC2 GAP and inhibits its activity on Rheb and the TSC1/2 complex formation, thus allowing Rheb to bind directly to activated mTORC1 [28, 87]. FoxO3 transcription factor (forkhead box) in its active conformation promotes autophagy through the gene transcription of LC3, Bnip3 (Bcl-2/adenovirus E1B 19 kDa-interacting protein 3), Vps34 and ULK1. One mechanism by which autophagy can inhibit Akt is by phosphorylation and the consequent inactivation of the transcription factor FoxO3, thereby impeding its translocation to the nucleus [91].

4.2.4 Regulation of Autophagy by Cellular Energy Charge

During periods of metabolic stress, autophagy activation is essential for cell viability. In mammalian cells, the reduction of ATP levels is detected by the kinase AMPK (5′-AMP-activated protein kinase). AMPK is activated by the upstream kinase LKB1 (Liver Kinase B1) when the ratio of ATP/AMP decreases. This activation induces phosphorylation and activation of TSC1/2, which inhibits the activity of mTORC1 through Rheb [29]. Autophagy activation through the negative regulation of mTOR results in an increased production of ATP by recycling nutrients. Additionally, the LKB1-AMPK pathway phosphorylates and activates p27kip1, an inhibitor of cdk (cyclin-dependent kinase), and induces a stop in the cell cycle, essential to prevent cell death and promote survival in response to the bioenergetic stress of nutrient deprivation [40]. The nitrosative stress-mediated nitric oxide (NO) inhibits autophagosome synthesis. NO inhibits the kinase IKKβ (inhibitor of nuclear factor kB kinase β), which leads to a drop in the activity of AMPK and TSC1/2 and the activation of mTORC1 [71]. Activation of IKKβ phosphorylates AMPK and JNK1 to stimulate autophagy in a process independent of nuclear factor NF-kB [15]. Cytoplasmic p53 inhibits autophagy, possibly through the AMPK/TSC/mTORC1 pathway. When low levels of glucose are detected, p53 is activated through AMPK mediated phosphorylation, thus inhibiting the activity of mTORC1 [80]. AMPK kinase is involved in hypoxia-induced autophagy. During hypoxia, mitochondrial respiration is altered causing a decrease in the ratio of ATP/AMP, which is detected by AMPK. AMPK activation results in the inhibition of mTORC1 through TSC2 [59].

4.3 Regulation of Autophagy by mTOR-Independent Pathways

In addition to the regulation of autophagy by mTORC1 and various signaling pathways and the downstream affects, several signaling pathways resulting in autophagy independent of mTORC1 and susceptible to chemical perturbations have been described.

4.3.1 Regulation of Autophagy by the Inositol Pathway

The inositol pathway is stimulated by PLC (phospholipase C), which hydrolyzes PtdIns(4,5)P_2 to form Ins(1,4,5)P_3 and DAG (diacylglycerol). Ins(1,4,5)P_3 operates as a second messenger and binds to its receptor (IP_3R) in the ER, releasing Ca^{2+} from intracellular stores, primarily cytoplasmic ER. The IP_3R pore driver comprises only 5 % of the receptor and is positioned at the C-terminus of the protein, whereas the ligand binding site for Ins(1,4,5)P_3 is located between amino acids 226–578 in the amino terminal [76]. Ins(1,4,5)P_3 is degraded by a 5′-phosphatase and IPPsdr (inositol polyphosphate 1-phosphatase) to form InsP, which is hydrolyzed by IMPase (inositol monophosphatase) and the free inositol required for signaling via this route. High intracellular concentrations of Ins(1,4,5)P_3 inhibit the synthesis of autophagosomes [70]. Mood stabilizing drugs, such as lithium, carbamazepine or valproic acid decrease intracellular inositol by inactivation of IMPase and facilitate the removal of substrates without the inhibition of mTORC1 [85].

4.3.2 Regulation of Autophagy by cAMP

It is likely that the regulation of autophagy through intracellular levels of Ins(1,4,5)P_3 is due to the release of calcium from the ER, through IP_3R. Elevated cytosolic calcium levels by activated calpain, which activates G protein coupled (Gαs) receptors, increases the activity of the AD (adenylate cyclase), thus directly influencing cAMP levels [60]. Elevated intracellular cAMP levels inhibit autophagy, and this response is mediated through Epac (exchange protein activated by cAMP directly). Epac activation through Rap2B in turn activates a G protein Ras family, which induces the hydrolysis of PtdIns(4,5)P_2 to Ins(1,4,5)P_3 through PLCε(191). Inhibition of AD activity at 2′5′ddA induces autophagy, and autofagosomal degradation increases by a route independent of mTORC1. Clonidine and tilmenidine, drug agonists for the imidazole-1(I1R) receptor, both induce autophagy by an mTOR-independent pathway, thereby reducing cAMP levels [85].

4.3.3 Regulation of Autophagy by JNK1/BECN1

BECN1 is a 60 kDa protein required for the activation key and has various bonding patterns that regulate autophagy. The post-translational modification of BECN1 along with its association with other proteins results in different PI3KC3 complexes that regulate autophagy. Proteins that are able to bind to BECN1 include hVps34, UVRAG, AMBRA1, Bif-1, Rubicon, IP3R, ATG14L/Barkor and Bcl-2 [27]. Autophagosome synthesis requires PI3CK3 activity, which is increased by the interaction of hVps34-BECN1 [67]. The BECN1-UVRAG and BECN1-ATG14L complexes are involved in the regulation of the early steps of autophagy by activating the formation of autophagosomes. However, it has been observed that the interaction of UVRAG with Rubicon inhibits autophagy by an arrest in autophagosome formation [90]. AMBRA1 is associated with ATG14L in the PI3KC3 complex and acts as a positive modulator of autophagy [22]. Bcl-2 is to date the only protein that inhibits autophagy by associating directly with BECN1. Interaction with Bcl-2 prevents BECN1 assembly in the hVps34 BECN1 complex and inhibits autophagy. Therefore, both proteins are subjected to various post-translational modifications for their interaction [13]. Phosphorylation of Bcl-2 at residues Thr69, Ser70 and Ser87 by JNK1 (c-Jun N-terminal kinase 1), in the absence of nutrients, allows BECN1 cleavage and subsequent activation of autophagy through the formation of the BECN1/hVps34 complex. Nutrient deficiency also inhibits mTORC1, but the expression of constitutively active JNK1 does not disrupt mTORC1 activity. Rapamycin, moreover, does not affect phosphorylation of Bcl-2 by JNK1, suggesting that the regulation of the autophagy routes through JNK1/BECN1/PI3KC3 and mTOR are possibly independent [69]. The mono-ubiquitination of Bcl-2 by the E3 ubiquitin kinase Parkin stabilizes the binding of Bcl-2 with BECN1, increases the half-life of Bcl-2 and decreases autophagy activation, suggesting that the mono-ubiquitination of Bcl-2 inhibits autophagy and increases the amount of Bcl-2 available for binding to BECN1 [1].

4.3.4 Control of Autophagy by Calcium

The increase in cytosolic Ca^{2+} levels has a complex effect on the regulation of autophagy, involving both autophagosome formation and fusion of autophagosomes with lysosomes [26]. Treatments using thapsigargin (an inhibitor of ER Ca^{2+}/Mg^{2+} ATPase) or ionomycin (Ca^{2+} ionophore) block the autophagic flux and slows the breakdown of the contents of autophagosomes [24]. Gsα protein regulates cAMP levels through the activation of the AD. Pharmacological inhibition of calpains (calcium-activated neutral proteinase) with calpastatin and calpeptin or by gene silencing increases autophagic flow without disruption of mTORC1. By contrast, activation of calpains 1 and 2 to open Ca^{2+} channels, or the overexpression of constitutively active calpain 2, inhibit the formation of autophagosomes [85].

Acknowledgements Rosa-Ana González-Polo was supported by a talent research contract (Junta de Extremadura, TA13009, Spain) and received research support from the Ministerio de Ciencia e Innovación, Spain (PI14/00170). Dr. José M. Fuentes received research support from the Ministerio de Ciencia e Innovación, Spain, CIBERNED (CB06/05/004), Consejería, Economía, Comercio e Innovación Junta de Extremadura (GR15045), Ministerio de Ciencia e Innovación, Spain (PI150034). This work is supported also by "Fondo Europeo de Desarrollo Regional" (FEDER), from European Union. We all thanks to FUNDESALUD for helpful assistance.

References

1. Abrahamsen H, Stenmark H, Platta HW (2012) Ubiquitination and phosphorylation of beclin 1 and its binding partners: tuning class III phosphatidylinositol 3-kinase activity and tumor suppression. FEBS Lett 586(11):1584–1591. doi:10.1016/j.febslet.2012.04.046
2. Agarraberes FA, Dice JF (2001) A molecular chaperone complex at the lysosomal membrane is required for protein translocation. J Cell Sci 114(Pt 13):2491–2499

3. Agarraberes FA, Terlecky SR, Dice JF (1997) An intralysosomal hsp70 is required for a selective pathway of lysosomal protein degradation. J Cell Biol 137(4):825–834. doi:10.1083/jcb.137.4.825
4. Ashford TP, Porter KR (1962) Cytoplasmic components in hepatic cell lysosomes. J Cell Biol 12:198–202
5. Baba M, Takeshige K, Baba N, Ohsumi Y (1994) Ultrastructural analysis of the autophagic process in yeast: detection of autophagosomes and their characterization. J Cell Biol 124(6): 903–913
6. Baehrecke EH (2005) Autophagy: dual roles in life and death? Nat Rev Mol Cell Biol 6(6):505–510. doi:10.1038/nrm1666
7. Behrends C, Sowa ME, Gygi SP, Harper JW (2010) Network organization of the human autophagy system. Nature 466(7302):68–76. doi:10.1038/nature09204
8. Bjorkoy G, Lamark T, Johansen T (2006) p62/SQSTM1: a missing link between protein aggregates and the autophagy machinery. Autophagy 2(2):138–139
9. Bursch W (2001) The autophagosomal-lysosomal compartment in programmed cell death. Cell Death Differ 8(6):569–581. doi:10.1038/sj.cdd.4400852
10. Bursch W, Ellinger A, Gerner C, Frohwein U, Schulte-Hermann R (2000) Programmed cell death (PCD). Apoptosis, autophagic PCD, or others? Ann N Y Acad Sci 926:1–12
11. Buytaert E, Callewaert G, Vandenheede JR, Agostinis P (2006) Deficiency in apoptotic effectors Bax and Bak reveals an autophagic cell death pathway initiated by photodamage to the endoplasmic reticulum. Autophagy 2(3):238–240
12. Ciechanover A (2005) Proteolysis: from the lysosome to ubiquitin and the proteasome. Nature Rev Mol Cell Biol 6(1):79–87. doi:10.1038/nrm1552
13. Ciechomska IA, Goemans GC, Skepper JN, Tolkovsky AM (2009) Bcl-2 complexed with Beclin-1 maintains full anti-apoptotic function. Oncogene 28(21):2128–2141. doi:10.1038/onc.2009.60
14. Colman RJ, Anderson RM, Johnson SC, Kastman EK, Kosmatka KJ, Beasley TM, Allison DB, Cruzen C, Simmons HA, Kemnitz JW, Weindruch R (2009) Caloric restriction delays disease onset and mortality in rhesus monkeys. Science 325(5937):201–204. doi:10.1126/science.1173635
15. Criollo A, Senovilla L, Authier H, Maiuri MC, Morselli E, Vitale I, Kepp O, Tasdemir E, Galluzzi L, Shen S, Tailler M, Delahaye N, Tesniere A, De Stefano D, Younes AB, Harper F, Pierron G, Lavandero S, Zitvogel L, Israel A, Baud V, Kroemer G (2010) The IKK complex contributes to the induction of autophagy. EMBO J 29(3):619–631. doi:10.1038/emboj.2009.364
16. Chiang HL, Dice JF (1988) Peptide sequences that target proteins for enhanced degradation during serum withdrawal. J Biol Chem 263(14):6797–6805
17. Chin RM, Fu X, Pai MY, Vergnes L, Hwang H, Deng G, Diep S, Lomenick B, Meli VS, Monsalve GC, Hu E, Whelan SA, Wang JX, Jung G, Solis GM, Fazlollahi F, Kaweeteerawat C, Quach A, Nili M, Krall AS, Godwin HA, Chang HR, Faull KF, Guo F, Jiang M, Trauger SA, Saghatelian A, Braas D, Christofk HR, Clarke CF, Teitell MA, Petrascheck M, Reue K, Jung ME, Frand AR, Huang J (2014) The metabolite alpha-ketoglutarate extends lifespan by inhibiting ATP synthase and TOR. Nature 510(7505):397–401. doi:10.1038/nature13264
18. Degtyarev M, De Maziere A, Orr C, Lin J, Lee BB, Tien JY, Prior WW, van Dijk S, Wu H, Gray DC, Davis DP, Stern HM, Murray LJ, Hoeflich KP, Klumperman J, Friedman LS, Lin K (2008) Akt inhibition promotes autophagy and sensitizes PTEN-null tumors to lysosomotropic agents. J Cell Biol 183(1):101–116. doi:10.1083/jcb.200801099
19. Dubouloz F, Deloche O, Wanke V, Cameroni E, De Virgilio C (2005) The TOR and EGO protein complexes orchestrate microautophagy in yeast. Mol Cell 19(1):15–26. doi:10.1016/j.molcel.2005.05.020
20. Feng Y, He D, Yao Z, Klionsky DJ (2014) The machinery of macroautophagy. Cell Res 24(1):24–41. doi:10.1038/cr.2013.168
21. Fengsrud M, Roos N, Berg T, Liou W, Slot JW, Seglen PO (1995) Ultrastructural and immunocytochemical characterization of autophagic vacuoles in isolated hepatocytes: effects of

vinblastine and asparagine on vacuole distributions. Exp Cell Res 221(2):504–519. doi:10.1006/excr.1995.1402
22. Fimia GM, Stoykova A, Romagnoli A, Giunta L, Di Bartolomeo S, Nardacci R, Corazzari M, Fuoco C, Ucar A, Schwartz P, Gruss P, Piacentini M, Chowdhury K, Cecconi F (2007) Ambra1 regulates autophagy and development of the nervous system. Nature 447(7148):1121–1125. doi:10.1038/nature05925
23. Fujiwara Y, Furuta A, Kikuchi H, Aizawa S, Hatanaka Y, Konya C, Uchida K, Yoshimura A, Tamai Y, Wada K, Kabuta T (2013) Discovery of a novel type of autophagy targeting RNA. Autophagy 9(3):403–409. doi:10.4161/auto.23002
24. Ganley IG, Wong PM, Gammoh N, Jiang X (2011) Distinct autophagosomal-lysosomal fusion mechanism revealed by thapsigargin-induced autophagy arrest. Mol Cell 42(6):731–743. doi:10.1016/j.molcel.2011.04.024
25. Gonzalez-Polo RA, Niso-Santano M, Ortiz-Ortiz MA, Gomez-Martin A, Moran JM, Garcia-Rubio L, Francisco-Morcillo J, Zaragoza C, Soler G, Fuentes JM (2007) Relationship between autophagy and apoptotic cell death in human neuroblastoma cells treated with paraquat: could autophagy be a "brake" in paraquat-induced apoptotic death? Autophagy 3(4):366–367
26. Gordon PB, Holen I, Fosse M, Rotnes JS, Seglen PO (1993) Dependence of hepatocytic autophagy on intracellularly sequestered calcium. J Biol Chem 268(35):26107–26112
27. He C, Levine B (2010) The Beclin 1 interactome. Curr Opin Cell Biol 22(2):140–149. doi:10.1016/j.ceb.2010.01.001
28. Inoki K, Li Y, Zhu T, Wu J, Guan KL (2002) TSC2 is phosphorylated and inhibited by Akt and suppresses mTOR signalling. Nat Cell Biol 4(9):648–657. doi:10.1038/ncb839
29. Inoki K, Zhu T, Guan KL (2003) TSC2 mediates cellular energy response to control cell growth and survival. Cell 115(5):577–590
30. Jewell JL, Russell RC, Guan KL (2013) Amino acid signalling upstream of mTOR. Nat Rev Mol Cell Biol 14(3):133–139. doi:10.1038/nrm3522
31. Jung CH, Ro SH, Cao J, Otto NM, Kim DH (2010) mTOR regulation of autophagy. FEBS Lett 584(7):1287–1295. doi:10.1016/j.febslet.2010.01.017
32. Kabeya Y, Mizushima N, Ueno T, Yamamoto A, Kirisako T, Noda T, Kominami E, Ohsumi Y, Yoshimori T (2000) LC3, a mammalian homologue of yeast Apg8p, is localized in autophagosome membranes after processing. EMBO J 19(21):5720–5728. doi:10.1093/emboj/19.21.5720
33. Kaushik S, Bandyopadhyay U, Sridhar S, Kiffin R, Martinez-Vicente M, Kon M, Orenstein SJ, Wong E, Cuervo AM (2011) Chaperone-mediated autophagy at a glance. J Cell Sci 124(Pt 4):495–499. doi:10.1242/jcs.073874
34. Kaushik S, Massey AC, Mizushima N, Cuervo AM (2008) Constitutive activation of chaperone-mediated autophagy in cells with impaired macroautophagy. Mol Biol Cell 19(5):2179–2192. doi:10.1091/mbc.E07-11-1155
35. Klionsky DJ (2005) The molecular machinery of autophagy: unanswered questions. J Cell Sci 118(Pt 1):7–18. doi:10.1242/jcs.01620
36. Klionsky DJ, Cuervo AM, Dunn WA Jr, Levine B, van der Klei I, Seglen PO (2007) How shall I eat thee? Autophagy 3(5):413–416. doi:4377
37. Klionsky DJ, Schulman BA (2014) Dynamic regulation of macroautophagy by distinctive ubiquitin-like proteins. Nat Struct Mol Biol 21(4):336–345. doi:10.1038/nsmb.2787
38. Kondo Y, Kondo S (2006) Autophagy and cancer therapy. Autophagy 2(2):85–90
39. Kuma A, Matsui M, Mizushima N (2007) LC3, an autophagosome marker, can be incorporated into protein aggregates independent of autophagy: caution in the interpretation of LC3 localization. Autophagy 3(4):323–328
40. Larrea MD, Liang J, Da Silva T, Hong F, Shao SH, Han K, Dumont D, Slingerland JM (2008) Phosphorylation of p27Kip1 regulates assembly and activation of cyclin D1-Cdk4. Mol Cell Biol 28(20):6462–6472. doi:10.1128/MCB.02300-07
41. Leroy E, Boyer R, Auburger G, Leube B, Ulm G, Mezey E, Harta G, Brownstein MJ, Jonnalagada S, Chernova T, Dehejia A, Lavedan C, Gasser T, Steinbach PJ, Wilkinson KD, Polymeropoulos

MH (1998) The ubiquitin pathway in Parkinson's disease. Nature 395(6701):451–452. doi:10.1038/26652
42. Levine B, Klionsky DJ (2004) Development by self-digestion: molecular mechanisms and biological functions of autophagy. Dev Cell 6(4):463–477
43. Loewith R, Jacinto E, Wullschleger S, Lorberg A, Crespo JL, Bonenfant D, Oppliger W, Jenoe P, Hall MN (2002) Two TOR complexes, only one of which is rapamycin sensitive, have distinct roles in cell growth control. Mol Cell 10(3):457–468
44. Loos B, Engelbrecht AM, Lockshin RA, Klionsky DJ, Zakeri Z (2013) The variability of autophagy and cell death susceptibility Unanswered questions. Autophagy 9(9):1270–1285. doi:10.4161/Auto.25560
45. LoPiccolo J, Blumenthal GM, Bernstein WB, Dennis PA (2008) Targeting the PI3K/Akt/mTOR pathway: effective combinations and clinical considerations. Drug Resist Updat 11(1–2):32–50. doi:10.1016/j.drup.2007.11.003
46. Madeo F, Pietrocola F, Eisenberg T, Kroemer G (2014) Caloric restriction mimetics: towards a molecular definition. Nat Rev Drug Discov 13(10):727–740. doi:10.1038/nrd4391
47. Marino G, Niso-Santano M, Baehrecke EH, Kroemer G (2014) Self-consumption: the interplay of autophagy and apoptosis. Nat Rev Mol Cell Biol 15(2):81–94. doi:10.1038/nrm3735
48. Martinez-Vicente M, Cuervo AM (2007) Autophagy and neurodegeneration: when the cleaning crew goes on strike. Lancet Neurol 6(4):352–361. doi:10.1016/S1474-4422(07)70076-5
49. Masiero E, Agatea L, Mammucari C, Blaauw B, Loro E, Komatsu M, Metzger D, Reggiani C, Schiaffino S, Sandri M (2009) Autophagy is required to maintain muscle mass. Cell Metab 10(6):507–515. doi:10.1016/j.cmet.2009.10.008
50. Matsunaga K, Saitoh T, Tabata K, Omori H, Satoh T, Kurotori N, Maejima I, Shirahama-Noda K, Ichimura T, Isobe T, Akira S, Noda T, Yoshimori T (2009) Two Beclin 1-binding proteins, Atg14L and Rubicon, reciprocally regulate autophagy at different stages. Nat Cell Biol 11(4):385–396. doi:10.1038/ncb1846
51. Meijer AJ (2008) Amino acid regulation of autophagosome formation. Methods Mol Biol 445:89–109. doi:10.1007/978-1-59745-157-4_5
52. Mijaljica D, Prescott M, Devenish RJ (2011) Microautophagy in mammalian cells: revisiting a 40-year-old conundrum. Autophagy 7(7):673–682
53. Mizushima N, Noda T, Yoshimori T, Tanaka Y, Ishii T, George MD, Klionsky DJ, Ohsumi M, Ohsumi Y (1998) A protein conjugation system essential for autophagy. Nature 395(6700):395–398. doi:10.1038/26506
54. Mizushima N, Yamamoto A, Hatano M, Kobayashi Y, Kabeya Y, Suzuki K, Tokuhisa T, Ohsumi Y, Yoshimori T (2001) Dissection of autophagosome formation using Apg5-deficient mouse embryonic stem cells. J Cell Biol 152(4):657–668
55. Nakai A, Yamaguchi O, Takeda T, Higuchi Y, Hikoso S, Taniike M, Omiya S, Mizote I, Matsumura Y, Asahi M, Nishida K, Hori M, Mizushima N, Otsu K (2007) The role of autophagy in cardiomyocytes in the basal state and in response to hemodynamic stress. Nat Med 13(5):619–624. doi:10.1038/nm1574
56. Nicklin P, Bergman P, Zhang B, Triantafellow E, Wang H, Nyfeler B, Yang H, Hild M, Kung C, Wilson C, Myer VE, MacKeigan JP, Porter JA, Wang YK, Cantley LC, Finan PM, Murphy LO (2009) Bidirectional transport of amino acids regulates mTOR and autophagy. Cell 136(3):521–534. doi:10.1016/j.cell.2008.11.044
57. Ohsumi Y (2001) Molecular dissection of autophagy: two ubiquitin-like systems. Nat Rev Mol Cell Biol 2(3):211–216. doi:10.1038/35056522
58. Pankiv S, Clausen TH, Lamark T, Brech A, Bruun JA, Outzen H, Overvatn A, Bjorkoy G, Johansen T (2007) p62/SQSTM1 binds directly to Atg8/LC3 to facilitate degradation of ubiquitinated protein aggregates by autophagy. J Biol Chem 282(33):24131–24145. doi:10.1074/jbc.M702824200
59. Papandreou I, Lim AL, Laderoute K, Denko NC (2008) Hypoxia signals autophagy in tumor cells via AMPK activity, independent of HIF-1, BNIP3, and BNIP3L. Cell Death Differ 15(10):1572–1581. doi:10.1038/cdd.2008.84

60. Parys JB, Decuypere JP, Bultynck G (2012) Role of the inositol 1,4,5-trisphosphate receptor/Ca2+-release channel in autophagy. Cell Commun Signal 10(1):17. doi:10.1186/1478-811X-10-17
61. Parzych KR, Klionsky DJ (2014) An overview of autophagy: morphology, mechanism, and regulation. Antioxid Redox Signal 20(3):460–473. doi:10.1089/ars.2013.5371
62. Petiot A, Ogier-Denis E, Blommaart EF, Meijer AJ, Codogno P (2000) Distinct classes of phosphatidylinositol 3′-kinases are involved in signaling pathways that control macroautophagy in HT-29 cells. J Biol Chem 275(2):992–998
63. Puissant A, Auberger P (2010) AMPK- and p62/SQSTM1-dependent autophagy mediate resveratrol-induced cell death in chronic myelogenous leukemia. Autophagy 6(5):655–657. doi:10.4161/auto.6.5.12126
64. Ravikumar B, Imarisio S, Sarkar S, O'Kane CJ, Rubinsztein DC (2008) Rab5 modulates aggregation and toxicity of mutant huntingtin through macroautophagy in cell and fly models of Huntington disease. J Cell Sci 121(Pt 10):1649–1660. doi:10.1242/jcs.025726
65. Ravikumar B, Rubinsztein DC (2004) Can autophagy protect against neurodegeneration caused by aggregate-prone proteins? Neuroreport 15(16):2443–2445
66. Rusten TE, Stenmark H (2009) How do ESCRT proteins control autophagy? J Cell Sci 122(Pt 13):2179–2183. doi:10.1242/jcs.050021
67. Salminen A, Kaarniranta K, Kauppinen A (2013) Beclin 1 interactome controls the crosstalk between apoptosis, autophagy and inflammasome activation: impact on the aging process. Ageing Res Rev 12(2):520–534. doi:10.1016/j.arr.2012.11.004
68. Salvador N, Aguado C, Horst M, Knecht E (2000) Import of a cytosolic protein into lysosomes by chaperone-mediated autophagy depends on its folding state. J Biol Chem 275(35):27447–27456. doi:10.1074/jbc.M001394200
69. Sarkar S (2013) Regulation of autophagy by mTOR-dependent and mTOR-independent pathways: autophagy dysfunction in neurodegenerative diseases and therapeutic application of autophagy enhancers. Biochem Soc Trans 41(5):1103–1130. doi:10.1042/BST20130134
70. Sarkar S, Floto RA, Berger Z, Imarisio S, Cordenier A, Pasco M, Cook LJ, Rubinsztein DC (2005) Lithium induces autophagy by inhibiting inositol monophosphatase. J Cell Biol 170(7):1101–1111. doi:10.1083/jcb.200504035
71. Sarkar S, Korolchuk VI, Renna M, Imarisio S, Fleming A, Williams A, Garcia-Arencibia M, Rose C, Luo S, Underwood BR, Kroemer G, O'Kane CJ, Rubinsztein DC (2011) Complex inhibitory effects of nitric oxide on autophagy. Mol Cell 43(1):19–32. doi:10.1016/j.molcel.2011.04.029
72. Schneider JL, Suh Y, Cuervo AM (2014) Deficient chaperone-mediated autophagy in liver leads to metabolic dysregulation. Cell Metab 20(3):417–432. doi:10.1016/j.cmet.2014.06.009
73. Schroeder S, Pendl T, Zimmermann A, Eisenberg T, Carmona-Gutierrez D, Ruckenstuhl C, Marino G, Pietrocola F, Harger A, Magnes C, Sinner F, Pieber TR, Dengjel J, Sigrist SJ, Kroemer G, Madeo F (2014) Acetyl-coenzyme A: a metabolic master regulator of autophagy and longevity. Autophagy 10(7):1335–1337. doi:10.4161/auto.28919
74. Shinojima N, Yokoyama T, Kondo Y, Kondo S (2007) Roles of the Akt/mTOR/p70S6K and ERK1/2 signaling pathways in curcumin-induced autophagy. Autophagy 3(6):635–637
75. Singh R, Xiang Y, Wang Y, Baikati K, Cuervo AM, Luu YK, Tang Y, Pessin JE, Schwartz GJ, Czaja MJ (2009) Autophagy regulates adipose mass and differentiation in mice. J Clin Invest 119(11):3329–3339. doi:10.1172/JCI39228
76. Skupin A, Falcke M (2008) The role of IP3R clustering in Ca2+ signalinG. Genome Inform 20:15–24
77. Stroupe C (2011) Autophagy: cells SNARE selves. Curr Biol 21(18):R697–R699. doi:10.1016/j.cub.2011.08.017
78. Suzuki K, Ohsumi Y (2007) Molecular machinery of autophagosome formation in yeast, Saccharomyces cerevisiae. FEBS Lett 581(11):2156–2161. doi:10.1016/j.febslet.2007.01.096
79. Tanida I, Ueno T, Kominami E (2004) LC3 conjugation system in mammalian autophagy. Int J Biochem Cell Biol 36(12):2503–2518. doi:10.1016/j.biocel.2004.05.009

80. Tasdemir E, Maiuri MC, Galluzzi L, Vitale I, Djavaheri-Mergny M, D'Amelio M, Criollo A, Morselli E, Zhu C, Harper F, Nannmark U, Samara C, Pinton P, Vicencio JM, Carnuccio R, Moll UM, Madeo F, Paterlini-Brechot P, Rizzuto R, Szabadkai G, Pierron G, Blomgren K, Tavernarakis N, Codogno P, Cecconi F, Kroemer G (2008) Regulation of autophagy by cytoplasmic p53. Nat Cell Biol 10(6):676–687. doi:10.1038/ncb1730
81. Terlecky SR, Dice JF (1993) Polypeptide import and degradation by isolated lysosomes. J Biol Chem 268(31):23490–23495
82. Todde V, Veenhuis M, van der Klei IJ (2009) Autophagy: principles and significance in health and disease. Biochim Biophys Acta 1792(1):3–13. doi:10.1016/j.bbadis.2008.10.016
83. Uemura T, Yamamoto M, Kametaka A, Sou YS, Yabashi A, Yamada A, Annoh H, Kametaka S, Komatsu M, Waguri S (2014) A cluster of thin tubular structures mediates transformation of the endoplasmic reticulum to autophagic isolation membrane. Mol Cell Biol 34(9):1695–1706. doi:10.1128/MCB.01327-13
84. Wada Y, Sun-Wada GH, Kawamura N (2013) Microautophagy in the visceral endoderm is essential for mouse early development. Autophagy 9(2):252–254. doi:10.4161/auto.22585
85. Williams A, Sarkar S, Cuddon P, Ttofi EK, Saiki S, Siddiqi FH, Jahreiss L, Fleming A, Pask D, Goldsmith P, O'Kane CJ, Floto RA, Rubinsztein DC (2008) Novel targets for Huntington's disease in an mTOR-independent autophagy pathway. Nat Chem Biol 4(5):295–305. doi:10.1038/nchembio.79
86. Yakhine-Diop SM, Bravo-San Pedro JM, Gomez-Sanchez R, Pizarro-Estrella E, Rodriguez-Arribas M, Climent V, Aiastui A, Lopez de Munain A, Fuentes JM, Gonzalez-Polo RA (2014) G2019S LRRK2 mutant fibroblasts from Parkinson's disease patients show increased sensitivity to neurotoxin 1-methyl-4-phenylpyridinium dependent of autophagy. Toxicology 324:1–9. doi:10.1016/j.tox.2014.07.001
87. Yang Q, Inoki K, Kim E, Guan KL (2006) TSC1/TSC2 and Rheb have different effects on TORC1 and TORC2 activity. Proc Natl Acad Sci U S A 103(18):6811–6816. doi:10.1073/pnas.0602282103
88. Yla-Anttila P, Vihinen H, Jokitalo E, Eskelinen EL (2009) 3D tomography reveals connections between the phagophore and endoplasmic reticulum. Autophagy 5(8):1180–1185
89. Young AR, Chan EY, Hu XW, Kochl R, Crawshaw SG, High S, Hailey DW, Lippincott-Schwartz J, Tooze SA (2006) Starvation and ULK1-dependent cycling of mammalian Atg9 between the TGN and endosomes. J Cell Sci 119(Pt 18):3888–3900. doi:10.1242/jcs.03172
90. Zhong Y, Wang QJ, Li X, Yan Y, Backer JM, Chait BT, Heintz N, Yue Z (2009) Distinct regulation of autophagic activity by Atg14L and Rubicon associated with Beclin 1-phosphatidylinositol-3-kinase complex. Nat Cell Biol 11(4):468–476. doi:10.1038/ncb1854
91. Zhou J, Liao W, Yang J, Ma K, Li X, Wang Y, Wang D, Wang L, Zhang Y, Yin Y, Zhao Y, Zhu WG (2012) FOXO3 induces FOXO1-dependent autophagy by activating the AKT1 signaling pathway. Autophagy 8(12):1712–1723. doi:10.4161/auto.21830
92. Zoncu R, Efeyan A, Sabatini DM (2011) mTOR: from growth signal integration to cancer, diabetes and ageing. Nat Rev Mol Cell Biol 12(1):21–35. doi:10.1038/nrm3025

Autophagy and Pattern Recognition Receptors

Christophe Viret and Mathias Faure

Abstract Autophagy is an evolutionarily conserved eukaryotic process that eliminates intracellular components through lysosomal degradation for recycling. It is a hierarchized multistep process that is involved in a multitude of cellular functions and its fine regulation is required for cell homeostasis as indicated by the various pathologies associated with autophagy dysfunctions. During the recent years, it was recognized that autophagy plays important roles in host defense against microbial infection as well. Although autophagy can markedly influence responses of the adaptive immune system, this chapter focuses on the relationship between autophagy and the innate arm of the mammalian immune system and more specifically between autophagy and the functioning of highly conserved receptors that recognize conserved molecular pattern rather than particular molecules among microbial components: the pattern recognition receptors (PRRs). Signaling pathways activated by the engagement of PRRs can both regulate autophagy to contribute to intracellular responses to infection, including through direct pathogen degradation, and be regulated, including negatively, by the autophagic activity/machinery of the cells. Such a reciprocal regulation between autophagy and PRRs optimizes efficient innate immune responses to intracellular infection and minimizes the occurrence of pathologies associated with uncontrolled immune responses. Thus, through close interconnections with innate immunity signaling pathways, autophagy represents an integrated component of autonomous cell defense mechanisms.

Abbreviations

AIM2	Absent in melanoma 2
ASC	Apoptosis-associated speck-like protein
Atg	Autophagy-related genes
CARD	Cytoplasmic C-terminal caspase activation and recruitment domain

C. Viret • M. Faure
CIRI, International Center for Infectiology Research, Inserm, U1111, CNRS UMR5308,
Université Lyon 1, Ecole Normale Supérieure de Lyon,
UCBL1 – 21, Avenue Tony Garnier, 69365 Lyon, France
e-mail: mathias.faure@inserm.fr

M.C. Maiuri, D. De Stefano (eds.), *Autophagy Networks in Inflammation*,
Progress in Inflammation Research, DOI 10.1007/978-3-319-30079-5_2

cGAMP	Cyclic GMP-AMP synthase (cGAS) pathway
DAI	DNA-dependent activator of IFN-regulatory factors
GABARAPs	γ-aminobutiric acid receptor-associated proteins
HCV	Hepatitis C virus
HSV	Herpes Simplex Virus-1
IFI	IFN-inducible protein 16
IFNAR	IFN-I receptor
IFN-Is	Interferons
IPS-1	IFN-β promoter stimulator 1 adaptor (also known as MAVS, Cardif or Visa)
IRF	Interferon regulatory factor-3 or -7
LIR	LC3-interacting region
LRR	Leucine-rich repeat
LRRFIP1	LRR in Flightless I interacting protein-1
MAP1LC3	Microtubule-associated protein 1 light chain 3
MBL	Mannan-Binding Lectin
MDA5	Melanoma differentiation-associated gene 5
MMR	Macrophage mannose receptor
MSR	Macrophage scavenger receptor or MARCO
NBR1	Neighbor of BRCA1 gene 1
NF-κB	Nuclear factor-κB
NLRs	Nod-like receptors
OAS	2′-5′-oligoadenylate synthase
PAMPS	Pathogen-associated molecular patterns
PRRs	Pattern recognition receptors
RIG-I	Retinoic acid-inducible gene I
RLR	RIG-I-like receptor
ROS	Reactive oxygen species
SQSTM1	Sequestosome 1-like receptors (SLRs)
STING	STimulator of INterferon Genes
TLRs	Toll-like receptors
TOR	Target of rapamycin
UB	Ubiquitin-binding

1 Introduction

1.1 Autophagy

Autophagy is a ubiquitous process in eukaryotic cells that culminates with the degradation of the cytoplasmic constituents within degradative compartments of lysosomal origin [52]. Under unperturbed steady state conditions, autophagy is active in all cells as it contributes to cell homeostasis through the continuous degradation and recycling of various intracellular elements larger than those handled by the proteasome such as protein aggregates or damaged/senescent organelles. There exist several forms of mammalian autophagy. The so-called chaperone-mediated autophagy relates to the unfolding and translocation of proteins bearing a particular motif into lysosomes prior to degradation. Microautophagy targets cytosolic elements to lysosomal degradation through direct invagination of the lysosomal membrane itself. Macroautophagy, thereafter referred to as autophagy, implies the sequestration of large cytoplasmic portions into vacuoles called autophagosomes that are characterized by a double membrane and ultimately fuse with lysosomal vacuoles. Under conditions of reduced nutrient resources or exposure to various stress factors, the induced autophagy that allows cell adaptation is non-selective in nature. In contrast, another type of autophagy directs the encapsulation and degradation of particular cargoes such as specific organelles. This selective form of autophagy relies on the engagement of specialized autophagy receptors [69, 90]). In either case, mammalian autophagy involves the so-called core autophagy machinery that includes multiple conserved Autophagy-related (Atg) genes. Autophagy is induced via the decoupling of a complex named the pre-initiation complex from the suppressive regulation of the target of rapamycin (TOR) Complex 1. This pre-initiation complex, that includes unc-51-like kinase 1 (ULK1), ATG13, ATG101 and FIP200 permits the activation of the class III phosphatidylinositol 3-kinase complex, comprising Beclin-1 and VPS34, which catalyses the assembly of an isolation membrane (or phagophore) originating from various sources including the endoplasmic reticulum, the mitochondrial membranes and possibly, additional input from other compartments [56]. VPS34 induces PI(3)P on the membrane of the phagophore, a step that promotes the recruitment of additional autophagy factors. The elongation and the closure of the phagophore engulf cytoplasmic materials to form the autophagosome. This formation is regulated by two ubiquitin-like conjugation systems. One is the ATG5-ATG12-ATG16L1 system assembled with the assistance of ATG7 and ATG10 and that promotes the stabilization/elongation of the developing autophagosome as well as the activation of the second system. This second system involves ATG4, ATG7, ATG3 and factors related to yeast ATG8 that include the γ-aminobutiric acid receptor-associated proteins (GABARAPs) and the microtubule-associated protein 1 light chain 3 (MAP1LC3) factors also called LC3s. The conjugation reaction leads to the stable association of phospatidyl-ethanolamine-lipidated forms of LC3/GABARAP molecules to the membrane of the developing autophagosome [70]. Such forms named LC3II are instrumental for the inclusion of multiple categories of cargoes into autophagosomes. Selective autophagy is indeed not restricted to natural constituents of the cell, it can also operate to target and eliminate microorganisms that enter the intracellular microenvironment, a process referred to as xenophagy [61]. Selective autophagy is promoted by ubiquitination of cargoes and engagement of adaptors that connect ubiquitinated materials (either damaged organelles or bacterial cells) to LC3/GABARAP factors. While the interaction with ubiquitin chains involves ubiquitin-binding (UB) domains, the interaction with LC3s/GABARAPs involves a so-called LC3-interacting region (LIR) [54, 94, 100]. Such adaptors molecules are referred to as sequestosome 1 (SQSTM1)-like receptors (SLRs). Presently, SLRs include SQSTM1/p62, neighbor of BRCA1 gene 1 (NBR1), nuclear domain

10 protein (NDP)52/Calcoco2, TAX1BP1, T6BP and optineurin [69, 90]. Ubiquinated bacterial cells are often recognized by combinations of SLRs for targeting to autophagosomes: SQSTM1/p62 and NDP52 (e.g. *Mycobacteria*); SQSTM1/p62, NDP52 and optineurin (e.g. *Salmonella*) or SQSTM1/p62, NBR1 and NDP52 (e.g. *Shigella*) [27, 72, 87, 113, 116, 117, 119, 124].

1.2 Pattern Recognition Receptors

The immune system evolved to protect the host from infection by eliminating pathogenic microorganisms. The innate component of the immune system allows the rapid activation of signaling pathways that lead to early and appropriate immune responses to resist infection and condition the development of adequate adaptive immune responses [68]. The triggering of these pathways, relies on the recognition of conserved motifs of microbial origin called pathogen-associated molecular patterns (PAMPS) recognized by specialized receptors [44]. Unlike rearranged antigen receptors of the adaptive immune system whose specificity is not predefined, innate immunity receptors are germline-encoded receptors that do not require gene rearrangement processes and are distributed in a non-clonal manner. Rather than particular molecules, they have evolved to primarily recognize invariant molecular patterns that are shared by large groups of microorganisms and are unlikely to vary without deleterious consequences for the persistence of microbes themselves, thus limiting the possibility that microbial mutations promotes escape from innate immunity recognition [44, 68]. This mode of recognition allows for the detection of huge numbers of microorganisms by the host by using a relatively restricted set of receptors. Globally, such receptors are termed pattern recognition receptors (PRRs). They can be secreted into biological fluids, expressed on cell surface, present within intracellular compartments or within the cytosol [45]. Families of secreted PRRs include Mannan-Binding Lectin (MBL), pentraxins, ficolins and collectins which can either function as opsonins on microbes, activate the classical complement pathway or activate the lectin pathway of complement [11, 38]. Certain cell surface PRRs can mediate phagocytosis of microbes. The macrophage mannose receptor (MMR) and the related DEC205 receptor of dendritic cells appear to have important functions as phagocytic receptors. Phagocytosis of pathogens can also involves scavenger receptors such as the macrophage scavenger receptor (MSR) or MARCO [80] as well as the C-type lectin receptor Dectin-1 which can also trigger oxidative burst or induce pro-inflammatory cytokines [12]. Among the best-studied membrane-associated PRRs are the Toll-like receptors (TLRs) which are made of a distal leucine-rich repeat (LRR) motif able to bind microbial ligands, associated to an intracellular TIR domain involved in signal transduction [2]. TLRs can sense microbial components as various as triacyl lipopeptides (TLR1), lipoteichoic acid (TLR2), double strand (ds) RNA (TLR3), lipopolysaccahide [79] (TLR4), flagellin (TLR5), diacyl lipopeptide (TLR6), single strand (ss) RNA (TLR7/8) and CpGDNA (TLR9). Among TLRs, TLR3/7/8 and 9 are located within endosomal compartments of the cell.

The signaling pathways of TLRs involve the recruitment of TIR domain-containing adaptor molecules including MyD88, TRIF, TIRAP or TRAM that lead to activation of transcription factors such as nuclear factor-κB (NF-κB) and AP-1 for induction of pro-inflammatory cytokines and chemokines or to activation of interferon regulatory factor (IRF)-3 or -7 for induction of anti-viral type I (α/β) interferons (IFN-Is) [60, 104]. IFN-I can be strongly induced following TLR3, 4, 7 and 9 engagement either via TRIF or MyD88 [49]. IFN-β binds the IFN-I receptor (IFNAR) according to an autocrine/paracrine-amplifying loop that induces high production of IFN-α which shares the same receptor. IFNAR signals through STAT1/2 factors that control the expression of a large number of anti-viral genes. Some viruses and bacteria can reach the intracellular milieu upon infection. Within cells, several PRRs are present in the cytosol. The protein kinase PKR reacts to viral dsRNA during viral infection/replication leading to neutralization of the eIF2α translation initiation factor that alters protein synthesis, activation of NF-κB and MAP kinases, induction of IFN-I genes and in some instances, to apoptosis of infected cells [17, 73]. Viral nucleic acids are also detected by the 2′-5′-oligoadenylate synthase (OAS) pathway and the OAS homologue cyclic GMP-AMP (cGAMP) synthase (cGAS) pathway. OAS and cGAS are cytosolic sensors of ds nucleic acids (dsRNA and DNA, respectively) that, upon engagement, generate 2′-5′-linked intermediates. 2′-5′-linked oligoadenylates induced by OASs lead to activation of the RNaseL hydrolase while 2′-5′-linked cGAMP induced by cGAS activates expression of *IFN-I* and other antiviral genes via the STimulator of INterferon Genes (STING) factor that is located on the endoplasmic reticulum membrane [40, 99]. Cytoplasmic C-terminal caspase activation and recruitment domain (CARD) helicases are a distinct family of sensors for viral RNAs. Thus, retinoic acid-inducible gene I (RIG)-I and melanoma differentiation-associated gene 5 (MDA5) are members of the RIG-I-like receptor (RLR) family of receptors that react to cytosolic RNAs (5′-triphosphorylated RNAs and long dsRNAs, respectively) via their RNA helicase domain and induce downstream signaling via their CARD domain upon conformational change. Their expression is upregulated during viral infection or IFN-I stimulation. Upon engagement, RIG-I/MDA5 recruit the IFN-β promoter stimulator 1 (IPS-1) adaptor (also known as MAVS, Cardif or Visa) that localizes to the mitochondrial outer membrane and can induce both inflammatory cytokines and type I IFN through the recruitment of TRAF6, TRAF3, TBK1/IKK kinases, NF-κB, IRF3 and IRF7 [105, 123].

Nod-like receptors (NLRs) represent another family of cytosolic PRRs [31, 41, 66]. They comprise a C-terminal ligand-binding LRR, a NACHT domain involved in oligomerization and an N-terminal domain involved in signaling. Among well-known NLRs are NOD1/2 receptors that sense peptidoglycans from bacteria and are instrumental for antibacterial defense. NOD2 was also found to recognize viral ssRNA. Upon engagement, NODs recruit the RICK/RIP2/CCK/Cardiac kinase leading to NF-κB or MAP kinases activation [101]. Other major NLRs are the NALP receptors that contain a PYD effector domain. Several NALPs are components of inflammasomes which are multimolecular complexes that function as platforms for conversion of procaspase-1 into active caspase-1 and the production of the proinflammatory cytokines IL-1β and IL-18 [66]. For instance, the NLRP3 inflammasome comprises NALP3, the apoptosis-associated speck-like protein (ASC), Cardinal and Caspase-1. The NALP receptors can be activated by bacteria. NALP3 reacts to *Staphylococcus aureus* and *Listeria monocytogenes* while other NLRs such as IPAF and NAIP5 (NLRC4 inflammasome) react to flagellins from *Salmonella typhimurium* and *Legionella pneumophila* [57, 115]. AIM2 (absent in melanoma 2) is another type of cytosolic PRR that reacts to cytosolic dsDNA via its HIN200 domain by recruiting ASC and activating caspase-1 to form a canonical inflammasome. AIM2 can sense DNA from vaccinia virus, cytomegalovirus and *Francisella tularensis*. Along with NLRP1 and NLRC4, AIM2 contributes to caspase-1 activation during *Listeria monocytogenes* infection [6, 57]. Besides AIM2, cytosolic dsDNA can be sensed by the DNA-dependent activator of IFN-regulatory factors (DAI) also called DLM-1/ZBP1 that recruits IRF3 and the TANK-binding kinase TBK1 for IFN-I production [103] or RNA polymerase III that converts DNA into 5′ppp RNA which can then bind RIG-I [1, 16]. Additional factors such as some DExD/H helicases, the IFN-inducible protein (IFI)16 and LRR in Flightless I interacting protein-1 (LRRFIP1) appear to be involved in cytosolic viral DNA detection to some extend [95].

1.3 Autophagy and Pattern Recognition Receptors

As described above, the infection of host cells by microorganisms is detected by the PRRs of the innate immune system that sense the presence of microbial components and, through adequate signaling pathways, induce appropriate cellular responses to resist infection. Since autophagy can also efficiently operate in defense reactions against infection through targeting and elimination of microorganisms, close interconnections have evolved between PRRs-induced pathways/responses and the autophagic activity/machinery of host cells. This chapter is an attempt to summarize the current knowledge on such reciprocal influences that occur in mammalian cells in terms of molecular mechanisms and functional consequences on immune responses to infection.

2 Regulation of Autophagy During PRRs Engagement

2.1 Toll-Like Receptors

Upon mycobacteria infection, TLR4 engagement by LPS was found to promote the selective autophagic activity that contributed to bacteria clearance through activation of VPS34-dependent formation of LC3 punctates and enhancement of bacteria targeting to autophagosomes. In human macrophages and a mouse macrophage cell line, such an induction was MyD88-independent and TRIF-dependent and involved the RIP1 and p38 downstream components [122]. In macrophages, autophagy-induced by TLR4 engagement involved the ubiquitination-related activation of Beclin-1 that requires its interaction with TRAF6 followed by Bcl-2 dissociation from Beclin-1. Beclin-1 ubiquitination can then be balanced by the deubiquitinating enzyme A20 that appears able to attenuate autophagy induction and NF-κB activation induced by TLR signaling [10, 96]. TRAF6 emerges thus as an important regulator of TLR signaling-induced autophagy. Another regulator appears to be the heat shock protein HSP90 whose stabilizing interaction with Beclin-1 was reported to be important for autophagy induction subsequent to TLR engagement or *Salmonella typhimurium* infection [121]. Optineurin plays an important role during TLR-4-induced autophagic degradation of bacteria. Optineurin phosphorylation by the TBK1 kinase promotes its binding to LC3 and the targeting of ubiquinated bacteria to autophagosomes where optineurin can colocalize with TBK1 and NDP52, that can also interact with TBK1 [119]. Thus, PRR-induced phosphorylation of autophagy receptors might represent a potent regulatory circuit to promote anti-microbial selective autophagy. In fact, TBK1 impacts the autophagy-mediated elimination of microbes in macrophages by regulating the maturation of autophagosomes [84]. Recruitment of TBK1 by the membrane trafficking regulator Rab8b promotes the assembly of the autophagic machinery and is also involved in the induction of anti-bacterial autophagic activity in macrophages by IL-1β. In RAW 264.7 mouse macrophages, autophagy can also be induced by agonists of TLR3 and TLR7 [25]. Thus, ssRNA and imiquimod that bind TLR7 are robust inducers of autophagosome formation (LC3 puncta formation and LC3I/II conversion). This was found to be a MyD88-dependent process that is able to result in destruction of intracellular bacteria independently of the ability of such bacteria to engage TLR7. In primary mouse macrophages however, such an induction was less marked than in RAW 264.7 cells. In the case of TLR3, Poly(I:C)-induced autophagy correlated with induction of IFN-β secretion but the latter is unlikely to act as an autophagy inducer since IFN-I do not induce autophagy in RAW 264.7 cells [35]. Because TLR3 signaling is TRIF-dependent and MyD88-independent, a role for TRIF in Poly(I:C)-induced autophagy is possible. Distinct microbes appear able to induce autophagy through TLR2 engagement with no predominant signaling pathway involved. TLR2 and the NOD/RIP2 pathway are involved in autophagy induction by *Listeria monocytogenes* infection in macrophages. The major downstream pathway involved is the extracellular regulated protein kinase (ERK) pathway [4]. TLR2 is also contributing to autophagy induction in RAW 264.7 mouse macrophages exposed to *Staphylococcus aureus*. Various signaling pathway were triggered upon such an infection but the crucial one for autophagy induction turned out to be the c-Jun N-terminal kinase (JNK) pathway with no role for the p38 or ERK pathways [28]. In macrophages, other TLRs are capable of triggering a Beclin-1/MyD88/TRIF-dependent autophagy. Those include TLR1, 3, 5, 6 and 7. Such TLRs engagement promotes the recruitment of Beclin-1 to the MyD88/TRIF signaling module while reducing its interaction with Bcl2 factor [98]. TLR4, 3, 7 and 8 can also induce autophagy in response to viral PAMPs [24, 98, 120]. In the presence of active vitamin D, the activation of macrophages through TLR8 engagement induces the expression of the human cathelicidin microbial peptide CAMP, which promotes the autophagic flux and leads to inhibition of human immunodeficiency virus type 1 (HIV-1) replication in infected human macrophages [14]. Upon LPS stimulation or exposure to

Salmonella typhimurium, macrophages release the high mobility group box 1 (HMGB1) protein, a factor normally associated to chromatin. This secretion is dependent on a functional inflammasome [58]. Upon translocation from the nucleus into the cytoplasm HMGB1 directly binds to Beclin-1 and displaces Bcl-2, acting as a positive regulator of the autophagic flux [110, 111]. Thus PAMPs stimulation of macrophages is susceptible to also modulate autophagy positively via endogenous factors such as HMGB1. In antigen presenting cells, TLR4 signaling is associated with the formation of ubiquitin-positive, LC3-positive cytosolic aggresome-like induced structures called ALIS. The conventional autophagy inducible by TLR4 itself can target ALIS with an important role for SQSTM1/p62: TLR4 signaling enhances SQSTM1/p62 expression and its recruitment to ALIS to promote the autophagic elimination of ALIS in a p38 and NrF2 activation dependent manner [32]. TLR adaptor proteins such as MyD88 and TRIF are themselves prone to constitutive aggregation, which may require control during TLR signaling. Such aggregation structures are associated with the recruitment of SQSTM1/p62 and HDAC6 factors to the MyD88 signaling complex along with TRAF6 [42]. SQSTM1/p62 and HDAC6 can in fact act as negative modulators of ligand-induced TLR signaling: upon TLR4 engagement, SQSTM1/p62 and HDAC6 recruitment can negatively regulate p38 and JNK activation. After TLR3 engagement, autophagy also selectively degrades TRAF6 as well as TRIF which is another TLR signaling factor prone to aggregation [33]. Such an autophagy was of a peculiar type since it involved neither Beclin-1 nor ATG5. TRIF and TRAF6 degradation involved NDP52 and was down modulated by the ubiquitin-editing enzyme A20.

2.2 NOD-Like Receptors

In human dendritic cells, NOD2 engagement by the bacterial wall component muramyl dipeptides (MDP) induces autophagosome formation that allows for bacterial control and enhancement of productive antigen presentation to CD4 T cells. In this case, ATG5, ATG7, ATG16L1 and RIPK-2 are required [19]. Dendritic cells from Crohn disease (CD) patients with CD-associated NOD2 or ATG16L1 risk alleles were indeed deficient in autophagy induction by MDP, bacterial handling and antigen presentation. NOD1 can also elicit the autophagic response to intracellular bacteria such as *Shigella flexneri* or *Listeria monocytogenes*. The underlying mechanism involves the recruitment of ATG16L1 to the plasma membrane at the site of bacterial entry with no role for the RIP2 adaptor or NF-κB. Cells from CD patients with NOD2 risk variants (such as NOD2L1007insC) were also unable to recruit ATG16L1 to the entry site and promote bacterial trafficking to autophagosomes, and CD patient cells with the CD-associated ATG16L1*300A allele showed impaired induction of autophagy upon exposure to MDP. Presumably, CD-associated mutations of NOD prevent the relocalization of ATG16L1 to the plasma membrane [114]. Hence, NOD1 and NOD2 sensors can activate autophagy in response to bacterial infection through engagement of ATG16L1 to promote pathogen clearance and immunity. In the context of infections, NOD2 signaling links autophagy to the NF-κB pathway. Upon binding of muramyl dipeptide, NOD2 transduces signals leading to NF-κB activation and autophagy. Other reports show that the reduced expression of Beclin-1 decreased the NOD2-dependent NF-κB activation [39]. The crosstalk between NF-κB and autophagy could be extended to other PRRs. Indeed, several TLRs can induce autophagy upon specific PAMPs recognition [25]. However, the role of NF-κB in this pathway has not yet been extensively studied. The understanding of the interplay between autophagy and NF-κB activation in the context of PRRs engagement requires further specific investigations. NLRs can negatively regulate autophagy. It was noticed that several NLRs, including NLRP3, NLRP4, NLRP10 and NLRC4 can interact with Beclin-1. The silencing of NLRP4 enhances the autophagic process including in the context of bacterial infection. Upon infection by *group A streptococcus*, NLRP4 was shown to recruit bacteria-containing phagosomes and dissociates from Beclin-1 enabling the initiation of Beclin-1-mediated autophagy. However, NLRP4 also interacts with class C VPS complexes resulting in inhibition of autophagosome and endosome maturation [47]. Whether others NLRs have similar suppressive influence on autophagy remains to be explored. The induction of AIM2 or NLRP3 inflammasomes in human macrophages promotes autophagosome formation in an ASC/caspase-1 independent manner that involves the Ras-like small G protein, RalB [97]. RalB activation recruits the Exo84 effector to promote autophagosome formation through the assembly of active ULK1 and Beclin-1 complexes needed for formation of isolation membranes [8]. Indeed, we will see latter that the autophagy induced by inflammatory signals acts back on inflammasome activity.

2.3 Other PRRs

CD46 is a cell surface glycoprotein present on all human nucleated cells. It functions as a complement regulatory protein and is involved in cytokine production and antigen presentation. In addition, it serves as a defense-inducing receptor for products from pathogen such as measles virus, adenovirus B/D, human herpes virus 6 and substrains of *group A Streptococcus* which was known to be targetable by autophagy [74]. In the recent years, it was found that CD46 crosslinking indeed induces autophagosome formation in an ATG5/ATG7-dependent fashion. This induction involved the scaffold protein GOPC that interacts with the Cyt-1 isoform of CD46 on one hand and with the Beclin-1/VPS34 complex on the other hand [46]. Autophagy induction also occurs upon recognition of Edmonston measles virus and emm6+ *group A Streptococcus. Group A Streptococcus* substrains unable to bind CD46 were slowly degraded relative to emm6+ substrains. Other microbes are susceptible to be degraded by autophagy via this pathway since CD46 appears able to bind C3b-opsonized pathogens. In the case of Edmonston measles virus, CD46-induced autophagy is rapid and transient; it precedes another phase of autophagy that is indeed beneficial to virus replication [89]. In the case of ubiquitin-mediated autophagy that targets Mycobacterium tuberculosis in macrophages, the cytosolic DNA sensor STING was found to be required for autophagy induction [117]. Upon infection, mycobacterial DNA becomes accessible to the STING-related signaling pathway due to permeabilization of the phagosomal membrane by the M. tuberculosis secretion system called ESX-1. This allows for ubiquitination of the bacteria. Along with TBK1 that functions in the IFN-I stimulatory pathway via IRF3 and IRF7 [43], the autophagy receptors SQSTM1/p62 and NDP52 were required to target bacteria to autophagosomes in this model. The cGAS sensor that normally triggers IFN-I production via the STING pathway upon cytosolic microbial DNA sensing, is in fact, able to interact with Beclin-1 [102]. The Beclin-1/cGAS interaction promotes autophagy through displacement of the autophagy inhibitor Rubicon and activation of the PI3K III pathway [62]. A role for STING in autophagy induction has also been reported in the case of dsDNA virus infection. In mouse myeloid cells infected with Herpes Simplex Virus (HSV)-1, the presence of DNA in the virion, but not viral replication, was essential for autophagy induction and sufficient to initiate LC3I/II conversion. This induction was dependent on the presence of STING [88]. In addition to HSV-1, human cytomegalovirus infection induces a rapid and sustained autophagy in fibroblasts that also involves viral DNA with no role for viral protein synthesis [67]. STING or other DNA sensor such as AIM2 could possibly play a role in such an induction. PKR can efficiently induce autophagy in response to dsRNA during viral infection [108] and is instrumental for the degradation of HSV-1 particles encapsulated within autophagosomes [77, 109]. Such a xenophagic process might represent an anti-viral mechanism to control viruses immediately after entry but has been indeed rarely observed. PKR can act upstream of Beclin-1 and regulate the activity of the Beclin-1/Vps34 complex. Accordingly, factors able to repress PKR activity such as STAT3, are susceptible to inhibit autophagy. Finally, PKR is also promoting autophagy in a different manner because through interactions with NLRs, it contributes to optimal inflammasome activity [63] which itself can be a pro-autophagic factor (see above). The Dectin-1 receptor that recognizes fungal cell wall components can trigger phagocytosis, reactive oxygen species (ROS) production and inflammatory cytokine production in macrophages. It can also induce the non-conventional secretion of inflammatory cytokines that depend on a marked autophagic activity and inflammasome activity induced via the Syk kinase pathway [76].

2.4 The Case of Sequestosome 1/p62-Like Receptors

Besides functioning as autophagy adaptors for ubiquinated bacterial cells in xenophagy, SLRs can initiate the generation of anti-microbial peptides through the processing of cysosolic precursors that are transferred into autophagic compartments prior to proteolytic conversion into microbicidal effectors. For instance, *Mycobacteria tuberculosis* can be subjected to killing within autolysosomes by antimicrobial peptides that derive from ubiquitinated cytosolic precursors recruited to the autolysosomal micro-environment of macrophages by SQSTM1/p62 [3, 86]. The delivery of antimicrobial peptide precursors to autophagolysosomes involves IFN-γ-inducible GTPases of the Gbp type. Gbp1 binds SQSTM1/p62-associated ubiquinated proteins for delivery to autophagosome membranes. Concomitantly, Gbp7 promotes the recruitment of Atg4b for membrane elongation and closure of the autophagosome [51]. Of note is the fact that Gbps are also mildly inducible by TLR engagement in macrophages [23]. Another autophagic role for SQSTM1/p62 is independent of ubiquitin and relates to anti-viral defense. Through direct interaction, SQSTM1/p62 can target viral proteins to autophagic degradation, a process called virophagy. This is the case for the capsid protein of the Sindbis virus. Although SQSTM1/p62 appears not to be the only factor involved in Sindbis virus capsid autophagic degradation, this observation indicates that autophagy adaptors can directly recognize viral determinants and induce their selective elimination. Although there are no other examples of such a direct interaction available at the moment, one could consider that this represents a situation where SLRs can effectively function as bona fide PRRs [78, 79]. A "protein-protein interaction"-based analysis revealed that more than one third of 44 factors of the "autophagy protein network" were reported to engage in interactions with RNA virus factors belonging to five RNA virus families [34]. The fraction of such interactions susceptible to be involved in virophagy remains to be investigated. During bacteria internalization, the membrane of the vacuole that surrounds the bacterial cell becomes disrupted leading to exposure of glycosylated molecules normally present on the cell surface, to the cytosolic environment. This causes the recruitment of galectins (such as galectin-3, -9 or -8) which can bind to the autophagic receptor NDP52 that is required for clearance of bacteria such as *Salmonella thyphimurium* [112].

3 Regulation of PRR-Related Signaling Pathways by Autophagy

3.1 Autophagy and NF-κB Activation

Besides being sensitive to modulations imposed by PRRs-induced signaling pathways, autophagy can significantly influence these pathways and regulate the associated immune responses. For instance, kinases that participate to NF-κB activation such as IKK or TAK1 are important for efficient induction of autophagy and conversely, autophagy is important for the full activation of NF-κB. Such mutual influence involves the suppressive role of IKK on phosphatidylinositol-3-kinase activity on one hand and the interaction of Beclin-1 with the TAB2/3 factors that cooperates with TAK1 on the other hand [18, 20–22]. In the context of infections, NOD2 signaling links autophagy to the NF-κB pathway. Upon MDP binding, NOD2-induced signaling leads to NF-κB activation and autophagy. NOD2-induced autophagy is independent of NF-κB [114] and in turn, positively influences NOD2-induced NF-κB activation since the down modulation of Beclin-1 reduces NOD2-dependent NF-κB activation [39].

3.2 Autophagy and Induction of Pro-inflammatory Cytokines

ATG16L1 can negatively regulate TLR-induced responses including inflammasome activation. ATG16L1 deficiency disturbs the recruitment of the ATG12/ATG5 complex to the isolation membrane leading to an inefficient LC3 conjugation to phosphatidylethanolamine and therefore a deficiency in autophagosome formation. Upon LPS stimulation of macrophages, ATG16L1 deficiency results in increased production of IL-1β and IL-18 inflammatory cytokines through TRIF-dependent, ROS-mediated activation of caspase-1 [92]. In mice lacking ATG16L1, sensitivity to chemically-induced colitis is exacerbated and can be attenuated by antibody-mediated neutralization of IL-1β and IL-18 emphasizing the crucial role of ATG16L1 in regulating intestinal inflammation possibly through suppression of ROS generation. In fact, in TLR agonist-treated macrophages, pro-IL-1β is specifically captured into autophagosomes where it becomes controlled by the autophagic activity of the cell. While autophagy enhancement led to pro-IL-1β degradation and blocked secretion, autophagy neutralization promoted the processing of IL-1β and its secretion in a NLRP3 and TRIF-dependent fashion [36]. The blockade of mitochondrial autophagy (mitophagy) causes the accumulation of damaged mitochondria and its associated release of ROS. This leads to assembly/activation of the NLRP3 inflammasome [126]. Thus, by contributing to the homeostasis of the mitochondrial pool, an unperturbed mitophagy/autophagy activity is important to avoid unwanted NLRP3 inflammasome activation by mitochondrial ROS associated with the accumulation of dysfunctional/senescent mitochondria. Indeed depletion of LC3B or Beclin-1 triggers high secretion level of IL-1β and IL-18 by macrophages due to accumulation of dysfunctional mitochondria and cytosolic translocation of mitochondrial DNA in response to LPS or ATP stimulation. Both ROS and NALP3 inflammasome contributes to the release of mitochondrial DNA in this context and further activates NLRP3 [75]. Autophagy can also limits inflammatory responses resulting from inflammasome activation in other ways. As mentioned above, the induction of AIM2 or NLRP3 inflammasomes can promotes autophagy in a RalB dependent manner [97]. The autophagy induced by such inflammatory signals in fact targets ubiquinated inflammasomes thereby limiting inflammatory cytokine secretion by the mean of inflammasome destruction. Accordingly, the inhibition of autophagy exacerbates inflammasome activity and stimulation of autophagy restrains it. Thus, inflammasome activation is associated with autophagic activity and the latter contributes to keep inflammation under control through the deconstruction of active inflammasomes. It is unknown whether there exists a crosstalk between autophagy and the NLRC4-dependent production of mature IL-1β that supports bacterial elimination in infected intestinal macrophages [30] or between autophagy and the NLRP3-NLRC4 cooperation that plays an important role in mouse defense against *Salmonella typhimurium* in vivo [13]. Dying autophagic cells are efficient activator of IL-1β production by macrophages due to NLRP3 inflammasome activation during phagocytosis. This activation required ATP release from dying autophagic cells and purinergic receptor activation on macrophages. Interestingly, while autophagic dying cells induced IL-1β production via NLRP3, they reduced some pro-inflammatory responses induced by TLR4 engagement such as TNFα, IL-6 and IL-8 production [5, 83]. Also, as mentioned above, LPS stimulation leads to IL-1 β secretion in macrophages lacking Beclin-1 or LC3B [75]. In macrophages with non-functional autophagy, LPS stimulation induces an enhanced level of abnormal/dys-functional mitochondria that increases the level of ROS production, activates the inflammasome, and induces caspase 1 and IL-1β secretion [75, 126]. Beclin-1 silencing is associated with IL-1β secretion in *M. tuberculosis* infected cells [81]. Induction of autophagy promotes unconventional secretion of IL-1β involving Rab8a and the Golgi protein GRASP [26]. These results argue for autophagy to negatively regulate inflammasome activation induced via innate immunity receptors by triggering degradation of inflammatory factors thus lowering the overall level of inflammation.

3.3 Autophagy and IFN-I Production

In mouse plasmacytoid dendritic cells infected by vesicular stomatitis virus (VSV) or Sendai virus, endosomal TLR7 appears to sense cytosolic intermediates of viral replication after autophagy-dependent transport into endosomes/lysosomes. Autophagy can therefore promote the delivery of cytosolic PAMPs to lysosomes, activating TLR7 signaling for induction of IL-12p40 and IFNα production [59]. Autophagy was also reported to be crucial for human immunodeficiency virus (HIV)-1-induced IFN-I production in human plasmacytoid dendritic cells, independently of viral replication [125]. Along the same line, autophagy induced upon respiratory syncytial virus (RSV) infection seemed necessary for optimal IFN-I production in murine dendritic cells derived from bone-marrow progenitors [71]. During herpes simplex virus 1 (HSV-1) infection of plasmacytoid dendritic cells, ATG5 contributed to TLR9-induced IFNα secretion but not that of IL-12 [106]. While autophagy may regulate positively IFN-I production in virally infected plasmacytoid dendritic cells, autophagy factors appear able to limit the production of IFN-I induced by RLRs. In the absence of ATG5 or ATG7, mouse cells can produce elevated amounts of IFN-Is in response to VSV infection. Conversely, ATG5 over expression inhibits IFN-I-inducing signaling. The ATG5-ATG12 complex is a direct and constitutive interactor of RIG-I and IPS-1 and appears to inhibit RLR signaling suggesting a homeostatic regulatory role for autophagy in RLR-related antiviral immune response. The ATG5-12-RIG-I-IPS-1 interactions are reinforced during VSV infection preventing RIG-I-inducible IFN-I production [48]. In fact, the enhanced RLR signaling and IFN-I production seen in autophagy-deficient, VSV-infected cells appears to imply an important role for ROS accumulation elicited by dysfunctional mitochondria and mitochondria-associated IPS1 [107]. Amplification of autophagy to inhibit IFN-I production is also a property of the hepatitis C virus (HCV). HCV infection elicits the so-called unfolded protein response [85] which promotes autophagy and is used by HCV to assist its RNA replication in hepatoma cells [50]. This mechanism involves IFN-I suppression. HCV-derived PAMPs induce RIG-I leading to IFN-β transcription that is antagonized by autophagy and UPR. The NS3/4A factor of HCV can both cleave IPS-1 limiting its ability to signal in infected cells, and interact with IRGM to promote autophagy [34]. Since both IRGM and IPS-1 are located at the level of mitochondria and can interact with ATG5, IRGM might be involved in NS3/4A activity [82]. As mentioned above, activation of the STING signaling pathway can promote autophagy. In turn, autophagy can regulate STING-dependent IFN-I responses. Thus, upon dsDNA detection, STING assembly in the Golgi apparatus and its translocation to the cytosol was found to be negatively modulated by ATG9a which regulates assembly of the STING-TBK1 complex leading to dampened IFN-I responses [91]. The STING signaling pathway of type I IFN induction can be negatively modulated by Beclin-1, which binds to cGAS and inhibits excessive cGAMP production during HSV-1 infection [62]. In addition, the cGAMP produced upon DNA recognition-mediated cGAS activation elicits the down modulation of STING recruitment through ULK1/ATG1-dependent phosphorylation after ULK1 activation, due to cyclic dinucelotides that alleviates ULK1 repression by the AMP activated protein kinase AMPK [55]. Thus, the Beclin-1-cGAS interaction may act as an attenuating factor of cGAS-driven IFN-I production. Hence, innate immunity signals that converge to both the STING pathway of IFN-I production and autophagy can be subjected to a regulatory feedback control initiated through engagement of autophagic factors such as ATG9a, Beclin-1 or ULK1, most likely driven by the need to avoid the deleterious consequences of IFN-I overproduction.

3.4 Pathogen-Orchestrated Modification of PRR Signaling by Autophagy

By manipulating the autophagic process, pathogens can escape innate immune responses and maintain their replication. Thus, the M45 factor of the murine cytomegalovirus interacts with a NF-κB regulatory factor called NEMO causing its degradation by autophagy and alteration of NF-κB-related antiviral genes induction [29]. In dendritic cells, HIV-1 inhibits autophagy rapidly after infection via its viral envelope protein. This inhibition interferes with innate immune response such as TLRs-induced TNF-α production [7]. A similar inhibition has been observed in the case of HCMV. While HCMV promotes the autophagic flux early after infection, it blocks autophagosome maturation at latter time points via de novo synthesis of the TRS1 viral factor, which interacts with Beclin-1. This phenomenon is distinct of the inhibition of antiviral PKR activity that also involves TRS1 engagement [15].

3.5 Autophagy and Scavenger Receptors

The influence of autophagy on the antibacterial phagocytic activity of mouse macrophages was studied in engineered mice lacking the ATG7 factor in the myeloid lineages. ATG7-deficient macrophages showed an early enhancement in bacterial internalization upon infection with Mycobacterium tuberculosis. Such macrophages displayed increased expression levels of the class A scavenger receptor MSR1 and MARCO through enhanced activity of the Nrf2 transcription factor due to defective autophagy and accumulation of SQSTM1/p62 which binds the Nrf2 repressor called KEAP1 [9]. These observations reveal an unexpected link between phagocytosis and autophagy and may suggest that, by regulating scavenger receptors expression, autophagy could represent a regulatory feedback control of phagocytosis.

4 PRRs and Non-anonical Autophagic Processes

TLR-signaling can also recruit factors of the autophagy machinery to promote the maturation of conventional phagosomes. Upon phagocytosis, latex particles bearing TLR agonists such as PAM3CSK4 (TLR2), LPS (TLR4) or Zymosan (TLR2), trigger the rapid recruitment of LC3 to phagosomes in a Beclin-1, ATG5 and ATG7-dependent manner in mouse RAW 264.7 macrophages [93]. Such a LC3 recruitment, which is independent of a preinitiation complex and is not associated with the presence of double membranes, and therefore does not involve conventional autophagosomes, assists the fusion of phagosomes with lysosomes. The addition of TLR agonists after phagocytosis of uncoated particles did not initiate LC3 recruitment to phagosomes indicating that phagosomal TLR signaling was crucial for this phenomenon. MyD88 played no role while p38 could be a contributing factor. This process whereby the engulfment of phagocytosed exogenous material is associated with the PRR signaling-dependent recruitment of autophagic factors onto the membrane of the phagosomal compartment to facilitate its fusion to lysosomes and ensure the rapid degradation of the cargo is named LC3-associated phagocytosis or LAP [37, 65, 93]. It is interesting to note that under such conditions of "beads phagocytosis", mitochondria accumulate around phagosomes containing the internalized beads when the beads carry ligands for TLR2/4 but not if they carry a ligand for TLR9 which is endosome-associated [118]. Whether this proximity reflects a role for mitochondria in LC3 recruitment and/or enhanced phagosome maturation (for instance through local ROS production), is not known. LC3II recruitment to phagosomes has also been observed in the case of Dectin-1-dependent macrophage activation. This was dependent on Syk, ROS and ATG5 [64]. Whether this phenomenon is related to the LAP process described above has not been studied.

5 Conclusions

The fact that some pathogens have evolved strategies to escape, or take advantage of, autophagy is by itself a strong indication that autophagy represents a major part of the autonomous defense mechanisms against intracellular infection in eukaryotes. As such, the autophagy machinery has developed intimate interconnections with signaling circuits induced by PRRs that sense the presence of microorganisms and rapidly trigger appropriate signaling pathways. Such interconnections translate into reciprocal functional influences between autophagic activity and innate immune responses: autophagy and immune responses induced by PRRs are integrated processes. Accordingly, defects in either PRR function or autophagy function can be associated to susceptibility to identical immunopathologies. For instance, mutations in autophagy genes such as ATG16L1, or in PRR genes such as NOD2, are both associated to susceptibility to the intestinal disorder CD that involves chronic inflammation and deregulated immune responses to invasive bacteria within the intestine. In the same vein, autophagy or PRR deficiency restricted to macrophages and granulocytes can confer susceptibility to infection by bacteria and parasites. Constitutive autophagy can be enhanced upon infection by intracellular pathogens via the engagement of PRRs. Such an induction can further promote PRRs-dependent signaling for IFN-I productions. Autophagy can also restrain IFN-I productions responses by down modulating the magnitude of the signaling pathways triggered by PRRs. A similar capacity for autophagy to down modulate PRR-induced signaling applies to pathogen-induced, TLRs-dependent production of pro-inflammatory cytokines. This is illustrated for instance by the capacity of autophagy to down regulate inflammasome activation. Hence, the reciprocal regulation between autophagy and PRRs appears to optimize innate immune responses to intracellular infection while limiting the occurrence of immune-pathologies that can be initiated by deregulated immune responses. Finally, we have seen that in many instances, both TLR-inducing pathways and autophagy are linked to/influenced by, mitochondrial homeostasis or mitochondria-related factors suggesting that the mitochondrial pool of the cell might represent a particularly important hub in the integrated relationship between autophagy and PRR signaling.

Acknowledgments Our work is supported by grants from Agence Nationale de la Recherche (ANR), Institut Universitaire de France (IUF, MF), INSERM, CNRS, FINOVI, Association François Aupetit and comité du Rhône de la Ligue National contre le Cancer. We acknowledge the framework of the LABEX ECOFECT (ANR-11-LABX-0042) of Université de Lyon, within the program Investissements d'Avenir (ANR-11-IDEX-0007) operated by the ANR, within which our work is performed.

References

1. Ablasser A, Bauernfeind F et al (2009) RIG-I-dependent sensing of poly(dA:dT) through the induction of an RNA polymerase III-transcribed RNA intermediate. Nat Immunol 10(10): 1065–1072
2. Akira S, Takeda K (2004) Toll-like receptor signalling. Nat Rev Immunol 4(7):499–511
3. Alonso S, Pethe K et al (2007) Lysosomal killing of Mycobacterium mediated by ubiquitin-derived peptides is enhanced by autophagy. Proc Natl Acad Sci U S A 104(14):6031–6036
4. Anand PK, Tait SW et al (2011) TLR2 and RIP2 pathways mediate autophagy of Listeria monocytogenes via extracellular signal-regulated kinase (ERK) activation. J Biol Chem 286(50):42981–42991

5. Ayna G, Krysko DV et al (2012) ATP release from dying autophagic cells and their phagocytosis are crucial for inflammasome activation in macrophages. PLoS One 7(6), e40069
6. Barber GN (2011) Innate immune DNA sensing pathways: STING, AIMII and the regulation of interferon production and inflammatory responses. Curr Opin Immunol 23(1):10–20
7. Blanchet FP, Moris A et al (2010) Human immunodeficiency virus-1 inhibition of immunoamphisomes in dendritic cells impairs early innate and adaptive immune responses. Immunity 32(5):654–669
8. Bodemann B, Orvedahl A et al (2011) RalB and the exocyst mediate the cellular starvation response by direct activation of autophagosome assembly. Cell 144:253–267
9. Bonilla DL, Bhattacharya A et al (2013) Autophagy regulates phagocytosis by modulating the expression of scavenger receptors. Immunity 39(3):537–547
10. Boone D, Turer E et al (2004) The ubiquitin-modifying enzyme A20 is required for termination of Toll-like receptor responses. Nat Immunol 5:1052–1060
11. Bottazzi B, Garlanda C et al (2006) Pentraxins as a key component of innate immunity. Curr Opin Immunol 18(1):10–15
12. Brown GD (2006) Dectin-1: a signalling non-TLR pattern-recognition receptor. Nat Rev Immunol 6(1):33–43
13. Broz P, Lamkanfi M, Mariathasan S, Dixit VM et al (2010) Redundant roles for inflammasome receptors NLRP3 and NLRC4 in host defense against Salmonella. J Exp Med 207:1745–1755
14. Campbell GR, Spector SA (2012) Vitamin D inhibits human immunodeficiency virus type 1 and Mycobacterium tuberculosis infection in macrophages through the induction of autophagy. PLoS Pathog 8(5), e1002689
15. Chaumorcel M, Lussignol M et al (2012) The human cytomegalovirus protein TRS1 inhibits autophagy via its interaction with Beclin 1. J Virol 86(5):2571–2584
16. Chiu YH, Macmillan JB et al (2009) RNA polymerase III detects cytosolic DNA and induces type I interferons through the RIG-I pathway. Cell 138(3):576–591
17. Clemens MJ, Elia A (1997) The double-stranded RNA-dependent protein kinase PKR: structure and function. J Interferon Cytokine Res 17(9):503–524
18. Comb WC, Hutti JE et al (2012) p85alpha SH2 domain phosphorylation by IKK promotes feedback inhibition of PI3K and Akt in response to cellular starvation. Mol Cell 45(6): 719–730
19. Cooney R, Baker J et al (2010) NOD2 stimulation induces autophagy in dendritic cells influencing bacterial handling and antigen presentation. Nat Med 16(1):90–97
20. Criollo A, Chereau F et al (2012) Autophagy is required for the activation of NFkappaB. Cell Cycle 11(1):194–199
21. Criollo A, Niso-Santano M et al (2011) Inhibition of autophagy by TAB2 and TAB3. EMBO J 30(24):4908–4920
22. Criollo A, Senovilla L et al (2010) The IKK complex contributes to the induction of autophagy. EMBO J 29(3):619–631
23. Degrandi D, Konermann C et al (2007) Extensive characterization of IFN-induced GTPases mGBP1 to mGBP10 involved in host defense. J Immunol 179(11):7729–7740
24. Delgado M, Singh S et al (2009) Autophagy and pattern recognition receptors in innate immunity. Immunol Rev 227(1):189–202
25. Delgado MA, Elmaoued RA et al (2008) Toll-like receptors control autophagy. EMBO J 27(7):1110–1121
26. Dupont N, Jiang S et al (2011) Autophagy-based unconventional secretory pathway for extracellular delivery of IL-1beta. EMBO J 30(23):4701–4711
27. Dupont N, Lacas-Gervais S et al (2009) Shigella phagocytic vacuolar membrane remnants participate in the cellular response to pathogen invasion and are regulated by autophagy. Cell Host Microbe 6(2):137–149
28. Fang L, Wu HM et al (2014) TLR2 mediates phagocytosis and autophagy through JNK signaling pathway in Staphylococcus aureus-stimulated RAW264.7 cells. Cell Signal 26(4): 806–814

29. Fliss PM, Jowers TP et al (2012) Viral mediated redirection of NEMO/IKKgamma to autophagosomes curtails the inflammatory cascade. PLoS Pathog 8(2), e1002517
30. Franchi L, Kamada N et al (2012) NLRC4-driven production of IL-1β discriminates between pathogenic and commensal bacteria and promotes host intestinal defense. Nat Immunol 13:449–456
31. Fritz JH, Ferrero RL et al (2006) Nod-like proteins in immunity, inflammation and disease. Nat Immunol 7(12):1250–1257
32. Fujita K, Maeda D et al (2011) Nrf2-mediated induction of p62 controls Toll-like receptor-4-driven aggresome-like induced structure formation and autophagic degradation. Proc Natl Acad Sci U S A 108(4):1427–1432
33. Funami K, Sasai M et al (2008) Homo-oligomerization is essential for Toll/interleukin-1 receptor domain-containing adaptor molecule-1-mediated NF-kappaB and interferon regulatory factor-3 activation. J Biol Chem 283(26):18283–18291
34. Gregoire IP, Richetta C et al (2011) IRGM is a common target of RNA viruses that subvert the autophagy network. PLoS Pathog 7(12), e1002422
35. Gutierrez MG, Master SS et al (2004) Autophagy is a defense mechanism inhibiting BCG and Mycobacterium tuberculosis survival in infected macrophages. Cell 119(6):753–766
36. Harris J, Hartman M et al (2011) Autophagy controls IL-1beta secretion by targeting pro-IL-1beta for degradation. J Biol Chem 286(11):9587–9597
37. Henault J, Martinez J et al (2012) Noncanonical autophagy is required for type I interferon secretion in response to DNA-immune complexes. Immunity 37(6):986–997
38. Holmskov U, Thiel S et al (2003) Collections and ficolins: humoral lectins of the innate immune defense. Annu Rev Immunol 21:547–578
39. Homer CR, Richmond AL et al (2010) ATG16L1 and NOD2 interact in an autophagy-dependent antibacterial pathway implicated in Crohn's disease pathogenesis. Gastroenterology 139(5):1630–1641, 1641.e1631–1632
40. Hornung V, Hartmann R et al (2014) OAS proteins and cGAS: unifying concepts in sensing and responding to cytosolic nucleic acids. Nat Rev Immunol 14(8):521–528
41. Inohara C et al (2005) NOD-LRR proteins: role in host-microbial interactions and inflammatory disease. Annu Rev Biochem 74:355–383
42. Into T, Inomata M et al (2010) Regulation of MyD88 aggregation and the MyD88-dependent signaling pathway by sequestosome 1 and histone deacetylase 6. J Biol Chem 285(46):35759–35769
43. Ishikawa H, Ma Z et al (2009) STING regulates intracellular DNA-mediated, type I interferon-dependent innate immunity. Nature 461(7265):788–792
44. Janeway CA Jr (1989) Approaching the asymptote? Evolution and revolution in immunology. Cold Spring Harb Symp Quant Biol 54(Pt 1):1–13
45. Janeway CA Jr, Medzhitov R (2002) Innate immune recognition. Annu Rev Immunol 20:197–216
46. Joubert PE, Meiffren G et al (2009) Autophagy induction by the pathogen receptor CD46. Cell Host Microbe 6(4):354–366
47. Jounai N, Kobiyama K et al (2011) NLRP4 negatively regulates autophagic processes through an association with beclin1. J Immunol 186:1646–1655
48. Jounai N, Takeshita F et al (2007) The Atg5 Atg12 conjugate associates with innate antiviral immune responses. Proc Natl Acad Sci U S A 104(35):14050–14055
49. Kawai T, Akira S (2006) Innate immune recognition of viral infection. Nat Immunol 7(2): 131–137
50. Ke PY, Chen SS (2011) Autophagy: a novel guardian of HCV against innate immune response. Autophagy 7(5):533–535
51. Kim B, Shenoy A et al (2011) A family of IFN-γ-inducible 65-kD GTPases protects against bacterial infection. Science 332:717–721
52. Klionsky DJ (2007) Autophagy: from phenomenology to molecular understanding in less than a decade. Nat Rev Mol Cell Biol 8(11):931–937

53. Klionsky DJ, Abdalla FC et al (2012) Guidelines for the use and interpretation of assays for monitoring autophagy. Autophagy 8(4):445–544
54. Klionsky DJ, Schulman BA (2014) Dynamic regulation of macroautophagy by distinctive ubiquitin-like proteins. Nat Struct Mol Biol 21(4):336–345
55. Konno H, Konno K et al (2013) Cyclic dinucleotides trigger ULK1 (ATG1) phosphorylation of STING to prevent sustained innate immune signaling. Cell 155(3):688–698
56. Lamb CA, Yoshimori T et al (2013) The autophagosome: origins unknown, biogenesis complex. Nat Rev Mol Cell Biol 14(12):759–774
57. Lamkanfi M, Dixit VM (2014) Mechanisms and functions of inflammasomes. Cell 157(5): 1013–1022
58. Lamkanfi M, Sarkar A et al (2010) Inflammasome-dependent release of the alarmin HMGB1 in endotoxemia. J Immunol 185(7):4385–4392
59. Lee HK, Lund JM et al (2007) Autophagy-dependent viral recognition by plasmacytoid dendritic cells. Science 315(5817):1398–1401
60. Lee MS, Kim YJ (2007) Signaling pathways downstream of pattern-recognition receptors and their cross talk. Annu Rev Biochem 76:447–480
61. Levine B (2005) Eating oneself and uninvited guests: autophagy-related pathways in cellular defense. Cell 120(2):159–162
62. Liang Q, Seo GJ et al (2014) Crosstalk between the cGAS DNA sensor and Beclin-1 autophagy protein shapes innate antimicrobial immune responses. Cell Host Microbe 15(2): 228–238
63. Lu B, Nakamura T et al (2012) Novel role of PKR in inflammasome activation and HMGB1 release. Nature 488(7413):670–674
64. Ma J, Becker C et al (2012) Dectin-1-triggered recruitment of light chain 3 protein to phagosomes facilitates major histocompatibility complex class II presentation of fungal-derived antigens. J Biol Chem 287(41):34149–34156
65. Martinez J, Almendinger J et al (2011) Microtubule-associated protein 1 light chain 3 alpha (LC3)-associated phagocytosis is required for the efficient clearance of dead cells. Proc Natl Acad Sci U S A 108(42):17396–17401
66. Martinon F, Tschopp J (2005) NLRs join TLRs as innate sensors of pathogens. Trends Immunol 26(8):447–454
67. McFarlane S, Aitken J et al (2011) Early induction of autophagy in human fibroblasts after infection with human cytomegalovirus or herpes simplex virus 1. J Virol 85(9):4212–4221
68. Medzhitov R, Janeway CA (1997) Innate immunity: the virtues of a nonclonal system of recognition. Cell 91(3):295–298
69. Mijaljica D, Nazarko TY et al (2012) Receptor protein complexes are in control of autophagy. Autophagy 8(11):1701–1705
70. Mizushima N, Yoshimori T et al (2011) The role of Atg proteins in autophagosome formation. Annu Rev Cell Dev Biol 27:107–132
71. Morris S, Swanson MS et al (2011) Autophagy-mediated dendritic cell activation is essential for innate cytokine production and APC function with respiratory syncytial virus responses. J Immunol 187(8):3953–3961
72. Mostowy S, Sancho-Shimizu V et al (2011) p62 and NDP52 proteins target intracytosolic Shigella and Listeria to different autophagy pathways. J Biol Chem 286:26987–26995
73. Munir M, Berg M (2013) The multiple faces of proteinkinase R in antiviral defense. Virulence 4(1):85–89
74. Nakagawa I, Amano A et al (2004) Autophagy defends cells against invading group A Streptococcus. Science 306(5698):1037–1040
75. Nakahira K, Haspel JA et al (2011) Autophagy proteins regulate innate immune responses by inhibiting the release of mitochondrial DNA mediated by the NALP3 inflammasome. Nat Immunol 12(3):222–230
76. Ohman T, Teirila L et al (2014) Dectin-1 pathway activates robust autophagy-dependent unconventional protein secretion in human macrophages. J Immunol 192(12):5952–5962

77. Orvedahl A, Alexander D et al (2007) HSV-1 ICP34.5 confers neurovirulence by targeting the Beclin 1 autophagy protein. Cell Host Microbe 1(1):23–35
78. Orvedahl A, MacPherson S et al (2010) Autophagy protects against Sindbis virus infection of the central nervous system. Cell Host Microbe 7(2):115–127
79. Orvedahl A, Sumpter R et al (2011) Image-based genome-wide siRNA screen identifies selective autophagy factors. Nature 480(7375):113–117
80. Pearson T, Greiner DL et al (2008) Creation of humanized mice to study human immunity. Curr Protoc Immunol Chapter 15:Unit 15.21
81. Peral de Castro C, Jones SA et al (2012) Autophagy regulates IL-23 secretion and innate T cell responses through effects on IL-1 secretion. J Immunol 189(8):4144–4153
82. Petkova DS, Viret C et al (2012) IRGM in autophagy and viral infections. Front Immunol 3:426
83. Petrovski G, Ayna G et al (2011) Phagocytosis of cells dying through autophagy induces inflammasome activation and IL-1beta release in human macrophages. Autophagy 7(3): 321–330
84. Pilli M, Arko-Mensah J et al (2012) TBK-1 promotes autophagy-mediated antimicrobial defense by controlling autophagosome maturation. Immunity 37:223–234
85. Plumet S, Duprex WP et al (2005) Dynamics of viral RNA synthesis during measles virus infection. J Virol 79(11):6900–6908
86. Ponpuak M, Davis AS et al (2010) Delivery of cytosolic components by autophagic adaptor protein p62 endows autophagosomes with unique antimicrobial properties. Immunity 32(3):329–341
87. Radtke AL, Delbridge LM et al (2007) TBK1 protects vacuolar integrity during intracellular bacterial infection. PLoS Pathog 3(3), e29
88. Rasmussen SB, Horan KA et al (2011) Activation of autophagy by alpha-herpesviruses in myeloid cells is mediated by cytoplasmic viral DNA through a mechanism dependent on stimulator of IFN genes. J Immunol 187(10):5268–5276
89. Richetta C, Gregoire IP et al (2013) Sustained autophagy contributes to measles virus infectivity. PLoS Pathog 9(9), e1003599
90. Rogov V, Dotsch V et al (2014) Interactions between autophagy receptors and ubiquitin-like proteins form the molecular basis for selective autophagy. Mol Cell 53(2):167–178
91. Saitoh T, Fujita N et al (2009) Atg9a controls dsDNA-driven dynamic translocation of STING and the innate immune response. Proc Natl Acad Sci U S A 106(49):20842–20846
92. Saitoh T, Fujita N et al (2008) Loss of the autophagy protein Atg16L1 enhances endotoxin-induced IL-1beta production. Nature 456(7219):264–268
93. Sanjuan MA, Dillon CP et al (2007) Toll-like receptor signalling in macrophages links the autophagy pathway to phagocytosis. Nature 450(7173):1253–1257
94. Shaid S, Brandts CH et al (2013) Ubiquitination and selective autophagy. Cell Death Differ 20(1):21–30
95. Sharma S, Fitzgerald KA (2011) Innate immune sensing of DNA. PLoS Pathog 7(4), e1001310
96. Shi C, Kehrl J (2010) TRAF6 and A20 regulate lysine 63-linked ubiquitination of Beclin-1 to control TLR4-induced autophagy. Sci Signal 3:ra42
97. Shi C, Shenderov K et al (2012) Activation of autophagy by inflammatory signals limits IL-1β production by targeting ubiquitinated inflammasomes for destruction. Nat Immunol 13(3):255–263
98. Shi CS, Kehrl JH (2008) MyD88 and Trif target Beclin 1 to trigger autophagy in macrophages. J Biol Chem 283(48):33175–33182
99. Silverman RH (2007) Viral encounters with 2′,5′-oligoadenylate synthetase and RNase L during the interferon antiviral response. J Virol 81(23):12720–12729
100. Stolz A, Ernst A et al (2014) Cargo recognition and trafficking in selective autophagy. Nat Cell Biol 16(6):495–501
101. Strober W, Murray PJ et al (2006) Signalling pathways and molecular interactions of NOD1 and NOD2. Nat Rev Immunol 6(1):9–20

102. Sun L, Wu J et al (2013) Cyclic GMP-AMP synthase is a cytosolic DNA sensor that activates the type I interferon pathway. Science 339(6121):786–791
103. Takaoka A, Wang Z et al (2007) DAI (DLM-1/ZBP1) is a cytosolic DNA sensor and an activator of innate immune response. Nature 448(7152):501–505
104. Takeda K, Kaisho T et al (2003) Toll-like receptors. Annu Rev Immunol 21:335–376
105. Takeuchi O, Akira S (2009) Innate immunity to virus infection. Immunol Rev 227(1):75–86
106. Tal MC, Iwasaki A (2009) Autophagy and innate recognition systems. Curr Top Microbiol Immunol 335:107–121
107. Tal MC, Sasai M et al (2009) Absence of autophagy results in reactive oxygen species-dependent amplification of RLR signaling. Proc Natl Acad Sci U S A 106(8):2770–2775
108. Talloczy Z, Jiang W et al (2002) Regulation of starvation- and virus-induced autophagy by the eIF2alpha kinase signaling pathway. Proc Natl Acad Sci U S A 99(1):190–195
109. Talloczy Z, Virgin HW et al (2006) PKR-dependent autophagic degradation of herpes simplex virus type 1. Autophagy 2(1):24–29
110. Tang D, Kang R et al (2010) HMGB1 release and redox regulates autophagy and apoptosis in cancer cells. Oncogene 29:5299–5310
111. Tang D, Kang R et al (2010) Endogenous HMGB1 regulates autophagy. J Cell Biol 190:881–892
112. Thurston T, Wandel M et al (2012) Galectin 8 targets damaged vesicles for autophagy to defend cells against bacterial invasion. Nature 482:414–418
113. Thurston TL, Ryzhakov G et al (2009) The TBK1 adaptor and autophagy receptor NDP52 restricts the proliferation of ubiquitin-coated bacteria. Nat Immunol 10(11):1215–1221
114. Travassos LH, Carneiro LA et al (2010) Nod1 and Nod2 direct autophagy by recruiting ATG16L1 to the plasma membrane at the site of bacterial entry. Nat Immunol 11(1):55–62
115. von Moltke J, Ayres JS et al (2013) Recognition of bacteria by inflammasomes. Annu Rev Immunol 31:73–106
116. von Muhlinen N, Akutsu M et al (2012) LC3C, bound selectively by a noncanonical LIR motif in NDP52, is required for antibacterial autophagy. Mol Cell 48:329–342
117. Watson R, Manzanillo P et al (2012) Extracellular M. tuberculosis DNA targets bacteria for autophagy by activating the host DNA-sensing pathway. Cell 150:803–815
118. West AP, Brodsky IE et al (2011) TLR signalling augments macrophage bactericidal activity through mitochondrial ROS. Nature 472(7344):476–480
119. Wild P, Farhan H et al (2011) Phosphorylation of the autophagy receptor optineurin restricts Salmonella growth. Science 333:228–233
120. Xie Z, Klionsky DJ (2007) Autophagosome formation: core machinery and adaptations. Nat Cell Biol 9(10):1102–1109
121. Xu C, Liu J et al (2011) Functional interaction of heat shock protein 90 and Beclin 1 modulates Toll-like receptor-mediated autophagy. FASEB J 25(8):2700–2710
122. Xu Y, Jagannath C et al (2007) Toll-like receptor 4 is a sensor for autophagy associated with innate immunity. Immunity 27(1):135–144
123. Yoneyama M, Fujita T (2008) Structural mechanism of RNA recognition by the RIG-I-like receptors. Immunity 29(2):178–181
124. Zheng Y, Shahnazari S et al (2009) The adaptor protein p62/SQSTM1 targets invading bacteria to the autophagy pathway. J Immunol 183:5909–5916
125. Zhou D, Kang KH et al (2012) Production of interferon alpha by human immunodeficiency virus type 1 in human plasmacytoid dendritic cells is dependent on induction of autophagy. J Infect Dis 205(8):1258–1267
126. Zhou R, Yazdi AS et al (2011) A role for mitochondria in NLRP3 inflammasome activation. Nature 469(7329):221–225

The Complex Crosstalk Between Autophagy and ROS Signalling Pathways

Kelly Airiau and Mojgan Djavaheri-Mergny

Abstract The homeostasis between the oxidant and antioxidant levels in cells is altered in several diseases including cancer, neurodegenerative diseases, and inflammatory disorders. Macroautophagy (hereafter referred to as autophagy) is a redox (reduction/oxidation)-sensitive process that results in degradation of cellular constituents such as proteins, lipids, and mitochondria through the lysosomal pathway. There is a complex and mutual relationship between pathways that control levels of reactive oxygen species (ROS) and autophagy. Autophagy is activated by various stimuli in cells and ROS are one of these autophagy inducers. The accumulation of ROS induces autophagy both by direct effect on the core autophagy machinery and by indirect influence on the components of the autophagy-regulatory signaling pathway. In turn, autophagy regulates the abundance of ROS in cells by promoting the clearance of damaged mitochondria and oxidized cellular substrates and by modulating activity of the detoxifying antioxidant systems. ROS are also involved in the initiation of inflammation, a process that required the secretion of several inflammatory mediators. Here, we will discuss the regulation of inflammatory responses by autophagy as a consequence of the interplay of autophagy and ROS signaling pathways.

Abbreviations

2-ME	2-methoxyestradiol
AMPK	5′-AMP-activated protein kinase
ARE	Antioxidant Response Element
ATG	Autophagy-related

K. Airiau • M. Djavaheri-Mergny (✉)
INSERM U1218 ACTION, Institut Bergonié, 229 Cours de l'Argonne, Bordeaux 33076, France

University of Bordeaux, Bordeaux, France
e-mail: mojgan.mergny@inserm.fr

M.C. Maiuri, D. De Stefano (eds.), *Autophagy Networks in Inflammation*, Progress in Inflammation Research, DOI 10.1007/978-3-319-30079-5_3

ATM	Serine-protein kinase ATM
ATP	Adenosine triphosphate
BCL2	B-cell CLL/lymphoma 2
BECN1	Autophagy related BCL2-interacting coiled-coil protein 1
BNIP3	BCL2/adenovirus E1B 19 kDa interacting protein 3
BNIP3	BNIP3 ligand
DAMP	Damage-associated molecular patterns
DNA	Deoxyribonucleic acid
FOXO3	Forkhead box O3
GSH	Glutathione
H_2O_2	Hydrogen peroxide
HIF1	Hypoxia Inducing Factor 1
HMGB1	High–mobility group box 1
HO	Heme oxygenase
IL	Interleukin
JNK	c-Jun N-terminal kinases
Keap1	Kelch-like ECH-associated protein 1
LC3	Microtubule-associated protein light chain 3
LPS	Lipopolysaccharide
MAPK	Mitogen-activated protein kinases
mETC	Mitochondrial electron transport chain
MnSOD	Manganese SuperOxide Dismutase
MTORC1	Mammalian target of rapamycin complex 1
NADPH	Nicotinamide adenine dinucleotide phosphate
NF-kB	Nuclear Factor-kappa B
NO	Nitric oxide
NOX	NADPH oxidase
NRF2	Nuclear Factor-like 2
$^1O_2$	Singlet oxygen
$O_2^{\cdot -}$	Superoxide anion
·OH	Hydroxyl radical
PAMP	Pathogen-associated molecular patterns
PERK	Protein kinase R-like endoplasmic reticulum kinase
PI3K	Phospho-inositol 3 kinase
ROS	Reactive Oxygen Species
SOD	Superoxide dismutase
SQSTM1(p62)	Sequestosome 1
TBI	Traumatic brain injury
TNFα	Tumor Necrosis Factor α
TRX	Thioredoxin
ULK1/2	Uncoordinated 51 -like kinases ½
UPS	Ubiquitin proteasome system
VPS34	Vacuolar Protein Sorting 34
α-KG	α-ketoglutarate

1 Introduction

1.1 Autophagy

The term autophagy encompasses several processes that are implicated in delivering cellular components to lysosomes for degradation; these include macroautophagy, chaperone-mediated autophagy, and microautophagy [29]. Macroautophagy (here after referred to as autophagy) occurs through a multi-step process that requires the participation of several proteins known as autophagy-related (ATG) proteins [30, 70]. Autophagy is induced by a variety of intrinsic and environmental stresses including nutrient and energy limiting conditions, ROS, hypoxia, and recognition of specific cargo such as damaged mitochondria and microorganisms [57, 65]. Upstream to ATG proteins, autophagy is regulated by several signaling pathways including those that directly control the ATG proteins such as MTORC1, AMPK, and VPS34 pathways [11, 41]. During the initiation of autophagy, MTORC1 is inhibited, which results in the activation of the ULK1/2 complex. Once activated, the ULK1/2 complex induces the phosphorylation of Atg13 and Fip200, subsequently resulting in the activation of Beclin 1, a key component of the class III phosphatidylinositol 3-kinase (VPS34) complex [11]. This step is required for the formation of a double-membrane structure, known as a phagophore, which further elongates to capture a portion of cytoplasm before forming a vesicle called the autophagosome; these are the initiation and nucleation steps. The elongation and the closure of the phagophore (the maturation step) require the recruitment of several proteins including two ubiquitin-like conjugation systems, Atg5-Atg12 and Atg8 (LC3)-phosphatidyl ethanolamine, to the phagophore. Eventually, the autophagosome matures into an autophagolysosome by fusion with a lysosome. As a consequence of this fusion, the captured components are degraded by lysosomal hydrolases, resulting in production of intracellular ATP and new pool of biomolecules [41] (Fig. 1). The autophagy process is critical for the adaptation of cells to stressful conditions such as nutrient and energy limitation [57, 65]. Autophagy also promotes the selective removal of aggregated proteins and damaged mitochondria that are accumulated in cells exposed to oxidative insults or other injuries, thereby limiting the abundance of potentially toxic components [2, 25, 59, 102]. In addition, autophagy eliminates microorganisms (bacteria, parasites, and viruses) that invade cells. Both autophagy and immune responses are regulated by ROS suggesting the functional interconnection between these two processes [19, 44, 60].

1.2 Reactive Oxygen Species

ROS are a variety of reactive molecules formed from partial reduction of an oxygen molecule. ROS include radical species such as superoxide anion ($O_2^{\cdot -}$), hydroxyl radical (·OH), and nitric oxide (NO) and non-radical ones such as hydrogen peroxide

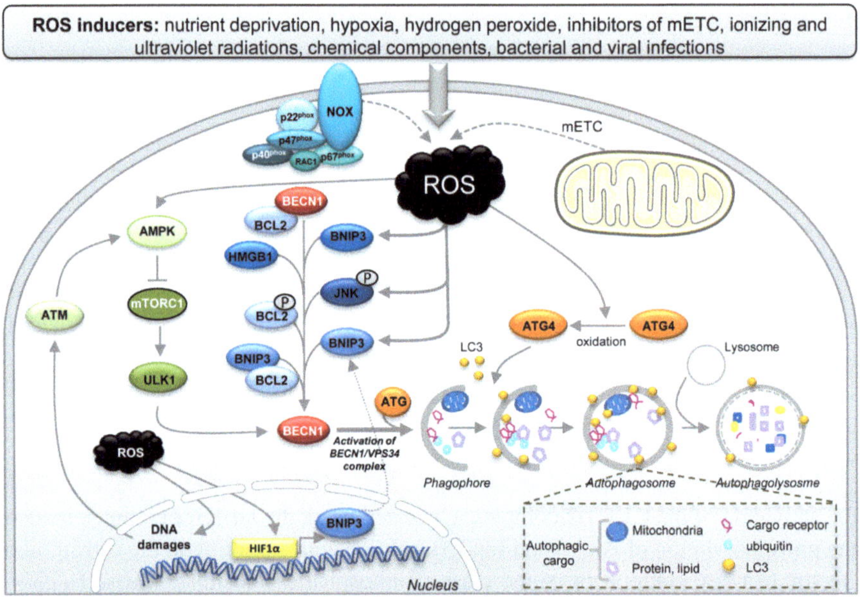

Fig. 1 Regulation of autophagy by reactive oxygen species. Autophagy is activated by ROS generated by intrinsic and environmental cellular cues. The major intracellular sources of ROS production are the mitochondrial electron transport chain (*mETC*) and the NADPH oxidase (*NOX*), a multi-proteins complex. Under starvation conditions, ROS can directly sense autophagy by inducing the oxidation of the essential autophagy protein ATG4, a process that maintains autophagy in its active form. ROS can also influence the activity of proteins that regulate autophagy. For example, ROS promotes the activation AMPK, which leads to autophagy induction as a consequence of the sequential inhibition of MTORC1 and the activation of ULK complexes. ROS-mediated DNA damage also promotes AMPK activation through an ATM-dependent mechanism. ROS stimulate autophagy by promoting the release the autophagic protein BECN 1 from its inhibitory interaction with Bcl-2 through two distinct redox mechanisms: (i) the competitive binding of either BNIP3 or HMGB1 to Bcl-2 and (ii) the activation of JNK-1, which mediates Bcl-2 phosphorylation

(H_2O_2) and singlet oxygen ($^1O_2$) [36, 43]. The biological properties of each of these species are governed by their chemical reactivity, their lifetimes in the particular biological environment, and their ability to cross the biological lipid membranes. Reactive oxygen species are constantly produced in cells under normal physiological conditions and in response to environmental injury and cellular stress conditions (e.g., mitochondrial respiration, phagocytosis, bacterial and viral infections, inflammatory cytokines, hypoxia, ultraviolet and ionizing radiations, and various pharmacological components) [17]. One of the main sources of endogenous ROS production is the mitochondrial respiratory electron transport chain (mETC) [48, 72]; electron leakage from oxidative phosphorylation generates superoxide anion, predominantly through the complexes I and III of mETC [90]. Superoxide anion is also generated through membrane-localized NADPH oxidase (NOX) enzymes in, for example, phagocytic cells activated upon microbial infection, a process that leads to killing of the bacteria that these cells engulf [80]. Superoxide anion is rapidly converted by superoxide dismutase (SOD1, SOD2) enzymes into hydrogen peroxide H_2O_2 [32], which is very harmful to cells because it may cross biological membranes and react with iron(II) (through the Fenton reaction) to generate the highly reactive hydroxyl radical .OH [36]. At the physiological level, ROS serve as second messengers to ensure cellular homeostasis by regulating various redox-sensitive signaling pathways, in particular those that govern cell growth, cell survival, differentiation, and immune and inflammatory processes [43, 81]. However, the excessive and aberrant production of ROS results in oxidative stress, a process that may lead to DNA oxidation, lipid peroxidation or protein oxidation, and mitochondria oxidative damage, all of which resulting in loss of cellular integrity when the cellular defense systems fail to repair or replace damaged components [87]. Cells have evolved several antioxidant defense mechanisms. These include enzymatic systems such as SODs, glutathione peroxidase, catalase, thioredoxin, heme oxygenase, and small molecules such as glutathione and NADPH that have ROS-scavenging functions [35]. Furthermore, the level of ROS is tightly controlled by activities of several transcription factors including NRF-2, NF-κB, and p53 that drive transcriptional activation of genes encoding proteins with antioxidant and detoxification functions [79, 87].

The ubiquitin proteasome system (UPS) and the lysosomal degradation pathways (macroautophagy and chaperone-mediated autophagy) protect cells against the toxic effects of oxidatively damaged components that accumulate during oxidative stress [50, 89]. Oxidized and aggregated proteins are degraded by both UPS and autophagy processes, but the degradation and recycling of damaged mitochondria in eukaryotes cells occurs only through autophagy. In this chapter, we will summarize the molecular interactions between macroautophagy and ROS. We will not discuss the roles of the proteasome or chaperone-mediated autophagy, two other major pathways involved in degradation of oxidized proteins.

2 Redox-Sensitive Signaling in the Regulation of Autophagy

2.1 ROS Regulatory Mechanisms That Impact Autophagy

A body of evidence indicates that mechanisms that sense ROS control autophagy under a variety of cellular contexts [22, 63, 85] (Fig. 1). The first evidence for such regulation came from study showing that nerve growth factor limitation induces the accumulation of ROS in neuronal cells leading to lipid peroxidation and the induction of autophagic cell death [51, 100]. Similarly, studies in yeast demonstrated that the mitochondrial ROS production induces lipid peroxidation of mitochondrial membranes, which elicits the activation of autophagy [52]. Subsequently, we revealed that direct addition of exogenous ROS, namely H_2O_2, promotes upregulation of both the essential autophagic protein Beclin 1 and autophagy in a range of cancer cell lines [26, 27]. A signaling role for ROS in autophagy execution has been also shown in several other ROS-generating settings, for instance, during nutrient and oxygen (hypoxia) limiting conditions [5, 8, 61, 86], in the context of inhibition the mitochondrial electron transport chain [15], in response to damage-associated molecular patterns (DAMPs); e.g., LPS [103], in response to bacterial and viral infections [44], in conditions of caspase inhibition [105], and in response to a range of oxidant agents and insults (e.g., arsenic trioxide, 2-methoxyestradiol, photodynamic therapy and ultraviolet radiation) [23, 49, 107].

ROS regulates autophagy not only in physiological conditions but also under various pathological contexts such as cancer, neurodegenerative diseases, aging, inflammatory-associated disorders, ischemia/reperfusion, and traumatic brain injury. In cancer cells, autophagy is stimulated in response to hypoxia (low oxygen) and to ROS produced by the stromal cells in the tumor microenvironment [8, 77] as well as anticancer agents that generate ROS [22]. Oxidative stress causes the intracellular accumulation of aggregate-prone proteins that are associated with cytotoxicity and can lead to degenerative disorders (including neurodegenerative diseases) in the absence of functional activity of the autophagy and the proteasome pathways [33, 40]. The activation of the inflammatory signaling cascade requires the generation of ROS, and autophagy mitigates inflammation and the development of inflammatory-associated pathologies because it eliminates damaged mitochondria [19, 34, 108]. In turn, evidence indicates that inflammasome activation block mitophagy [104]. Excessive ROS production promotes the activation of autophagy in myocardial injury in response to ischemia/reperfusion in the mouse heart [39]. Similarly, ROS participate in the induction of autophagy after traumatic brain injury in mice, a mechanism that contributes to neurologic outcomes [58].

The mitochondrial electron transport chain and the NADPH oxidase enzymes are two main cellular sources of ROS that are involved in the regulation of autophagy [23, 34, 44, 85]. A line of evidence indicates that nutrient limitation induces the production of mitochondrial ROS specially, hydrogen peroxide produced by rapid conversion of superoxide anion by the superoxide dismutase enzymes– a response responsible for the initiation of autophagy [86]. The activation of autophagy in this context results from direct oxidation of an autophagy core machinery protein, Atg4, which contributes to the vesicle elongation steps of autophagy by preventing the delipidation of the essential autophagic protein, LC3-II (Atg8-PE). Further studies indicated a major role of the anion superoxide in the induction of autophagy upon nutrient or metabolic substrate deprivation (e.g., glucose, pyruvate, glutamine, serum) [62]. In the same vein, the inhibition of mETC by rotenone or the mitochondrial antioxidant MnSOD by the anti-cancer agent 2-metoxyestrodial [14], induce the accumulation of superoxide anion, which in turn promotes autophagic cell death. Moreover, the contribution of mitochondrial ROS generation in activation of autophagy has been shown in cells subjected to hypoxic conditions [106]. During hypoxia, the activation of autophagy is dependent on the induction of expression of BNIP3 and BNIP3L proteins by HIF 1, a cardinal transcription factor necessary for adaption to hypoxic conditions [8, 67, 92]. BNIP3 is a BH3-domain protein that stimulates autophagy by direct binding to the autophagic protein LC3 or by interacting with Bcl-2, a mechanism that induces autophagy by inducing the dissociation of Beclin 1 from its inhibitor Bcl-2 [18]. Under moderate hypoxic conditions, AMPK, an energy sensor protein, elicits autophagy through inhibition of MTORC1. AMPK is also activated in cells upon hydrogen peroxide exposure, which results in glutathionylation of its reactive cysteines at positions 299 and 304 [109]. ATM, a kinase that coordinates the cellular response to DNA damage, is another redox sensitive protein that can promote autophagy through the AMPK pathway in cells under oxidative stress [1].

HMGB1, another redox-sensitive protein, also regulates autophagy [61]. HMGB1 contains three cysteines at positions 23, 45, and 106, and the redox-induced modifications of these cysteines influence its functions in different cellular processes [101]. Under oxidative stress, oxidation of cysteines at positions 23 and 45 activates autophagy by inducing the dissociation of Beclin 1 from its inhibitor Bcl-2, thus leading to autophagy induction. Other regulators of autophagy induction in response to ROS include Jun N-terminal kinase (JNK) and the endoplasmic reticulum stress sensor PERK [3, 98].

ROS can also be generated by NOX enzymes in multiple cellular compartments [7, 74]. This process occurs predominantly during pathogen killing by professional phagocytic cells [80]. Defective NOX2 activity causes the genetic disorder chronic granulomatous, which is characterized by increased susceptibility to pathogens [95]. Several pathogen-associated molecular patterns (PAMPs) (e.g., microbial pathogens) sense phagocytosis mediated by Toll-like or Fc gamma receptors through a mechanism that involves NOX activation, a process that ultimately results in pathogen killing [76]. During phagocytosis, NOX-mediated ROS generation leads to the recruitment of LC3 to the phagosomal membrane, thereby facilitating the fusion of phagosome with the lysosome [84]. It remains to be elucidated how ROS regulate the association of the autophagic components with the phagosomal membrane and which mechanisms are responsible for the fusion between phagosome and lysosome.

Autophagy is also regulated at the transcriptional level by a variety of redox-sensitive transcription factors including p53, HIF1, FOXO3, NRF-2, and NF-κB [79]. These transcription factors promote the transcription of Sestrins, BNIP3, LC3 and BNIP3, and p62/SQSTM1 (the autophagic cargo receptor), and Beclin 1, respectively, leading to sustained activation of autophagy [8, 12, 16, 47, 66]. Notably, ROS-mediated the induction of autophagy can operate to ensure cell survival or to promote cell death processes though mechanisms that have not been fully elucidated [4, 69]. The natures of ROS species, their cellular localizations, and the durations of the ROS-mediated signals are all factors that might influence the role of autophagy in the fates of cells subjected to oxidative injuries. The effectiveness of the antioxidant and the detoxifying processes are other parameters that impact the redox regulation of autophagy and its functional roles in cells exposed to ROS-generating conditions.

2.2 *Antioxidant-Mediated Regulatory Mechanisms That Control Autophagy*

Upon prolonged ROS accumulation, the levels of several antioxidant enzymes are upregulated as a result of the activation of specific transcription factors including NF-κB, NRF-2, and p53. Antioxidants have been shown to regulate autophagy both by

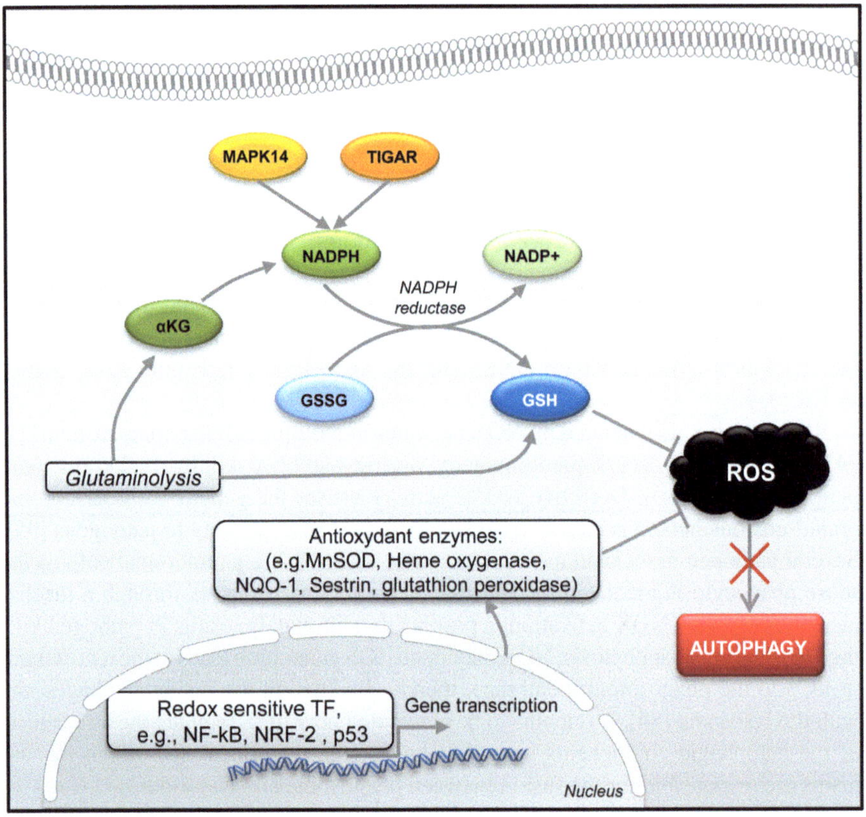

Fig. 2 Regulation of autophagy by the cellular antioxidant defense systems. Several redox sensitive transcription factors (*TF*), including NF-κB, NRF-2, and p53 can negatively regulate autophagy through their abilities to promote the transcriptional activation of specific antioxidant and detoxifying enzymes. ROS-induced autophagy is also inhibited by glutathione (*GSH*), a potent and major antioxidant molecule in cells. GSH is oxidized to form GSSG in response to a variety of oxidant sensors. The enzyme glutathione reductase catalyzes, in the presence of NADPH, the reduction of GSSG to GSH to maintain the cellular redox homeostasis. Under nutrient limiting conditions, NADPH levels are increased due to the activation of both TIGAR and the MAPK14, a mechanism that leads to decreased levels of ROS and the termination of the initial autophagy activation. Glutaminolysis (a process that is involved in the conversion of glutamine to α-ketoglutarate (*αKG*)) also has an inhibitory effect on autophagy by suppressing the accumulation of ROS and increasing the levels of GSH and NADPH

inhibiting basal levels of autophagy and by terminating the initial autophagy activation induced by ROS (Fig. 2).

NF-κB regulates autophagy through a redox-based mechanism. The activation of NF-κB in response to cytokines (e.g., TNFα, IL1β), Toll-like receptors, or increased levels of ROS elicits the transcription of a variety of target genes including those involved in the apoptotic and antioxidant responses (MnSOD, Heme oxygenase, and Thioredoxin) [6, 78, 93]. Loss of the NF-κB activity results in the accumulation of intracellular ROS, a process that participates in the apoptotic and necrotic cell death modalities [24]. We have previously shown that TNF induces autophagy in the absence of functional NF-κB in a manner that can be inhibited by antioxidants [26, 27]. Thus, NF-κB can repress both autophagy and apoptosis due to its ability to promote antioxidant responses. Whether or not the redox-regulating function of NF-κB depends on the duration, the abundance, or the nature of ROS generated in cells remains to be elucidated.

NRF-2 regulates autophagy upon induction of oxidative stress through the activation of transcription of genes encoding cellular defense enzymes [46, 54]. Under normal growth conditions, NRF-2 is repressed through sequestration within the cytosol by KEAP 1, a mechanism that facilitates the ubiquitination of NRF-2 and its subsequent degradation through proteasome pathway [55]. The interaction between NRF-2 and KEAP 1 is disrupted under oxidative stress conditions allowing the translocation of NRF-2 to the nucleus where it binds to the antioxidant response elements (AREs) located in the promoter regions of genes encoding several antioxidant enzymes [53]. Under oxidative stress, NRF-2 also promotes the transcription of the gene that encodes p62/SQSTM1, the autophagic cargo receptor [47]. Thus, NRF-2 has a dual role in the regulation of autophagy under oxidative stress conditions: on the one hand NRF-2-mediated induction of transcription of antioxidant driven genes that, ultimately, inhibits the initial activation of autophagy induced by ROS. On the other hand, NRF-2 facilitates termination of autophagy by enhancing the expression of p62/SQSTM1, which actively participates in the degradation of oxidized proteins by delivering them to the autophagosomes. In turn, p62/SQSTM1 has been shown to induce NRF-2 activation through inactivation of Keap1 [55]. NRF-2 and p62/SQSTM1, thereby form a positive feedback regulatory loop that controls autophagy in the presence of ROS.

TIGAR is another protein that exerts a negative impact on autophagy by reducing intracellular ROS levels [9]. TIGAR is a p53-inducible protein that drives a metabolic shift from glycolysis to the pentose phosphate pathway, which leads to the regeneration of NADPH, a key antioxidant involved in many reduction reactions including the reduction of oxidized form of glutathione to its reduced form GSH [10]. Loss of TIGAR leads to increases in ROS levels, activation of autophagy, and apoptotic cell death under nutrient limitation, metabolic stress, or exogenous addition of H_2O_2 [9]. Under nutrient limiting conditions, the activation of MAPK14/p38α also drives the production of NADPH as a result of a metabolic shift from glycolysis to the pentose phosphate pathway, thereby leading to the termination of the initial induction of autophagy [21]. The levels of NADPH and GSH are also increased by glutaminolysis, an enzymatic process that is responsible for the conversion of glutamine to α-ketoglutarate, an intermediate of the tricarboxylic acid cycle. As a consequence of glutaminolysis, the intracellular ROS levels decline, which result in autophagy inhibition [64]. Taken together, all these data revealed, thus, the influence of metabolic reprogramming (induced by TIGAR, glutaminolysis, and MAPK14) in the regulation of cellular redox status and autophagy, as well.

Interestingly, thiol-containing antioxidants (such as GSH and N-acetylcysteine) and vitamin E (a lipophilic antioxidant) inhibit autophagy in different cell lines as well as *in vivo* models such as in starved mice and in a zebrafish model of Huntington's disease [20, 94]. The thiol-containing antioxidants and vitamin E inhibit the JNK-1/Bcl-2 pathway and stimulate MTORC1, respectively. Taken together, all these studies indicate that the balance of antioxidant systems and the oxidant stressors has important influence on autophagic activity and on oxidative damage in cells [94].

3 Autophagy Limits ROS Accumulation and Oxidative Stress

ROS are clearly involved in autophagy induction, and a body of evidence indicates that autophagy, in turn, regulates the cellular redox status [4, 85]. There are at least three major mechanisms through which autophagy (and autophagy-regulatory proteins) modulates both ROS production and oxidative stress. These include (*i*) the selective degradation of damaged cellular components including damaged mitochondria and oxidized cellular substrates [2, 25, 34, 59, 102], (*ii*) the selective removal of catalase, an antioxidant enzyme [105], and (*iii*) the activation of antioxidant responses driven by the autophagy receptor p62/SQSTM1 [45].

Under moderately high levels of ROS, autophagy serves as a defense mechanism by removing damaged components to protect cells against oxidative stress. However, in some specific conditions, the induction of autophagy prolongs oxidative stress by selective degradation of the antioxidant enzyme catalase [105] or by not yet identified mechanism, thereby contributing to the execution of cell death process.

Autophagy is critical for the quality control of mitochondria and regulates mitochondria number in response to a variety of physiological and developmental signals [2, 25, 34, 59, 102]. In fact, autophagy is responsible for the degradation and recycling of superfluous and damaged mitochondria, the so-called mitophagy process. Mice harboring genetic defects in autophagy have tissues that accumulate high levels of ROS, abnormal mitochondria, lipid droplets, and aggregates-prone proteins [96]. Excessive ROS production induced by defective mitophagy causes DNA, protein, and lipid damages (including oxidative damages), which leads to persistent tissue damage, cell death and inflammation. Tissue damage and inflammation, in turn, increase the incidence of various pathologies such as neurodegenerative diseases, cancer, and the inflammatory-associated disorders [84, 96, 108]. In this sense, mice with genetic defect of autophagy in neural cells, accumulate abnormal mitochondria and aggregated protein in brain and manifest symptoms of neurodegenerative diseases as they age [38, 56, 100]. A piece of evidence suggests that tissue damage and inflammation resulting in autophagy defects enable the development and the progression of tumors as shown for hepatocellular carcinoma, lung adenocarcinomas, and lymphomas in animal models [91, 96]. The mechanisms though which these alterations lead to tumorigenesis are increased genomic instability, the aberrant accumulation of p62/SQSTM1, the failure to eliminate toxic aggregate–prone proteins, and the secretion of pro-inflammatory cytokines into the tumor microenvironment [97].

Interestingly, mice with defects in Atg16L1, a protein involved in mitophagy and autophagy, exhibit granule accumulation in intestinal Paneth cells, a similar phenotype to that of Crohn' disease, a major type of inflammatory bowel disease [13, 75]. Human genome-wide association studies have also linked several single-nucleotide polymorphisms (SNP) in *Atg16L1* to susceptibility to Crohn' disease in humans [37, 82]. However, the potential contribution of accumulation of damaged mitochondria and ROS to the development of Crohn' disease has not been experimentally confirmed.

Another link between ROS autophagy and inflammatory responses results from the redox regulation of inflammasome activity by autophagy [19, 34, 108]. The inflammasome is a multiprotein platform responsible for inflammatory processes that is activated following the recognition of pathogen-associated molecular pattern (PAMPs) or DAMPs by NOD-like receptor proteins [88]. Once activated, the inflammasome initiates the processing of caspase 1, which in turn promotes the maturation and the secretion of pro-inflammatory cytokines such as interleukin IL-1β and IL-18. A body of evidence suggests that ROS, mainly those generated by mitochondria, are required for inflammasome initiation [34, 108]. One important piece of evidence linking the ROS-dependent regulation of inflammasome with autophagy came from a study showing that autophagy inhibition results in the generation of mitochondrial ROS. This suggests that autophagy impedes inflammatory responses by preserving mitochondrial integrity. In agreement with this study, macrophages isolated from mice defective in autophagy (due to deficiencies in LC3, Atg7, or Atg16L1) display elevated levels of IL-1β and IL-18 secretion upon ATP stimulation; this supports the idea that defective autophagy results in inflammasome-mediated prolonged inflammatory responses that may ultimately contribute to tissue damage [19, 73, 83, 108].

Moreover, autophagy defends against oxidative stress as the process removes proteins and lipids that are oxidized by ROS [31, 42]. For example, autophagy is involved in the degradation of oxidized proteins under oxidative stress conditions in the plant *Arabidopsis thaliana* [99]. Autophagy is induced in keratinocytes exposed to ultraviolet A radiation. In keratinocytes from mice lacking *Atg7* (and therefore deficient in autophagy), UV-A radiation induces the accumulation of oxidized lipids as well as proteins aggregates containing the autophagy adaptor protein p62/SQSTM1 suggesting that autophagy protects cells against oxidative damage through a p62/SQSTM1-dependent mechanism [107]. p62/SQSTM1 is also a key regulator of NRF-2 and NF-κB, transcription factors involved in the cellular antioxidant responses [55, 71]. Intriguingly, the aggregation of p62/SQSTM1 due to a massive accumulation of ROS or in autophagy-deficient cells result in the loss of its functional role in NF-κB activation and leads to further ROS production [28, 68]. Thus, the aggregation of p62/SQSTM1 allows the switch of p62/SQSTM1 function from antioxidant to proxidant. As p62/SQSTM1 plays roles in several signaling pathways, it remains unclear whether the role of p62/SQSTM in redox modulation is dependent on only its function in autophagy or also on its control of other signaling pathways.

Taken together, all these studies reveal that the activation of autophagy by ROS constitutes often a negative feedback loop mechanism for keeping down the levels of ROS and oxidative stress in cells. This certainly has functional relevance in the prevention of tissue damage, cell death and inflammatory responses.

4 Conclusion

The aforementioned examples reveal the mutual and complex relationships among pathways that regulate the ROS levels and the autophagy signaling process. On the one hand, ROS serve as important autophagy activating signals in response to a variety of intrinsic and environmental cellular signals. Accordingly, the inhibition of ROS levels (e.g. using antioxidants) has been shown to impair the autophagic responses in several settings. On the other hand, autophagy is an important antioxidant defense mechanism under oxidative stress conditions since genetic defects in autophagy components lead to oxidative stress, tissue damage and inflammation in several animal models. These data suggest that defect in autophagy contribute to aging and to the development of several diseases (e.g., cancer, inflammatory disorders and neurodegenerative diseases). Although several cellular targets of ROS in the regulation of autophagy signaling process have been identified, further investigation is needed to identify other potential autophagy sensors (proteins, lipids, DNA) that can be targeted by ROS.

Moreover, the mechanisms through which autophagy is involved in the selective degradation of oxidized cellular substrates (e.g., damaged mitochondria, peroxidized lipids and oxidized proteins) are not well understood. Research in this area should identify new therapy concepts that may prevent the development of degenerative diseases linked to tissue damage and inflammation.

Acknowledgments This work was supported by funds from the following institutions: Institut National de la Santé et de la Recherche Médicale – INSERM, University of Bordeaux the Conseil Régional d'Aquitaine, Institut Bergonié, Ligue contre le Cancer – Comité de la Gironde, and INCa-DGOS-Inserm 6046.

References

1. Alexander A, Cai S-L, Kim J, Nanez A, Sahin M, MacLean KH, Inoki K, Guan K-L, Shen J, Person MD, Kusewitt D, Mills GB, Kastan MB, Walker CL (2010) ATM signals to TSC2 in the cytoplasm to regulate mTORC1 in response to ROS. Proc Natl Acad Sci U S A 107:4153–4158. doi:10.1073/pnas.0913860107
2. Ashrafi G, Schwarz TL (2013) The pathways of mitophagy for quality control and clearance of mitochondria. Cell Death Differ 20:31–42. doi:10.1038/cdd.2012.81
3. Avivar-Valderas A, Salas E, Bobrovnikova-Marjon E, Diehl JA, Nagi C, Debnath J, Aguirre-Ghiso JA (2011) PERK integrates autophagy and oxidative stress responses to promote survival during extracellular matrix detachment. Mol Cell Biol 31:3616–3629. doi:10.1128/MCB.05164-11

4. Azad MB, Chen Y, Gibson SB (2009) Regulation of autophagy by reactive oxygen species (ROS): implications for cancer progression and treatment. Antioxid Redox Signal 11:777–790. doi:10.1089/ARS.2008.2270
5. Azad MB, Chen Y, Henson ES, Cizeau J, McMillan-Ward E, Israels SJ, Gibson SB (2008) Hypoxia induces autophagic cell death in apoptosis-competent cells through a mechanism involving BNIP3. Autophagy 4:195–204
6. Baldwin AS (2012) Regulation of cell death and autophagy by IKK and NF-κB: critical mechanisms in immune function and cancer. Immunol Rev 246:327–345. doi: 10.1111/j.1600-065X.2012.01095
7. Bedard K, Krause K-H (2007) The NOX family of ROS-generating NADPH oxidases: physiology and pathophysiology. Physiol Rev 87:245–313. doi:10.1152/physrev.00044.2005
8. Bellot G, Garcia-Medina R, Gounon P, Chiche J, Roux D, Pouysségur J, Mazure NM (2009) Hypoxia-induced autophagy is mediated through hypoxia-inducible factor induction of BNIP3 and BNIP3L via their BH3 domains. Mol Cell Biol 29:2570–2581. doi:10.1128/MCB.00166-09
9. Bensaad K, Cheung EC, Vousden KH (2009) Modulation of intracellular ROS levels by TIGAR controls autophagy. EMBO J 28:3015–3026. doi:10.1038/emboj.2009.242
10. Bensaad K, Tsuruta A, Selak MA, Vidal MNC, Nakano K, Bartrons R, Gottlieb E, Vousden KH (2006) TIGAR, a p53-inducible regulator of glycolysis and apoptosis. Cell 126:107–120. doi:10.1016/j.cell.2006.05.036
11. Boya P, Reggiori F, Codogno P (2013) Emerging regulation and functions of autophagy. Nat Cell Biol 15:713–720. doi:10.1038/ncb2788
12. Budanov AV (2011) Stress-responsive sestrins link p53 with redox regulation and mammalian target of rapamycin signaling. Antioxid Redox Signal 15:1679–1690. doi:10.1089/ars.2010.3530
13. Cadwell K, Liu JY, Brown SL, Miyoshi H, Loh J, Lennerz JK, Kishi C, Kc W, Carrero JA, Hunt S, Stone CD, Brunt EM, Xavier RJ, Sleckman BP, Li E, Mizushima N, Stappenbeck TS, Virgin HW (2008) A key role for autophagy and the autophagy gene Atg16l1 in mouse and human intestinal Paneth cells. Nature 456:259–263. doi:10.1038/nature07416
14. Chen Y, McMillan-Ward E, Kong J, Israels SJ, Gibson SB (2008) Oxidative stress induces autophagic cell death independent of apoptosis in transformed and cancer cells. Cell Death Differ 15:171–182. doi:10.1038/sj.cdd.4402233
15. Chen Y, McMillan-Ward E, Kong J, Israels SJ, Gibson SB (2007) Mitochondrial electron-transport-chain inhibitors of complexes I and II induce autophagic cell death mediated by reactive oxygen species. J Cell Sci 120:4155–4166. doi:10.1242/jcs.011163
16. Copetti T, Demarchi F, Schneider C (2009) p65/RelA binds and activates the beclin 1 promoter. Autophagy 5:858–859
17. D'Autréaux B, Toledano MB (2007) ROS as signalling molecules: mechanisms that generate specificity in ROS homeostasis. Nat Rev Mol Cell Biol 8:813–824. doi:10.1038/nrm2256
18. Decuypere J-P, Parys JB, Bultynck G (2012) Regulation of the autophagic Bcl-2/beclin 1 interaction. Cells 1:284–312. doi:10.3390/cells1030284
19. Deretic V, Saitoh T, Akira S (2013) Autophagy in infection, inflammation and immunity. Nat Rev Immunol 13:722–737. doi:10.1038/nri3532
20. Desideri E, Filomeni G, Ciriolo MR (2012) Glutathione participates in the modulation of starvation-induced autophagy in carcinoma cells. Autophagy 8:1769–1781. doi:10.4161/auto.22037
21. Desideri E, Vegliante R, Cardaci S, Nepravishta R, Paci M, Ciriolo MR (2014) MAPK14/p38α-dependent modulation of glucose metabolism affects ROS levels and autophagy during starvation. Autophagy 10:1652–1665. doi:10.4161/auto.29456
22. Dewaele M, Maes H, Agostinis P (2010) ROS-mediated mechanisms of autophagy stimulation and their relevance in cancer therapy. Autophagy 6:838–854. doi:10.4161/auto.6.7.12113
23. Dewaele M, Martinet W, Rubio N, Verfaillie T, de Witte PA, Piette J, Agostinis P (2011) Autophagy pathways activated in response to PDT contribute to cell resistance against ROS damage. J Cell Mol Med 15:1402–1414. doi:10.1111/j.1582-4934.2010.01118.x

24. DiDonato JA, Mercurio F, Karin M (2012) NF-κB and the link between inflammation and cancer. Immunol Rev 246:379–400. doi:10.1111/j.1600-065X.2012.01099.x
25. Ding W-X, Yin X-M (2012) Mitophagy: mechanisms, pathophysiological roles, and analysis. Biol Chem 393:547–564. doi:10.1515/hsz-2012-0119
26. Djavaheri-Mergny M, Amelotti M, Mathieu J, Besançon F, Bauvy C, Codogno P (2007) Regulation of autophagy by NFkappaB transcription factor and reactives oxygen species. Autophagy 3:390–392
27. Djavaheri-Mergny M, Amelotti M, Mathieu J, Besançon F, Bauvy C, Souquère S, Pierron G, Codogno P (2006) NF-kappaB activation represses tumor necrosis factor-alpha-induced autophagy. J Biol Chem 281:30373–30382. doi:10.1074/jbc.M602097200
28. Duran A, Linares JF, Galvez AS, Wikenheiser K, Flores JM, Diaz-Meco MT, Moscat J (2008) The signaling adaptor p62 is an important NF-kappaB mediator in tumorigenesis. Cancer Cell 13:343–354. doi:10.1016/j.ccr.2008.02.001
29. Eskelinen E-L (2005) Maturation of autophagic vacuoles in mammalian cells. Autophagy 1:1–10
30. Feng Y, He D, Yao Z, Klionsky DJ (2014) The machinery of macroautophagy. Cell Res 24:24–41. doi:10.1038/cr.2013.168
31. Filomeni G, De Zio D, Cecconi F (2015) Oxidative stress and autophagy: the clash between damage and metabolic needs. Cell Death Differ 22:377–388. doi:10.1038/cdd.2014.150
32. Fridovich I (1995) Superoxide radical and superoxide dismutases. Annu Rev Biochem 64:97–112. doi:10.1146/annurev.bi.64.070195.000525
33. Ghavami S, Shojaei S, Yeganeh B, Ande SR, Jangamreddy JR, Mehrpour M, Christoffersson J, Chaabane W, Moghadam AR, Kashani HH, Hashemi M, Owji AA, Łos MJ (2014) Autophagy and apoptosis dysfunction in neurodegenerative disorders. Prog Neurobiol 112:24–49. doi:10.1016/j.pneurobio.2013.10.004
34. Green DR, Galluzzi L, Kroemer G (2011) Mitochondria and the autophagy-inflammation-cell death axis in organismal aging. Science 333:1109–1112. doi:10.1126/science.1201940
35. Gutteridge JMC, Halliwell B (2010) Antioxidants: molecules, medicines, and myths. Biochem Biophys Res Commun 393:561–564. doi:10.1016/j.bbrc.2010.02.071
36. Halliwell B, Cross CE (1994) Oxygen-derived species: their relation to human disease and environmental stress. Environ Health Perspect 102(Suppl 10):5–12
37. Hampe J, Franke A, Rosenstiel P, Till A, Teuber M, Huse K, Albrecht M, Mayr G, De La Vega FM, Briggs J, Günther S, Prescott NJ, Onnie CM, Häsler R, Sipos B, Fölsch UR, Lengauer T, Platzer M, Mathew CG, Krawczak M, Schreiber S (2007) A genome-wide association scan of nonsynonymous SNPs identifies a susceptibility variant for Crohn disease in ATG16L1. Nat Genet 39:207–211. doi:10.1038/ng1954
38. Hara T, Nakamura K, Matsui M, Yamamoto A, Nakahara Y, Suzuki-Migishima R, Yokoyama M, Mishima K, Saito I, Okano H, Mizushima N (2006) Suppression of basal autophagy in neural cells causes neurodegenerative disease in mice. Nature 441:885–889. doi:10.1038/nature04724
39. Hariharan N, Zhai P, Sadoshima J (2011) Oxidative stress stimulates autophagic flux during ischemia/reperfusion. Antioxid Redox Signal 14:2179–2190. doi:10.1089/ars.2010.3488
40. Harris H, Rubinsztein DC (2012) Control of autophagy as a therapy for neurodegenerative disease. Nat Rev Neurol 8:108–117. doi:10.1038/nrneurol.2011.200
41. He C, Klionsky DJ (2009) Regulation mechanisms and signaling pathways of autophagy. Annu Rev Genet 43:67–93. doi:10.1146/annurev-genet-102808-114910
42. Höhn A, Jung T, Grune T (2014) Pathophysiological importance of aggregated damaged proteins. Free Radic Biol Med 71:70–89. doi:10.1016/j.freeradbiomed.2014.02.028
43. Holmström KM, Finkel T (2014) Cellular mechanisms and physiological consequences of redox-dependent signalling. Nat Rev Mol Cell Biol 15:411–421. doi:10.1038/nrm3801
44. Huang J, Canadien V, Lam GY, Steinberg BE, Dinauer MC, Magalhaes MAO, Glogauer M, Grinstein S, Brumell JH (2009) Activation of antibacterial autophagy by NADPH oxidases. Proc Natl Acad Sci U S A 106:6226–6231. doi:10.1073/pnas.0811045106

45. Ishii T, Warabi E, Siow RCM, Mann GE (2013) Sequestosome1/p62: a regulator of redox-sensitive voltage-activated potassium channels, arterial remodeling, inflammation, and neurite outgrowth. Free Radic Biol Med 65:102–116. doi:10.1016/j.freeradbiomed.2013.06.019
46. Itoh K, Chiba T, Takahashi S, Ishii T, Igarashi K, Katoh Y, Oyake T, Hayashi N, Satoh K, Hatayama I, Yamamoto M, Nabeshima Y (1997) An Nrf2/small Maf heterodimer mediates the induction of phase II detoxifying enzyme genes through antioxidant response elements. Biochem Biophys Res Commun 236:313–322
47. Jain A, Lamark T, Sjøttem E, Bowitz Larsen K, Atesoh Awuh J, Øvervatn A, McMahon M, Hayes JD, Johansen T (2010) p62/SQSTM1 is a target gene for transcription factor NRF2 and creates a positive feedback loop by inducing antioxidant response element-driven gene transcription. J Biol Chem 285:22576–22591. doi:10.1074/jbc.M110.118976
48. Jensen PK (1966) Antimycin-insensitive oxidation of succinate and reduced nicotinamide-adenine dinucleotide in electron-transport particles. I. pH dependency and hydrogen peroxide formation. Biochim Biophys Acta 122:157–166
49. Kanzawa T, Kondo Y, Ito H, Kondo S, Germano I (2003) Induction of autophagic cell death in malignant glioma cells by arsenic trioxide. Cancer Res 63:2103–2108
50. Kiffin R, Christian C, Knecht E, Cuervo AM (2004) Activation of chaperone-mediated autophagy during oxidative stress. Mol Biol Cell 15:4829–4840. doi:10.1091/mbc.E04-06-0477
51. Kirkland RA, Adibhatla RM, Hatcher JF, Franklin JL (2002) Loss of cardiolipin and mitochondria during programmed neuronal death: evidence of a role for lipid peroxidation and autophagy. Neuroscience 115:587–602
52. Kissová I, Deffieu M, Samokhvalov V, Velours G, Bessoule J-J, Manon S, Camougrand N (2006) Lipid oxidation and autophagy in yeast. Free Radic Biol Med 41:1655–1661. doi:10.1016/j.freeradbiomed.2006.08.012
53. Kobayashi A, Kang M-I, Okawa H, Ohtsuji M, Zenke Y, Chiba T, Igarashi K, Yamamoto M (2004) Oxidative stress sensor Keap1 functions as an adaptor for Cul3-based E3 ligase to regulate proteasomal degradation of Nrf2. Mol Cell Biol 24:7130–7139. doi:10.1128/MCB.24.16.7130-7139.2004
54. Kobayashi M, Yamamoto M (2005) Molecular mechanisms activating the Nrf2-Keap1 pathway of antioxidant gene regulation. Antioxid Redox Signal 7:385–394. doi:10.1089/ars.2005.7.385
55. Komatsu M, Kurokawa H, Waguri S, Taguchi K, Kobayashi A, Ichimura Y, Sou Y-S, Ueno I, Sakamoto A, Tong KI, Kim M, Nishito Y, Iemura S, Natsume T, Ueno T, Kominami E, Motohashi H, Tanaka K, Yamamoto M (2010) The selective autophagy substrate p62 activates the stress responsive transcription factor Nrf2 through inactivation of Keap1. Nat Cell Biol 12:213–223. doi:10.1038/ncb2021
56. Komatsu M, Waguri S, Chiba T, Murata S, Iwata J, Tanida I, Ueno T, Koike M, Uchiyama Y, Kominami E, Tanaka K (2006) Loss of autophagy in the central nervous system causes neurodegeneration in mice. Nature 441:880–884. doi:10.1038/nature04723
57. Kroemer G, Mariño G, Levine B (2010) Autophagy and the integrated stress response. Mol Cell 40:280–293. doi:10.1016/j.molcel.2010.09.023
58. Lai Y, Hickey RW, Chen Y, Bayir H, Sullivan ML, Chu CT, Kochanek PM, Dixon CE, Jenkins LW, Graham SH, Watkins SC, Clark RSB (2008) Autophagy is increased after traumatic brain injury in mice and is partially inhibited by the antioxidant gamma-glutamylcysteinyl ethyl ester. J Cereb Blood Flow Metab 28:540–550. doi:10.1038/sj.jcbfm.9600551
59. Lemasters JJ (2005) Selective mitochondrial autophagy, or mitophagy, as a targeted defense against oxidative stress, mitochondrial dysfunction, and aging. Rejuvenation Res 8:3–5. doi:10.1089/rej.2005.8.3
60. Levine B, Mizushima N, Virgin HW (2011) Autophagy in immunity and inflammation. Nature 469:323–335. doi:10.1038/nature09782
61. Li G, Tang D, Lotze MT (2013) Ménage à trois in stress: DAMPs, redox and autophagy. Semin Cancer Biol 23:380–390. doi:10.1016/j.semcancer.2013.08.002

62. Li L, Chen Y, Gibson SB (2013) Starvation-induced autophagy is regulated by mitochondrial reactive oxygen species leading to AMPK activation. Cell Signal 25:50–65. doi:10.1016/j.cellsig.2012.09.020
63. Li L, Ishdorj G, Gibson SB (2012) Reactive oxygen species regulation of autophagy in cancer: implications for cancer treatment. Free Radic Biol Med 53:1399–1410. doi:10.1016/j.freeradbiomed.2012.07.011
64. Lorin S, Tol MJ, Bauvy C, Strijland A, Poüs C, Verhoeven AJ, Codogno P, Meijer AJ (2013) Glutamate dehydrogenase contributes to leucine sensing in the regulation of autophagy. Autophagy 9:850–860. doi:10.4161/auto.24083
65. Maiuri MC, Kroemer G (2015) Autophagy in stress and disease. Cell Death Differ 22:365–366. doi:10.1038/cdd.2014.236
66. Mammucari C, Milan G, Romanello V, Masiero E, Rudolf R, Del Piccolo P, Burden SJ, Di Lisi R, Sandri C, Zhao J, Goldberg AL, Schiaffino S, Sandri M (2007) FoxO3 controls autophagy in skeletal muscle in vivo. Cell Metab 6:458–471. doi:10.1016/j.cmet.2007.11.001
67. Mansfield KD, Guzy RD, Pan Y, Young RM, Cash TP, Schumacker PT, Simon MC (2005) Mitochondrial dysfunction resulting from loss of cytochrome c impairs cellular oxygen sensing and hypoxic HIF-α activation. Cell Metab 1:393–399. doi:10.1016/j.cmet.2005.05.003
68. Mathew R, Karp CM, Beaudoin B, Vuong N, Chen G, Chen H-Y, Bray K, Reddy A, Bhanot G, Gelinas C, Dipaola RS, Karantza-Wadsworth V, White E (2009) Autophagy suppresses tumorigenesis through elimination of p62. Cell 137:1062–1075. doi:10.1016/j.cell.2009.03.048
69. Mazure NM, Pouysségur J (2010) Hypoxia-induced autophagy: cell death or cell survival? Curr Opin Cell Biol 22:177–180. doi:10.1016/j.ceb.2009.11.015
70. Mizushima N, Yoshimori T, Ohsumi Y (2011) The role of Atg proteins in autophagosome formation. Annu Rev Cell Dev Biol 27:107–132. doi:10.1146/annurev-cellbio-092910-154005
71. Moscat J, Diaz-Meco MT (2009) p62 at the crossroads of autophagy, apoptosis, and cancer. Cell 137:1001–1004. doi:10.1016/j.cell.2009.05.023
72. Murphy MP (2009) How mitochondria produce reactive oxygen species. Biochem J 417:1–13. doi:10.1042/BJ20081386
73. Nakahira K, Haspel JA, Rathinam VA, Lee S-J, Dolinay T, Lam HC, Englert JA, Rabinovitch M, Cernadas M, Kim HP, Fitzgerald KA, Ryter SW, Choi AM (2011) Autophagy proteins regulate innate immune response by inhibiting NALP3 inflammasome-mediated mitochondrial DNA release. Nat Immunol 12:222–230. doi:10.1038/ni.1980
74. Nauseef WM (2008) Biological roles for the NOX family NADPH oxidases. J Biol Chem 283:16961–16965. doi:10.1074/jbc.R700045200
75. Nguyen HTT, Lapaquette P, Bringer M-A, Darfeuille-Michaud A (2013) Autophagy and Crohn's disease. J Innate Immun 5:434–443. doi:10.1159/000345129
76. Park J-B (2003) Phagocytosis induces superoxide formation and apoptosis in macrophages. Exp Mol Med 35:325–335
77. Pavlides S, Vera I, Gandara R, Sneddon S, Pestell RG, Mercier I, Martinez-Outschoorn UE, Whitaker-Menezes D, Howell A, Sotgia F, Lisanti MP (2011) Warburg meets autophagy: cancer-associated fibroblasts accelerate tumor growth and metastasis via oxidative stress, mitophagy, and aerobic glycolysis. Antioxid Redox Signal 16:1264–1284. doi:10.1089/ars.2011.4243
78. Perkins ND (2007) Integrating cell-signalling pathways with NF-kappaB and IKK function. Nat Rev Mol Cell Biol 8:49–62. doi:10.1038/nrm2083
79. Pietrocola F, Izzo V, Niso-Santano M, Vacchelli E, Galluzzi L, Maiuri MC, Kroemer G (2013) Regulation of autophagy by stress-responsive transcription factors. Semin Cancer Biol 23:310–322. doi:10.1016/j.semcancer.2013.05.008
80. Rada B, Hably C, Meczner A, Timár C, Lakatos G, Enyedi P, Ligeti E (2008) Role of Nox2 in elimination of microorganisms. Semin Immunopathol 30:237–253. doi:10.1007/s00281-008-0126-3
81. Ray PD, Huang B-W, Tsuji Y (2012) Reactive oxygen species (ROS) homeostasis and redox regulation in cellular signaling. Cell Signal 24:981–990. doi:10.1016/j.cellsig.2012.01.008
82. Rioux JD, Xavier RJ, Taylor KD, Silverberg MS, Goyette P, Huett A, Green T, Kuballa P, Barmada MM, Datta LW, Shugart YY, Griffiths AM, Targan SR, Ippoliti AF, Bernard E-J,

Mei L, Nicolae DL, Regueiro M, Schumm LP, Steinhart AH, Rotter JI, Duerr RH, Cho JH, Daly MJ, Brant SR (2007) Genome-wide association study identifies new susceptibility loci for Crohn disease and implicates autophagy in disease pathogenesis. Nat Genet 39:596–604. doi:10.1038/ng2032

83. Saitoh T, Fujita N, Jang MH, Uematsu S, Yang B-G, Satoh T, Omori H, Noda T, Yamamoto N, Komatsu M, Tanaka K, Kawai T, Tsujimura T, Takeuchi O, Yoshimori T, Akira S (2008) Loss of the autophagy protein Atg16L1 enhances endotoxin-induced IL-1beta production. Nature 456:264–268. doi:10.1038/nature07383
84. Sanjuan MA, Dillon CP, Tait SWG, Moshiach S, Dorsey F, Connell S, Komatsu M, Tanaka K, Cleveland JL, Withoff S, Green DR (2007) Toll-like receptor signalling in macrophages links the autophagy pathway to phagocytosis. Nature 450:1253–1257. doi:10.1038/nature06421
85. Scherz-Shouval R, Elazar Z (2011) Regulation of autophagy by ROS: physiology and pathology. Trends Biochem Sci 36:30–38. doi:10.1016/j.tibs.2010.07.007
86. Scherz-Shouval R, Shvets E, Fass E, Shorer H, Gil L, Elazar Z (2007) Reactive oxygen species are essential for autophagy and specifically regulate the activity of Atg4. EMBO J 26:1749–1760. doi:10.1038/sj.emboj.7601623
87. Schieber M, Chandel NS (2014) ROS function in redox signaling and oxidative stress. Curr Biol 24:R453–R462. doi:10.1016/j.cub.2014.03.034
88. Schroder K, Tschopp J (2010) The inflammasomes. Cell 140:821–832. doi:10.1016/j.cell.2010.01.040
89. Shang F, Taylor A (2011) Ubiquitin-proteasome pathway and cellular responses to oxidative stress. Free Radic Biol Med 51:5–16. doi:10.1016/j.freeradbiomed.2011.03.031
90. Starkov AA (2008) The role of mitochondria in reactive oxygen species metabolism and signaling. Ann N Y Acad Sci 1147:37–52. doi:10.1196/annals.1427.015
91. Takamura A, Komatsu M, Hara T, Sakamoto A, Kishi C, Waguri S, Eishi Y, Hino O, Tanaka K, Mizushima N (2011) Autophagy-deficient mice develop multiple liver tumors. Genes Dev 25:795–800. doi:10.1101/gad.2016211
92. Tracy K, Dibling BC, Spike BT, Knabb JR, Schumacker P, Macleod KF (2007) BNIP3 Is an RB/E2F target gene required for hypoxia-induced autophagy. Mol Cell Biol 27:6229–6242. doi:10.1128/MCB.02246-06
93. Trocoli A, Djavaheri-Mergny M (2011) The complex interplay between autophagy and NF-κB signaling pathways in cancer cells. Am J Cancer Res 1:629–649
94. Underwood BR, Imarisio S, Fleming A, Rose C, Krishna G, Heard P, Quick M, Korolchuk VI, Renna M, Sarkar S, García-Arencibia M, O'Kane CJ, Murphy MP, Rubinsztein DC (2010) Antioxidants can inhibit basal autophagy and enhance neurodegeneration in models of polyglutamine disease. Hum Mol Genet 19:3413–3429. doi:10.1093/hmg/ddq253
95. Van de Veerdonk FL, Dinarello CA (2014) Deficient autophagy unravels the ROS paradox in chronic granulomatous disease. Autophagy 10:1141–1142. doi:10.4161/auto.28638
96. White E (2012) Deconvoluting the context-dependent role for autophagy in cancer. Nat Rev Cancer 12:401–410. doi:10.1038/nrc3262
97. White E, Karp C, Strohecker AM, Guo Y, Mathew R (2010) Role of autophagy in suppression of inflammation and cancer. Curr Opin Cell Biol 22:212–217. doi:10.1016/j.ceb.2009.12.008
98. Xie C-M, Chan WY, Yu S, Zhao J, Cheng CHK (2011) Bufalin induces autophagy-mediated cell death in human colon cancer cells through reactive oxygen species generation and JNK activation. Free Radic Biol Med 51:1365–1375. doi:10.1016/j.freeradbiomed.2011.06.016
99. Xiong Y, Contento AL, Nguyen PQ, Bassham DC (2007) Degradation of oxidized proteins by autophagy during oxidative stress in Arabidopsis. Plant Physiol 143:291–299. doi:10.1104/pp.106.092106
100. Xue L, Fletcher GC, Tolkovsky AM (1999) Autophagy is activated by apoptotic signalling in sympathetic neurons: an alternative mechanism of death execution. Mol Cell Neurosci 14:180–198. doi:10.1006/mcne.1999.0780
101. Yang H, Lundbäck P, Ottosson L, Erlandsson-Harris H, Venereau E, Bianchi ME, Al-Abed Y, Andersson U, Tracey KJ, Antoine DJ (2011) Redox modification of cysteine residues regu-

lates the cytokine activity of High Mobility Group Box-1 (HMGB1). Mol Med 18:250–259. doi:10.2119/molmed.2011.00389
102. Youle RJ, Narendra DP (2011) Mechanisms of mitophagy. Nat Rev Mol Cell Biol 12:9–14. doi:10.1038/nrm3028
103. Yuan H, Perry CN, Huang C, Iwai-Kanai E, Carreira RS, Glembotski CC, Gottlieb RA (2009) LPS-induced autophagy is mediated by oxidative signaling in cardiomyocytes and is associated with cytoprotection. Am J Physiol Heart Circ Physiol 296:H470–H479. doi:10.1152/ajpheart.01051.2008
104. Yu J, Nagasu H, Murakami T, Hoang H, Broderick L, Hoffman HM, Horng T (2014) Inflammasome activation leads to Caspase-1-dependent mitochondrial damage and block of mitophagy. Proc Natl Acad Sci U S A 111:15514–15519. doi:10.1073/pnas.1414859111
105. Yu L, Wan F, Dutta S, Welsh S, Liu Z, Freundt E, Baehrecke EH, Lenardo M (2006) Autophagic programmed cell death by selective catalase degradation. Proc Natl Acad Sci U S A 103:4952–4957. doi:10.1073/pnas.0511288103
106. Zhang H, Bosch-Marce M, Shimoda LA, Tan YS, Baek JH, Wesley JB, Gonzalez FJ, Semenza GL (2008) Mitochondrial autophagy is an HIF-1-dependent adaptive metabolic response to hypoxia. J Biol Chem 283:10892–10903. doi:10.1074/jbc.M800102200
107. Zhao Y, Zhang C-F, Rossiter H, Eckhart L, König U, Karner S, Mildner M, Bochkov VN, Tschachler E, Gruber F (2013) Autophagy is induced by UVA and promotes removal of oxidized phospholipids and protein aggregates in epidermal keratinocytes. J Invest Dermatol 133:1629–1637. doi:10.1038/jid.2013.26
108. Zhou R, Yazdi AS, Menu P, Tschopp J (2011) A role for mitochondria in NLRP3 inflammasome activation. Nature 469:221–225. doi:10.1038/nature09663
109. Zmijewski JW, Banerjee S, Bae H, Friggeri A, Lazarowski ER, Abraham E (2010) Exposure to hydrogen peroxide induces oxidation and activation of AMP-activated protein kinase. J Biol Chem 285:33154–33164. doi:10.1074/jbc.M110.143685

Interplay Between Autophagy and Inflammasomes

Qiming Liang, Jianning Ge, and Xi Jin

Abstract The autophagy and inflammasome are two ancient innate immune pathways for controlling invading pathogens that are linked mutual regulation. In addition to controlling the cellular metabolic homeostasis through nutrient recycling, the autophagy "self-eating" process is also responsible for the degradation of damaged organelles, aggregated protein complexes, and pathogens to protect the integrity of the organism. As a cytosolic pathogen recognition receptor (PRR) complex, the inflammasome both induces and induced by autophagy through direct interaction with major autophagy proteins or through the effects of secondary molecules, such as mitochondrial reactive oxygen species and mitochondrial DNA. While the underlying molecular mechanisms of inflammasome activation and regulation are largely unknown, much of current knowledge has been established through investigation of the role of autophagy in innate immune response. Many of the newly uncovered links between autophagy and inflammasome have raised new questions about the mechanism controlling inflammasome function, which are highlighted in this chapter.

Abbreviations

8-OH-dG	Oxidized nucleoside 8-hydroxy-guanosine
ASC	Caspase recruitment domain adapter protein
BMDCs	Bone marrow derived dendritic cells
BMDMs	Bone marrow-derived macrophages
BRCC3	BRCA1-BRCA2-containing complex 3
BRISC	BRCC36-containing isopeptidase
DAMP	Danger-associated molecular pattern

Q. Liang (✉) • J. Ge
Department of Molecular Microbiology and Immunology, Keck School of Medicine, University of Southern California, 1501 San Pablo Street, ZNI 521, Los Angeles, CA 90033, USA
e-mail: liang4002@gmail.com

X. Jin
Department of Internal Medicine, University of Michigan, Ann Arbor, MI 48109, USA

M.C. Maiuri, D. De Stefano (eds.), *Autophagy Networks in Inflammation*, Progress in Inflammation Research, DOI 10.1007/978-3-319-30079-5_4

DHA	Omega-3 (ω3) fatty acid docosahexaenoic acid
EM	Electron microscope
FMF	Familial Mediterranean fever
IFI	Interferon-gamma-inducible gene
IL	Interleukin
LeTx	Anthrax lethal toxin
LRR	Leucine-rich repeat
LUBAC	Linear ubiquitin assembly complex
MAVS	Mitochondrial antiviral signaling
MDP	Muramyl dipeptide
MEFs	Mouse embryonic fibroblasts
mtDNA	Mitochondrial DNA
mtROS	Mitochondrial ROS
NACHT, NOD, NB-ARC	Adapter protein; central nucleotide-binding and oligomerization domain
NLR	Nod-like receptor
PAMP	Pathogen-associated molecular pattern
PRR	Pathogen recognition receptor
pyrin or CARD domain	Protein binding domain
Rip2	Receptor interacting protein 2
RLH	RIG-I-like helicase
TLR	Toll-like receptor
TUFM	Tu translation elongation factor
VDAC1	Voltage dependent anion channel 1
VSV	Vesicular stomatitis virus

1 Introduction of Inflammasome

Both mammals and plants rely on a group of pattern recognition receptors (PRRs) to detect the conserved pathogen-associated molecular pattern (PAMP) or danger-associated molecular pattern (DAMP), and mount an innate immune response to ensure a healthy organism. PAMPs are conserved features of bacterial and viral pathogens and DAMPs are byproducts of cell death or increased membrane permeability, which can arise due to invasion by a pathogen or by tissue damage [42, 94]. PRRs recognize various components of pathogens to initiate signaling transduction, which finally activates transcriptional factors to induce the production of inflammatory cytokines [90]. During the signaling transduction, the inflammasome has emerged as an important molecular protein complex which initiates proteolytic processing of the pro-inflammatory cytokines interleukin (IL)-1β and IL-18 into mature inflammatory cytokines. In addition, inflammasomes initiate pyroptotic cell death that may be independent of those cytokines. Inflammasomes are central to elicit innate immune responses against many pathogens, and are key components in the induction of host defenses following bacterial infection [19].

Inflammasome consists of an oligomerized Nod-like receptor (NLR), the apoptosis-associated speck-like protein containing a caspase recruitment domain (ASC) adapter protein, and an inflammatory caspase [Caspase-1, Caspase-4 or 5 (human), Caspase-11 (mouse)]. The NLR family of PRRs contains 22 members in humans and plays a central role in inflammasome formation. NLRs contain a protein binding domain (pyrin or CARD domain), a central nucleotide-binding and oligomerization domain (NACHT, NOD, NB-ARC), and a leucine-rich repeat (LRR) domain, which senses ligands or regulates NLR functions. Inflammasome complexes induce the cleavage of pro-Caspase-1 to the p10 and p20 subunit of active Caspase-1 enzyme, which in turn cleaves the pro-inflammatory cytokines [19]. Recently, it was shown that the Caspase-1 knockout mouse previously used to determine the role of Caspase-1 in inflammasome activation also had a truncation mutation in the Caspase-11 gene, which was subsequently shown to regulate non-canonical inflammasome activation leading to pyroptosis [49]. Although Caspase-11 is not required for inflammasome responses to some sterile simulants, such as K+ efflux, it is required for responses to gram-negative bacterial infections. Early work has hinted at potential molecular mechanisms of Caspase-11-mediated regulation of inflammasome activation, but additional work is required to determine what the targets of Caspase-11 are and how they might mediate pyroptosis. Phagocytes (such as macrophage and dendritic cell) are the only cell type known to contain inflammasomes and these same cells rely upon autophagy to monitor for the presence of pathogens. In this chapter, we will review recent discoveries of NLRP1, NLRP3, NLRP6, NLRP7, NLRP12, NLRC4, Pyrin and AIM2-mediated recognition of pathogens and discuss the cross-regulations between inflammasome and autophagy.

1.1 NLRP1 Inflammasome

NLRP1b, was reported as the primary mediator of mouse macrophage susceptibility to Anthrax lethal toxin (LeTx) based on the variable sensitivity displayed by inbred mouse strains to LeTx-induced macrophage necrosis [8]. Following work suggests that LeTx triggers the formation of a membrane-associated inflammasome complex including NLRP1, Caspase-1 and Caspase-11 in murine macrophages, resulting in cleavage of cytosolic Caspase-1 substrates and cell death [70]. Further study shows that the cleavage of NLRP1 by LeTx is required for the inflammasome activation, IL-1β release and macrophage pyroptosis induced by LeTx, because the uncleaved mutant of NLRP1b blocks this cleavage and also prevents the activation of Caspase-1 [12, 32]. As well, direct cleavage of NLRP1b itself is sufficient to induce inflammasome activation in the absence of LeTx, therefore NLRP1 is proposed to function as a sensor of protease activity and could conceivably detect a broader spectrum of pathogens [12]. Recent work shows that NLRP1 might also be a sensor to Toxoplasma gondii [26]. Oral infection with Toxoplasma triggers an inflammasome response which requires the effector Caspases-1 and 11, the adapter

ASC and NLRP1. However, N-terminal processing of NLRP1b, is not observed in response to Toxoplasma infection which suggests a novel mechanism that differs to the response to LeTx [26].

In addition to LeTx, muramyl dipeptide (MDP), a peptidoglycan constituent from both Gram-positive and Gram-negative bacteria can directly activate NLRP1 inflammasome [27]. Interestingly, LeTx shows a specificity only to mouse NLRP1. Unlike a single NLRP1 gene in human genome, mice have three homologous genes including Nlrp1a, Nlrp1b and Nlrp1c and only Nlrp1b encoded protein responses to LeTx while MDP can activate human NLRP1 but not mouse NLRP1 [8, 27]. The mechanism how MDP activates NLRP1 is still unknown even though it can directly bind to NLRP1. The structure study on NLRP1 suggests the binding of NLRP1 to MDP is not mediated by LRR domain [79]. Most recently, NLRP1b is reported to be activated when cells are deprived of glucose or treated with metabolic inhibitors [59]. The activation of NLRP1b in ATP depleted cells does not require N-terminus of NLRP1b which is necessary in LeTx-induced inflammasome activation. Further comparison of two alleles of NLRP1b that differed in their response to metabolic inhibitors leads to the finding that function to find domain (FIIND) of NLRP1b facilitates the detection of ATP depletion [69]. These results suggest that NLRP1b utilizes distinct regions as sensors to detect activating signals. Further investigations are still required to reveal the detailed mechanism by which NLRP1 get activated and what kind of signals NLRP1 senses.

1.2 NLRP3 Inflammasome

NLRP3 is the mostly well-studied NLR family member and can be activated by a wide spectrum of PAMPs and sterile DAMPs. NLRP3 activation usually is reflected at two levels: the first level is NF-κB signal-dependent transcriptional and translational upregulation of NLRP3 induced by Toll-like receptor (TLR) agonists and inflammatory cytokines [6] for the second level, NLRP3 mediates the formation of inflammasome protein complex with ASC and Caspase-1 upon NLRP3 "ligand" stimulation [84]. By far it is still unclear about the real NLRP3 ligand. An emerging evidences suggest that NLRP3 senses a broad range of pathogenic and stress-associated signal, including bacterial PAMPs such as *Staphylococcus aureus* [17], *Listeria monocytogenes* [62], *Neisseria gonorrhoeae* [22], and bacterial toxins [17, 34, 99], viruses such as Sendai virus [46], Influenza [2, 46], Adenovirus [67], and Encephalomyocarditis virus [74], Fungus such as *Candida albicans* [33], *Saccharomyces cerevisiae* [33], and protozoan like *Plasmodium* species [88], and DAMPs such as Extracellular ATP [62], Hyaluronan [104], Glucose [107], MSU [53], Amyloid-β [36], Skin irritants [92, 100], Imidazoquinoline compounds (R837, R848) [48], Silica [11, 21, 41], Asbestos [11, 21], and Alum [24, 30, 52, 58]. Structure diversity of NLRP3 stimuli indicates that the activation of NLRP3 inflammasome is not simply mediated by direct binding. It suggests that all these stimuli might converge on one or several common signaling cascade leading to NLRP3 inflammasome activation. Base on extensive and comprehensive studies, three major model of NLRP3 activation have been proposed: (1) ion flux model; (2) lysosomal rupture model; (3) reactive oxygen species (ROS) model.

In ion flux model, the changes of intracellular concentration of K^+ and Ca^{2+} are critical for the activation of NLRP3 inflammasome. Mitochondrial perturbation, the opening of a large membrane pore, ROS generation or a change in cell volume is not required for NLRP3 activation in ion flux model [65]. The permeation of the cell membrane to K^+ and Na^+ is proposed as a common event induced by all NLRP3 agonists. Extracellular potassium, as well as potassium channel inhibitor, inhibits NLRP3 activation in human monocytes while reduction of the intracellular K^+ concentration is sufficient to activate NLRP3 [54, 71]. Recently an important role of intracellular calcium level in NLRP3 inflammasome activation has been reported [55, 66, 81]. Increase of the intracellular Ca^{2+} concentration activates NLRP3 inflammasome while blockage of Ca^{2+} mobilization shows an inhibitory effect on NLRP3 inflammasome assembly and activation. Increased extracellular calcium also activates the NLRP3 inflammasome through signaling mediated by G protein-coupled calcium sensing receptor (CaSR) and GPRC6A. Phosphatidyl–inositol/Ca^{2+} pathway is shown to contribute to the NLRP3 inflammasome activation based on the study using phospholipase C (PLC) inhibitor [55, 81]. In addition to extracellular calcium, Ca^{2+} release from the endoplasmic reticulum amplifies NLRP3 inflammasome activation which indicates a connection between the ER stress and NLRP3 inflammasome activation [55, 81]. However, it is not known how the movement of calcium activates NLRP3 inflammasome. One previous study shows TAK1 phosphorylation following an elevation of cytosolic calcium level. As well, loss of IL-1β release is also shown after inhibition of TAK1 phosphorylation or TAK1 knockdown. More information is needed to validate the TAK1 mediated NLRP3 inflammasome activation model [14, 91]. It is not known whether the flux of both potassium and calcium are required for the activation of NLRP3. The precise mechanism under this model still needs further investigation.

Lysosomal rupture is also proposed as a mechanism by which NLRP3 inflammasome is activated by crystalline or particulate structures. Uptake of these agonists such as MSU, uric acid crystals, silica, asbestos, malarial hemozoin, hydroxyapatite, amyloid-β, and alum, results in the damage of lysosome and release of lysosomal content in to cytosol [43, 84]. NLRP3 inflammasome activation is inhibited in the presence of inhibitors of the lysosomal protease Cathepsin B [36, 41]. However, there is no obvious change in NLRP3 inflammasome activation in Cathepsin B-deficient macrophage treated by NLRP3 agonists [20].

The generation of ROS is proposed as the third model for NLRP3 activation [11, 18, 21]. ROS production is a common event resulted from pathogen infection, injury as well as environmental stress, and inhibition of ROS by antioxidants blocks inflammasome activation induced by NLRP3 agonists' treatment [11, 18, 21, 33, 71, 88]. NADPH oxidases and mitochondria are the two major routes that produce intracellular ROS. Activation of NLRP3 inflammasome is reduced upon suppression of NADPH oxidase common subunit p22 [21]. However, it is still controversial about the roles of NADPH oxidase in NLRP3 inflammasome activation as deficiency of NADPH oxidase does not affect NLRP3 activation in both human and mouse cells [5, 97]. Later study implicates a role of thioredoxin-interacting protein (TXNIP), an intracellular ROS sensor, in activation of NLRP3 inflammasome via a direct binding mechanism [107]. Recent study indicates that mitochondria themselves are more pivotal than ROS in NLRP3 inflammasome activation [106]. And another work suggests that mitochondria DNA (mtDNA) is reuiqred for the activation of NLRP3 even though the mechanism is still not clear [68].

1.3 NLRP6 Inflammasome

NLRP6 interacts with ASC and forms an inflammasome by overexpression. Nlrp6 deficient mice revealed that it is essential for restricting commensal bacteria to maintain homeostasis through promoting IL-18 release. Nlrp6 deficient mice are also resistant to infection with *Listeria monocytogenes*, *Salmonella typhimurium* and *Escherichia coli* due to an enhanced production of MAPK- and NF-κB-dependent cytokines [4, 25]. However, the ligands specifically sensed by NLRP6 are currently not known.

1.4 NLRP7 Inflammasome

NLRP7 is located on the long arm of human chromosome 19 at q13.42, but no mouse orthologous gene exists. Nlrp2 is the closest relative and NLRP2 is located adjacent to NLRP7 on human chromosome 19. NLRP7 is abundantly expressed in bone marrow, thymus, spleen, testis, and nervous system, and its transcription is upregulated upon LPS and IL-1β stimulation, suggesting a role in inflammation and host defense [50]. Overexpression of NLRP7 inhibits inflammation by several mechanisms, including direct binding and inhibition of inflammasome components, impairing transcription of pro-IL-1β and modulating the trafficking and release of IL-1β. However most of these studies were performed in a reconstitution system and the finding will require further confirmation. Using a siRNA-based screening, Stehlik's group identified that NLRP7 is an intracellular sensor for bacterial acylated lipoproteins, which is required for IL-1β release [50]. While TLR2 heterodimers are responsible for mediating NF-κB activation and subsequent transcription of pro-IL-1β in response to bacterial acylated lipoproteins, NLRP7 is specifically essential for bacterial acylated lipoprotein-mediated Caspase-1 activation and maturation of IL-1β and IL-18 by forming a large, high molecular weight complex including ASC and Caspase-1. Furthermore, NLRP7 is crucial for restricting growth of intracellular bacteria, including *Staphylococcus aureus* and *Listeria monocytogenes* in human macrophages. Thus, there is compelling evidence that NLRP7 acts as a direct or indirect cytosolic sensor of microbial acylated lipopeptides during infection.

1.5 NLRP12 Inflammasome

NLRP12 is highly expressed in myeloid-lineage cells, and there is emerging evidence that NLRP12 is able to function as an inflammasome component. Lien's group characterized that NLRP12 inflammasome is an important regulator of IL-1β and IL-18 release during *Yersinia pestis* infection, and NLRP12-deficient mice were more susceptible to *Yersinia pestis* challenge. NLRP12 binds to ASC and required for Caspase-1 activation, although NLRP3 is also required for this inflammasome. The ligand for the NLRP12 inflammasome activation is unknown, but the *Yersinia pestis* type III secretion system is required, suggesting that some secreted bacterial proteins may be directly recognized by NLRP12 and activates NLRP12 inflammasome. NLRP12 directed interferon-γ production via induction of IL-18, but had minimal effect on signaling to the transcription factor NF-κB upon *Yersinia pestis* infection [98]. However, other studies have suggested that NLRP12 as a negative regulator of colon inflammation and tumorigenesis in a DSS colitis model, and to dendritic cell recruitment, by impairing NF-κB activation [1, 3]. The effects of NLRP12 to host innate immunity may be cell type specific and stimuli dependent.

1.6 NLRC4 Inflammasome

NLRC4 is stimulated by intracellular flagellin and type III secretion systems from bacteria, including *Salmonella typhimurium*, *Shigella flexneri*, and *Escherichia coli* [64]. NLRC4 utilizes ASC [61], NALP (human) or Naip5 and Naip2 (mouse) adaptor proteins depending on the bacterial stimulus [51, 106]. While NLRC4 can directly detect the N-terminus of flagellin, Naip5 association is required to respond to the C-terminus of flagellin. The Naip2 adaptor protein detects a rod component of type III secretion systems and binds NLRC4 to activate inflammasome formation [51, 106]. The ASC adapter protein is important but not essential for all types of NLRC4 inflammasomes, as ASC deficient Bone marrow-derived macrophages (BMDMs) have a partial defect in Caspase-1 activity and IL-1β secretion upon flagellin simulation, suggesting that ASC enhances the activity of the NLRC4 inflammasome [9, 51, 64]. However, ASC binds NLRC4 and ASC deficient BMDMs have a complete defect in Caspase-1 cleavage and NLRC4 inflammasome foci formation upon infection with *Salmonella typhimurium* and *Legionella pneumophila* [61, 75]. As a further complication, Caspase-1 dependent cell death, or pyroptosis, is intact in ASC deficient BMDMs upon infection with *Pseudomonas aeruginosa*, *Listeria monocytogenes*, or *Shigella flexneri* [31, 83, 93]. Thus, the role of ASC in NLRC4 activation might depend on the context of inflammasome stimulation. Activation of NLRC4 during *Salmonella typhimurium* infection also requires phosphorylation at serine 533 (S533) between the NACHT and LRR domains, which is likely targeted for phosphorylation by the PKCδ kinase [77]. This is the first evidence of the phosphorylation-mediated regulation of inflammasome and it will be of interest to determine whether additional NLRs are regulated by phosphorylation and dephosphorylation.

1.7 Pyrin Inflammasome

Pyrin (also named as TRIM20, MEF, MEFV and FMF) is conserved in human and mice, and mutations in Pyrin gene are associated with the human autoinflammatory disease familial Mediterranean fever (FMF). Several studies indicate that Pyrin can interact with inflammasome adaptor ASC and induce inflammatory Caspase-1 activation in monocytic cells [103, 105]. Recently, Shao's group identified that Pyrin inflammasome detects inactivating modifications of host Rho GTPase by diverse bacterial toxins and infections, including *Clostridium difficile* glucosylating cytotoxin TcdB, FIC-domain adenylyltransferase effectors from *Vibrio parahaemolyticus* and *Histophilus somni*, ADP-ribosylating *Clostridium botulinum* C3 toxin and *Burkholderia cenocepacia* infection. Although diverse in the chemical form, all the modifications take place on different residues within or around the GTPase switch I region. Loss of the Pyrin inflammasome causes elevated intra-macrophage growth of *Burkholderia cenocepacia* and diminished lung inflammation in mice [103, 105]. Pyrin does not appear to directly interact with modified Rho, suggesting an indirect activation by the virulence activity of bacterial pathogens. It is interesting to further investigate the molecular linkage between modified Rho and Pyrin in future.

1.8 AIM2 Inflammasome

In humans, four AIM2-like receptors (IFI16, IFIX, MNDA, and AIM2) have been annotated, while 13 genes are predicted to exist in mice, which are referred to as γ-interferon-inducible genes (IFI). All ALRs, except for p202, contain an N-terminal PYD and one or two C-terminal partially conserved HIN200 DNA-binding domains [84]. Although the AIM2 is structurally unique from NLRs, it functions as a cytosolic PRR involved in inflammasome activation. AIM2 directly binds to cytosolic double stranded DNA through the HIN200 domain and forms an inflammasome in response to bacterial and viral infection, including *Francisella tularensis*, *Listeria monocytogenes*, Vaccinia virus and murine cytomegalovirus. BMDMs from AIM2−/− mice are deficient in pro-Caspase-1, pro-IL-1β, and pro-IL-18 processing after infection with *Francisella tularensis* and *Listeria monocytogenes*; thus emphasizing the importance of inflammasome activation against bacteria that replicate intracellularly [28, 40, 78].

2 Autophagy Inhibits Inflammasome Activation

Deletion or depletion of autophagy by genetic or pharmacological means has a clear enhancing effect on IL-1β secretion upon inflammasome stimulation or lipopolysaccharide (LPS) treatment conditions. In 2008, Saitoh et al. first provided the evidence that autophagy can modulate the activation of inflammasomes [82]. In their study, ATG16L1-deficience significantly enhances the activation of inflammasomes in response to LPS in macrophages. Although the nature of the inflammasome scaffold was not determined, their study indicated that NLRP3 inflammasome activation is dependent on K^+ efflux and ROS [82]. In line with what Saitoh et al. finding, Zhou et al. also reported that autophagy inhibitor 3-methyladenine (3-MA) treatment elevates NLRP3 inflammasomes activation in THP-1 macrophages [106]. Besides 3-MA treatment, Beclin-1, Atg7 or Atg5 silencing all led to inflammasome activation, which further established the regulatory role of autophagy in inflammasome activation. A more recent study showed that the exposure of macrophage cells to omega-3 (ω3) fatty acid docosahexaenoic acid (DHA) impeded inflammasome activation [101]. However, the ATG7 deficient cells were partially resistant to suppression effects of DHA. Besides macrophages, experiments conducted in human ARPE-19 cells also show that inflammasome was activated due to the inhibition of autophagy by bafilomycin A1 [72]. Therefore, it is a universal fact that autophagy inhibits inflammasome activation, which applies to different cell lines, mice, as well as different autophagy silencing approaches. Although it is widely accepted that autophagy regulates inflammasome activation, the underlying mechanisms are still under debates. Three key words have been placed on the hot spot, which ROS, mtDNA and inflammasome degradation. In the following contexts, they will be mainly highlighted.

2.1 ROS

Besides inflammasome activation, blockade of autophagy also leads to ROS accumulation, particularly mitochondrial ROS (mtROS). This suggests that when damaged mitochondria were not cleared by mitophagy (mitochondria autophagy), they released mtROS, which stimulated NLRP3 inflammasomes. If all traditional NLRP3 stimulants induce mitochondrial damage, the mtROS model would present a unified mechanism for the activation of NLRP3 by a structurally diverse set of stimulants. The mtROS model of NLRP3 inflammasome activation is further supported by the colocalization of NLRP3 with mitochondrial markers, suggesting that the inflammasome activation site is physically located near the source of mtROS [108]. Mitochondria function is specifically important for mtROS production and IL-1β secretion because shRNA-mediated depletion of voltage dependent anion channel 1 (VDAC1), which is important for mitochondrial function and mtROS production, reduces Caspase-1 cleavage and IL-1β secretion in response to sterile NLRP3 stimuli [89]. Notably, this effect is not observed for NLRC4 or AIM2 stimuli, making mtROS a specific requirement of NLRP3 inflammasome activation, although the role of mtROS in additional inflammasomes must also be examined further [87, 108]. High levels of Bcl-2, which partially blocks VDAC, reduces mtROS production, and blocks apoptosis, also reduces NLRP3 inflammasome activation in BMDMs and an immortalized human macrophage cell line [87, 108]. In contrast, it has previously been shown that a Bcl-2 over-expression in the human THP-1 macrophage cell line has lower levels of Caspase-1 activity and IL-1β secretion in response to MDP and ATP stimulation [10, 108], making it likely that the role of Bcl-2 may differ between cell lines or that MDP and ATP stimulation is unique from other NLRP3 stimulants. A major advance in understanding the molecular mechanism of the mtROS-mediated activation of NLRP3 was achieved in experiments using mice deficient for autophagophore formation and elongation due to knockout of LC3 ($LC3^{-/-}$) or the depletion of Beclin-1 ($Beclin1^{+/-}$) [68]. BMDMs from these mice had elevated Caspase-1 activation and IL-1β secretion upon stimulation of the NLRP3 inflammasome with ATP, which causes K^+ efflux in the cell through the P2X7 channel [47].

The mitochondria-targeted antioxidant Mito-TEMPO, a scavenger specific for mitochondrial ROS, leads to lower level of mtROS and subsquently less secretion of IL-1β in both wild-type macrophages and macrophages with depletion of autophagic proteins. Defective mitochondrial oxidative phosphorylation (OXPHOS) and elevated ROS result in increased NLRP3 inflammasome in bone marrow derived dendritic cells (BMDCs) from fibromyalgia patients. More interestingly, ROS also has been shown able to up-regulate autophagy. Thus, it is possible that under different circumstances, the interplay between ROS and autophagy is in a dynamic event to keep the balance of host innate immunity against pathogen invasion. Also, ROS activation of autophagy may represent a negative feedback regulation to limit ROS-induced excess inflammations.

2.2 *mtDNA*

Although the detail mechanisms have been intensively investigated, how mtROS leads to the inflammasome activation is still elusive. During NLRP3 inflammasome activation by ATP treatment, an increase in swollen and damaged mitochondria was visible by electron microscope (EM), which results in the mtDNA release from mitochondria [47]. Remarkably, BMDMs lacking mitochondrial DNA (mtDNA), called ρ0 cells, were unable to secrete IL-1β in response to NLRP3 stimuli and DNase I treatment reduced Caspase-1 activation and IL-1β secretion in normal BMDMs, suggesting that mtDNA is involved in NLRP3 inflammasome activation [47, 87]. In a series of elegantly designed experiments, it was revealed that mtDNA is released to the cytosol upon stimulation with ATP in an mtROS and NLRP3-dependent manner [47], indicating that Caspase-1 activation is downstream from NLRP3-mediated translocation of mtDNA to the cytosol. Since mtDNA is released to the cytosol in response to ATP is impaired in BMDMs from either ASC or NLRP3 knockout mice, the inflammasome itself is likely involved in mtDNA release [47]. This is in contrast to a more recent model proposed by Shimada et al. in which NLRP3 is activated after it directly binds to mtDNA released upon mitochondrial damage due to apoptosis triggered by NLRP3 stimulants [87]. Shimada et al. argue that mtDNA was not detected in NLRP3 KO BMDM upon ATP treatment because the mtDNA was degraded in the absence of NLRP3-binding, although this possibility has not been tested. Intriguingly, oxidized nucleoside 8-hydroxy-guanosine (8-OH-dG), a marker for oxidized mtDNA, was detectable in endogenous NLRP3 immunoprecipitations and was even capable of blocking IL-1β production when it was added in excess to BMDM by competitively binding to endogenous NLRP3 [87]. This strongly supports a role for oxidized mtDNA in NLRP3 inflammasome activation, although it is not clear whether NLRP3 may also facilitate mtDNA release before binding.

2.3 *Autophagy Degrades Inflammasomes*

In addition to its role as a negative regulator of NLRP3-inflammasome activation, autophagy also negatively regulates inflammasomes through the newly discovered autophagy-dependent degradation of inflammasome components and IL-1β [38, 86]. Studying the negative regulation of inflammasome activation is important for understanding how this potent signal is "turned off" to avoid acute tissue damage. Not surprisingly, many NLRs and autophagy factors are linked to autoimmune and autoinflammatory diseases [77, 85], which are characterized by high levels of inflammatory cytokines. As a negative regulatory mechanism of inflammasome activation, pro-IL-1β is targeted to autophagosomes for degradation in response to TLR stimulation [38]. Specifically, it has been shown that IL-1β is sequestered in the LC3-positive autophagosomes upon TLR stimulation and pro-IL-1β protein levels decreased when autophagy was induced by rapamycin [38]. This suggests that TLR stimulation induces both pro-IL-1β expression and degradation by autophagy, thereby limiting the amount of available pro-IL-1β protein in the absence of NLR stimulation. A recent report suggests that mature IL-1β uses the autophagy machinery for secretion in a noncanonical secretory pathway [23]. However, more evidence is required to vigorously evaluate this hypothesis.

Inflammasomes are also negatively regulated by autophagy upstream from IL-1β secretion. It has been shown that a portion of ASC-containing inflammasomes is redirected towards autophagosomes and autophagolysosomes upon NLRP3 or AIM2 stimulation in THP-1 and primary human macrophages. Mechanistically, ASC localization to autophagosomes is dependent on both Beclin-1 and p62, a protein that specifically recruits ubiquitinated proteins to autophagosomes for degradation [86]. ASC and ASC-containing inflammasome complexes could be recruited to autophagosomes by p62 since the K63-ubiquitination of ASC is detectable upon AIM2 stimulation [86]. However, many new questions will need to be answered to elucidate the molecular mechanism of this potential negative regulation of inflammasomes. The ubiquitin ligase that modifies ASC and the trigger for ubiquitination are unknown. However, ASC protein levels did not change upon AIM2 stimulation and AIM2 levels actually increased at the timepoints in this study, making the significance of ASC localization to the autophagosome unclear. Perhaps the degradation of ASC-containing inflammasomes is balanced by increased protein expression, or proportionally very few inflammasomes are recruited to autophagosomes under stimulation conditions. Since experimental evidence is lacking, further investigation into the molecular mechanisms, regulation, and purpose of targeting inflammasomes to autophagosomes is imperative for determining how autophagy may work as a potential 'off switch' for activated inflammasomes. Another type of ubiquitination, linear unibiqination, has been characterized on ASC recently. The linear ubiquitination of ASC is mediated by linear ubiquitin assembly complex (LUBAC) and essential for NLRP3 inflammasome activation [80]. However, how different types of ubiquitin chains target ASC to autophagy or activation is still unclear.

In addition to ASC, NLRP3 is also ubiquitinated, although binding to p62 has not been reported. So it is not known whether NLRP3 can be independently recruited to autophagosomes [45, 60, 76]. NLRP3 is ubiquitinated in the LRR domain with a mix of lysine 63 (K63) and lysine 48 (K48) linked chains by unknown ubiquitin ligases. Ubiquitinated NLRP3 is inactive because its deubiquitination is required for inflammasome activation and is triggered by TLR activation, mtROS, and ATP [76]. The deubiquitinase responsible for NLRP3 activation, BRCA1-BRCA2-containing complex 3 (BRCC3), is a member of the BRCC36-containing isopeptidase (BRISC) deubiquitination complex, which has been suggested to specifically cleave K63-linked ubiquitin chains and not K48-linked chains [15, 16, 29], making the mechanism of K48-linked ubiquitin removal unclear. Although K48-linked ubiquitinated proteins are often targeted for proteasomal degradation, MG132 proteasomal inhibitor treatment did not affect ubiquitinated NLRP3 protein levels in the presence of a BCC3 inhibitor [76], suggesting that ubiquitination of NLRP3 does not lead to proteasomal degradation of NLRP3. Thus, the mechanism of inhibition of NLRP3 by ubiquitination remains unknown.

3 NLRs Upregulate Autophagy

The positive and negative regulation of inflammasomes by autophagy is complemented by NLR-mediated control of autophagy. NOD2 is a positive regulator of autophagy. NOD2 utilizes the receptor interacting protein 2 (Rip2) adapter protein kinase to form a protein complex resembling the inflammasome, called a nodosome, which forms upon stimulation with MDP. Instead of activating Caspase-1, the NOD2 nodosome activates NF-κB, resulting in inflammatory cytokine production [7]. Both *NOD2* and *Atg16L1* have been identified by several groups as the genes with risk alleles for Crohn's disease, a chronic inflammatory bowel disease characterized by high levels of inflammatory cytokines [13, 37, 63]. A T300A mutation in the N-terminus of the WD repeats of Atg16L1 is associated with Crohn's disease in the presence or absence of NOD2 LRR domain point mutations R702W, G908R, or L1007C [7, 77]. Recently, NOD2 and Atg16L1 have been functionally linked in a mechanism of bacterial pathogen clearance, which may explain why they are both genetically linked to Crohn's disease [39, 96]. NOD2 induces autophagophore formation upon MDP stimulation by binding and recruiting Atg16L1 to the bacterial entry site of the plasma membrane to initiate autophagic destruction of the bacterial pathogen. Strikingly, dendritic cells from Crohn's patients with *NOD2* or *Atg16L1* variant alleles have normal autophagy levels in response to a TLR1/2 ligand, but have significantly reduced autophagy levels in response to MDP, a NOD2 stimulant. Consequently, cells from Crohn's patients have a defect in lysosomal destruction of bacterial pathogens [39]. The resulting higher load of bacteria in cells from Crohn's patients may contribute to higher levels of inflammatory cytokines, including IL-6 and IL-1β, which are detected in Crohn's patients' tissues. The increase of IL-1β cytokine production in Crohn's patients could be due to either increased pro-IL-1β protein levels or increased Caspase-1 activity. Consistent with a role for NOD2 in Crohn's disease, the enhanced IL-1β secretion in Crohn's patients' cells in response to MDP is due to increased pro-IL1β mRNA levels and not due to Caspase-1 activity, which remains unchanged when compared to healthy patients [73]. NOD2 is constitutively expressed in immune cells and inducible expressed in intestinal epithelial cells upon pro-inflammatory stimulation [7, 35]. Thus, a potential mechanism for the trigger that controls NOD2-mediated autophagy could be the induction of NOD2 expression. Since RIP2 is ubiquitinated and phosphorylated to regulate NOD2 activation, additional post-transcriptional mechanisms may also regulate NOD2 protein levels [9, 95]. As a potential mechanism for disease pathogenesis, NOD2-mediated autophagy may serve as a model for future studies since many additional autophagy and NLR proteins are also linked to chronic inflammatory diseases with unknown mechanisms. In addition to NOD2, NLRX1 also positive regulates autophagy. NLRX1, a mitochondria NLR protein, promotes autophagy during viral infection through an interaction with mitochondrial Tu translation elongation factor (TUFM) [56]. TUFM likely increases autophagy through its interaction with the Atg5-Atg12 complex, which is an essential component of the vesicle elongation step of autophagy and is unable to bind NLRX1 on its own. Although it is not clear how the NLRX1/TUFM interaction with Atg5-Atg12 elevates autophagy, the resulting decrease in vesicular stomatitis virus (VSV) production in NLRX1 and Atg5 knockout mouse embryonic fibroblasts (MEFs) suggests that NLRX1 and autophagy are proviral factor during VSV infection [56]. NLRX1 knockout MEFs had a larger decrease in VSV replication than $Atg5^{-/-}$ MEFs, which might be due to the proposed negative regulation of RIG-I-like helicase (RLH) signaling through the mitochondrial antiviral signaling (MAVS) adapter protein by NLRX1 [57, 102]. The function has been called into question because different methods of knocking out NLRX1 in mice have given conflicting results. Several other NLR members, including NLRC4, NLRP3, NLRP4, and NLRP10, interact with Beclin-1. In particular, NLRP4 showed a strong binding affinity to Beclin1 among these four and knockdown NLRP4 resulted in upregulation of the autophagic process under both physiological conditions and invasive bacterial infections [44]. Besides, NLRP4 physically associates with the class C vacuolar protein-sorting complex, thereby negatively regulating the maturation step of the autophagosome and endosome.

4 Conclusions

The intersection of inflammasomes and autophagy is an exciting area of research with many unanswered questions that should be addressed by future investigation. Of course, for every host defense there is an opposing pathogen offense, making it likely that bacterial or viral proteins may exist that target or exploit the newly discovered links between autophagy and inflammasome regulation to avoid detection or destruction. Thus, as we uncover new mechanisms of regulation between the ancient innate immune pathways of inflammasomes and autophagy, we may also find novel host-pathogen interactions that may be targeted therapeutically. Moreover, treatments for chronic inflammatory diseases in which NLRs and autophagy proteins are implicated will rely on these findings as well.

References

1. Allen IC, Lich JD, Arthur JC, Jania CM, Roberts RA, Callaway JB, Tilley SL, Ting JP (2012) Characterization of NLRP12 during the development of allergic airway disease in mice. PLoS One 7:e30612
2. Allen IC, Scull MA, Moore CB, Holl EK, McElvania-TeKippe E, Taxman DJ, Guthrie EH, Pickles RJ, Ting JP (2009) The NLRP3 inflammasome mediates in vivo innate immunity to influenza A virus through recognition of viral RNA. Immunity 30:556–565
3. Allen IC, Wilson JE, Schneider M, Lich JD, Roberts RA, Arthur JC, Woodford RM, Davis BK, Uronis JM, Herfarth HH et al (2012) NLRP12 suppresses colon inflammation and tumorigenesis through the negative regulation of noncanonical NF-kappaB signaling. Immunity 36:742–754
4. Anand PK, Malireddi RK, Lukens JR, Vogel P, Bertin J, Lamkanfi M, Kanneganti TD (2012) NLRP6 negatively regulates innate immunity and host defence against bacterial pathogens. Nature 488:389–393
5. Bauer C, Duewell P, Mayer C, Lehr HA, Fitzgerald KA, Dauer M, Tschopp J, Endres S, Latz E, Schnurr M (2010) Colitis induced in mice with dextran sulfate sodium (DSS) is mediated by the NLRP3 inflammasome. Gut 59:1192–1199
6. Bauernfeind FG, Horvath G, Stutz A, Alnemri ES, MacDonald K, Speert D, Fernandes-Alnemri T, Wu J, Monks BG, Fitzgerald KA et al (2009) Cutting edge: NF-kappaB activating pattern recognition and cytokine receptors license NLRP3 inflammasome activation by regulating NLRP3 expression. J Immunol 183:787–791
7. Billmann-Born S, Lipinski S, Böck J, Till A, Rosenstiel P, Schreiber S (2011) The complex interplay of NOD-like receptors and the autophagy machinery in the pathophysiology of Crohn disease. Eur J Cell Biol 90:593–602
8. Boyden ED, Dietrich WF (2006) Nalp1b controls mouse macrophage susceptibility to anthrax lethal toxin. Nat Genet 38:240–244
9. Broz P, von Moltke J, Jones JW, Vance RE, Monack DM (2010) Differential requirement for Caspase-1 autoproteolysis in pathogen-induced cell death and cytokine processing. Cell Host Microbe 8:471–483
10. Bruey JM, Bruey-Sedano N, Luciano F, Zhai D, Balpai R, Xu C, Kress CL, Bailly-Maitre B, Li X, Osterman A et al (2007) Bcl-2 and Bcl-XL regulate proinflammatory caspase-1 activation by interaction with NALP1. Cell 129:45–56
11. Cassel SL, Eisenbarth SC, Iyer SS, Sadler JJ, Colegio OR, Tephly LA, Carter AB, Rothman PB, Flavell RA, Sutterwala FS (2008) The Nalp3 inflammasome is essential for the development of silicosis. Proc Natl Acad Sci U S A 105:9035–9040

12. Chavarria-Smith J, Vance RE (2013) Direct proteolytic cleavage of NLRP1B is necessary and sufficient for inflammasome activation by anthrax lethal factor. PLoS Pathog 9:e1003452
13. Cho JH (2008) The genetics and immunopathogenesis of inflammatory bowel disease. Nat Rev Immunol 8:458–466
14. Compan V, Baroja-Mazo A, Lopez-Castejon G, Gomez AI, Martinez CM, Angosto D, Montero MT, Herranz AS, Bazan E, Reimers D et al (2012) Cell volume regulation modulates NLRP3 inflammasome activation. Immunity 37:487–500
15. Cooper EM, Boeke JD, Cohen RE (2010) Specificity of the BRISC deubiquitinating enzyme is not due to selective binding to Lys63-linked polyubiquitin. J Biol Chem 285:10344–10352
16. Cooper EM, Cutcliffe C, Kristiansen TZ, Pandey A, Pickart CM, Cohen RE (2009) K63-specific deubiquitination by two JAMM/MPN+ complexes: BRISC-associated Brcc36 and proteasomal Poh1. EMBO J 28:621–631
17. Craven RR, Gao X, Allen IC, Gris D, Bubeck Wardenburg J, McElvania-Tekippe E, Ting JP, Duncan JA (2009) Staphylococcus aureus alpha-hemolysin activates the NLRP3-inflammasome in human and mouse monocytic cells. PLoS One 4:e7446
18. Cruz CM, Rinna A, Forman HJ, Ventura AL, Persechini PM, Ojcius DM (2007) ATP activates a reactive oxygen species-dependent oxidative stress response and secretion of proinflammatory cytokines in macrophages. J Biol Chem 282:2871–2879
19. Davis BK, Wen H, Ting JP (2011) The inflammasome NLRs in immunity, inflammation, and associated diseases. Annu Rev Immunol 29:707–735
20. Dostert C, Guarda G, Romero JF, Menu P, Gross O, Tardivel A, Suva ML, Stehle JC, Kopf M, Stamenkovic I et al (2009) Malarial hemozoin is a Nalp3 inflammasome activating danger signal. PLoS One 4:e6510
21. Dostert C, Petrilli V, Van Bruggen R, Steele C, Mossman BT, Tschopp J (2008) Innate immune activation through Nalp3 inflammasome sensing of asbestos and silica. Science 320:674–677
22. Duncan JA, Gao X, Huang MT, O'Connor BP, Thomas CE, Willingham SB, Bergstralh DT, Jarvis GA, Sparling PF, Ting JP (2009) Neisseria gonorrhoeae activates the proteinase cathepsin B to mediate the signaling activities of the NLRP3 and ASC-containing inflammasome. J Immunol 182:6460–6469
23. Dupont N, Jiang S, Pilli M, Ornatowski W, Bhattacharya D, Deretic V (2011) Autophagy-based unconventional secretory pathway for extracellular delivery of IL-1beta. EMBO J 30:1–11
24. Eisenbarth SC, Colegio OR, O'Connor W, Sutterwala FS, Flavell RA (2008) Crucial role for the Nalp3 inflammasome in the immunostimulatory properties of aluminium adjuvants. Nature 453:1122–1126
25. Elinav E, Strowig T, Kau AL, Henao-Mejia J, Thaiss CA, Booth CJ, Peaper DR, Bertin J, Eisenbarth SC, Gordon JI et al (2011) NLRP6 inflammasome regulates colonic microbial ecology and risk for colitis. Cell 145:745–757
26. Ewald SE, Chavarria-Smith J, Boothroyd JC (2014) NLRP1 is an inflammasome sensor for Toxoplasma gondii. Infect Immun 82:460–468
27. Faustin B, Lartigue L, Bruey JM, Luciano F, Sergienko E, Bailly-Maitre B, Volkmann N, Hanein D, Rouiller I, Reed JC (2007) Reconstituted NALP1 inflammasome reveals two-step mechanism of caspase-1 activation. Mol Cell 25:713–724
28. Fernandes-Alnemri T, Yu J-W Datta P, Wu J, Alnemri ES (2009) AIM2 activates the inflammasome and cell death in response to cytoplasmic DNA. Nature 458:509–513
29. Finley D (2009) Recognition and processing of ubiquitin-protein conjugates by the proteasome. Annu Rev Biochem 78:477–513
30. Franchi L, Nunez G (2008) The Nlrp3 inflammasome is critical for aluminium hydroxide-mediated IL-1beta secretion but dispensable for adjuvant activity. Eur J Immunol 38:2085–2089
31. Franchi L, Stoolman J, Kanneganti TD, Verma A, Ramphal R, Nunez G (2007) Critical role for Ipaf in Pseudomonas aeruginosa-induced caspase-1 activation. Eur J Immunol 37:3030–3039
32. Frew BC, Joag VR, Mogridge J (2012) Proteolytic processing of Nlrp1b is required for inflammasome activity. PLoS Pathog 8:e1002659

33. Gross O, Poeck H, Bscheider M, Dostert C, Hannesschlager N, Endres S, Hartmann G, Tardivel A, Schweighoffer E, Tybulewicz V et al (2009) Syk kinase signalling couples to the Nlrp3 inflammasome for anti-fungal host defence. Nature 459:433–436
34. Gurcel L, Abrami L, Girardin S, Tschopp J, van der Goot FG (2006) Caspase-1 activation of lipid metabolic pathways in response to bacterial pore-forming toxins promotes cell survival. Cell 126:1135–1145
35. Gutierrez O, Pipaon C, Inohara N, Fontalba A, Ogura Y, Prosper F, Nunez G, Fernandez-Luna JL (2002) Induction of Nod2 in myelomonocytic and intestinal epithelial cells via nuclear factor-kappa B activation. J Biol Chem 277:41701–41705
36. Halle A, Hornung V, Petzold GC, Stewart CR, Monks BG, Reinheckel T, Fitzgerald KA, Latz E, Moore KJ, Golenbock DT (2008) The NALP3 inflammasome is involved in the innate immune response to amyloid-beta. Nat Immunol 9:857–865
37. Hampe J, Franke A, Rosenstiel P, Till A, Teuber M, Huse K, Albrecht M, Mayr G, De La Vega FM, Briggs J et al (2007) A genome-wide association scan of nonsynonymous SNPs identifies a susceptibility variant for Crohn disease in ATG16L1. Nat Genet 39:207–211
38. Harris J, Hartman M, Roche C, Zeng SG, O'Shea A, Sharp FA, Lambe EM, Creagh EM, Golenbock DT, Tschopp J et al (2011) Autophagy controls IL-1 secretion by targeting Pro-IL-1 for degradation. J Biol Chem 286:9587–9597
39. Homer CR, Kabi A, Marina-García N, Sreekumar A, Nesvizhskii AI, Nickerson KP, Chinnaiyan AM, Núñez G, McDonald C (2012) A dual role for receptor-interacting protein kinase 2 (RIP2) kinase activity in nucleotide-binding oligomerization domain 2 (NOD2)-dependent autophagy. J Biol Chem 287:25565–25576
40. Hornung V, Ablasser A, Charrel-Dennis M, Bauernfeind F, Horvath G, Caffrey DR, Latz E, Fitzgerald KA (2009) AIM2 recognizes cytosolic dsDNA and forms a caspase-1-activating inflammasome with ASC. Nature 458:514–518
41. Hornung V, Bauernfeind F, Halle A, Samstad EO, Kono H, Rock KL, Fitzgerald KA, Latz E (2008) Silica crystals and aluminum salts activate the NALP3 inflammasome through phagosomal destabilization. Nat Immunol 9:847–856
42. Janeway CA, Medzhitov R (2002) Innate immune recognition. Annu Rev Immunol 20: 197–216
43. Jin C, Frayssinet P, Pelker R, Cwirka D, Hu B, Vignery A, Eisenbarth SC, Flavell RA (2011) NLRP3 inflammasome plays a critical role in the pathogenesis of hydroxyapatite-associated arthropathy. Proc Natl Acad Sci U S A 108:14867–14872
44. Jounai N, Kobiyama K, Shiina M, Ogata K, Ishii KJ, Takeshita F (2011) NLRP4 negatively regulates autophagic processes through an association with beclin1. J Immunol 186: 1646–1655
45. Juliana C, Fernandes-Alnemri T, Kang S, Farias A, Qin F, Alnemri ES (2012) Non-transcriptional priming and deubiquitination regulate NLRP3 inflammasome activation. J Biol Chem 287:36617–36622
46. Kanneganti TD, Body-Malapel M, Amer A, Park JH, Whitfield J, Franchi L, Taraporewala ZF, Miller D, Patton JT, Inohara N et al (2006) Critical role for Cryopyrin/Nalp3 in activation of caspase-1 in response to viral infection and double-stranded RNA. J Biol Chem 281: 36560–36568
47. Kanneganti TD, Lamkanfi M, Kim YG, Chen G, Park JH, Franchi L, Vandenabeele P, Nunez G (2007) Pannexin-1-mediated recognition of bacterial molecules activates the cryopyrin inflammasome independent of Toll-like receptor signaling. Immunity 26:433–443
48. Kanneganti TD, Ozoren N, Body-Malapel M, Amer A, Park JH, Franchi L, Whitfield J, Barchet W, Colonna M, Vandenabeele P et al (2006) Bacterial RNA and small antiviral compounds activate caspase-1 through cryopyrin/Nalp3. Nature 440:233–236
49. Kayagaki N, Warming S, Lamkanfi M, Vande Walle L, Louie S, Dong J, Newton K, Qu Y, Liu J, Heldens S et al (2011) Non-canonical inflammasome activation targets caspase-11. Nature 479:117–121
50. Khare S, Dorfleutner A, Bryan NB, Chawon Y, Radian AD, de Almeida L, Rojanasakul Y, Stehlik C (2012) An NLRP7-containing inflammasome mediates recognition of microbial lipopeptides in human macrophages. Immunity 36:464–476

51. Kofoed EM, Vance RE (2011) Innate immune recognition of bacterial ligands by NAIPs determines inflammasome specificity. Nature 477:592–595
52. Kool M, Petrilli V, De Smedt T, Rolaz A, Hammad H, van Nimwegen M, Bergen IM, Castillo R, Lambrecht BN, Tschopp J (2008) Cutting edge: alum adjuvant stimulates inflammatory dendritic cells through activation of the NALP3 inflammasome. J Immunol 181:3755–3759
53. Kummer JA, Broekhuizen R, Everett H, Agostini L, Kuijk L, Martinon F, van Bruggen R, Tschopp J (2007) Inflammasome components NALP 1 and 3 show distinct but separate expression profiles in human tissues suggesting a site-specific role in the inflammatory response. J Histochem Cytochem 55:443–452
54. Lamkanfi M, Mueller JL, Vitari AC, Misaghi S, Fedorova A, Deshayes K, Lee WP, Hoffman HM, Dixit VM (2009) Glyburide inhibits the Cryopyrin/Nalp3 inflammasome. J Cell Biol 187:61–70
55. Lee GS, Subramanian N, Kim AI, Aksentijevich I, Goldbach-Mansky R, Sacks DB, Germain RN, Kastner DL, Chae JJ (2012) The calcium-sensing receptor regulates the NLRP3 inflammasome through Ca2+ and cAMP. Nature 492:123–127
56. Lei Y, Wen H, Yu Y, Taxman DJ, Zhang L, Widman DG, Swanson KV, Wen K-W, Damania B, Moore CB et al (2012) The mitochondrial proteins NLRX1 and TUFM form a complex that regulates type I interferon and autophagy. Immunity 36:933–946
57. Levinsohn JL, Newman ZL, Hellmich KA, Fattah R, Getz MA, Liu S, Sastalla I, Leppla SH, Moayeri M (2012) Anthrax lethal factor cleavage of nlrp1 is required for activation of the inflammasome. PLoS Pathog 8:e1002638
58. Li H, Willingham SB, Ting JP, Re F (2008) Cutting edge: inflammasome activation by alum and alum's adjuvant effect are mediated by NLRP3. J Immunol 181:17–21
59. Liao KC, Mogridge J (2013) Activation of the Nlrp1b inflammasome by reduction of cytosolic ATP. Infect Immun 81:570–579
60. Lopez-Castejon G, Luheshi NM, Compan V, High S, Whitehead RC, Flitsch SL, Kirov A, Prudovsky I, Swanton E, Brough D (2012) Deubiquitinases regulate the activity of caspase-1 and IL-1beta secretion via assembly of the inflammasome. J Biol Chem 288:2721–2733
61. Mariathasan S, Newton K, Monack DM, Vucic D, French DM, Lee WP, Roose-Girma M, Erickson S, Dixit VM (2004) Differential activation of the inflammasome by caspase-1 adaptors ASC and Ipaf. Nature 430:213–218
62. Mariathasan S, Weiss DS, Newton K, McBride J, O'Rourke K, Roose-Girma M, Lee WP, Weinrauch Y, Monack DM, Dixit VM (2006) Cryopyrin activates the inflammasome in response to toxins and ATP. Nature 440:228–232
63. Massey D, Parkes M (2007) Common pathways in Crohn's disease and other inflammatory diseases revealed by genomics. Gut 56:1489–1492
64. Miao EA, Alpuche-Aranda CM, Dors M, Clark AE, Bader MW, Miller SI, Aderem A (2006) Cytoplasmic flagellin activates caspase-1 and secretion of interleukin 1beta via Ipaf. Nat Immunol 7:569–575
65. Munoz-Planillo R, Kuffa P, Martinez-Colon G, Smith BL, Rajendiran TM, Nunez G (2013) K(+) efflux is the common trigger of NLRP3 inflammasome activation by bacterial toxins and particulate matter. Immunity 38:1142–1153
66. Murakami T, Ockinger J, Yu J, Byles V, McColl A, Hofer AM, Horng T (2012) Critical role for calcium mobilization in activation of the NLRP3 inflammasome. Proc Natl Acad Sci U S A 109:11282–11287
67. Muruve DA, Petrilli V, Zaiss AK, White LR, Clark SA, Ross PJ, Parks RJ, Tschopp J (2008) The inflammasome recognizes cytosolic microbial and host DNA and triggers an innate immune response. Nature 452:103–107
68. Nakahira K, Haspel JA, Rathinam VA, Lee SJ, Dolinay T, Lam HC, Englert JA, Rabinovitch M, Cernadas M, Kim HP et al (2011) Autophagy proteins regulate innate immune responses by inhibiting the release of mitochondrial DNA mediated by the NALP3 inflammasome. Nat Immunol 12:222–230
69. Neiman-Zenevich J, Liao KC, Mogridge J (2014) Distinct regions of NLRP1B are required to respond to anthrax lethal toxin and metabolic inhibition. Infect Immun 82:3697–3703

70. Nour AM, Yeung YG, Santambrogio L, Boyden ED, Stanley ER, Brojatsch J (2009) Anthrax lethal toxin triggers the formation of a membrane-associated inflammasome complex in murine macrophages. Infect Immun 77:1262–1271
71. Petrilli V, Papin S, Dostert C, Mayor A, Martinon F, Tschopp J (2007) Activation of the NALP3 inflammasome is triggered by low intracellular potassium concentration. Cell Death Differ 14:1583–1589
72. Pippo N, Korkmaz A, Hytti M, Kinnunen K, Salminen A, Atalay M, Kaarniranta K, Kauppinen A (2014) Decline in cellular clearance systems induces inflammasome signaling in human ARPE-19 cells. Biochim Biophys Acta 1843:3038–3046
73. Plantinga TS, Crişan TO, Oosting M, van de Veerdonk FL, de Jong DJ, Philpott DJ, van der Meer JW, Girardin SE, Joosten LA, Netea MG (2011) Crohn's disease-associated ATG16L1 polymorphism modulates pro-inflammatory cytokine responses selectively upon activation of NOD2. Gut 60(9):1229–1235
74. Poeck H, Bscheider M, Gross O, Finger K, Roth S, Rebsamen M, Hannesschlager N, Schlee M, Rothenfusser S, Barchet W et al (2010) Recognition of RNA virus by RIG-I results in activation of CARD9 and inflammasome signaling for interleukin 1 beta production. Nat Immunol 11:63–69
75. Proell M, Gerlic M, Mace PD, Reed JC, Riedl SJ (2013) The CARD plays a critical role in ASC foci formation and inflammasome signaling. Biochem J 499(3):613–621
76. Py BF, Kim MS, Vakifahmetoglu-Norberg H, Yuan J (2012) Deubiquitination of NLRP3 by BRCC3 critically regulates inflammasome activity. Mol Cell 49:331–338
77. Ramjeet M, Hussey S, Philpott DJ, Travassos LH (2010) "Nodophagy": new crossroads in Crohn disease pathogenesis. Gut Microbes 1:307–315
78. Rathinam VAK, Jiang Z, Waggoner SN, Sharma S, Cole LE, Waggoner L, Vanaja SK, Monks BG, Ganesan S, Latz E et al (2010) The AIM2 inflammasome is essential for host defense against cytosolic bacteria and DNA viruses. Nat Immunol 11:395–402
79. Reubold TF, Hahne G, Wohlgemuth S, Eschenburg S (2014) Crystal structure of the leucine-rich repeat domain of the NOD-like receptor NLRP1: implications for binding of muramyl dipeptide. FEBS Lett 588:3327–3332
80. Rodgers MA, Bowman JW, Fujita H, Orazio N, Shi M, Liang Q, Amatya R, Kelly TJ, Iwai K, Ting J et al (2014) The linear ubiquitin assembly complex (LUBAC) is essential for NLRP3 inflammasome activation. J Exp Med 211:1333–1347
81. Rossol M, Pierer M, Raulien N, Quandt D, Meusch U, Rothe K, Schubert K, Schoneberg T, Schaefer M, Krugel U et al (2012) Extracellular Ca2+ is a danger signal activating the NLRP3 inflammasome through G protein-coupled calcium sensing receptors. Nat Commun 3:1329
82. Saitoh T, Fujita N, Jang MH, Uematsu S, Yang BG, Satoh T, Omori H, Noda T, Yamamoto N, Komatsu M et al (2008) Loss of the autophagy protein Atg16L1 enhances endotoxin-induced IL-1beta production. Nature 456(7219):264–268
83. Sauer JD, Pereyre S, Archer KA, Burke TP, Hanson B, Lauer P, Portnoy DA (2011) Listeria monocytogenes engineered to activate the Nlrc4 inflammasome are severely attenuated and are poor inducers of protective immunity. Proc Natl Acad Sci U S A 108:12419–12424
84. Schroder K, Tschopp J (2010) The inflammasomes. Cell 140(6):821–832
85. Shaw PJ, McDermott MF, Kanneganti TD (2011) Inflammasomes and autoimmunity. Trends Mol Med 17:57–64
86. Shi CS, Shenderov K, Huang NN, Kabat J, Abu-Asab M, Fitzgerald KA, Sher A, Kehrl JH (2012) Activation of autophagy by inflammatory signals limits IL-1beta production by targeting ubiquitinated inflammasomes for destruction. Nat Immunol 13:255–263
87. Shimada K, Crother TR, Karlin J, Dagvadorj J, Chiba N, Chen S, Ramanujan VK, Wolf AJ, Vergnes L, Ojcius DM et al (2012) Oxidized mitochondrial DNA activates the NLRP3 inflammasome during apoptosis. Immunity 36:401–414
88. Shio MT, Eisenbarth SC, Savaria M, Vinet AF, Bellemare MJ, Harder KW, Sutterwala FS, Bohle DS, Descoteaux A, Flavell RA et al (2009) Malarial hemozoin activates the NLRP3 inflammasome through Lyn and Syk kinases. PLoS Pathog 5:e1000559

89. Shoshan-Barmatz V, Ben-Hail D (2012) VDAC a multi-functional mitochondrial protein as a pharmacological target. Mitochondrion 12:24–34
90. Song DH, Lee JO (2012) Sensing of microbial molecular patterns by Toll-like receptors. Immunol Rev 250:216–229
91. Sutterwala FS, Haasken S, Cassel SL (2014) Mechanism of NLRP3 inflammasome activation. Ann N Y Acad Sci 1319:82–95
92. Sutterwala FS, Ogura Y, Zamboni DS, Roy CR, Flavell RA (2006) NALP3: a key player in caspase-1 activation. J Endotoxin Res 12:251–256
93. Suzuki T, Franchi L, Toma C, Ashida H, Ogawa M, Yoshikawa Y, Mimuro H, Inohara N, Sasakawa C, Núñez G (2007) Differential regulation of caspase-1 activation pyroptosis and autophagy via Ipaf and ASC in Shigella-infected macrophages. PLoS Pathog 3:e111
94. Takeuchi O, Akira S (2010) Pattern recognition receptors and inflammation. Cell 140: 805–820
95. Tigno-Aranjuez JT, Abbott DW (2012) Ubiquitination and phosphorylation in the regulation of NOD2 signaling and NOD2-mediated disease. Biochim Biophys Acta 1823:2022–2028
96. Travassos LH, Carneiro LAM, Ramjeet M, Hussey S, Kim Y-G, Magalhães JG, Yuan L, Soares F, Chea E, Le Bourhis L et al (2010) Nod1 and Nod2 direct autophagy by recruiting ATG16L1 to the plasma membrane at the site of bacterial entry. Nat Immunol 11:55–62
97. van Bruggen R, Koker MY, Jansen M, van Houdt M, Roos D, Kuijpers TW, van den Berg TK (2010) Human NLRP3 inflammasome activation is Nox1-4 independent. Blood 115: 5398–5400
98. Vladimer GI, Weng D, Paquette SWM, Vanaja SK, Rathinam VAK, Aune MH, Conlon JE, Burbage JJ, Proulx MK, Liu Q et al (2012) The NLRP12 inflammasome recognizes Yersinia pestis. Immunity 37:96–107
99. Warren SE, Mao DP, Rodriguez AE, Miao EA, Aderem A (2008) Multiple Nod-like receptors activate caspase 1 during Listeria monocytogenes infection. J Immunol 180:7558–7564
100. Watanabe H, Gaide O, Petrilli V, Martinon F, Contassot E, Roques S, Kummer JA, Tschopp J, French LE (2007) Activation of the IL-1beta-processing inflammasome is involved in contact hypersensitivity. J Invest Dermatol 127:1956–1963
101. Williams-Bey Y, Boularan C, Vural A, Huang NN, Hwang IY, Shan-Shi C, Kehrl JH (2014) Omega-3 free fatty acids suppress macrophage inflammasome activation by inhibiting NF-kappaB activation and enhancing autophagy. PLoS One 9:e97957
102. Xia X, Cui J, Wang HY, Zhu L, Matsueda S, Wang Q, Yang X, Hong J, Songyang Z, Chen ZJ et al (2011) NLRX1 negatively regulates TLR-induced NF-kappaB signaling by targeting TRAF6 and IKK. Immunity 34:843–853
103. Xu H, Yang J, Gao W, Li L, Li P, Zhang L, Gong YN, Peng X, Xi JJ, Chen S et al (2014) Innate immune sensing of bacterial modifications of Rho GTPases by the Pyrin inflammasome. Nature 513:237–241
104. Yamasaki K, Muto J, Taylor KR, Cogen AL, Audish D, Bertin J, Grant EP, Coyle AJ, Misaghi A, Hoffman HM et al (2009) NLRP3/cryopyrin is necessary for interleukin-1beta (IL-1beta) release in response to hyaluronan an endogenous trigger of inflammation in response to injury. J Biol Chem 284:12762–12771
105. Yang J, Xu H, Shao F (2014) The immunological function of familial Mediterranean fever disease protein Pyrin. Sci China Life Sci 57(12):1156–1161
106. Zhao Y, Yang J, Shi J, Gong YN, Lu Q, Xu H, Liu L, Shao F (2011) The NLRC4 inflammasome receptors for bacterial flagellin and type III secretion apparatus. Nature 477:596–600
107. Zhou R, Tardivel A, Thorens B, Choi I, Tschopp J (2010) Thioredoxin-interacting protein links oxidative stress to inflammasome activation. Nat Immunol 11:136–140
108. Zhou R, Yazdi AS, Menu P, Tschopp J (2010) A role for mitochondria in NLRP3 inflammasome activation. Nature 469:221–225

What Is the Pathobiology of Inflammation to Cell Death? Apoptosis, Necrosis, Necroptosis, Autophagic Cell Death, Pyroptosis, and NETosis

Rui Kang and Daolin Tang

Abstract Cell death and immunity are two evolutionarily-conserved processes that maintain homeostasis under changing conditions in the internal and external environment. Although these processes utilize fundamentally different machinery, cell death and immunity are highly interconnected and share a number of critical modifiers. Inflammation, the body's important immune response to injuries or infections, is a complex process involving various types of immune cells and signaling molecules. Different types of cell death including apoptosis, necrosis, necroptosis, autophagic cell death, pyroptosis, and NETosis can lead to the development of different immune and inflammatory responses including either immunogenic cell death (ICD) or tolerogenic cell death (TCD). The molecular mechanisms of ICD and TCD are beginning to be elucidated and have critical implications for the treatment of various acute and chronic diseases. In particular, damage-associated molecular patterns (DAMPs), endogenous molecules released during cell death and tissue injury, exhibit cytokine and chemokine activities in the regulation of the balance between ICD and TCD. In this chapter, recent advances in our understanding of the relationship between cell death, inflammation, and DAMPs are reviewed.

Abbreviations

ACD	Accidental cell death
AIF	Apoptosis-inducing factor
AIM2	Absent in melanoma 2
ASC	Apoptosis-associated speck-like protein containing a CARD
ATG	Autophagy-related

R. Kang (✉) • D. Tang (✉)
Department of Surgery, University of Pittsburgh Cancer Institute, University of Pittsburgh, Pittsburgh, PA, USA
e-mail: kangr@upmc.edu; tangd2@upmc.edu

M.C. Maiuri, D. De Stefano (eds.), *Autophagy Networks in Inflammation*, Progress in Inflammation Research, DOI 10.1007/978-3-319-30079-5_5

ATP	Adenosine triphosphate
Bak	Bcl-2 homologous antagonist/killer
Bax	bcl-2-like protein 4
BCL10	B-cell CLL/lymphoma 10
Bcl-2	B-cell lymphoma 2
Bcl-xL	B-cell lymphoma-extra large
BECN1	Beclin 1
Bid	BH3 interacting-domain death agonist
CTLs	Cytotoxic T lymphocytes
DAMPs	Damage-associated molecular patterns
DCC	Colorectal carcinoma
DRs	Death receptors
ENDOG	Endonuclease G
FADD	Fas-Associated protein with Death Domain
HMGB1	High mobility group box 1
HSPs	Heat shock proteins
IBD	Inflammatory bowel disease
ICD	Immunogenic cell death
IFN	Interferon
IL	Interleukin
LAMP	Lysosomal-associated membrane protein
LAP	LC3-associated phagocytosis
LC3	Microtubule-associated protein 1 light chain 3
LPS	Lipopolysaccharide
MAPK	Mitogen-activated protein kinases
MDA5	Melanoma differentiation-associated protein 5
MLKL	Mixed lineage kinase domain-like protein
MPO	Myeloperoxidase
NADPH	Nicotinamide adenine dinucleotide phosphate-oxidase
NCCD	Nomenclature Committee on Cell Death
NDP52/CALCOCO2	Nuclear dot protein 52
NETs	Neutrophil extracellular traps
NF-κB	Nuclear factor-κB
NK	Natural killer
NLRC4	NLR family CARD domain-containing protein 4
NLRP1	NLR family pyrin domain containing 1
NLRs	NOD-like receptors
Omi/HTRA2	HtrA serine peptidase 2
PAMPs	Pathogen-associated molecular patterns
PARP-1	Poly ADP-ribose polymerase 1
PD4	Peptidylarginine deiminase 4
PIK3C3	Phosphatidylinositol 3-kinase, catalytic subunit type 3
PIK3R4	Phosphoinositide-3-kinase, regulatory subunit 4
PKB/AKT	Protein kinase B
PMA	Phorbol myristate acetate

PRRs	Pattern recognition receptors
PtdIns3K	Phosphatidylinositol 3-kinase
PUMA	p53 upregulated modulator of apoptosis
RAGE	The receptor for advanced glycation end products
RB1CC1/FIP200	RB1-Inducible Coiled-Coil 1/FAK Family Kinase-Interacting Protein of 200 kDa
RCD	Regulated cell death
RIG-1	Retinoic acid-inducible gene 1
RIP3/RIPK3	Receptor-interacting serine-threonine kinase 3
RLRs	RIG-I-like receptors
ROS	Reactive oxygen species
SMAC/DIABLO	Second mitochondria-derived activator of caspases
SQSTM1/p62	Sequestosome 1
STAT3	Signal transducer and activator of transcription 3
T3SS	Type III secretion system
TAX1BP1	Human T-cell leukemia virus type I binding protein 1
TCD	Tolerogenic cell death
TIM3	T-cell immunoglobulin mucin 3
TLRs	Toll-like receptors
TMEM173/STING	Transmembrane protein 173
TNF	Tumor necrosis factor
TNFR1	TNF receptor 1
TRAIL	TNF-related apoptosis-inducing ligand
TUNEL	Terminal deoxynucleotidyl transferase dUTP nick end labeling
ULK1	UNC-51-like kinase 1
UNC5A	unc-5 homolog A
VPS34	Vacuolar protein sorting 34
WIPI1	WD repeat domain phosphoinositide-interacting protein 1
ZBP1/DAI	Z-DNA-binding protein 1.

1 Introduction

Cell death plays a fundamental role in various physiological and pathological processes. It is not only a universal feature of normal development and aging, but also firmly established in the pathogenesis and treatment of human disease. Although different types of cell death exhibit different morphological, biochemical, functional, and immunological characteristics, these forms of cell death may coexist and be related mechanistically. In early 1970s [80, 137], cell death was first morphologically classified and cell death was divided into (1) type I cell death, namely apoptosis, exhibiting morphologic features of membrane blebbing, cell shrinkage, fragmentation of nucleus and chromosomal DNA, and chromatin condensation; (2) type II cell death, namely autophagy, characterized by lack of chromatin

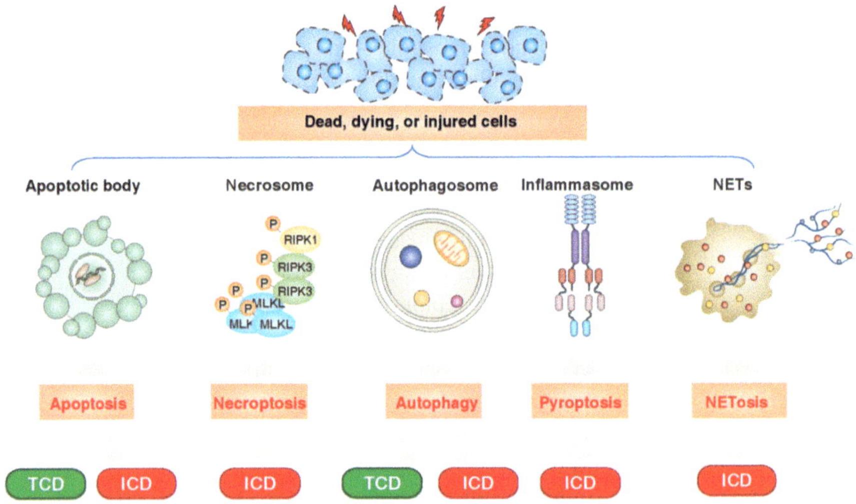

Fig. 1 Immunological characteristics of cell death

condensation and extensive cytoplasmic vacuolation involving swelling of organelles; and (3) type III cell death, namely necrosis, displaying plasma membrane rupture, release of cytoplasmic constituents, and moderate chromatin condensation. In 2007, 2009, 2012, and 2014, *Cell Death and Differentiation* published serial recommendations of the Nomenclature Committee on Cell Death (NCCD) for cell death modalities, classification of cell death, molecular definitions of cell death subroutines, and essential versus accessory aspects of cell death, respectively [45–47, 85]. These guidelines and recommendations will be helpful in understanding the biology of cell death in human health and diseases including cancer, autoimmune disorders, neurodegenerative diseases, ischemic and inflectional diseases, and aging.

According to NCCD recommendations, cell death is generally divided into accidental cell death (ACD) or regulated cell death (RCD) [45]. The onset of ACD is extremely fast and does not usually involve specific molecular machinery. In contrast, RCD processes such as apoptosis, autophagic cell death, necroptosis, pyroptosis, and NETosis are dynamic and regulated by specific molecular machinery. The biochemical processes underlying genetically-controlled RCD is complex and requires unique initiator, effector, and signaling pathways. For example, caspase [129] and autophagy-related (ATG) protein [113] are essential regulators for apoptosis and autophagy, respectively.

With respect to immunological characteristics, cell death can be divided into two distinct types: immunogenic cell death (ICD) and tolerogenic cell death (TCD) (Fig. 1) [52]. The activity and category of molecules released from dead/dying cells and the manner of clearance of dead/dying cells by neighboring cells differentiate ICD and TCD [130, 187]. In particular, damage-associated molecular patterns (DAMPs) are endogenous molecules released by dead, dying, or injured cells that function as danger signals to mediate inflammatory and immunological responses after abnormal cell death [151]. Besides mediating ICD, DAMPs also contribute to TCD through redox modification, suggesting a context-dependent role for DAMPs in cell death and inflammation [63]. Here, we briefly introduce the process of inflammation and the notion and category of DAMPs, and focus on the pathobiology of inflammation to different cell death types.

2 Inflammatory Process in Disease

Inflammation is induced when innate immune cells detect infection or tissue injury. Inflammation can be divided into acute inflammation or chronic inflammation. Acute inflammation occurs faster over minutes, hours, and days, while chronic inflammation occurs over a longer period of time. The principal causes of acute inflammation include microbial infections (e.g., bacterial, viral, fungal, and parasitic infection), physical agents (e.g., trauma, irradiation, burns, or excessive cooling), chemical agents (e.g., corrosives, acids, alkalis, and reducing agents), foreign bodies (e.g., splinters, dirt, sutures, and crystal deposits), tissue necrosis (e.g., ischemic infarction) and hypersensitivity reactions [131]. Chronic inflammation can be caused by pathogenic infection (e.g., bacterial and viral), environmental antigens (e.g., pollen grains and mold spores), autoimmune reaction, or persistent activation of inflammatory mediators [19]. Clinical characteristics of acute inflammation include heat (Latin: *calor*), redness (Latin: *rubor*), swelling (Latin: *tumor*), pain (Latin: *dolor*), and loss of function (Latin: *functio laesa*). The Roman encyclopedist Aulus Cornelius Celsus documented the first four cardinal signs of inflammation 2,000 years ago, whereas Rudolf Virchow, a German physician who is known as the "father of modern pathology," added loss of function as the fifth cardinal sign of inflammation in the nineteenth century. Clearly, inflammation is normally a localized, protective host immunity response to eliminate microbial infection and promote repair of damaged tissue that is detected by the presentation of "stranger signals" and "danger signals" to pattern recognition receptors [40]. In contrast, excessive, inappropriate, or uncontrolled inflammation has been implicated in a number of acute and chronic human diseases including cancer through cytokine storm, immunity dysfunction, and tissue damage [25]. The net effect of inflammation in certain pathologic conditions will be determined by whether the disadvantages outweigh the advantages or the advantages outweigh the disadvantages [59]. Thus, inflammation is a double-edged sword in a number of pathological processes.

The acute inflammatory response is a complex process involving a cascade mechanism that mainly includes a vascular reaction and a cellular response [109, 146]. The vascular system is the first system to respond to an injury. The vascular reaction includes (1) increased vascular caliber that promotes blood flow and (2) structural and functional changes in the microvasculature that increase permeability, the leakage of plasma proteins, and emigration of leukocytes from the microcirculation. In addition, endothelial cells are activated, resulting in increased adhesion of leukocytes and transendothelial migration of the leukocytes. Granulocytes, monocytes, and lymphocytes are major leukocytes that take part in the inflammatory cascade as well as consequent repair and injury. Among them, neutrophils, eosinophils, and monocytes can migrate to the site of infection and act as phagocytes to engulf and digest cellular debris and microorganisms. In addition, several leukocytes release not only enzymatic granules to kill pathogenic invaders, but also inflammatory mediators (e.g., cytokines and chemokines) to sustain the inflammatory response. In many cases, granulocytes contribute to acute inflammation, whereas monocytes and lymphocytes are responsible for chronic inflammation. Of note, induction of cell death may be of the utmost importance in inflammation and immunity. Type of cell death not only controls the development, differentiation, and activity of immune cells, but also the strategy employed by immune cells such as leukocytes to remove unwanted cells from the body [24, 82, 140].

3 Stranger and Danger Signals: PAMPs and DAMPs

One of the major functions of the immune system is to detect and distinguish a wide variety of stranger and danger signals during stress. Infection from invasive pathogens and sterile inflammation from tissue injury have different origins and mediators in the initiation of the immune response [12]. The exogenous stranger signals are referred to pathogen-associated molecular patterns (PAMPs) that are the components of microorganisms [67]. Lipopolysaccharide (LPS), the prototypical PAMP, is a major component of the Gram-negative bacterial membrane. Other PAMPs include microbial nucleic acids (e.g., DNA, dsRNA, ssRNA, and 5′-triphosphate RNA) and molecular structures associated with microbial envelopes (e.g., peptidoglycans, lipoteichoic acid, flagellin, and glycosylphosphatidylinositol) [2]. In contrast, the endogenous danger signals are referred to damage-associated molecular patterns (DAMPs), which are cell-derived molecules from self [107, 139]. Most DAMPs are proteins from the nucleus (e.g., high mobility group box 1 [HMGB1] and histone), cytosol (e.g., S100 and heat shock proteins [HSPs]), mitochondria (e.g., mitochondrial transcription factor A and formyl peptides), and plasma (e.g., C3a, C4a, and C5a). Several non-protein DAMPs have been identified, such as adenosine triphosphate (ATP), uric acid crystals, hyaluronan, heparin sulfate, RNA, genomic DNA, and mitochondrial DNA. Recognition and activation of PAMPs and DAMPs are mediated by a significant number of pattern recognition receptors (PRRs) that are widely expressed not only in immune cells but also in non-immune cells. These PRRs include Toll-like receptors (TLRs), NOD-like receptors (NLRs), RIG-I-like receptors (RLRs), absent in melanoma 2 (AIM2)-like receptors, the receptor for advanced glycation end products (RAGE), and DNA/RNA sensors. As so-called "Signal 0 s," PAMPs and DAMPs can activate multiple inflammatory signaling pathways (e.g., nuclear factor-κB [NF-κB], inflammasome, signal transducer and activator of transcription 3 [STAT3], and mitogen-activated protein kinases [MAPK]) that originate from these PRRs [151].

This interplay between autophagy and DAMPs in cell injury, adaption, and death has recently been studied in several cells. DAMPs have a normal function inside the cells of origin. In healthy cells, endogenous DAMPs (e.g., HMGB1, histone, and HSPs) usually contribute to sustaining cellular homeostasis. Knockout of HMGB1 and HSPs decreases autophagy and increases cell death in response to stress [121, 152, 153]. Moreover, autophagy plays a context-dependent role in the regulation of DAMP release and degradation in inflammation and death [182]. Many questions remain about how feedback loops between autophagy and DAMPs allow cells to maintain homeostasis in different cell death types.

4 HMGB1 as the Prototypic DAMP

HMGB1 is the prototypic DAMP and has been established to play an essential role in the regulation of infection, sterile inflammation, and other inflammation-associated diseases such as cancer [6, 74, 78, 154]. HMGB1 is normally a nuclear architecture factor with DNA chaperone activity that contributes to nucleosome and genomic stability [51]. In addition to the nucleus, HMGB1 is located in the cytosol, mitochondria, plasma, and extracellular space. HMGB1 can be either actively secreted by immune cells in response to PAMPs or passively released by death pathways including necrosis, apoptosis, autophagy, pyroptosis, and NETosis (Fig. 2). HMGB1 is a late mediator of sepsis during infection [170] and an early mediator of tissue injury during sterile inflammation [160]. Although oxidative stress may be a common mechanism that regulates HMGB1 release [155], several mechanisms have been proposed to define HMGB1 release in response to death. For example, HMGB1 released during necrosis and necroptosis is positively regulated by receptor-interacting serine-threonine kinase 3 (RIP3/RIPK3), poly [ADP-ribose] polymerase 1 (PARP-1), and cathepsin [33, 89, 189], whereas HMGB1 released during apoptosis in dendritic cells is positively regulated by caspase3/7 and mitochondrial reactive oxygen species (ROS) production [79]. In addition, the ATG5-mediated autophagy and pathogenic autophagic cell death pathway is required for HMGB1 secretion from cancer cells, fibroblasts, and macrophages in response to toxin, starvation, and LPS [36, 152, 158]. Inhibition of caspase 1 activity can diminish HMGB1 release and the inflammatory response during PAMP- and DAMP-induced pyroptosis [73, 103]. The release of HMGB1 during NETosis is regulated by autophagy [72, 112], suggesting interplay between autophagy and death in the regulation of DAMP release.

The redox status of HMGB1 may decide whether cell death is ICD or TCD [149]. Reduced HMGB1 with cytokine and chemokine activity has been shown to contribute to induction of the inflammatory signaling pathway and autophagy [150, 167]. In contrast, oxidized HMGB1 loses immune activity at the late stage of the inflammatory response [167] and facilitates TCD during apoptosis [79]. Apart from this, the activity of extracellular HMGB1 may be regulated by its receptors and cleavage. For instance, binding with TLRs and RAGE increases the immune activity of HMGB1, whereas binding with CD24 and T-cell immunoglobulin mucin 3 (TIM3) diminishes

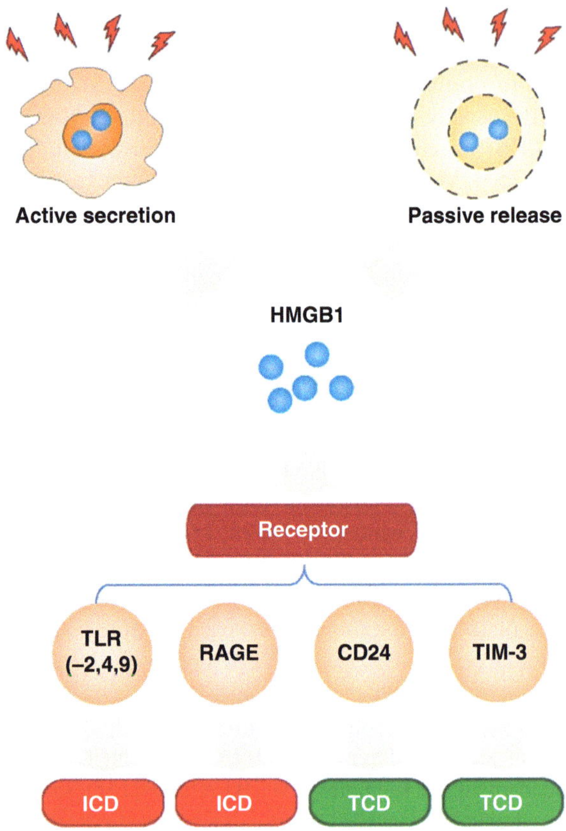

Fig. 2 Receptor-mediated activity of HMGB1

the immune activity of HMGB1 (Fig. 2) [20, 156]. The synergistic effect of HMGB1 with other substances such as LPS, DNA, histone, and IgG has been shown to modulate the host inflammatory response [161]. Collectively, HMGB1 is an inducer, sensor, mediator, and effector of inflammation during infection and tissue injury.

5 Apoptosis and Inflammation

5.1 The Process of Apoptosis

In 1972, the pathologists Kerr, Wyllie, and Currie coined the term “apoptosis” for regulated cell death during the development of an organism that proceeds through active, controlled morphological changes [80]. In addition to playing a physiological role, death by apoptosis facilitates loss of cells in a variety of pathological conditions such as DNA damage and accumulation of misfolded proteins and infection. Apoptosis is triggered through three major “extrinsic,” “intrinsic,” and “cytolytic” pathways [39]. Molecular crosstalk between extrinsic, intrinsic, and cytolytic apoptosis pathways is a highly-regulated process.

The extrinsic pathway is mediated by activation of pro-apoptotic transmembrane death receptors (DRs) when they recognize corresponding pro-apoptotic ligands [8]. The DRs are members of the tumor necrosis factor (TNF) receptor gene superfamily, including the Fas receptor (FasR/CD95), TNF receptor 1 (TNFR1), lymphotoxin receptor, DR3, and DR4/DR5. Apart from DRs, the receptors unc-5 homolog A (UNC5A) and deleted in colorectal carcinoma (DCC) have been demonstrated to mediate the alternative extrinsic pro-apoptotic pathway [110]. The signaling-mediated crosstalk between DRs, UNC5A, and DCC remains unclear.

The intrinsic pathway is mediated by mitochondria [169]. The mitochondrial pathways of apoptosis are usually triggered by signals from DNA damage, chemotherapy, irradiation, and cell-survival factor depletion. These stimuli can cause structural and functional changes of the inner mitochondrial membrane, which then induce the opening of mitochondrial permeability transition pores, loss of mitochondrial transmembrane potential, and release of mitochondrial apoptosis-inducing factor [147]. Cytochrome c [100], second mitochondria-derived activator of caspases (SMAC/DIABLO) [34], and HtrA serine peptidase 2 (Omi/HTRA2) [163] are released from mitochondria into the cytosol at the early stage of apoptosis and can activate effector caspases to induce apoptosis. Omi/HTRA2 also promotes the caspase-independent apoptosis pathway through cleavage of cytoskeletal proteins [57]. In addition, apoptosis-inducing factor (AIF) [27, 145] and endonuclease G (ENDOG) [96] is released from mitochondria to the nucleus at the later stage of apoptosis, which promotes apoptosis in a caspase-independent manner. The pro-(e.g., Bax, Bak, Bid, and PUMA) and anti-(e.g., Bcl-2 and Bcl-xL) apoptotic Bcl-2 family members are critical regulators of the mitochondrial apoptotic pathway through controlling mitochondrial transmembrane potential [180].

The cytolytic pathway is mediated by cytotoxic T lymphocytes (CTLs) and natural killer (NK) cells and is an important defense mechanism that kills viral and cancer cells [159, 162]. The pore-forming protein perforin is the one of main effector molecules of CTLs and NK cells. Perforin associated with granzymes is the major component of cytotoxic granules. Once a target cell is recognized, the perforin pores open and allow perforin/granzymes to enter the target cell and induce apoptosis. This perforin/granzyme-mediated cytolytic pathway requires classical components of the mitochondrial cell-death pathway such as effector caspases and pro-apoptotic Bcl-2 family members (e.g., Bid).

5.2 *The Pathobiology of Inflammation to Apoptosis*

Apoptotic cell death is generally recognized as a non-immunogenic, non-inflammatory process. Firstly, apoptotic cells have an intact plasma membrane and don't rapidly release their intracellular contents. Secondly, apoptosis of inflammatory cells, including epithelial cells and dendritic cells, can diminish the production and activity of pro-inflammatory cytokines and DAMPs such as HMGB1 and interleukin (IL)-1β [66, 79]. Thirdly, phagocytic clearance of apoptotic cells plays a significant role in the resolution of inflammation [58]. Phagocytes include macrophages, neutrophils, monocytes, and dendritic cells [42]. The initiation of successful phagocytosis generally depends on four steps [125, 126]: (1) recruitment of phagocytes to the site of infection and injury when apoptotic cells release "find-me" signals; (2) phagocyte-mediated engulfment of apoptotic cells through upregulation of "eat-me" signals and downregulation of "don't eat me" signals; (3) the maturation of phagosomes and subsequent fusion with lysosomes to generate phagolysosomes, in which cell corpses are degraded, and (4) suppression or initiation of the innate immune response, depending on the stimuli. In this way, phagocytosis protects tissue against harmful exposure to the inflammatory contents of dying cells, promotes tissue repair and wound healing, and prevents autoimmunity. Apart from phagocytes, autophagy has the ability to directly clean apoptotic cells during embryonic development, suggesting a physiological role of autophagy in development [123]. Oxidative stress-mediated interplay between autophagy and phagocytosis has been important for the clearance of bacterial and apoptotic cells [168].

Apoptosis may be not only a TCD process, but also an ICD process arising from an increased production of apoptosis and/or impaired clearances. Apoptotic DNA fragmentation is a key feature of apoptosis. Excessive apoptosis can directly lead to release of nuclear DAMPs including DNA, histone, and HMGB1 [10, 21]. Interestingly, DNA and histone may promote phagocytosis [53], whereas HMGB1 is a negative regulator of phagocytosis [98]. Extracellular HMGB1 can bind phosphatidylserine in apoptotic neutrophils or integrin $\alpha_v\beta_3$ in phagocytic macrophages, which inhibits the phagocytosis of apoptotic cells [44, 98]. These findings suggest that HMGB1 can inhibit phagocytosis-mediated anti-inflammatory responses during apoptosis. Increased apoptosis of professional phagocytes may lead to impaired clearance of apoptotic cells. Accumulating evidence suggests that certain DAMPs are powerful immunological adjuvants that contribute to ICD-mediated anticancer therapy [63, 84, 88]. ICD involves several steps. Firstly, DAMPs such as calreticulin [119], HMGB1 [7], HSP70/HSP90 [49, 142], and ATP [50] are exposed to cell membrane or released into the extracellular space from proapoptotic, postapoptotic, and/or necrotic cells. Secondly, PPRs such as CD91, TLR4, and P2X7 are present on dendritic cells and can recognize these DAMPs. Finally, this recognition between DAMPs and PRRs has multiple significant functional roles, including the clearance of dying cells, presentation of tumor antigens, and production of inflammasome-dependent proinflammatory cytokines. These models should also allow assessment of other inflammation-associated diseases.

6 Necrosis and Inflammation

6.1 The Process of Necrosis

Necrosis has been long considered an accidental, passive, and type III cell death which lacks the morphological characteristics of apoptosis or autophagy and follows overwhelming stress such as heat shock, osmotic shock, hypoxic injury, pressure disruption, mechanical stress, and freeze-thawing [188]. It recently became clear that necrosis is not only accidental, but also regulated and programmed [47, 165]. Recently, the term "necroptosis" has been used to describe the process of programmed necrosis. Necroptosis can be triggered in multiple cell types by ligands for DR, Toll-like receptors (e.g., TLR-2, -3, -4, -5 and -9), DNA, and/or RNA sensors (e.g., Z-DNA-binding protein 1 [ZBP1/DAI], RIG-1 and melanoma differentiation-associated protein 5 [MDA5]) and interferon (IFN) receptors [70, 166]. In particular, TNFα and TNF-related apoptosis-inducing ligand (TRAIL), the same ligands that activate death receptor-mediated apoptosis, induce necroptosis when the activation of caspase-8 is inhibited in genetic (e.g., knockout of caspase-8 or FADD) and pharmacological (e.g., ZVAD-FMK) ways [71, 172]. Like necrosis, necroptosis is characterized by early loss of plasma membrane integrity, leakage of intracellular contents, and organelle swelling. Although they lack typical apoptotic morphological characteristics, necroptotic cells are usually terminal deoxynucleotidyl transferase dUTP nick end labeling (TUNEL)-positive. ATP depletion, accumulated ROS, and increased permeability of cellular membranes contribute to necrosis and necroptosis. At the molecular level, RIP1/RIPK1, RIP3/RIPK3, and mixed lineage kinase domain-like protein (MLKL) are three major regulators and mediators of necroptosis [56, 115, 118, 144]. The function of RIP1 kinase, an important upstream regulator of necroptosis, can be regulated intricately by several posttranslational modification including phosphorylation and ubiquitination [22]. Following RIP1/RIPK1 activation, it binds to RIP3/RIPK3 and then forms necrosome, a functional complex required for necroptosis [95, 166]. MLKL, the downstream component of necroptosis, can be phosphorylated by RIP3/RIPK3 and then recruited to the necrosome through its interaction with RIP3/RIPK3 [115, 144]. Finally, MLKL translocates to the plasma and cytoplasmic membranes, where it modulates ion channel activities and Ca^{2+} influx to lead to necrosis [17]. Collectively, necroptosis is a programmed and actively-regulated process under specific conditions.

6.2 *The Pathobiology of Inflammation to Necrosis*

Much has been learned about the critical role of necrosis in inflammation. The release of intracellular contents after cellular membrane damage is the cause of a sterile inflammatory response in necrosis and necroptosis. Many DAMPs, including protein and non-protein DAMPs released from accidental necrotic cells, have been described. For example, nuclear DAMPs including HMGB1, histones, and genomic DNA have been observed in necrotic tissues and serum from patients with inflammatory diseases [21, 74, 75, 135]. Mitochondrial DAMPs, including mitochondrial DNA, are emerging biomarkers of infection and tissue ischemia-reperfusion injury [183]. Several DAMPs including HMGB1, S100A9, IL-33, and mitochondrial DNA have been associated with necroptotic conditions in vitro and in vivo [37, 83, 89], but the mechanisms behind these immune effects remain unclear and require further investigation. Necroptosis in the absence of caspase might affect the immunological activity of released DAMPs in inflammatory disease. In addition, many productions (e.g., TNFα and IFN) from inflammatory cells such as macrophages and dendritic cells can cause tissue necrosis and necroptosis, suggesting a feed-forward loop between inflammation and necrosis.

More importantly, inhibition of necroptosis by both pharmacological inhibitors (e.g., necrostatin-1 and necrosulfonamide) and genetic knockout of critical regulators (e.g., RIP13/RIPK1, RIP3/RIPK3, and MLKL) could reduce, delay, or prevent the development of pathology in mouse models of numerous inflammation-associated diseases [26, 28, 128, 144, 148]. These diseases include acute ischemia reperfusion injury in the brain, heart, retina, and kidney; traumatic brain injury; stroke; retinal detachment; age-related macular degeneration; necrotizing pancreatitis; atherosclerosis; inflammatory bowel disease; acute liver failure and injury; steatohepatitis; vaccinia virus infection; acute peritonitis; systemic inflammatory response syndrome; and transplant rejection. Thus, interference with necroptosis might provide therapeutic benefits to human diseases involving inflammation and cell death [97, 186].

7 Autophagy and Inflammation

7.1 The Process of Autophagy

The term "autophagy" was coined in 1963 by Christian de Duve, who used electron microscopy to observe the ultrastructure in rat liver hepatocytes with autophagic vacuolization [176]. Currently, autophagy is a lysosome-dependent degradation system important in sustaining cellular homeostasis [23]. At least three major types of autophagy have been identified: (1) Macroautophagy is characterized by the formation and maturation of several membrane structures including the phagophore, autophagosome, and autolysosme, which allow the delivery of a large number of different cargos such as cytosolic components, proteins, and organelles into the lysosome for degradation. (2) Chaperone-mediated autophagy is responsible for the degradation of proteins carrying KFERQ-like motif. This process requires HSC70 chaperone-mediated recognition and subsequent lysosomal-associated membrane protein (LAMP)-2A-mediated translocation across lysosomes for degradation. (3) Microautophagy is mediated by the direct engulfment of cytoplasmic components into the lysosome.

Macroautophagy (hereafter referred to as autophagy), the most well-studied type, is primarily regulated by ATG proteins [177]. These ATG proteins can form several functional complexes involved in different stages of autophagy [9, 41]. In mammals, these complexes include: (1) UNC-51-like kinase 1 (ULK1) kinase complex containing the core proteins ULK1 (or ULK2), ATG13, and RB1CC1/FIP200, which is required for the induction of autophagosome formation [62, 69]; (2) the Phosphatidylinositol 3-kinase (PtdIns3K) complex containing the core proteins PIK3C3/VPS34, PIK3R4/p150, BECN1 and ATG14, which is required for induction of the nucleation of the phagophore; (3) ATG9 and its cycling system containing the core proteins ATG9, ATG2, and WIPI1/2, which contributes to the elongation of the phagophore; (4) The ATG12–ATG5 conjugation system containing the core proteins ATG5, ATG12, ATG7, ATG10, and ATG16L1, which is required for phagophore expansion by membrane addition; and (5) The Ubl LC3 conjugation system containing the core proteins microtubule-associated protein 1 light chain 3 (LC3, a homolog of yeast Atg8), ATG3, ATG4, and ATG7, which is also required for phagophore expansion by membrane addition. However, the role of ATG proteins at the stage of autolysosome formation, degradation, and efflux remains unknown.

Induction of autophagy is an important cellular stress response that enables cells to catabolize unused proteins and damaged organelles for recycling and reuse to promote survival [87, 116]. The term "autophagic cell death" was originally defined as a morphological change of a type II cell death accompanied by large-scale autophagic vacuolization of the cytoplasm [86]. In addition to being associated with the appearance of autophagy, cell death can be trigged by abnormal autophagy [101]. Recent studies indicate that excessive or uncontrolled levels of autophagy can kill cells under some conditions, especially in the absence of an intact apoptosis pathway [104]. A recent study suggests that Draper, the *Drosophila melanogaster* orthologue of the Caenorhabditis elegans engulfment receptor CED-1, functions to separate autophagy associated with cell death from autophagy leading to cell survival [108]. In particular, "autosis" was identified as a novel form of autophagy-dependent and non-apoptotic cell death by Beth Levine in 2013 [102]. Different from other cell death pathways, autosis has several special morphological features such as focal plasma membrane rupture, enhanced cell substrate adhesion, focal ballooning of the perinuclear space, and dilation and fragmentation of the endoplasmic reticulum [102]. This process requires the core autophagy machinery and plasma membrane Na^+/K^+-ATPase in response to autophagy-inducing peptides, starvation, and hypoxia-ischemia [102]. It remains to be determined whether Na^+/K^+-ATPase is a casual factor in autophagic cell death.

7.2 *The Pathobiology of Inflammation to Autophagy*

Autophagy-mediated anti-inflammatory responses are involved in several different pathways [30–32, 93, 94, 114]: (1) Direct elimination of microorganisms by xenophagy. Xenophagy is a form of selective autophagy that directly takes up and degrades cytosolic invasive pathogens (such as viruses, bacteria, and protozoa) [3]. This process requires specific adaptor proteins such as sequestosome 1 (SQSTM1/p62), neighbor of Brca1 gene (NBR1), nuclear dot protein 52 (NDP52/CALCOCO2), human T-cell leukemia virus type I binding protein 1 (TAX1BP1), and optineurin to recognize molecular tags present on invading microorganisms and translocate to a double membrane autophagosome. In vivo studies from autophagy-deficient mice have confirmed the importance of xenophagy in the clearance of pathogens and prevention against infection. (2) Direct elimination of microorganisms by LC3-associated phagocytosis (LAP). LAP is an autophagic-phagocytosis "hybrid" process with characteristics of phagocytosis and autophagy. LAP utilizes autophagic machinery to selectively degrade pathogens and cellular corpses from apoptotic, necrotic, and necroptotic cells in a single-membrane phagosome [105]. This process requires certain ATG proteins such as LC3, BECN1, PI3KC3, ATG5, and ATG7, but not ULK1 [105]. In addition, MORN2 (MORN repeat-containing protein) has recently been identified to promote the recruitment of LC3 to phagosomes in LAP [1]. (3) Suppression of inflammasome-dependent IL-1β activation. Knockout of ATG16L1, ATG7, and ATG5 in mice increases infection-mediated sepsis and inflammasome-dependent IL-1β activation [133]. Mitophagy is a process in which mitochondria are targeted for degradation via the autophagy pathway [179]. Mitophagy deficiency results in the accumulation of damaged mitochondria and increased ROS production. This in turn activates inflammasome-dependent IL-1β release as well as Type 1 IFN production [117, 120, 185]. In addition, autophagy inhibits inflammasome activation by direct degradation of pro-IL-1β and SQSTM1/p62 stability [55, 91]. Autophagy deficiency leads to HMGB1-DNA complex-induced AIM2 inflammasome activation, suggesting a potential role of autophagy in inhibition of the activity of DAMP complex [99]. (4) Suppression of calpain-dependent IL-1α activation. Macrophages from autophagy-defective mice secrete high levels of IL-1α in a calpain-dependent manner. Interestingly, inflammasome activation is not required for this process [18]. (5) Suppression of the NF-κB pathway. B-cell CLL/lymphoma 10 (BCL10)-containing complex is an important regulator of NF-κB activation by T- and B-cell receptors. Autophagy not only inhibits BCL10-containing complex formation, but also directly degrades BCL10 to reduce NF-κB activation in antigen-activated T cells [122]. (6) Degradation of HMGB1. Several Chinese herbs (Tanshinone IIA sodium sulfonate and epigallocatechin gallate) with anti-inflammatory activity can induce autophagy-dependent HMGB1 degradation in macrophages [184]. (7) Inhibition of TMEM173/STING-dependent type I IFN production. LC3 and ATG9a can co-localize with transmembrane protein 173 (TMEM173/STING) and inhibit type I IFN production after dsDNA stimulation [132]. In addition, ULK1-mediated TMEM173/STING phosphorylation is a negative regulator of type I IFN production [81].

Autophagy also has the ability to promote proinflammatory cytokine expression and release in immune cells: (1) Affection of PRR-mediated inflammation signaling. As ligands of PRRs, PAMPs and DAMPs can induce autophagy, which in turn triggers negative or positive feedback control of the immune response by regulating cytokine release [151]. For example, autophagy promotes TLR9 signaling in B cells and increases type I IFN production in plasmacytoid dendritic cells [134]. (2) Secretion of immune mediators, including IL-1β and DAMPs. In addition to inhibition of IL-1β release, autophagy as well as LAP also promotes IL-1β release [36]. These findings suggest a complex role of autophagy in the regulation of IL-1β processes and production. The secretion of DAMPs is mediated by a non-classical or unconventional secretory route in which they don't use the endoplasmic reticulum-to-Golgi membrane pathway [38, 48]. Autophagy has been shown to increase release of DAMPs (e.g., HMGB1, ATP, and DNA) in immune and non-immune cells by unconventional secretory pathways [36, 152, 171].

In addition, interplay between autophagy and other cell death pathways such as apoptosis, necroptosis, and NETosis may influence the inflammatory response. In many cases, autophagy inhibits apoptosis by degradation of activated caspase, whereas apoptosis blocks autophagy by cleavage of ATGs such as ATG5, BECN1, and ATG16L [77, 104]. In contrast, autophagy deficiency in inflammatory cells contributes to apoptosis and necrosis and worsens efferocytosis [157]. Another example is that autophagy promotes monosodium urate crystal-induced HMGB1 release by NETosis [72, 112] In addition, positively-regulated necroptosis by autophagy may contribute to DAMP release and inflammatory cell death in endothelial and cancer cells [13].

8 Pyroptosis and Inflammation

8.1 The Process of Pyroptosis

The term "pyroptosis" was originally coined to indicate proinflammatory (="pyro") programmed cell death ("=ptosis") during bacterial infection in macrophages [15]. Currently, it refers to a regulated form of cell death in immune cells mediated by the activation of inflammasome [11]. Inflammasome, a multi-protein oligomer, is activated by PAMPs (e.g., LPS and bacterial flagellin) and DAMPs (e.g., ATP, uric acid, DNA, and RNA) and triggers the activation of caspase-1 and caspase-11 and finally the release of proinflammatory cytokines (e.g., IL-1β, IL-18, IL-33, and HMGB1) [11, 136]. Morphological characteristics of pyroptosis include cytoplasmic swelling, DNA fragmentation, and pore formation. In contrast to apoptosis, the formation of pyroptotic DNA fragmentation in pyroptosis is independent of caspase-activated DNase. Inflammasomes are divided into NLRs (e.g., NLRP1, NLRP3, and NLRC4) and non-NLRs (e.g., AIM2) [106, 136]. NLRs have common structural features with a nucleotide-binding domain and a C-terminus leucine-rich repeat. NLRP1 is activated by bacterial toxins and its activity is negatively regulated by anti-apoptotic Bcl-2 proteins such as Bcl-2 and Bcl-xL. NLRP3 is the most-studied inflammasome and becomes activated by extracellular ATP, crystalline, monosodium urate, alum, silica, asbestos, low intracellular potassium, or high intracellular ROS concentrations [92]. NLRC4 is activated by Gram-negative bacteria possessing a functional T3SS or T4SS, such as *Salmonella typhimurium*, *Legionella pneumophila*, and *Pseudomonas aeruginosa*. In non-NLR inflammasomes, AIM2 is activated by from cytosolic bacterial, viral, and host dsDNA [61]. The direct interaction between AIM2 and NLRs has not been established in disease.

8.2 The Pathobiology of Inflammation to Pyroptosis

An important innate immune effector mechanism, pyroptosis facilitates the clearance of invading intracellular pathogens. Several microorganisms (e.g., *Salmonella enterica Typhimurium*, *Legionella pneumophila*, *Burkholderia thailandensis*, the influenza virus, and *Shigella*) and their components can directly induce the activation of caspase-1 and inflammasome-mediated IL-1β and IL-18 production, which limit systemic infection. In contrast, increasing evidence indicates that inflammasome contributes to the pathogenesis of several inflammatory diseases such as diabetes, inflammatory bowel disease (IBD), rheumatoid arthritis, atherosclerosis, and pancreatitis. For example, NALP3 inflammasome-mediated IL-1β production from infiltrated macrophages in the pancreas can cause death of pancreatic β cells and subsequent diabetes [68]. Moreover, ROS production during apoptosis or mitophagy deficiency in pancreatic β cells also accelerates NALP3 inflammasome activation and IL-1β production, as well as the expression of chemotactic factors [141, 173]. This further worsens immune-cell infiltration and pancreatic β cell damage. NLRP3-mediated caspase-1 activation is unregulated in adipose-tissue and facilitates insulin signaling [164]. NLRP3- and NLRC4-dependent inflammasomes contribute to intestinal host defense against microbiota through promoting epithelial cell repair [43]. In contrast, activation of inflammasome promotes dextran sodium sulfate-induced colitis as well as colon cancer associated with IBD by inducing the excessive production of proinflammatory cytokines [181]. Fatty acids induce sterile inflammation in atherosclerosis partly by NLRP3-induced IL-1β-production [35]. Genetic deletion of caspase-1, ASC, and NLRP3 protects against experimental acute pancreatitis and chronic obesity-induced pancreatic damage [60], suggesting that inflammasome initiates inflammation in pancreatic injury.

The regulatory mechanisms of inflammasome activation is extremely complex and facilitate a balanced inflammasome-mediated immune response in disease [124]. Double-stranded RNA-dependent protein kinase (PKR) is a serine/threonine protein kinase that is activated by autophosphorylation after binding to dsRNA. PKR activation is implicated in inflammation and immune dysfunction through its regulation of several inflammation (e.g., mitogen-activated protein kinases, interferon regulatory factor 3, NF-κB), apoptosis, and autophagy pathways. Moreover, activation of PKR is implicated in the crosstalk between inflammasome, DAMP release, and cell death [76]. The release of HMGB1 is significantly decreased in macrophages from $PKR^{-/-}$ mice in response to multiple pyroptosis-associated stimuli such as ATP, monosodium urate, adjuvant aluminium, and live Escherichia coli [103]. As a newly-identified inflammasome component, PKR can sustain the structure and function of NLRP3, NLRP1, AIM2, and NLRC4 during inflammasome activation. Given that autophagy generally acts as an inhibitor of inflammasome activation, the mechanism of inflammasome components, including PKR coordinating autophagy and pyroptosis in inflammatory cell responses to infection, remains to be explored. In addition, HMGB1 can be specifically processed to create an active A-box peptide by caspase-1, but not other caspases (-2, -3, -5, -7, -9 or -11) [90].

9 NETosis and Inflammation

9.1 The Process of NETosis

Neutrophils, the most common type of white blood cell, provide the first line of defense of the innate immune system that protects the host from infection. Several mechanisms have been proposed to explain the activity and function of neutrophils in infection. Neutrophils have been well-demonstrated to kill and digest pathogens by phagocytosis and respiratory burst [138]. Neutrophils can not only engulf pathogens, but also kill them outside of the cell by neutrophil extracellular traps (NETs). NETs are the process by which neutrophils release extracellular web-like DNA-protein structures to capture and kill pathogens and then prevent the spread of infection. This process was first observed in neutrophils following treatment with PMA or interleukin-8 (IL-8) by Volker Brinkmann, Arturo Zychlinsky and colleagues in 2004 [16]. We now know that it is a form of regulated cell death, subsequently coined "NETosis" [143, 178]. In addition to NETosis, vesicular secretion is also responsible for neutrophils rapidly expelling their nuclear contents. NETosis, distinct from either necrosis or apoptosis, occurs not only in neutrophils, but also in other non-neutrophils including endothelial and cancer cells [29, 127]. NETs are generated in response to a number of pathological, physiological, and pharmacological stimuli such as pro-inflammatory stimuli (e.g., LPS, IL-8, and TNF), infections (e.g., microorganisms and pathogens), hypoxia, phorbol esters, or calcium ionophores.

The signaling cascade that triggers NETosis is known to involve several key steps, including the generation of ROS by NADPH oxidase, translocation of the granular enzymes neutrophil elastase and myeloperoxidase (MPO) to the nucleus, and the production of histone citrullination by peptidylarginine deiminase 4 (PD4) [14]. This dynamic process is positively regulated by several kinases such as protein kinase C, MAPK, and protein kinase B (PKB/AKT). Of note, PD4-mediated citrullination of H3 is often observed during NET formation and is a widely-used marker for NETosis monitoring in vitro. PD4 knockout mice exhibit impaired NETosis and are highly susceptible to some severe skin infections, but not lung infection and arthritis [14, 54]. These findings suggest that PD4 may play a tissue-dependent role in the regulation of NETosis. PD4-independent NETosis may exist and needs further identification [111].

9.2 *The Pathobiology of Inflammation to NETosis*

Apoptosis is essential for neutrophil functional shutdown; this process ensures that the cellular membrane remains intact and therefore prevents the release of granule contents and oxygen metabolites to damage tissue. Phagocytosis of apoptotic neutrophils by tissue macrophages is a mechanism to resolve inflammation. Chronic inflammation may result from an impaired ability of macrophages to induce phagocytosis of apoptotic neutrophils.

In addition, aberrant NETosis formation and impaired NET degradation may cause release and activation of inflammatory nuclear DAMPs (e.g., histone, DNA, and HMGB1) and granule components in tissue injury [127]. The release of these molecules during NETosis is positively-regulated by autophagy [72, 112], whereas DNase1 enzyme can degrade NETs. Once released, nuclear DAMPs induce pro-inflammatory and toxic responses in vivo and in vitro [174, 175]. Activation of TLRs (e.g., TLR2, TLR4, and TLR9) and NLRP3 inflammasome mediate the activity of nuclear DAMPs [4, 64, 65, 174]. NETosis-associated DAMP release and tissue damage have been demonstrated to be involved in the pathologies of several inflammatory and autoimmune disorders such as systemic lupus erythematosus, rheumatoid arthritis, asthma, vessel vasculitis, and psoriasis [14]. Circulating levels of DNA, histones, HMGB1, and nucleosomes may be useful candidate biomarkers in human diseases [5, 21].

10 Conclusions

The inflammatory response to cell death is complex and is likely dependent on stimuli, cell type, stage, and interplay between cells and their environment. This complexity needs further study to define settings in which cell death acts potently to either promote or inhibit the inflammatory response. In many cases, apoptosis and autophagy are forms of TCD, whereas necrosis/necroptosis, pyroptosis, and NETosis are forms of ICD (Fig. 1). In some cases, apoptosis and autophagy can cause ICD (Fig. 1). The category of DAMPs released from different cell death pathways and their activity, function, modification, as well as receptors will decide whether the cell death is TCD or ICD. Autophagy generally is an important stress response promoting cell survival, although autophagic cell death is recognized now in some cases. Aberrant cell death in immune cells or surrounding cells may limit or amplify the inflammatory response. In the future, biochemical and structural studies are needed to explore the underlying molecular mechanisms of DAMP release and how that death influences surrounding tissues and global immune responses. An improved understanding of the molecular mechanism of the inflammatory response has led to important advances in the treatment of chronic and acute diseases. However, inflammation is a dynamic process which is orchestrated by many signaling molecules, and targeting one or a few may not be enough.

Acknowledgments We apologize to the researchers who were not referenced due to space limitations. We thank Christine Heiner (Department of Surgery, University of Pittsburgh) for her critical reading of the manuscript. This work was supported by the National Institutes of Health (NIH) (R01CA160417 to D.T.) and the American Association for Cancer Research (AACR) (13-20-25-TANG to D.T.).

Conflict of Interest The authors declare no conflicts of interest.

References

1. Abnave P, Mottola G, Gimenez G et al (2014) Screening in planarians identifies MORN2 as a key component in LC3-associated phagocytosis and resistance to bacterial infection. Cell Host Microbe 16:338–350
2. Akira S, Hemmi H (2003) Recognition of pathogen-associated molecular patterns by TLR family. Immunol Lett 85:85–95

3. Alexander DE, Leib DA (2008) Xenophagy in herpes simplex virus replication and pathogenesis. Autophagy 4:101–103
4. Allam R, Darisipudi MN, Tschopp J et al (2013) Histones trigger sterile inflammation by activating the NLRP3 inflammasome. Eur J Immunol 43:3336–3342
5. Allam R, Kumar SV, Darisipudi MN et al (2014) Extracellular histones in tissue injury and inflammation. J Mol Med 92:465–472
6. Andersson U, Tracey KJ (2011) HMGB1 is a therapeutic target for sterile inflammation and infection. Annu Rev Immunol 29:139–162
7. Apetoh L, Ghiringhelli F, Tesniere A et al (2007) Toll-like receptor 4-dependent contribution of the immune system to anticancer chemotherapy and radiotherapy. Nat Med 13:1050–1059
8. Ashkenazi A, Dixit VM (1999) Apoptosis control by death and decoy receptors. Curr Opin Cell Biol 11:255–260
9. Behrends C, Sowa ME, Gygi SP et al (2010) Network organization of the human autophagy system. Nature 466:68–76
10. Bell CW, Jiang W, Reich CF 3rd et al (2006) The extracellular release of HMGB1 during apoptotic cell death. Am J Physiol Cell Physiol 291:C1318–C1325
11. Bergsbaken T, Fink SL, Cookson BT (2009) Pyroptosis: host cell death and inflammation. Nat Rev Microbiol 7:99–109
12. Bianchi ME (2007) DAMPs, PAMPs and alarmins: all we need to know about danger. J Leukoc Biol 81:1–5
13. Bonapace L, Bornhauser BC, Schmitz M et al (2010) Induction of autophagy-dependent necroptosis is required for childhood acute lymphoblastic leukemia cells to overcome glucocorticoid resistance. J Clin Invest 120:1310–1323
14. Branzk N, Papayannopoulos V (2013) Molecular mechanisms regulating NETosis in infection and disease. Semin Immunopathol 35:513–530
15. Brennan MA, Cookson BT (2000) Salmonella induces macrophage death by caspase-1-dependent necrosis. Mol Microbiol 38:31–40
16. Brinkmann V, Reichard U, Goosmann C et al (2004) Neutrophil extracellular traps kill bacteria. Science 303:1532–1535
17. Cai Z, Jitkaew S, Zhao J et al (2014) Plasma membrane translocation of trimerized MLKL protein is required for TNF-induced necroptosis. Nat Cell Biol 16:55–65
18. Castillo EF, Dekonenko A, Arko-Mensah J et al (2012) Autophagy protects against active tuberculosis by suppressing bacterial burden and inflammation. Proc Natl Acad Sci U S A 109:E3168–E3176
19. Cavanagh MM, Weyand CM, Goronzy JJ (2012) Chronic inflammation and aging: DNA damage tips the balance. Curr Opin Immunol 24:488–493
20. Chen GY, Tang J, Zheng P et al (2009) CD24 and Siglec-10 selectively repress tissue damage-induced immune responses. Science 323:1722–1725
21. Chen R, Kang R, Fan XG et al (2014) Release and activity of histone in diseases. Cell Death Dis 5:e1370. doi:10.1038/cddis.2014.337
22. Cho YS, Challa S, Moquin D et al (2009) Phosphorylation-driven assembly of the RIP1-RIP3 complex regulates programmed necrosis and virus-induced inflammation. Cell 137:1112–1123
23. Choi AM, Ryter SW, Levine B (2013) Autophagy in human health and disease. N Engl J Med 368:651–662
24. Christenson K, Thoren FB, Bylund J (2012) Analyzing cell death events in cultured leukocytes. Methods Mol Biol 844:65–86
25. Coussens LM, Werb Z (2002) Inflammation and cancer. Nature 420:860–867
26. Dannappel M, Vlantis K, Kumari S et al (2014) RIPK1 maintains epithelial homeostasis by inhibiting apoptosis and necroptosis. Nature 513:90–94
27. Daugas E, Susin SA, Zamzami N et al (2000) Mitochondrio-nuclear translocation of AIF in apoptosis and necrosis. FASEB J 14:729–739
28. Degterev A, Huang Z, Boyce M et al (2005) Chemical inhibitor of nonapoptotic cell death with therapeutic potential for ischemic brain injury. Nat Chem Biol 1:112–119

29. Demers M, Wagner DD (2014) NETosis: a new factor in tumor progression and cancer-associated thrombosis. Semin Thromb Hemost 40:277–283
30. Deretic V (2011) Autophagy in immunity and cell-autonomous defense against intracellular microbes. Immunol Rev 240:92–104
31. Deretic V, Levine B (2009) Autophagy, immunity, and microbial adaptations. Cell Host Microbe 5:527–549
32. Deretic V, Saitoh T, Akira S (2013) Autophagy in infection, inflammation and immunity. Nat Rev Immunol 13:722–737
33. Ditsworth D, Zong WX, Thompson CB (2007) Activation of poly(ADP)-ribose polymerase (PARP-1) induces release of the pro-inflammatory mediator HMGB1 from the nucleus. J Biol Chem 282:17845–17854
34. Du C, Fang M, Li Y et al (2000) Smac, a mitochondrial protein that promotes cytochrome c-dependent caspase activation by eliminating IAP inhibition. Cell 102:33–42
35. Duewell P, Kono H, Rayner KJ et al (2010) NLRP3 inflammasomes are required for atherogenesis and activated by cholesterol crystals. Nature 464:1357–1361
36. Dupont N, Jiang S, Pilli M et al (2011) Autophagy-based unconventional secretory pathway for extracellular delivery of IL-1beta. EMBO J 30:4701–4711
37. Duprez L, Takahashi N, Van Hauwermeiren F et al (2011) RIP kinase-dependent necrosis drives lethal systemic inflammatory response syndrome. Immunity 35:908–918
38. Eder C (2009) Mechanisms of interleukin-1beta release. Immunobiology 214:543–553
39. Elmore S (2007) Apoptosis: a review of programmed cell death. Toxicol Pathol 35:495–516
40. Eming SA, Krieg T, Davidson JM (2007) Inflammation in wound repair: molecular and cellular mechanisms. J Invest Dermatol 127:514–525
41. Feng Y, He D, Yao Z et al (2014) The machinery of macroautophagy. Cell Res 24:24–41
42. Flannagan RS, Jaumouille V, Grinstein S (2012) The cell biology of phagocytosis. Annu Rev Pathol 7:61–98
43. Franchi L, Kamada N, Nakamura Y et al (2012) NLRC4-driven production of IL-1beta discriminates between pathogenic and commensal bacteria and promotes host intestinal defense. Nat Immunol 13:449–456
44. Friggeri A, Yang Y, Banerjee S et al (2010) HMGB1 inhibits macrophage activity in efferocytosis through binding to the alphavbeta3-integrin. Am J Physiol Cell Physiol 299: C1267–C1276
45. Galluzzi L, Bravo-San Pedro JM, Vitale I et al (2015) Essential versus accessory aspects of cell death: recommendations of the NCCD 2015. Cell Death Differ 22:58–73
46. Galluzzi L, Maiuri MC, Vitale I et al (2007) Cell death modalities: classification and pathophysiological implications. Cell Death Differ 14:1237–1243
47. Galluzzi L, Vitale I, Abrams JM et al (2012) Molecular definitions of cell death subroutines: recommendations of the Nomenclature Committee on Cell Death 2012. Cell Death Differ 19:107–120
48. Gardella S, Andrei C, Ferrera D et al (2002) The nuclear protein HMGB1 is secreted by monocytes via a non-classical, vesicle-mediated secretory pathway. EMBO Rep 3:995–1001
49. Garg AD, Krysko DV, Vandenabeele P et al (2012) Hypericin-based photodynamic therapy induces surface exposure of damage-associated molecular patterns like HSP70 and calreticulin. Cancer Immunol Immunother 61:215–221
50. Ghiringhelli F, Apetoh L, Tesniere A et al (2009) Activation of the NLRP3 inflammasome in dendritic cells induces IL-1beta-dependent adaptive immunity against tumors. Nat Med 15:1170–1178
51. Goodwin GH, Sanders C, Johns EW (1973) A new group of chromatin-associated proteins with a high content of acidic and basic amino acids. Eur J Biochem 38:14–19
52. Green DR, Ferguson T, Zitvogel L et al (2009) Immunogenic and tolerogenic cell death. Nat Rev Immunol 9:353–363
53. Gullstrand B, Lefort MH, Tyden H et al (2012) Combination of autoantibodies against different histone proteins influences complement-dependent phagocytosis of necrotic cell material

by polymorphonuclear leukocytes in systemic lupus erythematosus. J Rheumatol 39:1619–1627
54. Hahn S, Giaglis S, Chowdhury CS et al (2013) Modulation of neutrophil NETosis: interplay between infectious agents and underlying host physiology. Semin Immunopathol 35: 439–453
55. Harris J, Hartman M, Roche C et al (2011) Autophagy controls IL-1beta secretion by targeting pro-IL-1beta for degradation. J Biol Chem 286:9587–9597
56. He S, Wang L, Miao L et al (2009) Receptor interacting protein kinase-3 determines cellular necrotic response to TNF-alpha. Cell 137:1100–1111
57. Hegde R, Srinivasula SM, Zhang Z et al (2002) Identification of Omi/HtrA2 as a mitochondrial apoptotic serine protease that disrupts inhibitor of apoptosis protein-caspase interaction. J Biol Chem 277:432–438
58. Hochreiter-Hufford A, Ravichandran KS (2013) Clearing the dead: apoptotic cell sensing, recognition, engulfment, and digestion. Cold Spring Harb Perspect Biol 5:a008748. doi:10.1101/cshperspect.a008748
59. Hohlfeld R, Kerschensteiner M, Meinl E (2007) Dual role of inflammation in CNS disease. Neurology 68:58–63; discussion S91–S96
60. Hoque R, Sohail M, Malik A et al (2011) TLR9 and the NLRP3 inflammasome link acinar cell death with inflammation in acute pancreatitis. Gastroenterology 141:358–369
61. Hornung V, Ablasser A, Charrel-Dennis M et al (2009) AIM2 recognizes cytosolic dsDNA and forms a caspase-1-activating inflammasome with ASC. Nature 458:514–518
62. Hosokawa N, Hara T, Kaizuka T et al (2009) Nutrient-dependent mTORC1 association with the ULK1-Atg13-FIP200 complex required for autophagy. Mol Biol Cell 20:1981–1991
63. Hou W, Zhang Q, Yan Z et al (2013) Strange attractors: DAMPs and autophagy link tumor cell death and immunity. Cell Death Dis 4:e966. doi:10.1038/cddis.2013.493
64. Huang H, Chen HW, Evankovich J et al (2013) Histones activate the NLRP3 inflammasome in Kupffer cells during sterile inflammatory liver injury. J Immunol 191:2665–2679
65. Huang H, Evankovich J, Yan W et al (2011) Endogenous histones function as alarmins in sterile inflammatory liver injury through Toll-like receptor 9 in mice. Hepatology 54:999–1008
66. Iyer SS, Accardi CJ, Ziegler TR et al (2009) Cysteine redox potential determines pro-inflammatory IL-1beta levels. PLoS One 4:e5017. doi:10.1371/journal.pone.0005017
67. Janeway CA Jr (1989) Approaching the asymptote? Evolution and revolution in immunology. Cold Spring Harb Symp Quant Biol 54(Pt 1):1–13
68. Jourdan T, Godlewski G, Cinar R et al (2013) Activation of the Nlrp3 inflammasome in infiltrating macrophages by endocannabinoids mediates beta cell loss in type 2 diabetes. Nat Med 19:1132–1140
69. Jung CH, Jun CB, Ro SH et al (2009) ULK-Atg13-FIP200 complexes mediate mTOR signaling to the autophagy machinery. Mol Biol Cell 20:1992–2003
70. Kaczmarek A, Vandenabeele P, Krysko DV (2013) Necroptosis: the release of damage-associated molecular patterns and its physiological relevance. Immunity 38:209–223
71. Kaiser WJ, Upton JW, Long AB et al (2011) RIP3 mediates the embryonic lethality of caspase-8-deficient mice. Nature 471:368–372
72. Kambas K, Mitroulis I, Apostolidou E et al (2012) Autophagy mediates the delivery of thrombogenic tissue factor to neutrophil extracellular traps in human sepsis. PLoS One 7:e45427. doi:10.1371/journal.pone.0045427
73. Kamo N, Ke B, Ghaffari AA et al (2013) ASC/caspase-1/IL-1beta signaling triggers inflammatory responses by promoting HMGB1 induction in liver ischemia/reperfusion injury. Hepatology 58:351–362
74. Kang R, Chen R, Zhang Q et al (2014) HMGB1 in health and disease. Mol Aspects Med 40:1–116
75. Kang R, Lotze MT, Zeh HJ et al (2014) Cell death and DAMPs in acute pancreatitis. Mol Med 20:466–477
76. Kang R, Tang D (2012) PKR-dependent inflammatory signals. Sci Signal 5:pe47. doi:10.1126/scisignal.2003511

77. Kang R, Zeh HJ, Lotze MT et al (2011) The Beclin 1 network regulates autophagy and apoptosis. Cell Death Differ 18:571–580
78. Kang R, Zhang Q, Zeh HJ 3rd et al (2013) HMGB1 in cancer: good, bad, or both? Clin Cancer Res 19:4046–4057
79. Kazama H, Ricci JE, Herndon JM et al (2008) Induction of immunological tolerance by apoptotic cells requires caspase-dependent oxidation of high-mobility group box-1 protein. Immunity 29:21–32
80. Kerr JF, Wyllie AH, Currie AR (1972) Apoptosis: a basic biological phenomenon with wide-ranging implications in tissue kinetics. Br J Cancer 26:239–257
81. Konno H, Konno K, Barber GN (2013) Cyclic dinucleotides trigger ULK1 (ATG1) phosphorylation of STING to prevent sustained innate immune signaling. Cell 155:688–698
82. Kono H, Rock KL (2008) How dying cells alert the immune system to danger. Nat Rev Immunol 8:279–289
83. Kovalenko A, Kim JC, Kang TB et al (2009) Caspase-8 deficiency in epidermal keratinocytes triggers an inflammatory skin disease. J Exp Med 206:2161–2177
84. Kroemer G, Galluzzi L, Kepp O, Zitvogel L (2013) Immunogenic cell death in cancer therapy. Annu Rev Immunol 31:51–72
85. Kroemer G, Galluzzi L, Vandenabeele P et al (2009) Classification of cell death: recommendations of the Nomenclature Committee on Cell Death 2009. Cell Death Differ 16:3–11
86. Kroemer G, Levine B (2008) Autophagic cell death: the story of a misnomer. Nat Rev Mol Cell Biol 9:1004–1010
87. Kroemer G, Marino G, Levine B (2010) Autophagy and the integrated stress response. Mol Cell 40:280–293
88. Krysko DV, Garg AD, Kaczmarek A et al (2012) Immunogenic cell death and DAMPs in cancer therapy. Nat Rev Cancer 12:860–875
89. Lau A, Wang S, Jiang J et al (2013) RIPK3-mediated necroptosis promotes donor kidney inflammatory injury and reduces allograft survival. Am J Transplant 13:2805–2818
90. Leblanc PM, Doggett TA, Choi J et al (2014) An immunogenic peptide in the a-box of HMGB1 reverses apoptosis-induced tolerance through RAGE. J Biol Chem 289:7777–7786
91. Lee J, Kim HR, Quinley C et al (2012) Autophagy suppresses interleukin-1beta (IL-1beta) signaling by activation of p62 degradation via lysosomal and proteasomal pathways. J Biol Chem 287:4033–4040
92. Leemans JC, Cassel SL, Sutterwala FS (2011) Sensing damage by the NLRP3 inflammasome. Immunol Rev 243:152–162
93. Levine B (2005) Eating oneself and uninvited guests: autophagy-related pathways in cellular defense. Cell 120:159–162
94. Levine B, Mizushima N, Virgin HW (2011) Autophagy in immunity and inflammation. Nature 469:323–335
95. Li J, McQuade T, Siemer AB et al (2012) The RIP1/RIP3 necrosome forms a functional amyloid signaling complex required for programmed necrosis. Cell 150:339–350
96. Li LY, Luo X, Wang X (2001) Endonuclease G is an apoptotic DNase when released from mitochondria. Nature 412:95–99
97. Linkermann A, Green DR (2014) Necroptosis. N Engl J Med 370:455–465
98. Liu G, Wang J, Park YJ et al (2008) High mobility group protein-1 inhibits phagocytosis of apoptotic neutrophils through binding to phosphatidylserine. J Immunol 181:4240–4246
99. Liu L, Yang M, Kang R et al (2014) HMGB1-DNA complex-induced autophagy limits AIM2 inflammasome activation through RAGE. Biochem Biophys Res Commun 450:851–856
100. Liu X, Kim CN, Yang J et al (1996) Induction of apoptotic program in cell-free extracts: requirement for dATP and cytochrome c. Cell 86:147–157
101. Liu Y, Levine B (2014) Autosis and autophagic cell death: the dark side of autophagy. Cell Death Differ 22:367–376
102. Liu Y, Shoji-Kawata S, Sumpter RM Jr et al (2013) Autosis is a Na+, K+-ATPase-regulated form of cell death triggered by autophagy-inducing peptides, starvation, and hypoxia-ischemia. Proc Natl Acad Sci U S A 110:20364–20371

103. Lu B, Nakamura T, Inouye K et al (2012) Novel role of PKR in inflammasome activation and HMGB1 release. Nature 488:670–674
104. Maiuri MC, Zalckvar E, Kimchi A et al (2007) Self-eating and self-killing: crosstalk between autophagy and apoptosis. Nat Rev Mol Cell Biol 8:741–752
105. Martinez J, Almendinger J, Oberst A et al (2011) Microtubule-associated protein 1 light chain 3 alpha (LC3)-associated phagocytosis is required for the efficient clearance of dead cells. Proc Natl Acad Sci U S A 108:17396–17401
106. Martinon F, Mayor A, Tschopp J (2009) The inflammasomes: guardians of the body. Annu Rev Immunol 27:229–265
107. Matzinger P (1994) Tolerance, danger, and the extended family. Annu Rev Immunol 12:991–1045
108. McPhee CK, Logan MA, Freeman MR et al (2010) Activation of autophagy during cell death requires the engulfment receptor Draper. Nature 465:1093–1096
109. Medzhitov R (2008) Origin and physiological roles of inflammation. Nature 454:428–435
110. Mehlen P, Bredesen DE (2011) Dependence receptors: from basic research to drug development. Sci Signal 4:mr2. doi:10.1126/scisignal.2001521
111. Metzler KD, Goosmann C, Lubojemska A et al (2014) A myeloperoxidase-containing complex regulates neutrophil elastase release and actin dynamics during NETosis. Cell Rep 8:883–896
112. Mitroulis I, Kambas K, Chrysanthopoulou A et al (2011) Neutrophil extracellular trap formation is associated with IL-1beta and autophagy-related signaling in gout. PLoS One 6:e29318. doi:10.1371/journal.pone.0029318
113. Mizushima N, Yoshimori T, Ohsumi Y (2011) The role of Atg proteins in autophagosome formation. Annu Rev Cell Dev Biol 27:107–132
114. Munz C (2009) Enhancing immunity through autophagy. Annu Rev Immunol 27:423–449
115. Murphy JM, Czabotar PE, Hildebrand JM et al (2013) The pseudokinase MLKL mediates necroptosis via a molecular switch mechanism. Immunity 39:443–453
116. Murrow L, Debnath J (2013) Autophagy as a stress-response and quality-control mechanism: implications for cell injury and human disease. Annu Rev Pathol 8:105–137
117. Nakahira K, Haspel JA, Rathinam VA et al (2010) Autophagy proteins regulate innate immune responses by inhibiting the release of mitochondrial DNA mediated by the NALP3 inflammasome. Nat Immunol 12:222–230
118. Newton K, Dugger DL, Wickliffe KE et al (2014) Activity of protein kinase RIPK3 determines whether cells die by necroptosis or apoptosis. Science 343:1357–1360
119. Obeid M, Tesniere A, Ghiringhelli F et al (2007) Calreticulin exposure dictates the immunogenicity of cancer cell death. Nat Med 13:54–61
120. Oka T, Hikoso S, Yamaguchi O et al (2012) Mitochondrial DNA that escapes from autophagy causes inflammation and heart failure. Nature 485:251–255
121. Park MA, Curiel DT, Koumenis C et al (2008) PERK-dependent regulation of HSP70 expression and the regulation of autophagy. Autophagy 4:364–367
122. Paul S, Kashyap AK, Jia W et al (2012) Selective autophagy of the adaptor protein Bcl10 modulates T cell receptor activation of NF-kappaB. Immunity 36:947–958
123. Qu X, Zou Z, Sun Q et al (2007) Autophagy gene-dependent clearance of apoptotic cells during embryonic development. Cell 128:931–946
124. Rathinam VA, Vanaja SK, Fitzgerald KA (2012) Regulation of inflammasome signaling. Nat Immunol 13:333–342
125. Ravichandran KS (2010) Find-me and eat-me signals in apoptotic cell clearance: progress and conundrums. J Exp Med 207:1807–1817
126. Ravichandran KS (2011) Beginnings of a good apoptotic meal: the find-me and eat-me signaling pathways. Immunity 35:445–455
127. Remijsen Q, Kuijpers TW, Wirawan E et al (2011) Dying for a cause: NETosis, mechanisms behind an antimicrobial cell death modality. Cell Death Differ 18:581–588
128. Rickard JA, O'Donnell JA, Evans JM et al (2014) RIPK1 regulates RIPK3-MLKL-driven systemic inflammation and emergency hematopoiesis. Cell 157:1175–1188

129. Riedl SJ, Shi Y (2004) Molecular mechanisms of caspase regulation during apoptosis. Nat Rev Mol Cell Biol 5:897–907
130. Rock KL, Kono H (2008) The inflammatory response to cell death. Annu Rev Pathol 3:99–126
131. Ryan GB, Majno G (1977) Acute inflammation. A review. Am J Pathol 86:183–276
132. Saitoh T, Fujita N, Hayashi T et al (2009) Atg9a controls dsDNA-driven dynamic translocation of STING and the innate immune response. Proc Natl Acad Sci U S A 106:20842–20846
133. Saitoh T, Fujita N, Jang MH et al (2008) Loss of the autophagy protein Atg16L1 enhances endotoxin-induced IL-1beta production. Nature 456:264–268
134. Saitoh T, Satoh T, Yamamoto N et al (2011) Antiviral protein Viperin promotes Toll-like receptor 7- and Toll-like receptor 9-mediated type I interferon production in plasmacytoid dendritic cells. Immunity 34:352–363
135. Scaffidi P, Misteli T, Bianchi ME (2002) Release of chromatin protein HMGB1 by necrotic cells triggers inflammation. Nature 418:191–195
136. Schroder K, Tschopp J (2010) The inflammasomes. Cell 140:821–832
137. Schweichel JU, Merker HJ (1973) The morphology of various types of cell death in prenatal tissues. Teratology 7:253–266
138. Segal AW (2005) How neutrophils kill microbes. Annu Rev Immunol 23:197–223
139. Seong SY, Matzinger P (2004) Hydrophobicity: an ancient damage-associated molecular pattern that initiates innate immune responses. Nat Rev Immunol 4:469–478
140. Serhan CN, Savill J (2005) Resolution of inflammation: the beginning programs the end. Nat Immunol 6:1191–1197
141. Shimada K, Crother TR, Karlin J et al (2012) Oxidized mitochondrial DNA activates the NLRP3 inflammasome during apoptosis. Immunity 36:401–414
142. Spisek R, Charalambous A, Mazumder A et al (2007) Bortezomib enhances dendritic cell (DC)-mediated induction of immunity to human myeloma via exposure of cell surface heat shock protein 90 on dying tumor cells: therapeutic implications. Blood 109:4839–4845
143. Steinberg BE, Grinstein S (2007) Unconventional roles of the NADPH oxidase: signaling, ion homeostasis, and cell death. Sci STKE 2007:pe11
144. Sun L, Wang H, Wang Z et al (2012) Mixed lineage kinase domain-like protein mediates necrosis signaling downstream of RIP3 kinase. Cell 148:213–227
145. Susin SA, Lorenzo HK, Zamzami N et al (1999) Molecular characterization of mitochondrial apoptosis-inducing factor. Nature 397:441–446
146. Tabas I, Glass CK (2013) Anti-inflammatory therapy in chronic disease: challenges and opportunities. Science 339:166–172
147. Tait SW, Green DR (2010) Mitochondria and cell death: outer membrane permeabilization and beyond. Nat Rev Mol Cell Biol 11:621–632
148. Takahashi N, Vereecke L, Bertrand MJ et al (2014) RIPK1 ensures intestinal homeostasis by protecting the epithelium against apoptosis. Nature 513:95–99
149. Tang D, Billiar TR, Lotze MT (2012) A Janus tale of two active high mobility group box 1 (HMGB1) redox states. Mol Med 18:1360–1362
150. Tang D, Kang R, Cheh CW et al (2010) HMGB1 release and redox regulates autophagy and apoptosis in cancer cells. Oncogene 29:5299–5310
151. Tang D, Kang R, Coyne CB et al (2012) PAMPs and DAMPs: signal 0 s that spur autophagy and immunity. Immunol Rev 249:158–175
152. Tang D, Kang R, Livesey KM et al (2010) Endogenous HMGB1 regulates autophagy. J Cell Biol 190:881–892
153. Tang D, Kang R, Livesey KM et al (2011) High-mobility group box 1 is essential for mitochondrial quality control. Cell Metab 13:701–711
154. Tang D, Kang R, Zeh HJ 3rd et al (2010) High-mobility group box 1 and cancer. Biochim Biophys Acta 1799:131–140
155. Tang D, Kang R, Zeh HJ et al (2011) High-mobility group box 1, oxidative stress, and disease. Antioxid Redox Signal 14:1315–1335

156. Tang D, Lotze MT (2012) Tumor immunity times out: TIM-3 and HMGB1. Nat Immunol 13:808–810
157. Teplova I, Lozy F, Price S et al (2013) ATG proteins mediate efferocytosis and suppress inflammation in mammary involution. Autophagy 9:459–475
158. Thorburn J, Horita H, Redzic J et al (2009) Autophagy regulates selective HMGB1 release in tumor cells that are destined to die. Cell Death Differ 16:175–183
159. Trapani JA, Smyth MJ (2002) Functional significance of the perforin/granzyme cell death pathway. Nat Rev Immunol 2:735–747
160. Tsung A, Sahai R, Tanaka H et al (2005) The nuclear factor HMGB1 mediates hepatic injury after murine liver ischemia-reperfusion. J Exp Med 201:1135–1143
161. Urbonaviciute V, Furnrohr BG, Weber C et al (2007) Factors masking HMGB1 in human serum and plasma. J Leukoc Biol 81:67–74
162. van den Brink MR, Burakoff SJ (2002) Cytolytic pathways in haematopoietic stem-cell transplantation. Nat Rev Immunol 2:273–281
163. van Loo G, van Gurp M, Depuydt B et al (2002) The serine protease Omi/HtrA2 is released from mitochondria during apoptosis. Omi interacts with caspase-inhibitor XIAP and induces enhanced caspase activity. Cell Death Differ 9:20–26
164. Vandanmagsar B, Youm YH, Ravussin A et al (2011) The NLRP3 inflammasome instigates obesity-induced inflammation and insulin resistance. Nat Med 17:179–188
165. Vanden Berghe T, Linkermann A, Jouan-Lanhouet S et al (2014) Regulated necrosis: the expanding network of non-apoptotic cell death pathways. Nat Rev Mol Cell Biol 15:135–147
166. Vandenabeele P, Galluzzi L, Vanden Berghe T et al (2010) Molecular mechanisms of necroptosis: an ordered cellular explosion. Nat Rev Mol Cell Biol 11:700–714
167. Venereau E, Casalgrandi M, Schiraldi M et al (2012) Mutually exclusive redox forms of HMGB1 promote cell recruitment or proinflammatory cytokine release. J Exp Med 209:1519–1528
168. Vernon PJ, Tang D (2013) Eat-me: autophagy, phagocytosis, and reactive oxygen species signaling. Antioxid Redox Signal 18:677–691
169. Wang C, Youle RJ (2009) The role of mitochondria in apoptosis*. Annu Rev Genet 43:95–118
170. Wang H, Bloom O, Zhang M et al (1999) HMG-1 as a late mediator of endotoxin lethality in mice. Science 285:248–251
171. Wang Y, Martins I, Ma Y et al (2013) Autophagy-dependent ATP release from dying cells via lysosomal exocytosis. Autophagy 9:1624–1625
172. Welz PS, Wullaert A, Vlantis K et al (2011) FADD prevents RIP3-mediated epithelial cell necrosis and chronic intestinal inflammation. Nature 477:330–334
173. Wen H, Ting JP, O'Neill LA (2012) A role for the NLRP3 inflammasome in metabolic diseases – did Warburg miss inflammation? Nat Immunol 13:352–357
174. Xu J, Zhang X, Monestier M et al (2011) Extracellular histones are mediators of death through TLR2 and TLR4 in mouse fatal liver injury. J Immunol 187:2626–2631
175. Xu J, Zhang X, Pelayo R et al (2009) Extracellular histones are major mediators of death in sepsis. Nat Med 15:1318–1321
176. Yang Z, Klionsky DJ (2010) Eaten alive: a history of macroautophagy. Nat Cell Biol 12:814–822
177. Yang Z, Klionsky DJ (2010) Mammalian autophagy: core molecular machinery and signaling regulation. Curr Opin Cell Biol 22:124–131
178. Yipp BG, Petri B, Salina D et al (2012) Infection-induced NETosis is a dynamic process involving neutrophil multitasking in vivo. Nat Med 18:1386–1393
179. Youle RJ, Narendra DP (2011) Mechanisms of mitophagy. Nat Rev Mol Cell Biol 12:9–14
180. Youle RJ, Strasser A (2008) The BCL-2 protein family: opposing activities that mediate cell death. Nat Rev Mol Cell Biol 9:47–59
181. Zaki MH, Lamkanfi M, Kanneganti TD (2011) The Nlrp3 inflammasome: contributions to intestinal homeostasis. Trends Immunol 32:171–179

182. Zhang Q, Kang R, Zeh HJ 3rd et al (2013) DAMPs and autophagy: cellular adaptation to injury and unscheduled cell death. Autophagy 9:451–458
183. Zhang Q, Raoof M, Chen Y et al (2010) Circulating mitochondrial DAMPs cause inflammatory responses to injury. Nature 464:104–107
184. Zhang Y, Li W, Zhu S et al (2012) Tanshinone IIA sodium sulfonate facilitates endocytic HMGB1 uptake. Biochem Pharmacol 84:1492–1500
185. Zhou R, Yazdi AS, Menu P et al (2010) A role for mitochondria in NLRP3 inflammasome activation. Nature 469:221–225
186. Zhou W, Yuan J (2014) Necroptosis in health and diseases. Semin Cell Dev Biol 35C:14–23
187. Zitvogel L, Kepp O, Kroemer G (2010) Decoding cell death signals in inflammation and immunity. Cell 140:798–804
188. Zong WX, Thompson CB (2006) Necrotic death as a cell fate. Genes Dev 20:1–15
189. Zou J, Kawai T, Tsuchida T et al (2013) Poly IC triggers a cathepsin D- and IPS-1-dependent pathway to enhance cytokine production and mediate dendritic cell necroptosis. Immunity 38:717–728

Autophagy in Chronic Inflammation

Johan Arnold, Diane Murera, and Frédéric Gros

Abstract Autophagy is a catabolic process consisting in the formation of cytoplasmic vacuoles, fusing with lysosomes and leading to the degradation of their content. Part of the autophagy machinery is also involved in specialized forms of endocytosis and vesicle trafficking. The role of autophagy, initially described as a response to energetic stress, has now been extended to other stress signals like tissue damage and infection. Autophagy is indeed deeply involved in the regulation of inflammation and in the biology of immune cells. Autophagy regulates cell metabolism and integrates it to the elimination of microorganisms, to the fine-tuning of inflammation and to the activation of the adaptive immune system. The inflammatory response aims at controlling pathogen invasion and at initiating tissue repair. If unrestricted, inflammation can become chronic and be the source of the so-called autoinflammatory and autoimmune pathologies. These complex disorders result from a combination between genetic and environmental factors. A clear genetic link between Crohn's disease and autophagy deregulation has been demonstrated. Autophagy deregulation provoked by environmental triggers like nutrient excess or by aging, are also linked to low-grade inflammation observed during metabolic syndrome, especially in the case of type II diabetes and atherosclerosis. Both genetic causes and environmental triggers could also link autophagy deregulation to autoimmune pathologies like rheumatoid arthritis, systemic lupus erythematosus and multiple sclerosis. The emerging causality between autophagy deregulation and chronic inflammation, subject of intense studies as it could lead to new therapeutic options, will be described in this chapter.

J. Arnold • D. Murera
CNRS UPR3572- Immunopathologie et Chimie Thérapeutique, Institut de Biologie Moléculaire et Cellulaire, Université de Strasbourg, Strasbourg, France

F. Gros (✉)
CNRS UPR3572- Immunopathologie et Chimie Thérapeutique, Institut de Biologie Moléculaire et Cellulaire, Université de Strasbourg, Strasbourg, France

Frédéric GROS, Institut de Biologie Moléculaire et Cellulaire, UPR3572 CNRS, Immunopathologie et Chimie Thérapeutique,
15 rue René Descartes, Strasbourg 67000, France
e-mail: f.gros@ibmc-cnrs.unistra.fr

M.C. Maiuri, D. De Stefano (eds.), *Autophagy Networks in Inflammation*, Progress in Inflammation Research, DOI 10.1007/978-3-319-30079-5_6

Abbreviations

AMPK AMP	Activated protein kinase
APC	Antigen presenting cells
ATG	Autophagy-related genes
ATP	Adenosine tri-phosphate
CD	Crohn's disease
CDS	Cytosolic DNA sensor
CMA	Chaperone mediated autophagy
DCs	Dendritic cells
DSS	Dextran sodium sulphate
EAE	Experimental autoimmune encephalomyelitis
ER	Endoplasmic reticulum
FFA	Free fatty acids
FoxO	Forkhead homeobox type protein O
GWAS	Genome-wide analysis studies
HFD	High fat diet
HM	Hypomorphic
HMGB1	High-mobility group 1 protein
IAPP	Islet amyloid peptide
IBD	Inflammatory bowel disease
IFN	Interferon
ILC	Innate lymphoid cells
IRGM	Immunity-related GTPase M
LAP	LC3-associated phagocytosis
LC3	Light chain 3 standing for microtubule-associated protein 1 light chain 3B
LDL	Low density lipoproteins
LPS	Lipopolysaccharide
MAMP	Microbe-associated molecular pattern
MHC	Major histocompatibility complex
MiR	Micro RNA
MS	Multiple sclerosis
NET	Neutrophil extracellular trap
NLRP3	NACHT, LRR and PYD domains-containing protein 3: IL, interleukin
NOD	Nucleotide oligomerization domain
NRV	Norovirus
PBMC	Peripheral blood mononuclear cells
pDCs	Plasmacytoid dendritic cells
PFKFB3	6-phosphofructo-2-kinase/fructose-2,6-biphosphatase 3
PRR	Pattern recognition receptors
RA	Rheumatoid arthritis
RASF	RA synovial fibroblasts
RLR	Retinoic acid induced gene (RIG)-like receptors

ROS	Reactive oxygen species
SLE	Systemic lupus erythematosus
SNP	Single nucleotide polymorphism
STING	CDS-activated proteins stimulator of IFN gene
TBK1	TRAF family member-associated nuclear factor-κB activator-binding kinases abbreviated
TCR	T cell receptor
TEC	Thymic epithelial cell
Th	t helper
TIID	Type II diabetes
TLR	Toll-like receptor
TNF	Tumour necrosis factor
TRAF	TNF receptor-associated factor
Treg	Regulatory T cell
UPR	Unfolded protein response
VSMC	Vascular smooth muscle cells

1 Autophagy in the Regulation of Inflammation

Inflammation is a physiological process, shared by higher animal eukaryotes, triggered by several stress signals. First, entry of microorganisms in otherwise sterile tissues, changes in the composition of the bacterial flora on epithelia, for example in the gut, can initiate inflammation. Secondly, inflammation can also be triggered by cytokines, themselves inducible by infection, or by danger signals, like the ones delivered by intracellular components released in the extracellular milieu (nuclear components like high mobility group 1 (HMGB1) protein or adenosine triphosphate (ATP) release). Oxygen stress induced by intra or extracellular reactive oxygen species (ROS) is also a potent stimulant of inflammation. Regulated inflammation leads to the initiation of an immune response aiming at controlling infection, and to tissue repair. Inflammation and the subsequent immune response, which can further participate in inflammation, must be down regulated at the end of the process. If not, chronic inflammation can lead to disorders linked to aberrant tissue remodelling, excessive cell death and tissue damage, sometimes associated to an autoimmune reaction. Inflammation is thus at the crossroads between metabolic-stress, control or elimination of pathogens by the immune response, and tissue homeostasis.

Autophagy is a process linked to self-digestion by cells of their own components via lysosomal degradation. Several forms of autophagy coexist in animal cells. Chaperone-mediated autophagy (CMA), and microautophagy both allow direct translocation of cytosolic material inside lysosomes. The relations of CMA and microautophagy with inflammation are plausible although not yet proven and will thus not be discussed in this chapter.

Macroautophagy, the best-characterized form of autophagy, will be hereafter called autophagy. It involves the formation of double membrane vesicles fusing with lysosomes leading to degradation of their content. The proteins encoded by autophagy-related genes (*ATG*), play a major role in canonical autophagy but also in other processes like endocytosis and vesicle trafficking [17]. Autophagy was initially described as a catabolic mechanism involved in metabolic stress response, like deprivation of amino acids. It is becoming increasingly clear that in higher eukaryotes autophagy is also involved in other cellular stress responses, like ROS reaction [26], hypoxia [22], unfolded protein response (UPR) and endoplasmic reticulum (ER)-stress [20], genotoxic stress [90] and pathogen recognition [85]. All these responses are main actors in the inflammation process and sometimes impact the subsequent immune response. It is thus not surprising that autophagy is a fundamental player in the inflammation process.

A very ancestral function of autophagy, beyond its role as sensor of energy stress, is probably the elimination of pathogens, particularly intracellular ones. This very particular aspect of autophagy is called xenophagy, when autophagic machinery directly engulfs pathogens or facilitates their translocation and degradation into lysosomes [23]. Autophagy is triggered in that case by metabolic stress induced by pathogen invasion, or by direct recognition of microbe associated molecular patterns (MAMP) via pattern recognition receptors (PRR). In that case autophagy is induced upon inflammation, and contributes to its resolution, by eliminating pathogens.

Interestingly, autophagy is tightly linked to mitochondrial homeostasis. Mitochondria are issued from ancestral proteobacterias, having adopted intracellular life in compartmented cells. It is a very seducing concept that evolution, initially dedicating autophagy to controll invasive pathogens, drove this degradative pathway toward the regulation of symbiotic organelle homeostasis. With respect to the regulation of inflammation, the maintenance of well functioning mitochondria is of great interest. First, balanced autophagic activity limits ROS produced by damaged mitochondria [98], dampening pro inflammatory stimulus. Secondly, mitochondria removal leads to the degradation of PRR associated to their membrane, like retinoic acid induced gene-like receptors (RLR). As a consequence, inhibition of autophagy can lead to hyper-responsiveness to cytosolic double stranded RNA in terms of type I interferon (IFN) secretion [34].

Autophagy also regulates the recognition of cytosolic DNA. In certain circumstances, mitochondrial DNA can be released in the cytosol, especially of PRR-stimulated macrophages [59]. In the previsouly cited publication, an unidentified cytosolic DNA sensor (CDS) leads to the activation of the NACHT, LRR and PYD domains-containing protein 3 (NALP3) inflammasome and to the subsequent production of the pro-inflammatory cytokines interleukin (IL)-1ß and IL-18. Another publication reported a contribution of toll-like receptor (TLR) 9 to that respect [63].

Considering the growing literature on this subject, autophagy is more widely linked to the regulation of cytosolic DNA recognition. Interestingly, not only mitochondria associated anti-viral PRR can be regulated by autophagy machinery. ATG9a protein activity and microtubule associated protein light chain 3, often abbreviated as LC3, both associate with the CDS-activated proteins stimulator of IFN gene (STING) and prevents its trafficking from ER to compartments containing tumour necrosis factor (TNF) receptor-associated factor (TRAF) family member-associated nuclear factor-κB activator-binding kinases, abbreviated as TBK1 [80]. This non-canonical autophagic function of ATG9a contributes to limit inflammatory responses toward cytosolic DNA.

Type I IFNs are secreted early after viral infection by both immune and non-immune cells. Autophagy, as introduced above, can also regulate pro-inflammatory cytokine secretion, especially in phagocytes. One of the first reports showing a link between inflammasome and the autophagy machinery was published by Saitoh and colleagues [81]. They showed that *Atg16L1*-deficient macrophages secreted higher amounts of IL-1ß after TLR stimulation than controls. They showed that this deregulated secretion was linked to a higher activity of the inflammasome. Kehrl's group showed more recently that the assembled inflammasome was subjected to ubiquitination and targeted to autophagy-related lysosomal degradation via sequestosome1/p62 binding [82]. It was also shown that ATG16L1 activity might indirectly control TRAF6 level via p62, then downregulating the intensity of IL1-ß receptor signalling [43]. Autophagy also limits inflammasome activity by preventing ROS production by damaged mitochondria [98]. Interestingly, a recent publication showed that during viral infection by influenza, receptor-interacting serine-threonine kinase 2 and nucleotide oligomerization domain 2 (NOD2) mediated degradation of damaged mitochondria by autophagy, limiting the activation of NLRP3 inflammasome by intracellular ROS [50].

On the onset of an immune response, innate immunity is the main first actor in the induction of immunity. Although the subsequent adaptive immune response is dependent on this first wave of pro-inflammatory signals, adaptive immune cells can also contribute to the maintenance of inflammation by the cytokines they secrete and by the tissue damage they induce. In certain circumstances, in addition to the initial inflammation, an antigen, sometimes encoded by self-genetic information, can be recognized by cognate T or B lymphocytes. These immune reactions mediated by antibodies or by cytotoxic cells, accompanying inflammation, can be responsible for tissue degradation. Autophagy can contribute to the abnormal activation of the adaptive immune system and to the maintenance of inflammation. First autophagy contributes to endogenous and exogenous/antigen presentation to T cells [79], and thus probably also to autoantigen presentation. Autophagy machinery as discussed later, may play a role for dead cell clearance limiting the access to autoantigen [54]. Moreover, autophagy is a key player in the regulation of lymphocyte survival and activation [71] and by this mean is suspected to regulate inflammation related to the activity of adaptive immunity.

Considering the increasing evidence that autophagy is a key player in immunity, it does not seem surprising now that the link between human immune diseases and autophagy deregulation is emerging.

2 Autophagy and Crohn's Disease

The first era of identification of candidate genes for autoinflammatory and autoimmune diseases, before genome-wide analysis studies (GWAS), strongly relied on linkage disequilibrium studies. One of them identified three variants of *NOD2* genes, strongly associated with Crohn's disease development (CD, [30]): one frameshift mutation, also identified by another independent study [62], and two nonsense variations, altering the leucine-rich repeat domain. In several independent studies, *NOD2* polymorphisms, are strongly linked to the development of inflammatory bowel disease [97]. The most common mutated variants are characteristic of patients suffering from CD rather than ulcerative colitis. IBD in general and CD in particular, are related to hyper inflammation caused by environmental factors like an abnormal commensal flora or abnormal reaction against the flora. Genetic predispositions participate in the inflammation process. Loss-of-function mutations of NOD2 lead to a decrease of IL-6, IL-8 and TNF-α response against cytosolic microbial products probably implying an impaired clearance of pathogens [61, 89]. This diminished inflammatory acute response is suspected to lead paradoxically to a chronic immune response at term. This could be due to pathogen overload linked to changes in the composition of the flora, but also to skewing of the adaptive immune response towards excessive T helper (Th)1/Th17 patterns. Indeed, loss of function of NOD2 mutants in peripheral blood mononuclear cells (PBMC) from CD patients showed impaired IL-10 production leading to the hypothesis that NOD2 participates to the global down regulation of Th cell activity [61]. In the same line, reduced regulatory T (Treg) cell numbers and survival have been found in CD patients with NOD2 loss of function mutation [74]. NOD2 is also highly expressed by innate lymphoid cell (ILC) population from the intestine. These ILC may contribute to homeostasis via their cytolytic activity against infected cells but also by the production of regulatory cytokines like IL-22, known to decrease inflammatory symptoms in a mouse model of colitis. Finally NOD2 mutations are linked to a decrease in antibacterial peptides secretion by Paneth cells in response to bacterial invasion [92]. This decrease in defensin production is eager to contribute to the increased bacterial load in the intestine, favouring hyper inflammation.

GWAS greatly contributed to the identification of new candidate genes for complex chronic inflammatory diseases, often of polygenic origin. One of the first genetic evidence pointing out a role for autophagy machinery in chronic inflammation was found in the context of CD. These studies confirmed *NOD2* as a candidate gene for CD development as single nucleotide polymorphisms (SNPs) were found to be strongly associated with the disease. Strikingly, these studies identified SNPs in the *ATG16L1* gene region, strongly correlated with CD development [6, 25, 77]. The more frequent coding variant of ATG16L1 leads to a substitution of a threonine to an alanine (T300A) and is located at the vicinity of the WD-repeat domain. Another candidate gene was also described: the immune related GTPase M [56, 67]. The latter protein is described as an inducer of xenophagy, particularly important in IFN-γ induced response towards mycobacteria and viruses [69]. Interestingly, one of the exonic variants for IRGM, although conservative in terms of amino acid sequence, leads to an increased sensitivity to down-regulation by micro RNA (MiR) 196 that is overexpressed in CD [8].

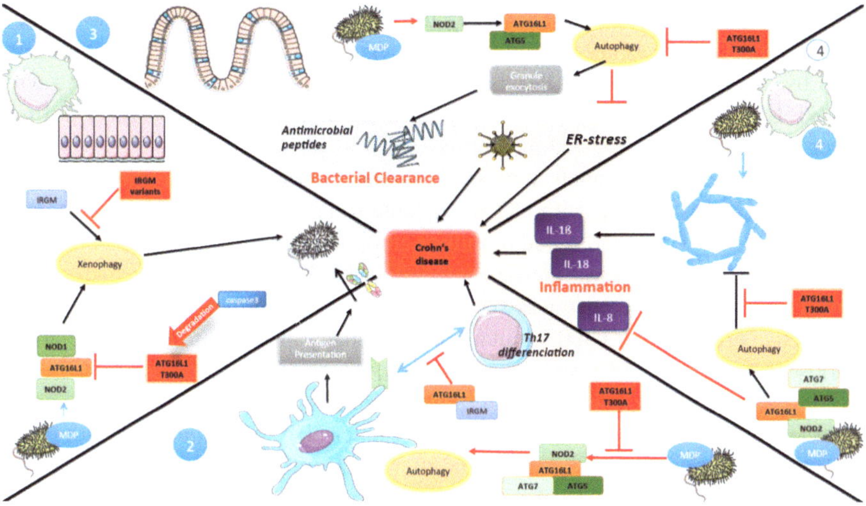

Fig. 1 Possible involvement of autophagy in CD development. *1* During bacterial infection of phagocytes or epithelial cells, ATG16L1 is recruited to the plasma membrane and can associate with NOD1 or NOD2. This induces the elimination of the bacteria through xenophagy. NOD-dependent xenophagy is compromised when ATG16L1^{T300A} protein variant is expressed. ATG16L1^{T300A} variant is more sensitive to degradation by caspase 3 and interacts less with NOD1/NOD2. The *IRGM*, identified as a candidate gene in CD development, encodes a protein known to also induce xenophagy in response to mycobacterial infection. IRGM variants may also be involved in xenophagy impairment and thus in decreased bacterial clearance. *2* MDP induces autophagy in a NOD2-dependent manner, in APCs. Autophagy proteins ATG16L1, ATG5 and ATG7 are implicated in antigen presentation via MHC-II molecules. NOD2 and ATG16L1^{T300A} variants are associated to decreased antigen presentation of bacteria associated antigens. This could contribute to abnormal regulation of adaptive immunity and control of the commensal flora. ATG16L1 and IRGM deficiencies are correlated with increased duration of synaptic contact between T cells and dendritic cells, skewing T cell polarization towards Th17 pro-inflammatory phenotype. *3* NOD2 loss of function mutations lead to a decreased secretion of antibacterial peptides by Paneth cells located in the gut epithelium, inhibiting bacterial clearance thus contributing to a hyper inflammatory environment. Autophagy proteins ATG16L1 and ATG5 are involved in granule exocytosis by Paneth cells, in response to NOD2 stimulation, contributing to bacterial clearance in the gut. ATG16L1^{T300A} polymorphism is linked to reduced antimicrobial peptide secretion, particularly under ER-stress and/or viral infection. *4* ATG16L1 deficiency in macrophages is associated with an increased pro-inflammatory cytokine (IL-1ß and IL18) secretion dependent on inflammasome activation after PRR stimulation. Autophagy proteins ATG5, ATG7 and ATG16L1 are known to downregulate inflammasome activity. ATG16L1 deficient and ATG16L1^{T300A} mice are more prone to activate inflammasome, suggesting the importance of autophagy in inflammation response regulation in CD. An autophagy-independent role for ATG16L1 is the control of IL-8 secretion in response to NOD receptor stimulation. Abbreviations: *ATG* autophagy-related genes, *NOD* nucleotide oligomerization domain, *IRGM* immune related GTPase, *CD* Crohn's disease, *MDP* muramyldipeptide, *MHC-II* major histocompatibility complex class II, *IL* interleukin, *PRR* pattern recognition receptors

From this discovery, autophagy was a matter of great interest in the field of CD research. Interestingly, several studies showed common potential relations between ATG16L1 polymorphisms and NOD2 loss-of-function. First, autophagy is directly linked to the elimination of pathogens as it is involved in xenophagy (Fig. 1, frame 1). The first study showing a link between NOD receptors and autophagy was published by Travassos and colleagues [88]. They first found that muramyldipeptide (MDP), ligand for NOD2, was able to induce autophagic activity. Moreover, during infection by invasive bacteria, NOD1 is recruited at the plasma membrane along with ATG16L1. This core-ATG protein favoured the elimination of *Shigella flexneri* after its capture at the plasma membrane, and was physically associated with both NOD1 and NOD2. Of much interest, the authors showed that the most common NOD2 mutant associated with CD was unable to lead to ATG16L1 recruitment at the plasma membrane and failed to induce autophagy. In addition, the T300A variant of ATG16L1 protein was inefficient to induce LC3 punctate structure in response to MDP while no difference was seen under rapamycin stimulation. This suggests a selective effect of the T300A mutant on xenophagy in regard to other specialized forms of autophagy. This very elegant study, proposing a rationale for a role of autophagy, in conjunction with NOD2 in the control of gut flora, was followed by another study published by Cooney and colleagues [19]. In conjunction with insufficient elimination of invasive bacteria after NOD stimulation, they propose that ATG16L1 mutations in CD impair the regulatory function of antigen presenting cells (APC; Fig. 1, frame 2). Their study on dendritic cells (DCs) cells showed that MDP stimulation induced autophagy, through a NOD2-dependent mechanism. This increase of LC3-decorated membrane load, needed the canonical core-machinery proteins ATG5 and ATG7, and promoted presentation by major histocompatibility complex (MHC) class II molecules of *Salmonella enterica*-associated antigens to T cells. They unambiguously show that the most common NOD2 and the ATG16L1^{T300A} variants, associated to CD, also led to impaired presentation in the same assay. Thus in addition to eliciting the efficacy of xenophagy, the authors propose that the ATG16L1-related autophagy induced by NOD2 also prime adaptive immune response to control gut pathogens. With respect to antigen presentation, autophagy could also regulate the activation of cognate T cells by APCs, via regulation of the stability of the immune synapse. Wildenberg and collaborators showed that DCs derived from patients carrying the *ATG16L1*^{T300A} allele induced more stable synapses with T cells during antigen presentation, favouring Th17 cell differentiation [94]. They confirm the possible involvement of ATG16L1 in this phenomenon and suggest IRGM as also implicated, by knock down experiments. Autophagy deregulation could thus impair the control of gut flora both by innate and adaptive immunity.

Another interesting point linking NOD2 with ATG16L1 in CD pathogenesis, is their described role in intestinal epithelium homeostasis and especially in Paneth cell function (Fig. 1, frame 3). Pr Virgin's team generated mice bearing a hypomorphic (*HM*) allele of *Atg16L1* by gene trap strategy [9]. As ATG16L1 complete deficiency is lethal in mice, this new model allowed studying *in vivo* the consequences of a diminished expression of this essential autophagy gene. The authors compared *Atg16L1*^{HM} mice with transgenic mice harbouring a conditional deletion of *Atg5* in cells expressing CRE recombinase, under the control of the Villin promoter, restricting autophagy deficiency to the intestinal epithelium. This work showed that both ATG5 and ATG16L1, and thus part of autophagic machinery, were integral to granule exocytosis by Paneth cells. Moreover, the adipocytokines leptin and adiponectin are increased at the transcriptional level in Paneth cells from *Atg16L1*^{HM} mice. Interestingly, the production of these cytokines is increased in patients suffering from CD. The same study validated the abnormal granule production or cytoplasmic localization in biological samples from patients carrying the T300A risk allele. The impact of this most common variant found in CD on Paneth cell function was confirmed by Xavier's team, using a mouse model knock-in for *Atg16L1* locus, with the T300A allele [41]. Cadwell and colleagues published 2 years after their initial description of *Atg16L1*^{HM} mice, that when they re-derived breeders from embryos in an enhanced barrier animal facility, they did not find any Paneth cell defect in the progeny, contrary to what they had previously described [10]. Most interestingly, virus infection by CR6 strains of norovirus (NRV) leading to a persistent intestinal infection, and present in the former conventional animal facility, recapitulated previously observes granule abnormalities in Paneth cells. This virus plus susceptibility gene interaction necessary for Paneth cell abnormalities, did however not spontaneously induces colitis, pointing out a role for another environmental trigger, additional to the genetic background and viral infection. The authors thus provoked colitis by chemical-induced injury after dextran sodium sulphate (DSS) treatment. They described that *Atg16L1*^{HM} mice, infected by NRV CR6 strain exhibited aberrant response to DSS in the colon leading to atrophy of the mucosa. Inhibiting TNF-α and IFN-γ, which is quite relevant considering CD characteristic inflammation and its common treatment, reduced tissue damage. Other stimulus like ER-stress, seems to be able to trigger intestinal inflammation. Using mice deficient for *Xbp1* specifically in the intestinal epithelium, Blumberg's team showed that ER-stress induced autophagy in this tissue [1]. Invalidation of *Atg16L1* simultaneously with *Xpb1* deletion in intestinal epithelium led to enteritis originating from Paneth cells. Two reports investigated in details autophagic activity in Paneth cells from CD patients [15, 86]. They paradoxically found an increased autophagic activity in Paneth cells. An aberrant distribution of granules in the cytoplasm was nevertheless confirmed, associated to detectable crinophagy, mechanism targeting secretory granules toward autolysosomes. From these studies it seems that a deregulated or aberrantly targeted autophagy, rather than simply diminished autophagy, contributes to abnormal granule exocytosis in CD.

Aside impaired bacterial clearance resulting from defective xenophagy, adaptive immunity, and antibacterial peptide secretion, several studies investigated the potential role that autophagy deregulation could play on inflammatory mediators (Fig. 1, frame 4). As mentioned earlier, autophagy limits inflammasome activation. In the previously cited publication by Saitoh and colleagues [81], *Atg16L1* deficiency was shown to favour increased IL-1ß and IL-18 secretions by macrophages in response to lipopolysaccharide (LPS). Foetal liver chimeric mice grafted with hematopoietic cells deficient for ATG16L1, exhibited more severe DSS-induced colitis than wild type counterparts. This phenomenon was at least in part dependent on IL-1ß and IL-18 secretions.

The role of inflammasome in the development of CD is controversial, as IL-1ß secretion is also believed to be necessary for tissue repair. As a matter of fact, DSS-induced colitis, partly dependent on inflammasome activation, does not totally recapitulate CD inflammation. Moreover, NLRP3 variants linked to the development of CD, are correlated to low mRNA levels [91] arguing against an increased inflammasome activity during CD. It remains however possible that IL-1ß secretion above homeostatic level contributes to CD inflammation when autophagy is impaired. Mihai Netea's group thus showed that PBMCs from patients carrying the *ATG16L1*^{T300A} allele, expressed higher levels of IL-1ß after NOD2 stimulation [70]. This increased production did not seem in that case correlated with pro-IL1ß processing by caspase 1. Lassen and colleagues proved that *Atg16L1*^{T300A} mice were more susceptible to IL-1ß-linked inflammation induced by bacteria. This suggests again a role for inflammasome activation in the regulation of response towards intestinal pathogens or commensal flora [41].

Interestingly, this latter work showed that the T300A allele modulated ATG16L1 expression by rendering it more susceptible to degradation by caspase 3. A study published the same year also describes enhanced degradation of ATG16L1^{T300A} protein variant by caspase 3 [58]. This loss of stability is responsible for enhanced pro-inflammatory cytokine secretion by macrophages and defective clearance of Y. enterocolitica in the intestine. This mechanism is caspase-dependent as suppression of the mutation-associated caspase cleavage site abolishes hyper inflammation.

Thus apart from its major role in regulating xenophagy, antigen presentation and inflammasome activity, ATG16L1 expression level seems in general linked to pro-inflammatory cytokine secretion. A recent study showed that this regulatory role could be partly played by autophagy-independent mechanisms. Sorbara and colleagues studied in details the interaction between NOD1 and NOD2 receptors with ATG16L1 [84]. They showed that an ATG16L1 protein incapable to induce autophagy was still able to down regulate cytokine production in response to NOD1 and NOD2 stimulations. They further show that the ATG16L1^{T300A} contributes to increase IL-8 signalling in response to *Shigella* infection, in an autophagy independent manner. This points out again a role for ATG proteins in regulating inflammation independently from canonical autophagy as proved for the regulation of STING-TBK1 association by ATG9a or of RLR store level by ATG5-ATG12 conjugates.

3 Autophagy Inflammation Associated to Metabolic Syndrome

Metabolic syndrome consists in a combination of clinical factors linked to a deregulated metabolism. It includes obesity, high insulin level, hypertension, and high cholesterol load. These parameters can lead to pathologies tightly linked to inflammation, like type II diabetes (TIID), atherosclerosis and other cardiovascular diseases. According to the central role autophagy plays on both regulation of metabolism and inflammation, great attention has been paid to this field.

Indeed, autophagy is known be downregulated under high energetic diet, increased in frequency with modern western style nutrition. Autophagy also decreases with age, when metabolic syndrome prevalence increases (Fig. 2, frame 1).

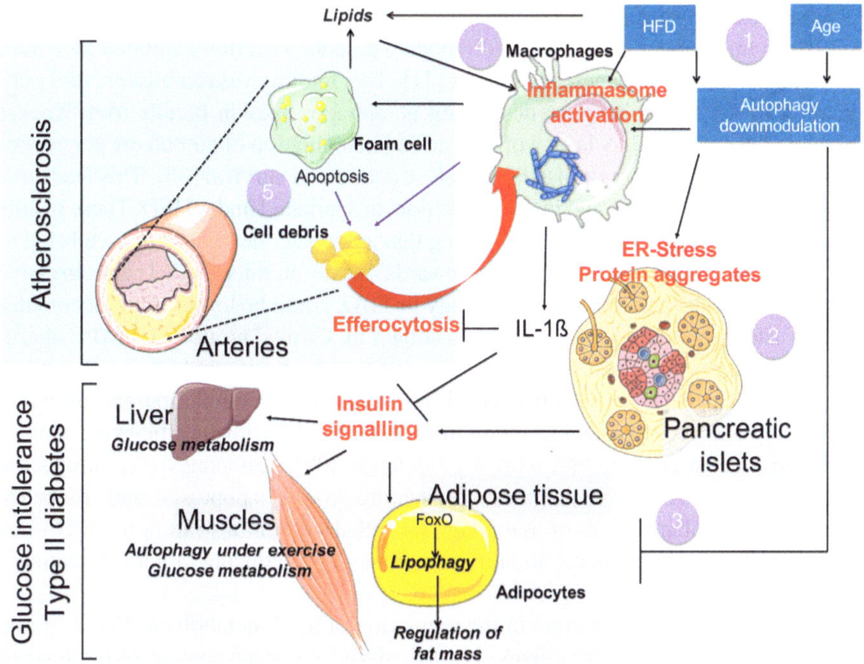

Fig. 2 Implications of autophagy in the metabolic syndrome. *1* Autophagy activity declines with age and can be downregulated by HFD. *2* Autophagy directly contributes to islet beta survival and function by limiting accumulation of protein aggregates and ER-stress. *3* Lipophagy, a form of autophagy involved in lipid droplets elimination, regulating fat mass and lipid release, could also be impaired by an autophagic activity decline. Moreover autophagic activity, specially induced during exercise, is important for glucose tolerance in muscles. *4* Inflammasome activity in macrophages is increased by HFD, including by the direct impact of lipid uptake, like fatty acids and cholesterol. This inflammatory activity is further increased by autophagy decline. IL-1ß released in this context can interfere with insulin signaling, contributing to glucose intolerance. *5* During atherosclerotic plaque formation, inflammation contributes to the recruitment of macrophages that can differentiate into foam cells. Autophagy impairment contributes to inflammation, recruitment of inflammatory cells and foam cell apoptosis. The debris generated by dead cells and the lipids released in the intima can further enhance inflammation. IL-1ß, over-produced by macrophages when autophagy is impaired, can also inhibit efferocytosis, which contributes to the accumulation of debris. Abbreviations: *HFD* high fat diet, *ER* endoplasmic reticulum, *IL* interleukin

These statements led to the hypothesis that autophagy impairment could participate in the development of pathologies associated to metabolic syndrome.

A pioneer study in the field of TIID showed that ß-cells from rat under high fat diet (HFD), exhibited abnormal ubiquitin protein aggregates [36]. Interestingly, these aggregates could be reproduced *in vitro* in a pancreatic cell line, under high glucose concentration treatment, and were shown to be dependent on oxidative stress. These aggregates, not related with aggresomes as their formation are actin and microtubule independent, are increased after 3-methyladenine treatment, an inhibitor of autophagy. Autophagy may thus be involved in their clearance (Fig. 2, frame 2).

This suggested protective role of autophagy was confirmed by another study in mice. C57BL/6 mice under HFD and db/db mouse, models for TIID, exhibited increased autophagic activity in pancreatic ß-cells [21]. This finding was reminiscent of a publication by Li and colleagues describing autophagosomes in ß-cells from Zucker diabetic fatty rats [46]. A later work described that ablation of autophagy genes specifically in ß-cells, compromised islet cell structure and survival [35]. This leads to a loss of glucose tolerance and insulin secretion, exacerbated under HFD. These results were confirmed by another study showing that autophagy deficiency in ß-cells led to ER-stress correlated with progression towards obesity in mice [73]. The progressive decreased activity of protective autophagy in TIID, could be linked to the accumulation of islet amyloid peptide (IAPP), as shown by Costes' group [78]. IAPP aggregates, co-expressed with insulin, are associated with obesity and block autophagic flux. The direct link *in vivo,* between increased IAPP accumulation and impaired autophagy was recently demonstrated in a mouse model with specific expression of IAPP in ß-cells, concomitant with *Atg7* deletion [39]. Autophagy thus allows the degradation of IAPP aggregates, preventing ß-cell apoptosis and diabetes. Interestingly, enhancement of autophagy by trehalose administration to HFD mice improved glucose tolerance, validating autophagy as a seducing target for treatment of TIID.

Autophagy is also important in the regulation of lipid metabolism (Fig. 2, frame 3). A mechanism called lipophagy, is particularly important to control the load of lipid droplets in the adipose tissue. Ciriolo's team showed that the forkhead homeobox type protein O1 (FoxO1) was activated under nutrient restriction in a murine adipocyte cell line [44]. FoxO transcription factors are related to the regulation of lysosomal processes in general, and autophagy in particular, at the transcriptional level. In this context, FoxO1 activation led to improved lysosomal degradation of lipid droplets via lipophagy. The drug metformin used as a treatment for TIID, and known to induce autophagy, led to the same effect. Thus autophagy regulates fat mass, lipid metabolism and the release of free fatty acids (FFA) by adipocytes.

The importance of autophagy on the onset of diabetes is not restricted to adipocytes and pancreatic islet cells. Mobilization of energy from muscles is also involved in TIID development and autophagy plays also here a regulatory role (Fig. 2, frame 3). Beth Levine's group demonstrated that preventing Beclin1-induced autophagy upon starvation and exercise, compromised glucose metabolism in skeletal and cardiac muscles, predisposing to glucose intolerance [28].

In addition to the increasing evidence about the protective role of autophagy on pancreatic ß-cell survival under stress, several studies highlighted a direct link between autophagy impairment and chronic inflammation characteristic of diabetes (Fig. 2, frame 4). IL-1ß is known to participate in insulin resistance, by directly inhibiting Akt signalling after insulin receptor stimulation, and by TNF-α, also known to limit insulin effect. The uptake of the saturated fatty acid palmitate, has been shown to induce NALP3 inflammasome activation in macrophages via NADPH oxidase activation and ROS generation [93]. Inflammasome activation in HFD regimen is here shown to induce insulin resistance *in vivo*. In homeostatic condition, the AMP-activated protein kinase (AMPK) activity can limit ROS generation by favouring ß-oxidation of

FFA. Strikingly, palmitate treatment in addition to LPS stimulation led to a decrease in AMPK activation, and inhibits autophagic activity. This decrease is suspected to impair mitophagy and to favour mitochondrial ROS release as mentioned above, participating in the overexpression of IL-1ß.

In contrast, an uncontrolled activation of autophagy could have deleterious effects on inflammation, in non-immune cells. The mouse beta cell line INS-1(823/13) treated with palmitate, over-expressed cathepsin B in an *Atg7*-dependent manner, cathepsin B being a lysosomal protease responsible for increased pro-inflammatory cytokine expression [45]. Inflammasome activation in that case may be favoured by excessive autophagy, leading to IL-1ß production that contributes to cell stress and limits insulin secretion in response to glucose stimulation. Thus autophagy must be tightly regulated to prevent excessive inflammasome activation and IL-1ß production.

The majority of the previously cited studies, focused on one cell type either immune or non-immune. A recent report showed that global decrease of autophagic activity modelled by *Atg7*$^{+/-}$ mice, led to low-grade inflammation associated to TIID after crossing with ob/ob mice [48]. This could explain why the risk to develop TIID increases with age when autophagic activity declines, and validate systemic autophagy modulation as a valuable therapeutic strategy.

Another manifestation of the metabolic syndrome tightly linked to inflammation is atherosclerosis (Fig. 2, frame 5). HFD can lead to the accumulation of lipids under the arterial epithelium, in a region called intima. This region then expands while the activation of the epithelium leads to cytokine secretion that attracts monocytes, which will further differentiate into macrophages. Native or oxidized lipids, low density lipoproteins (LDL), are then uptaken via scavenger receptors by macrophages. Macrophages then accumulate, perpetuating inflammation by cytokine or ROS release, which will further oxidize lipids. Macrophages can also differentiate into foam cells, containing elevated stores of lipids. These cells are prone to apoptosis and necrosis, generating debris that will also contribute to fuelling inflammation.

As endothelial cells, vascular smooth muscle cells (VSMC) are also directly sensitive to inflammatory cytokines. Jia and colleagues showed that TNF-α activated autophagy in VSMC isolated from atherosclerotic plaques [33]. In this context autophagy is suspected to participate in plaque instability by contributing to stress and cell death. Interestingly, autophagosome formation and LC3 processing are detected in cells of the intima of atherosclerotic plaques, including macrophages [52, 53]. To delineate the potential roles of autophagy activation in the plaque, two studies generated and described mouse models deficient for *Atg5* in macrophages, on a pro-atherosclerosis background (apolipoprotein null mice or LDL-receptor deficient mice; [47, 75]). As expected, ATG5-deficient macrophages secreted larger amounts of IL-1ß after LPS stimulation than wild-type counterparts. Of note co-incubation with cholesterol crystals, abundant components of the plaque also activated the inflammasome, probably via impairment of lysosomal degradation, which impacts autophagy. The inflammation induced by cytokines released from activated macrophages attracts immune cells and is accompanied by an increase in cell death.

Efferocytosis, a phagocytic process dedicated to the elimination of cell debris, is also frequently described as defective in atherosclerotic plaques. Inhibited autophagy, in the work of Liao and collaborators, leads to sensitization of macrophages to cell death after treatment with an oxisterol found in plaques and known to induce ER-stress. At the same time oxidative stress is increased in macrophages deficient for autophagy, rendering them less sensitive to clearance by surrounding phagocytes.

Autophagy is also important for lipid degradation by lysosomes in foam cells and for the subsequent cholesterol efflux [65]. In this context autophagy is induced in macrophages by the regulation of the ataxia telangiectasia mutated (Atm)-mammalian target of rapamycin pathway [42]. In line with this observation, impaired autophagy in advanced plaques, could contribute to foam cell apoptosis, lipid release, and cell debris accumulation, again contributing to inflammation. Autophagy can thus protect cells from death or contribute to their elimination although the latter mechanism is not totally understood. Autophagy in phagocytes also contributes to efferocytosis and defects in this specialized LC3-assisted phagocytosis (LAP) could also be involved in plaque formation [54].

Thus autophagy appears as an induced protective mechanism against plaque formation. Its progressive down regulation may participate to plaque evolution although the causes are not totally understood. Progressive inhibition of autophagy with age by increase of MiR-216a expression in endothelial cells may be a causal factor [57]. Aside from aging, and MiR expression, complex genetic background could participate in atherosclerosis susceptibility. A better understanding of the role of autophagy in metabolic disorders in general, partly linked to life-style and nutrition, could improve the existing treatments by limiting chronic inflammation.

4 Autophagy and Autoimmune Chronic Inflammation

Autoinflammatory diseases are chronic inflammations, clearly involving immune reactions toward autoantigens, *i.e.* molecules encoded by the self-genome. These disorders, as other auto inflammatory conditions, are mostly linked to a combination between environmental factors and genetic background. Both innate and adaptive immunity deregulations are prone to trigger autoimmunity. Pro-inflammatory background is eager to favour autoimmunity occurrence, and here again autophagy is central.

As for CD, GWAS designated new candidate genes for the development of autoimmunity (Fig. 3, frame 1). Among them several were linked to the development of systemic lupus erythematosus (SLE). This systemic autoimmune disease is characterized by the production of autoantibodies directed against nuclear auto antigens. Antibody deposits lead to chronic inflammation in several tissues like skin, kidney, cardiovascular system, and nervous system. In 2008, SNPs in *ATG5* locus were linked to the development of SLE [27]. Other studies confirmed this potential association in Asian population [99] while other failed to identify *ATG5* polymorphisms in a Finnish cohort of patients [32].

The functional relevance of SNPs identified in *ATG5* locus is not proven. One study demonstrated that one allelic variant more frequent in SLE patients was

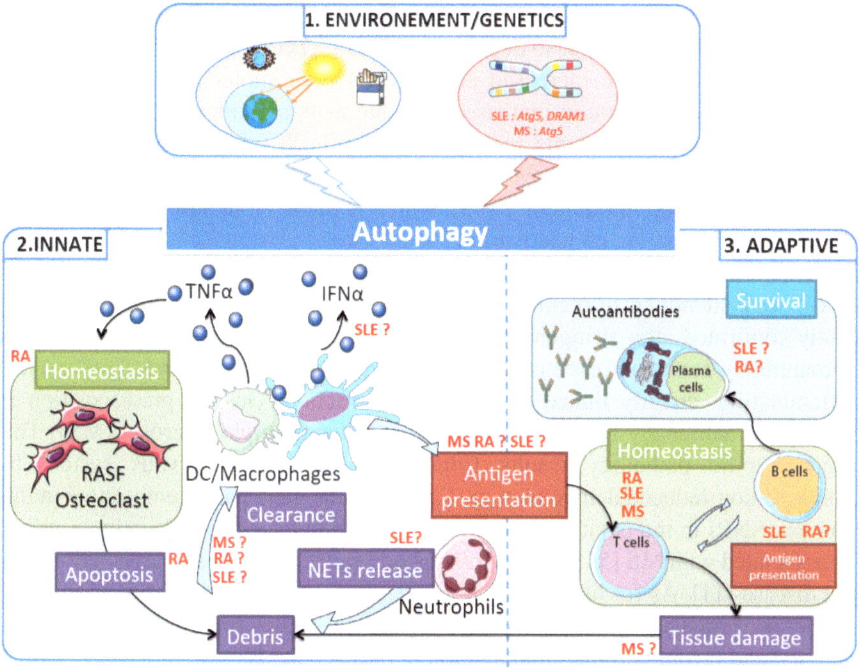

Fig. 3 Mechanisms linking autophagy deregulation to autoimmune disorders. *1* Environmental factors and genetic background could contribute to the deregulation of autophagy. Polymorphisms of unknown functional relevance have been linked to SLE and MS susceptibility. Autophagy deregulation can affect both innate and adaptive immune responses. *2* Autophagy could be involved in the deregulation of innate immunity. Autophagy regulates IFN-I production in response to nucleic acid-containing antigens, which is relevant to SLE. Autophagy is induced in RASF upon TNF-α stimulation. TNF-α is overexpressed in RA. Autophagy in response to TNF-α contributes to survival of RASF and activates osteoclastogenesis, leading to bone resorption. Autophagy contributes to clearance of dead cells by macrophages. The latter mechanism is defective in SLE and its impairment could also contribute to fuel inflammation in RA and MS. Excessive NETs release by neutrophils, mechanism dependent on autophagy, could also lead to the accumulation of debris containing nucleic acids, eliciting inflammation in SLE. Autophagy impacts antigen presentation by macrophages and DCs and could participate in autoantigen presentation in RA and SLE. This has been shown for the EAE model. Autophagy is also involved in the presentation of citrullinated epitopes, frequent in RA, to T cells. *3* Apart from antigen presentation, defects of autophagy, intrinsic to the adaptive immune system, could contribute to autoimmune disorders. Autophagy is implicated in T lymphocyte homeostasis and in the survival of memory B cells and plasma cells. This could contribute to abnormal autoreactive T cell survival in RA, SLE and in MS leading to tissue damage. In RA and SLE, it could contribute to the chronic secretion of pathogenic autoantibodies. Abbreviations: *Atg* autophagy-related gene, *DRAM1* DNA damaged-related autophagy modulator 1, *IFN* interferon, *SLE* systemic lupus erythematosus, *MS* multiple sclerosis, *RA* rheumatoid arthritis, *IFN* interferon, *TNF* tumour necrosis factor, *RA* rheumatoid arthritis, *RASF* RA synovial fibroblasts, *NET* neutrophil extracellular trap, *DC* dendritic cells, *EAE* experimental autoimmune encephalomyelitis

associated with increased *ATG5* mRNA expression [99]. Interestingly, another study focusing on asthma, also described SNPs in *ATG5* locus associated with the pathology, and with an increased activity of the promoter [51]. Correlatively, *ATG5* mRNA expression is increased in acute asthma. Although the direct link between *ATG5* allelic variants and its expression in SLE is not established, other genes regulating autophagy have been identified in GWAS such as DNA-damage regulated autophagy modulator 1 [95], which is involved in autophagy induction upon genetic stress via p53 activation.

Genetic links with other autoimmune diseases are less clear. Polymorphisms on *ATG5* were identified in rheumatoid arthritis (RA) context but could not be definitely confirmed after stringent statistical correction [64]. RA is characterized by autoantibody secretion responsible for systemic manifestations and T cell-related inflammation, directly linked to cartilage destruction. Genetic predisposition in relation to autophagy has also been suggested for multiple sclerosis (MS). This organ-specific pathology results from damages in the nervous system by inflammation-induced demyelination. T cells are the main pathogenic actors in this context. SNPs in the putative *ATG5* promoter region have been recently described as associated with MS [55] while another study failed to prove any association with the disease [11]. As for lupus, even if direct genetic predisposition linked to *ATGs* cannot be formally proved for RA and MS, indirect causes could impair autophagic activity, and favour chronic inflammation in these contexts.

Systemic pathologies like SLE and RA are linked to aberrant production of, and/or reaction towards, type I IFN and TNF-α respectively (Fig. 3, frame 2). Given the important relationship between ATG5/ATG12 and IFN-I secretion, by regulation of RLR and TLR availability for their ligands, it is possible that impairment of autophagic activity contributes to increased IFN-I production. Plasmacytoid dendritic cells (pDC) are important producers of IFN-α, especially via recognition by the intracellular sensor TLR9. Interestingly, a form of LAP has been linked in these cells to translocation of antibody-associated DNA complexes after recognition via Fc receptors [29]. Autophagy machinery in this context was shown to contribute to translocation of endocytosed DNA to TLR9 positive compartments, inducing IFN-α secretion. A deregulation in this traffic route could be relevant in SLE as insufficient clearance of nuclear debris, associated to antibodies, are thought to trigger and/or to sustain inflammation. Aside from insufficient clearance of debris resulting from apoptotic cells, DNA can also be released from activated neutrophils in neutrophil-extracellular traps (NETs). When produced in excess, they could contribute to provide TLR9 ligands phagocytosed by pDCs [40]. Interestingly, autophagy has been shown to participate in NET release [76]. Future studies should be done on lupus animal models or with SLE patient's samples to assess a potential link between deregulated autophagy and IFN-I secretion.

In contrast more experimental results have been obtained regarding the interplays between TNF-α, autophagy and RA. TNF-α is an inducer of autophagy in several cell types as vascular smooth muscle cells [33], skeletal muscle cells [38], epithelial cells but also in immune cells like macrophages [5]. TNF-α has been shown to induce autophagy in RA synovial fibroblasts (RASF) isolated from patients [18]. These cells are central to the development of cartilage inflammation thanks to the inflammatory cytokines and growth factors they secrete. These results are in line with other studies [37, 83], which showed that autophagy is induced under ER-stress and further increased by TNF-α. Autophagy protects fibroblasts from cell death, in concert with CCAAT/-enhancer-binding protein homologous protein under-expression, and probably contributes to their abnormal survival and secretion of growth factors. The microbial product LPS in complexes with damage associated molecular pattern HMGB1, are known to trigger experimental arthritis in mice. In the human pathology, these complexes could result from cell debris on a non-sterile inflammation site. LPS-HMGB1 complexes favour the differentiation and survival of RASF, concomitant with autophagy activation [72], again pointing out a protective role of autophagy on RASF and thus on the maintenance of inflammation. Autophagic vacuoles and expression of Beclin-1 and ATG7 are also increased in osteoclasts from RA patients [49]. Autophagy in this work was shown to help osteoclastogenesis and contributes to bone-resorption in a TNFα-dependent manner.

SLE and RA have in common the chronic generation of cellular debris at the site of inflammation. Adequate clearance of cell remnants is thought to be central to the prevention of autoimmunity. Numerous mouse models deficient for apoptotic cell clearance are prone to lupus-like pathologies. A specialized form of LAP, involving LC3A, has been shown to be implicated in the elimination of dead cells by macrophages [54]. Invalidation of this pathway leads to increased pro-inflammatory cytokine production by macrophages. The *in vivo* relevance of this observation and the potential link with human SLE are still to be demonstrated.

The existence of a pro-inflammatory background contributes to break the tolerance against auto antigens. A chronic inflammatory microenvironment is eager to activate APCs that become abnormally able to prime T cells against self-peptides. Autophagy plays an important role in antigen presentation, both by MHC class I and II molecules. Autophagy has been shown to be necessary for the presentation of endogenous self-peptides by thymic epithelial cells (TECs). Mice with autophagy-deficient thymus [60] or conditionally deleted for *Atg5* specifically in TECs [2], exhibit abnormal central selection of T cells leading to an autoimmune phenotype consisting on colitis. Defects in central tolerance are not formally proved in the majority of autoimmune diseases. It remains however possible that defects in autophagy activity in the thymus, associated for example with age (when inflammatory diseases become more frequent), contributes to skew T cell repertoire towards autoimmunity. Moreover, defects in Treg generation and survival are associated with colitis, an autoimmune phenotype observed in the study published by Nedjic and colleagues [60]. It is thus possible that autophagy defects in the thymus contribute to abnormal peripheral tolerance.

A role for autophagy in autoantigen presentation in the periphery is also plausible. Indeed, autophagy contributes to the presentation of cytosolic epitopes to CD4 T cells, including self-epitopes, which could be quite relevant for autoantigen presentation. A recent study showed that specific deletion of autophagy in DCs limits the development of experimental autoimmune encephalomyelitis (EAE), a murine model for human MS [7]. The improvement of the clinical score was correlated with a decrease in CD4 T cell priming. Although not yet proven in the case of RA, autophagy could contribute to autoepitope presentation. Unanues' group showed that autophagy in APCs, i.e. DCs, macrophages and B lymphocytes, contributed to the presentation of citrullinated epitopes [31]. Interestingly citrullinated peptides are common antigens in RA, and aberrant autophagy could contribute to generate such autoantigens. Autophagic activity has been shown to be upregulated in B cells from SLE patients and mouse models for lupus [16]. Increased autophagy could then contributes to autoantigen presentation, including citrullinated epitopes, but to date, no experimental proof has been provided.

The initial peripheral break of tolerance in autoimmune diseases leads to abnormal autoreactive lymphocyte survival (Fig. 3, frame 3). Here again, deregulated autophagy can contribute to the chronicity of autoimmune inflammation. Autophagy plays an important role in T lymphocyte survival and polarization. The first study suggesting a link between deregulated autophagy in lymphocytes was performed in T cells from MS patients. ATG5 expression was found increased in T cells isolated from EAE mice and MS patients [4]. A work performed in our laboratory identified autophagy deregulation in T cells from both mouse models for lupus and SLE patients [24]. The autophagic vacuole load in T cells was mainly observable under T cell receptor (TCR) related stimulation in mice and was increasing with age, contrary to control mice. An increase in autophagic compartments in SLE T cells was confirmed by three other studies [3, 12, 16]. Interestingly the study by Alessandri and colleagues showed an increase in the autophagosome-associated marker LC3, especially in naive CD4 T cells suggesting a predisposed deregulation. Our study including induction of systemic acute inflammation in normal mice showed that the activation of T cells in this context, was not sufficient to increase autophagy. Pierdomonici's group concludes that accumulation of autophagosomes is due to a blockade of autophagy rather than increased autophagy induction. We cannot formally exclude this hypothesis as our observation could results from disequilibrium between induction and degradation of autophagic vacuoles. However in our setting, the blockade was not total as LC3 was still accumulated after treatment with lysosomal protease inhibitors. Moreover Alessandri and colleagues use starvation or treatment with autologous serum from SLE patients, as triggers of autophagy. Interestingly, they show that SLE serum can induce autophagy in normal T cells, reminiscent of another study identifying the pro-autophagic impact of SLE serum on a neuron cell line [87], but not in SLE T cells. It is possible however that SLE T cells cannot further increase autophagy under metabolic demand or when re-exposed to their already stimulating environment. It is also possible that other pathways like the TCR pathway, relevant for auto reactive T cell activation, can contribute to elevate LC3 levels. In any case, further investigation is needed to discriminate at which level the deregulation occurs, and if an increase in autophagosome generation or a decrease in degradation, or both, could be involved. Furthermore, metabolic versus antigen-induced autophagic stimulations should be distinguished as their regulations and outcomes may be different.

The regulation of autophagy in T cells from RA has also been recently studied. Weyand's team described a deficiency in glucose metabolism in CD4 T cells from

RA, in response to TCR stimulation [96]. This was associated with insufficient 6-phosphofructo-2-kinase/fructose-2,6-bisphosphatase 3 (PFKFB3) induction. This impairment in energy mobilization could explain part of the apoptosis-prone phenotype of RA T cells. Interestingly, autophagy was also impaired by PFKFB3 deficiency, contributing to the difficulty to mobilize energy during activation.

Thus autophagy could play dual roles in autoreactive T cells, contributing to their abnormal survival when increased and sensitizing to apoptosis when impaired. In the context of MS, inhibiting autophagy in T cells from autoimmune diseases could be a therapeutic option as shown in the study by Kovacs and colleagues. They demonstrate that mice with specific deletion of Beclin1 in T cells are less sensitive to EAE induction. This was correlated with increased cell death, especially for Th1 cells, pointing out an interesting regulatory role of autophagy on Th cell polarization. The induction of apoptosis in the absence of autophagy could be related, in this experimental model to increased stores of pro-caspases.

In RA, and SLE, autoantibody secretion is part of the pathology. Deregulation of B lymphocyte homeostasis is thus one typical feature of these systemic autoimmune diseases. Clarke and colleagues were the first to describe autophagy increase in B cells from lupus prone mouse model NZB/W, and from SLE patients [16]. Two major studies showed a role for autophagy in memory B cell and plasma cell survival [14, 68]. An increase in autophagy could contribute to the abnormal survival of autoreactive B lymphocytes and autoantibody secreting plasma cells. It thus appears that modulation of autophagy in lymphocytes could be a beneficial strategy to limit auto reactive lymphocyte survival. Moreover, autophagy is suspected to play a role in the relocalization of DNA containing antigenic complexes toward TLR9 positive endosomes [13]. Autophagy machinery could thus contribute to B cell hyperactivity against nuclear antigens. Studies on mouse models with autoimmune-prone backgrounds, also deficient for autophagy in lymphocytes, could help decipher the role of ATGs in the development of the pathology.

Of much interest, therapies like rapamycin, hydroxychloroquine or P140 peptide [66], are known to modulate autophagic activity. It would be interesting to investigate if such therapeutic effects are actually linked to modulation of autophagy in immune cells like APCs or lymphocytes.

5 Conclusions

Autophagy is a physiological response, at the crossroads between energy sensing, and reaction to stress induced by tissue damage and/or by infection. This mechanism is thus a master integrator of both innate and adaptive immunity to the surrounding environment, by regulation of inflammation. Both genetic and environmental factors could contribute to the deregulation of autophagy. Allelic variations on autophagy genes seem to be strongly related to the susceptibility to develop autoinflammatory diseases like CD. Other polymorphisms on *ATGs* suggest a role for autophagy deregulation in autoimmune diseases, like SLE, RA and MS, although the functional relevance of these variants remains to be fully determined. Environmental factors like infection, or changes in metabolic equilibrium linked to diet, or genetically programmed like aging, can contribute to provide a low-grade inflammatory environment, prone to trigger chronic inflammation. Autophagy involvement in preventing inflammation linked to these three causes, linked to TIID and atherosclerosis, is now clearly demonstrated. Apart from RA, MS, SLE and CD, other inflammatory pathologies could also imply autophagy deregulation. Interestingly, therapeutic molecules modulating the autophagy process have shown efficacy in autoinflammatory diseases like metformin for TIID, rapamycin, hydroxychloroquine and P140 peptide, in autoimmune diseases. A better understanding about the precise roles of autophagy in chronic inflammation will help design new molecules, and new therapeutic approaches, to treat these complex diseases.

Acknowledgments The authors, all working in Pr Sylviane's Muller group, would like to thank her for scientific and material support, as well as for fruitful discussions. Research in SM's laboratory is financially supported by the French Centre National de la Recherche Scientifique and the Laboratory of Excellence Medalis (ANR-10-LABX-0034), Initiative of Excellence (IdEx), Strasbourg University. DM is a recipient of a PhD fellowship from Fond National de Recherche, Luxembourg. JA is a recipient of a PhD grant from Association de Recherche contre le Cancer (ARC)

We thank the authours of Servier Medical arts, providing precious frames used in our figures.

Conflicts of Interests The authors declare no conflicts of interests

References

1. Adolph TE, Tomczak MF, Niederreiter L, Ko HJ, Bock J, Martinez-Naves E, Glickman JN, Tschurtschenthaler M, Hartwig J, Hosomi S, Flak MB, Cusick JL, Kohno K, Iwawaki T, Billmann-Born S, Raine T, Bharti R, Lucius R, Kweon MN, Marciniak SJ, Choi A, Hagen SJ, Schreiber S, Rosenstiel P, Kaser A, Blumberg RS (2013) Paneth cells as a site of origin for intestinal inflammation. Nature 503(7475):272–276. doi:10.1038/nature12599
2. Aichinger M, Wu C, Nedjic J, Klein L (2013) Macroautophagy substrates are loaded onto MHC class II of medullary thymic epithelial cells for central tolerance. J Exp Med 210(2):287–300. doi:10.1084/jem.20122149
3. Alessandri C, Barbati C, Vacirca D, Piscopo P, Confaloni A, Sanchez M, Maselli A, Colasanti T, Conti F, Truglia S, Perl A, Valesini G, Malorni W, Ortona E, Pierdominici M (2012) T lymphocytes from patients with systemic lupus erythematosus are resistant to induction of autophagy. FASEB J 26(11):4722–4732. doi:10.1096/fj.12-206060
4. Alirezaei M, Fox HS, Flynn CT, Moore CS, Hebb AL, Frausto RF, Bhan V, Kiosses WB, Whitton JL, Robertson GS, Crocker SJ (2009) Elevated ATG5 expression in autoimmune demyelination and multiple sclerosis. Autophagy 5(2):152–158
5. Andrade RM, Wessendarp M, Gubbels M-J, Striepen B, Subauste CS (2006) CD40 induces macrophage anti–Toxoplasma gondii activity by triggering autophagy-dependent fusion of pathogen-containing vacuoles and lysosomes. J Clin Invest 116(9):2366–2377. doi:10.1172/JCI28796
6. Barrett JC, Hansoul S, Nicolae DL, Cho JH, Duerr RH, Rioux JD, Brant SR, Silverberg MS, Taylor KD, Barmada MM, Bitton A, Dassopoulos T, Datta LW, Green T, Griffiths AM, Kistner EO, Murtha MT, Regueiro MD, Rotter JI, Schumm LP, Steinhart AH, Targan SR, Xavier RJ,

Libioulle C, Sandor C, Lathrop M, Belaiche J, Dewit O, Gut I, Heath S, Laukens D, Mni M, Rutgeerts P, Van Gossum A, Zelenika D, Franchimont D, Hugot J-P, de Vos M, Vermeire S, Louis E, Cardon LR, Anderson CA, Drummond H, Nimmo E, Ahmad T, Prescott NJ, Onnie CM, Fisher SA, Marchini J, Ghori J, Bumpstead S, Gwilliam R, Tremelling M, Deloukas P, Mansfield J, Jewell D, Satsangi J, Mathew CG, Parkes M, Georges M, Daly MJ (2008) Genome-wide association defines more than 30 distinct susceptibility loci for Crohn's disease. Nat Genet 40(8):955–962

7. Bhattacharya A, Parillon X, Zeng S, Han S, Eissa NT (2014) Deficiency of autophagy in dendritic cells protects against experimental autoimmune encephalomyelitis. J Biol Chem 289(38):26525–26532. doi:10.1074/jbc.M114.575860
8. Brest P, Lapaquette P, Souidi M, Lebrigand K, Cesaro A, Vouret-Craviari V, Mari B, Barbry P, Mosnier JF, Hebuterne X, Harel-Bellan A, Mograbi B, Darfeuille-Michaud A, Hofman P (2011) A synonymous variant in IRGM alters a binding site for miR-196 and causes deregulation of IRGM-dependent xenophagy in Crohn's disease. Nat Genet 43(3):242–245. doi:10.1038/ng.762
9. Cadwell K, Liu JY, Brown SL, Miyoshi H, Loh J, Lennerz JK, Kishi C, Kc W, Carrero JA, Hunt S, Stone CD, Brunt EM, Xavier RJ, Sleckman BP, Li E, Mizushima N, Stappenbeck TS, Virgin HW (2008) A key role for autophagy and the autophagy gene Atg16l1 in mouse and human intestinal Paneth cells. Nature 456(7219):259–263. doi:10.1038/nature07416
10. Cadwell K, Patel KK, Maloney NS, Liu TC, Ng AC, Storer CE, Head RD, Xavier R, Stappenbeck TS, Virgin HW (2010) Virus-plus-susceptibility gene interaction determines Crohn's disease gene Atg16L1 phenotypes in intestine. Cell 141(7):1135–1145. doi:10.1016/j.cell.2010.05.009
11. Cai PP, Wang HX, Zhuang JC, Liu QB, Zhao GX, Li ZX, Wu ZY (2014) Variants of autophagy-related gene 5 are associated with neuromyelitis optica in the Southern Han Chinese population. Autoimmunity 47(8):563–6. doi:10.3109/08916934.2014.929668
12. Caza TN, Fernandez DR, Talaber G, Oaks Z, Haas M, Madaio MP, Lai ZW, Miklossy G, Singh RR, Chudakov DM, Malorni W, Middleton F, Banki K, Perl A (2014) HRES-1/Rab4-mediated depletion of Drp1 impairs mitochondrial homeostasis and represents a target for treatment in SLE. Ann Rheumat Dis 73(10):1888–1897. doi:10.1136/annrheumdis-2013-203794
13. Chaturvedi A, Dorward D, Pierce SK (2008) The B cell receptor governs the subcellular location of Toll-like receptor 9 leading to hyperresponses to DNA-containing antigens. Immunity 28(6):799–809. doi:10.1016/j.immuni.2008.03.019
14. Chen M, Hong MJ, Sun H, Wang L, Shi X, Gilbert BE, Corry DB, Kheradmand F, Wang J (2014) Essential role for autophagy in the maintenance of immunological memory against influenza infection. Nat Med 20(5):503–510. doi:10.1038/nm.3521
15. Ciccia F, Accardo-Palumbo A, Rizzo A, Guggino G, Raimondo S, Giardina A, Cannizzaro A, Colbert RA, Alessandro R, Triolo G (2014) Evidence that autophagy, but not the unfolded protein response, regulates the expression of IL-23 in the gut of patients with ankylosing spondylitis and subclinical gut inflammation. Ann Rheumat Dis 73(8):1566–1574. doi:10.1136/annrheumdis-2012-202925
16. Clarke AJ, Ellinghaus U, Cortini A, Stranks A, Simon AK, Botto M, Vyse TJ (2014) Autophagy is activated in systemic lupus erythematosus and required for plasmablast development. Ann Rheumat Dis. doi:10.1136/annrheumdis-2013-204343
17. Codogno P, Mehrpour M, Proikas-Cezanne T (2012) Canonical and non-canonical autophagy: variations on a common theme of self-eating? Nat Rev Mol Cell Biol 13(1):7–12. doi:10.1038/nrm3249
18. Connor AM, Mahomed N, Gandhi R, Keystone EC, Berger SA (2012) TNFalpha modulates protein degradation pathways in rheumatoid arthritis synovial fibroblasts. Arthritis Res Ther 14(2):R62. doi:10.1186/ar3778
19. Cooney R, Baker J, Brain O, Danis B, Pichulik T, Allan P, Ferguson DJ, Campbell BJ, Jewell D, Simmons A (2010) NOD2 stimulation induces autophagy in dendritic cells influencing bacterial handling and antigen presentation. Nat Med 16(1):90–97. doi:10.1038/nm.2069

20. Deegan S, Saveljeva S, Gorman AM, Samali A (2013) Stress-induced self-cannibalism: on the regulation of autophagy by endoplasmic reticulum stress. Cell Mol Life Sci 70(14):2425–2441. doi:10.1007/s00018-012-1173-4
21. Ebato C, Uchida T, Arakawa M, Komatsu M, Ueno T, Komiya K, Azuma K, Hirose T, Tanaka K, Kominami E, Kawamori R, Fujitani Y, Watada H (2008) Autophagy is important in islet homeostasis and compensatory increase of beta cell mass in response to high-fat diet. Cell Metab 8(4):325–332. doi:10.1016/j.cmet.2008.08.009
22. Goda N, Kanai M (2012) Hypoxia-inducible factors and their roles in energy metabolism. Int J Hematol 95(5):457–463. doi:10.1007/s12185-012-1069-y
23. Gomes LC, Dikic I (2014) Autophagy in antimicrobial immunity. Mol Cell 54(2):224–233. doi:10.1016/j.molcel.2014.03.009
24. Gros F, Arnold J, Page N, Decossas M, Korganow AS, Martin T, Muller S (2012) Macroautophagy is deregulated in murine and human lupus T lymphocytes. Autophagy 8(7):1113–1123. doi:10.4161/auto.20275
25. Hampe J, Franke A, Rosenstiel P, Till A, Teuber M, Huse K, Albrecht M, Mayr G, De La Vega FM, Briggs J, Gunther S, Prescott NJ, Onnie CM, Hasler R, Sipos B, Folsch UR, Lengauer T, Platzer M, Mathew CG, Krawczak M, Schreiber S (2007) A genome-wide association scan of nonsynonymous SNPs identifies a susceptibility variant for Crohn disease in ATG16L1. Nat Genet 39(2):207–211. doi:10.1038/ng1954
26. Harijith A, Ebenezer DL, Natarajan V (2014) Reactive oxygen species at the crossroads of inflammasome and inflammation. Front Physiol 5:352. doi:10.3389/fphys.2014.00352
27. Harley JB, Alarcon-Riquelme ME, Criswell LA, Jacob CO, Kimberly RP, Moser KL, Tsao BP, Vyse TJ, Langefeld CD (2008) Genome-wide association scan in women with systemic lupus erythematosus identifies susceptibility variants in ITGAM, PXK, KIAA1542 and other loci. Nat Genet 40 (2):204–210. doi:http://www.nature.com/ng/journal/v40/n2/suppinfo/ng.81_S1.html
28. He C, Bassik MC, Moresi V, Sun K, Wei Y, Zou Z, An Z, Loh J, Fisher J, Sun Q, Korsmeyer S, Packer M, May HI, Hill JA, Virgin HW, Gilpin C, Xiao G, Bassel-Duby R, Scherer PE, Levine B (2012) Exercise-induced BCL2-regulated autophagy is required for muscle glucose homeostasis. Nature 481(7382):511–515. doi:10.1038/nature10758
29. Henault J, Martinez J, Riggs JM, Tian J, Mehta P, Clarke L, Sasai M, Latz E, Brinkmann MM, Iwasaki A, Coyle AJ, Kolbeck R, Green DR, Sanjuan MA (2012) Noncanonical autophagy is required for type I interferon secretion in response to DNA-immune complexes. Immunity 37(6):986–997. doi:10.1016/j.immuni.2012.09.014
30. Hugot JP, Chamaillard M, Zouali H, Lesage S, Cezard JP, Belaiche J, Almer S, Tysk C, O'Morain CA, Gassull M, Binder V, Finkel Y, Cortot A, Modigliani R, Laurent-Puig P, Gower-Rousseau C, Macry J, Colombel JF, Sahbatou M, Thomas G (2001) Association of NOD2 leucine-rich repeat variants with susceptibility to Crohn's disease. Nature 411(6837):599–603. doi:10.1038/35079107
31. Ireland JM, Unanue ER (2011) Autophagy in antigen-presenting cells results in presentation of citrullinated peptides to CD4 T cells. J Exp Med 208(13):2625–2632. doi:10.1084/jem.20110640
32. Jarvinen TM, Hellquist A, Zucchelli M, Koskenmies S, Panelius J, Hasan T, Julkunen H, D'Amato M, Kere J (2012) Replication of GWAS-identified systemic lupus erythematosus susceptibility genes affirms B-cell receptor pathway signalling and strengthens the role of IRF5 in disease susceptibility in a Northern European population. Rheumatology 51(1):87–92. doi:10.1093/rheumatology/ker263
33. Jia G, Cheng G, Gangahar DM, Agrawal DK (2006) Insulin-like growth factor-1 and TNF-[alpha] regulate autophagy through c-jun N-terminal kinase and Akt pathways in human atherosclerotic vascular smooth cells. Immunol Cell Biol 84(5):448–454
34. Jounai N, Takeshita F, Kobiyama K, Sawano A, Miyawaki A, Xin K-Q, Ishii KJ, Kawai T, Akira S, Suzuki K, Okuda K (2007) The Atg5–Atg12 conjugate associates with innate antiviral immune responses. Proc Natl Acad Sci U S A 104(35):14050–14055. doi:10.1073/pnas.0704014104
35. Jung HS, Chung KW, Won Kim J, Kim J, Komatsu M, Tanaka K, Nguyen YH, Kang TM, Yoon K-H, Kim J-W, Jeong YT, Han MS, Lee M-K, Kim K-W, Shin J, Lee M-S (2008) Loss

of autophagy diminishes pancreatic β cell mass and function with resultant hyperglycemia. Cell Metab 8(4):318–324. doi:http://dx.doi.org/10.1016/j.cmet.2008.08.013
36. Kaniuk NA, Kiraly M, Bates H, Vranic M, Volchuk A, Brumell JH (2007) Ubiquitinated-protein aggregates form in pancreatic beta-cells during diabetes-induced oxidative stress and are regulated by autophagy. Diabetes 56(4):930–939. doi:10.2337/db06-1160
37. Kato M, Ospelt C, Gay RE, Gay S, Klein K (2014) Dual role of autophagy in stress-induced cell death in rheumatoid arthritis synovial fibroblasts. Arthritis Rheumatol 66(1):40–48. doi:10.1002/art.38190
38. Keller CW, Fokken C, Turville SG, Lunemann A, Schmidt J, Munz C, Lunemann JD (2011) TNF-alpha induces macroautophagy and regulates MHC class II expression in human skeletal muscle cells. J Biol Chem 286(5):3970–3980. doi:10.1074/jbc.M110.159392
39. Kim J, Cheon H, Jeong YT, Quan W, Kim KH, Cho JM, Lim YM, Oh SH, Jin SM, Kim JH, Lee MK, Kim S, Komatsu M, Kang SW, Lee MS (2014) Amyloidogenic peptide oligomer accumulation in autophagy-deficient beta cells induces diabetes. J Clin Invest 124(8):3311–3324. doi:10.1172/JCI69625
40. Lande R, Ganguly D, Facchinetti V, Frasca L, Conrad C, Gregorio J, Meller S, Chamilos G, Sebasigari R, Riccieri V, Bassett R, Amuro H, Fukuhara S, Ito T, Liu YJ, Gilliet M (2011) Neutrophils activate plasmacytoid dendritic cells by releasing self-DNA-peptide complexes in systemic lupus erythematosus. Sci Trans Med 3(73):73ra19. doi:10.1126/scitranslmed.3001180
41. Lassen KG, Kuballa P, Conway KL, Patel KK, Becker CE, Peloquin JM, Villablanca EJ, Norman JM, Liu TC, Heath RJ, Becker ML, Fagbami L, Horn H, Mercer J, Yilmaz OH, Jaffe JD, Shamji AF, Bhan AK, Carr SA, Daly MJ, Virgin HW, Schreiber SL, Stappenbeck TS, Xavier RJ (2014) Atg16L1 T300A variant decreases selective autophagy resulting in altered cytokine signaling and decreased antibacterial defense. Proc Natl Acad Sci U S A 111(21):7741–7746. doi:10.1073/pnas.1407001111
42. Le Guezennec X, Brichkina A, Huang YF, Kostromina E, Han W, Bulavin DV (2012) Wip1-dependent regulation of autophagy, obesity, and atherosclerosis. Cell Metab 16(1):68–80. doi:10.1016/j.cmet.2012.06.003
43. Lee J, Kim HR, Quinley C, Kim J, Gonzalez-Navajas J, Xavier R, Raz E (2012) Autophagy suppresses interleukin-1beta (IL-1beta) signaling by activation of p62 degradation via lysosomal and proteasomal pathways. J Biol Chem 287(6):4033–4040. doi:10.1074/jbc.M111.280065
44. Lettieri Barbato D, Tatulli G, Aquilano K, Ciriolo MR (2013) FoxO1 controls lysosomal acid lipase in adipocytes: implication of lipophagy during nutrient restriction and metformin treatment. Cell Death Dis 4:e861. doi:10.1038/cddis.2013.404
45. Li S, Du L, Zhang L, Hu Y, Xia W, Wu J, Zhu J, Chen L, Zhu F, Li C, Yang S (2013) Cathepsin B contributes to autophagy-related 7 (Atg7)-induced nod-like receptor 3 (NLRP3)-dependent proinflammatory response and aggravates lipotoxicity in rat insulinoma cell line. J Biol Chem 288(42):30094–30104. doi:10.1074/jbc.M113.494286
46. Li X, Zhang L, Meshinchi S, Dias-Leme C, Raffin D, Johnson JD, Treutelaar MK, Burant CF (2006) Islet microvasculature in islet hyperplasia and failure in a model of type 2 diabetes. Diabetes 55(11):2965–2973. doi:10.2337/db06-0733
47. Liao X, Sluimer JC, Wang Y, Subramanian M, Brown K, Pattison JS, Robbins J, Martinez J, Tabas I (2012) Macrophage autophagy plays a protective role in advanced atherosclerosis. Cell Metab 15(4):545–553. doi:10.1016/j.cmet.2012.01.022
48. Lim YM, Lim H, Hur KY, Quan W, Lee HY, Cheon H, Ryu D, Koo SH, Kim HL, Kim J, Komatsu M, Lee MS (2014) Systemic autophagy insufficiency compromises adaptation to metabolic stress and facilitates progression from obesity to diabetes. Nat Commun 5:4934. doi:10.1038/ncomms5934
49. Lin NY, Beyer C, Giessl A, Kireva T, Scholtysek C, Uderhardt S, Munoz LE, Dees C, Distler A, Wirtz S, Kronke G, Spencer B, Distler O, Schett G, Distler JH (2013) Autophagy regulates TNFalpha-mediated joint destruction in experimental arthritis. Ann Rheum Dis 72(5):761–768. doi:10.1136/annrheumdis-2012-201671
50. Lupfer C, Thomas PG, Anand PK, Vogel P, Milasta S, Martinez J, Huang G, Green M, Kundu M, Chi H, Xavier RJ, Green DR, Lamkanfi M, Dinarello CA, Doherty PC, Kanneganti TD

(2013) Receptor interacting protein kinase 2-mediated mitophagy regulates inflammasome activation during virus infection. Nat Immunol 14(5):480–488. doi:10.1038/ni.2563
51. Martin LJ, Gupta J, Jyothula SSSK, Butsch Kovacic M, Biagini Myers JM, Patterson TL, Ericksen MB, He H, Gibson AM, Baye TM, Amirisetty S, Tsoras AM, Sha Y, Eissa NT, Hershey GKK (2012) Functional variant in the autophagy-related 5 gene promotor is associated with childhood asthma. PLoS One 7(4):e33454. doi:10.1371/journal.pone.0033454
52. Martinet W, De Bie M, Schrijvers DM, De Meyer GR, Herman AG, Kockx MM (2004) 7-ketocholesterol induces protein ubiquitination, myelin figure formation, and light chain 3 processing in vascular smooth muscle cells. Arterioscler Thromb Vasc Biol 24(12):2296–2301. doi:10.1161/01.ATV.0000146266.65820.a1
53. Martinet W, De Meyer GR (2009) Autophagy in atherosclerosis: a cell survival and death phenomenon with therapeutic potential. Circ Res 104(3):304–317. doi:10.1161/CIRCRESAHA.108.188318
54. Martinez J, Almendinger J, Oberst A, Ness R, Dillon CP, Fitzgerald P, Hengartner MO, Green DR (2011) Microtubule-associated protein 1 light chain 3 alpha (LC3)-associated phagocytosis is required for the efficient clearance of dead cells. Proc Natl Acad Sci U S A 108(42):17396–17401. doi:10.1073/pnas.1113421108
55. Mayes MD, Bossini-Castillo L, Gorlova O, Martin JE, Zhou X, Chen WV, Assassi S, Ying J, Tan FK, Arnett FC, Reveille JD, Guerra S, Teruel M, Carmona FD, Gregersen PK, Lee AT, Lopez-Isac E, Ochoa E, Carreira P, Simeon CP, Castellvi I, Gonzalez-Gay MA, Spanish Scleroderma G, Zhernakova A, Padyukov L, Alarcon-Riquelme M, Wijmenga C, Brown M, Beretta L, Riemekasten G, Witte T, Hunzelmann N, Kreuter A, Distler JH, Voskuyl AE, Schuerwegh AJ, Hesselstrand R, Nordin A, Airo P, Lunardi C, Shiels P, van Laar JM, Herrick A, Worthington J, Denton C, Wigley FM, Hummers LK, Varga J, Hinchcliff ME, Baron M, Hudson M, Pope JE, Furst DE, Khanna D, Phillips K, Schiopu E, Segal BM, Molitor JA, Silver RM, Steen VD, Simms RW, Lafyatis RA, Fessler BJ, Frech TM, Alkassab F, Docherty P, Kaminska E, Khalidi N, Jones HN, Markland J, Robinson D, Broen J, Radstake TR, Fonseca C, Koeleman BP, Martin J (2014) Immunochip analysis identifies multiple susceptibility loci for systemic sclerosis. Am J Human Genet 94(1):47–61. doi:10.1016/j.ajhg.2013.12.002
56. McCarroll SA, Huett A, Kuballa P, Chilewski SD, Landry A, Goyette P, Zody MC, Hall JL, Brant SR, Cho JH, Duerr RH, Silverberg MS, Taylor KD, Rioux JD, Altshuler D, Daly MJ, Xavier RJ (2008) Deletion polymorphism upstream of IRGM associated with altered IRGM expression and Crohn's disease. Nat Genet 40(9):1107–1112. doi:10.1038/ng.215
57. Menghini R, Casagrande V, Marino A, Marchetti V, Cardellini M, Stoehr R, Rizza S, Martelli E, Greco S, Mauriello A, Ippoliti A, Martelli F, Lauro R, Federici M (2014) MiR-216a: a link between endothelial dysfunction and autophagy. Cell Death Dis 5:e1029. doi:10.1038/cddis.2013.556
58. Murthy A, Li Y, Peng I, Reichelt M, Katakam AK, Noubade R, Roose-Girma M, DeVoss J, Diehl L, Graham RR, van Lookeren Campagne M (2014) A Crohn's disease variant in Atg16l1 enhances its degradation by caspase 3. Nature 506(7489):456–462. doi:10.1038/nature13044
59. Nakahira K, Haspel JA, Rathinam VA, Lee SJ, Dolinay T, Lam HC, Englert JA, Rabinovitch M, Cernadas M, Kim HP, Fitzgerald KA, Ryter SW, Choi AM (2011) Autophagy proteins regulate innate immune responses by inhibiting the release of mitochondrial DNA mediated by the NALP3 inflammasome. Nat Immunol 12(3):222–230. doi:10.1038/ni.1980
60. Nedjic J, Aichinger M, Emmerich J, Mizushima N, Klein L (2008) Autophagy in thymic epithelium shapes the T-cell repertoire and is essential for tolerance. Nature 455(7211):396–400. doi:10.1038/nature07208
61. Netea MG, Kullberg BJ, de Jong DJ, Franke B, Sprong T, Naber TH, Drenth JP, Van der Meer JW (2004) NOD2 mediates anti-inflammatory signals induced by TLR2 ligands: implications for Crohn's disease. Eur J Immunol 34(7):2052–2059. doi:10.1002/eji.200425229
62. Ogura Y, Bonen DK, Inohara N, Nicolae DL, Chen FF, Ramos R, Britton H, Moran T, Karaliuskas R, Duerr RH, Achkar JP, Brant SR, Bayless TM, Kirschner BS, Hanauer SB, Nunez G, Cho JH (2001) A frameshift mutation in NOD2 associated with susceptibility to Crohn's disease. Nature 411(6837):603–606. doi:10.1038/35079114

63. Oka T, Hikoso S, Yamaguchi O, Taneike M, Takeda T, Tamai T, Oyabu J, Murakawa T, Nakayama H, Nishida K, Akira S, Yamamoto A, Komuro I, Otsu K (2012) Mitochondrial DNA that escapes from autophagy causes inflammation and heart failure. Nature 485(7397):251–255. doi:http://www.nature.com/nature/journal/v485/n7397/abs/nature10992.html-supplementary-information
64. Orozco G, Eyre S, Hinks A, Bowes J, Morgan AW, Wilson AG, Wordsworth P, Steer S, Hocking L; UKRAG consortium, Thomson W, Worthington J, Barton A (2011) Study of the common genetic background for rheumatoid arthritis and systemic lupus erythematosus. Ann Rheum Dis 70(3):463–468. doi:10.1136/ard.2010.137174
65. Ouimet M, Franklin V, Mak E, Liao X, Tabas I, Marcel YL (2011) Autophagy regulates cholesterol efflux from macrophage foam cells via lysosomal acid lipase. Cell Metab 13(6):655–667. doi:10.1016/j.cmet.2011.03.023
66. Page N, Gros F, Schall N, Briand JP, Muller S (2011) A therapeutic peptide in lupus alters autophagic processes and stability of MHCII molecules in MRL/lpr B cells. Autophagy 7(5):539–540
67. Parkes M, Barrett JC, Prescott NJ, Tremelling M, Anderson CA, Fisher SA, Roberts RG, Nimmo ER, Cummings FR, Soars D, Drummond H, Lees CW, Khawaja SA, Bagnall R, Burke DA, Todhunter CE, Ahmad T, Onnie CM, McArdle W, Strachan D, Bethel G, Bryan C, Lewis CM, Deloukas P, Forbes A, Sanderson J, Jewell DP, Satsangi J, Mansfield JC, Wellcome Trust Case Control C, Cardon L, Mathew CG (2007) Sequence variants in the autophagy gene IRGM and multiple other replicating loci contribute to Crohn's disease susceptibility. Nat Genet 39(7):830–832. doi:10.1038/ng2061
68. Pengo N, Scolari M, Oliva L, Milan E, Mainoldi F, Raimondi A, Fagioli C, Merlini A, Mariani E, Pasqualetto E, Orfanelli U, Ponzoni M, Sitia R, Casola S, Cenci S (2013) Plasma cells require autophagy for sustainable immunoglobulin production. Nat Immunol 14(3):298–305. doi:10.1038/ni.2524
69. Petkova DS, Viret C, Faure M (2012) IRGM in autophagy and viral infections. Front Immunol 3:426. doi:10.3389/fimmu.2012.00426
70. Plantinga TS, Crisan TO, Oosting M, van de Veerdonk FL, de Jong DJ, Philpott DJ, van der Meer JW, Girardin SE, Joosten LA, Netea MG (2011) Crohn's disease-associated ATG16L1 polymorphism modulates pro-inflammatory cytokine responses selectively upon activation of NOD2. Gut 60(9):1229–1235. doi:10.1136/gut.2010.228908
71. Puleston DJ, Simon AK (2014) Autophagy in the immune system. Immunology 141(1):1–8. doi:10.1111/imm.12165
72. Qin Y, Chen Y, Wang W, Wang Z, Tang G, Zhang P, He Z, Liu Y, Dai SM, Shen Q (2014) HMGB1-LPS complex promotes transformation of osteoarthritis synovial fibroblasts to a rheumatoid arthritis synovial fibroblast-like phenotype. Cell Death Dis 5:e1077. doi:10.1038/cddis.2014.48
73. Quan W, Hur KY, Lim Y, Oh SH, Lee JC, Kim KH, Kim GH, Kim SW, Kim HL, Lee MK, Kim KW, Kim J, Komatsu M, Lee MS (2012) Autophagy deficiency in beta cells leads to compromised unfolded protein response and progression from obesity to diabetes in mice. Diabetologia 55(2):392–403. doi:10.1007/s00125-011-2350-y
74. Rahman MK, Midtling EH, Svingen PA, Xiong Y, Bell MP, Tung J, Smyrk T, Egan LJ, Faubion WA Jr (2010) The pathogen recognition receptor NOD2 regulates human FOXP3+ T cell survival. J Immunol 184(12):7247–7256. doi:10.4049/jimmunol.0901479
75. Razani B, Feng C, Coleman T, Emanuel R, Wen H, Hwang S, Ting JP, Virgin HW, Kastan MB, Semenkovich CF (2012) Autophagy links inflammasomes to atherosclerotic progression. Cell Metab 15(4):534–544. doi:10.1016/j.cmet.2012.02.011
76. Remijsen Q, Berghe TV, Wirawan E, Asselbergh B, Parthoens E, De Rycke R, Noppen S, Delforge M, Willems J, Vandenabeele P (2011) Neutrophil extracellular trap cell death requires both autophagy and superoxide generation. Cell Res 21(2):290–304. doi:http://www.nature.com/cr/journal/v21/n2/suppinfo/cr2010150s1.html
77. Rioux JD, Xavier RJ, Taylor KD, Silverberg MS, Goyette P, Huett A, Green T, Kuballa P, Barmada MM, Datta LW, Shugart YY, Griffiths AM, Targan SR, Ippoliti AF, Bernard EJ, Mei

L, Nicolae DL, Regueiro M, Schumm LP, Steinhart AH, Rotter JI, Duerr RH, Cho JH, Daly MJ, Brant SR (2007) Genome-wide association study identifies new susceptibility loci for Crohn disease and implicates autophagy in disease pathogenesis. Nat Genet 39(5):596–604. doi:10.1038/ng2032
78. Rivera JF, Costes S, Gurlo T, Glabe CG, Butler PC (2014) Autophagy defends pancreatic beta cells from human islet amyloid polypeptide-induced toxicity. J Clin Invest 124(8):3489–3500. doi:10.1172/JCI71981
79. Romao S, Gannage M, Munz C (2013) Checking the garbage bin for problems in the house, or how autophagy assists in antigen presentation to the immune system. Semin Cancer Biol 23(5):391–396. doi:10.1016/j.semcancer.2013.03.001
80. Saitoh T, Fujita N, Hayashi T, Takahara K, Satoh T, Lee H, Matsunaga K, Kageyama S, Omori H, Noda T, Yamamoto N, Kawai T, Ishii K, Takeuchi O, Yoshimori T, Akira S (2009) Atg9a controls dsDNA-driven dynamic translocation of STING and the innate immune response. Proc Natl Acad Sci U S A 106(49):20842–20846. doi:10.1073/pnas.0911267106
81. Saitoh T, Fujita N, Jang MH, Uematsu S, Yang BG, Satoh T, Omori H, Noda T, Yamamoto N, Komatsu M, Tanaka K, Kawai T, Tsujimura T, Takeuchi O, Yoshimori T, Akira S (2008) Loss of the autophagy protein Atg16L1 enhances endotoxin-induced IL-1beta production. Nature 456(7219):264–268. doi:10.1038/nature07383
82. Shi CS, Shenderov K, Huang NN, Kabat J, Abu-Asab M, Fitzgerald KA, Sher A, Kehrl JH (2012) Activation of autophagy by inflammatory signals limits IL-1beta production by targeting ubiquitinated inflammasomes for destruction. Nat Immunol 13(3):255–263. doi:10.1038/ni.2215
83. Shin YJ, Han SH, Kim DS, Lee GH, Yoo WH, Kang YM, Choi JY, Lee YC, Park SJ, Jeong SK, Kim HT, Chae SW, Jeong HJ, Kim HR, Chae HJ (2010) Autophagy induction and CHOP under-expression promotes survival of fibroblasts from rheumatoid arthritis patients under endoplasmic reticulum stress. Arthritis Res Ther 12(1):R19. doi:10.1186/ar2921
84. Sorbara MT, Ellison LK, Ramjeet M, Travassos LH, Jones NL, Girardin SE, Philpott DJ (2013) The protein ATG16L1 suppresses inflammatory cytokines induced by the intracellular sensors Nod1 and Nod2 in an autophagy-independent manner. Immunity 39(5):858–873. doi:10.1016/j.immuni.2013.10.013
85. Tang D, Kang R, Coyne CB, Zeh HJ, Lotze MT (2012) PAMPs and DAMPs: signal 0 s that spur autophagyandimmunity.ImmunolRev249(1):158–175.doi:10.1111/j.1600-065X.2012.01146.x
86. Thachil E, Hugot JP, Arbeille B, Paris R, Grodet A, Peuchmaur M, Codogno P, Barreau F, Ogier-Denis E, Berrebi D, Viala J (2012) Abnormal activation of autophagy-induced crinophagy in Paneth cells from patients with Crohn's disease. Gastroenterology 142(5):1097–1099 e1094. doi:10.1053/j.gastro.2012.01.031
87. Towns R, Kabeya Y, Yoshimori T, Guo C, Shangguan Y, Hong S, Kaplan M, Klionsky DJ, Wiley JW (2005) Sera from patients with type 2 diabetes and neuropathy induce autophagy and colocalization with mitochondria in SY5Y cells. Autophagy 1(3):163–170
88. Travassos LH, Carneiro LA, Ramjeet M, Hussey S, Kim YG, Magalhaes JG, Yuan L, Soares F, Chea E, Le Bourhis L, Boneca IG, Allaoui A, Jones NL, Nunez G, Girardin SE, Philpott DJ (2010) Nod1 and Nod2 direct autophagy by recruiting ATG16L1 to the plasma membrane at the site of bacterial entry. Nat Immunol 11(1):55–62. doi:10.1038/ni.1823
89. van Heel DA, Ghosh S, Butler M, Hunt KA, Lundberg AM, Ahmad T, McGovern DP, Onnie C, Negoro K, Goldthorpe S, Foxwell BM, Mathew CG, Forbes A, Jewell DP, Playford RJ (2005) Muramyl dipeptide and toll-like receptor sensitivity in NOD2-associated Crohn's disease. Lancet 365(9473):1794–1796. doi:10.1016/S0140-6736(05)66582-8
90. Vessoni AT, Filippi-Chiela EC, Menck CF, Lenz G (2013) Autophagy and genomic integrity. Cell Death Differ 20(11):1444–1454. doi:10.1038/cdd.2013.103
91. Villani A-C, Lemire M, Fortin G, Louis E, Silverberg MS, Collette C, Baba N, Libioulle C, Belaiche J, Bitton A, Gaudet D, Cohen A, Langelier D, Fortin PR, Wither JE, Sarfati M, Rutgeerts P, Rioux JD, Vermeire S, Hudson TJ, Franchimont D (2009) Common variants in the NLRP3 region contribute to Crohn's disease susceptibility. Nat Genet 41(1):71–76. doi:http://www.nature.com/ng/journal/v41/n1/suppinfo/ng.285_S1.html

92. Wehkamp J, Harder J, Weichenthal M, Schwab M, Schäffeler E, Schlee M, Herrlinger KR, Stallmach A, Noack F, Fritz P, Schröder JM, Bevins CL, Fellermann K, Stange EF (2004) NOD2 (CARD15) mutations in Crohn's disease are associated with diminished mucosal α-defensin expression. Gut 53(11):1658–1664. doi:10.1136/gut.2003.032805
93. Wen H, Gris D, Lei Y, Jha S, Zhang L, Huang MT, Brickey WJ, Ting JP (2011) Fatty acid-induced NLRP3-ASC inflammasome activation interferes with insulin signaling. Nat Immunol 12(5):408–415. doi:10.1038/ni.2022
94. Wildenberg ME, Vos AC, Wolfkamp SC, Duijvestein M, Verhaar AP, Te Velde AA, van den Brink GR, Hommes DW (2012) Autophagy attenuates the adaptive immune response by destabilizing the immunologic synapse. Gastroenterology 142(7):1493–1503 e1496. doi:10.1053/j.gastro.2012.02.034
95. Yang W, Tang H, Zhang Y, Tang X, Zhang J, Sun L, Yang J, Cui Y, Zhang L, Hirankarn N, Cheng H, Pan HF, Gao J, Lee TL, Sheng Y, Lau CS, Li Y, Chan TM, Yin X, Ying D, Lu Q, Leung AM, Zuo X, Chen X, Tong KL, Zhou F, Diao Q, Tse NK, Xie H, Mok CC, Hao F, Wong SN, Shi B, Lee KW, Hui Y, Ho MH, Liang B, Lee PP, Cui H, Guo Q, Chung BH, Pu X, Liu Q, Zhang X, Zhang C, Chong CY, Fang H, Wong RW, Sun Y, Mok MY, Li XP, Avihingsanon Y, Zhai Z, Rianthavorn P, Deekajorndej T, Suphapeetiporn K, Gao F, Shotelersuk V, Kang X, Ying SK, Zhang L, Wong WH, Zhu D, Fung SK, Zeng F, Lai WM, Wong CM, Ng IO, Garcia-Barcelo MM, Cherny SS, Shen N, Tam PK, Sham PC, Ye DQ, Yang S, Zhang X, Lau YL (2013) Meta-analysis followed by replication identifies loci in or near CDKN1B, TET3, CD80, DRAM1, and ARID5B as associated with systemic lupus erythematosus in Asians. Am J Human Genet 92(1):41–51. doi:10.1016/j.ajhg.2012.11.018
96. Yang Z, Fujii H, Mohan SV, Goronzy JJ, Weyand CM (2013) Phosphofructokinase deficiency impairs ATP generation, autophagy, and redox balance in rheumatoid arthritis T cells. J Exp Med 210(10):2119–2134. doi:10.1084/jem.20130252
97. Yao Q (2013) Nucleotide-binding oligomerization domain containing 2: structure, function, and diseases. Semin Arthritis Rheum 43(1):125–130. doi:http://dx.doi.org/10.1016/j.semarthrit.2012.12.005
98. Zhou R, Yazdi AS, Menu P, Tschopp J (2011) A role for mitochondria in NLRP3 inflammasome activation. Nature 469(7329):221–225. doi:10.1038/nature09663
99. Zhou XJ, Lu XL, Lv JC, Yang HZ, Qin LX, Zhao MH, Su Y, Li ZG, Zhang H (2011) Genetic association of PRDM1-ATG5 intergenic region and autophagy with systemic lupus erythematosus in a Chinese population. Ann Rheum Dis 70(7):1330–1337. doi:10.1136/ard.2010.140111

Xenophagy: Autophagy in Direct Pathogen Elimination

Eun-Kyeong Jo, Jin Kyung Kim, and Seungmin Hwang

Abstract Autophagy is an intracellular catabolic pathway that delivers unwanted cytoplasmic materials to the lysosome for degradation. The degradative autophagy pathway has been shown to play essential roles not only in cellular homeostasis but also in cell-autonomous immune defense against invading pathogens. Recent data further showed that the contribution of autophagy to the host defense is more than simple degradation of the invaders, indicating novel nondegradative roles of the autophagy-related genes. In this chapter, we review recent advances in the studies on the role of the autophagy pathway/genes and selective autophagy receptors in defending the host against microbial infection. In addition, the crosstalk between pathogen recognition receptors, especially the toll-like receptors, and the autophagy pathway during pathogen infections is discussed. Understanding the role of autophagy in controlling intracellular pathogens will lead to a better chance of developing new therapeutic treatments for infectious diseases.

Abbreviations

3-MA	3-methyladenine
AMPK	AMP-activated protein kinase
ATGs	Autophagy-related genes
Bcl-2	B-cell lymphoma 2
FcγR	Fcγ receptor
FIP200	Focal Adhesion Kinase interacting protein of 200 kDa
GABARAP	γ-aminobutyric acid receptor-associated protein

E.-K. Jo (✉) • J.K. Kim
Department of Microbiology, Infection Signaling Network Research Center, College of Medicine, Chungnam National University, 6 Munhwa-dong, Jungku, Daejeon 301-747, South Korea
e-mail: hayoungj@cnu.ac.kr; pcjlovesh6@naver.com

S. Hwang
Department of Pathology, The University of Chicago, 924 E. 57th Street BSLC R110, Chicago, IL 60637, USA
e-mail: shwang@bsd.uchicago.edu

M.C. Maiuri, D. De Stefano (eds.), *Autophagy Networks in Inflammation*, Progress in Inflammation Research, DOI 10.1007/978-3-319-30079-5_7

GABARAP-like 1	GABARAPL1
GAS	Group A *Streptococcus*
GATE-16	Golgi-associated ATPase enhancer of 16 kDa
HECT	Homologous to the E6-AP Carboxyl Terminus
HSV	Herpes simplex virus
IL	Interleukin
KEAP1	Kelch-like ECH-associated protein 1
LC3	Microtubule-associated protein 1 light chain 3
LGALS8	Sugar receptor galectin-8
LIR	LC3-interacting region
MAVS	Mitochondrial antiviral signaling protein
miRNA	MicroRNA
MTOR	Mechanistic target of rapamycin
MyD88	Myeloid differentiation primary response 88
NADPH	Nicotinamide adenine dinucleotide phosphate
NBR1	Neighbor Of BRCA1 Gene 1
NDP52	Nuclear dot protein 52 kDa
NF-κB	Nuclear factor-κB
NLRC4	NLR family CARD domain-containing protein 4
NLRP3	NOD-like receptor family, pyrin domain containing 3
NLRX1	Nucleotide-binding oligomerization domain, leucine rich repeat containing X1
NOX2	NADPH Oxydase 2
NRF2	Nuclear factor erythroid 2-related factor
OPTN	Receptor optineurin
p62/SQSTM1	Sequestosome 1
PAI-2	Plasminogen activator inhibitor type 2
PPARGC1A	Peroxisome proliferator-activated receptor-gamma, coactivator 1α
Rheb	Ras homolog enriched in brain
RIG-1	Retinoic acid-inducible gene 1
ROS	Reactive oxygen species
SCV	Salmonella-containing vacuole
SLO	Pore-forming toxin streptolysin O
SMURF1	SMAD specific E3 ubiquitin protein ligase 1
TBK1	TANK-binding kinase
TLRs	Toll-like receptors
TNF	Tumor necrosis factor
TRIF	TIR-domain-containing adapter-inducing interferon-β
TUFM	Tu translation elongation factor, mitochondrial, also known as EF-TuMT
UBA	Ubiquitin-associated domain
ULK1	UNC-51-like kinase1
VDR	Vitamin D receptor.

1 Introduction

Autophagy is an evolutionarily conserved pathway that delivers cytoplasmic materials to the lysosome for degradation [44]. Depending on how the materials are delivered, the autophagy pathway can be categorized as chaperone-mediated autophagy (direct translocation across the lysosomal membrane through chaperone proteins), microautophagy (direct engulfment by the lysosome), or macroautophagy. Macroautophagy is the major autophagy pathway, which sequesters and transports a portion of cytosol to the lysosome in a double-membrane bound vesicle called autophagosome [44]. Any cytoplasmic material can be the cargo of the autophagosome. Through this catabolic pathway, the cells maintain homeostasis and meet the demand for essential metabolites under various cellular and metabolic stress conditions, like nutrient deprivation and accumulation of damaged organelles [51]. Thus, autophagy generally functions as a survival mechanism maintaining the integrity of cells by recycling essential metabolites and clearing cellular debris [23]. The critical role of the autophagy pathway has been demonstrated in a wide array of physiologic and pathologic conditions, such as immunity, inflammation, cell survival and death, lifespan extension, cell differentiation, etc [43, 44, 77]. Macroautophagy (henceforth autophagy) can target bulk cytoplasmic materials nonselectively or specific substrates selectively for lysosomal degradation [38]. Invading pathogens in cytoplasm can be the cargo of the selective autophagy pathway, which is mediated through a subset of autophagy receptors including p62/SQSTM1, NBR1, NDP52, and Optineurin. These are all ubiquitin-binding proteins that selectively link the ubiquitinated autophagic cargos to the autophagy machinery for the delivery to the lysosome and subsequent degradation of the cargos [40, 69]. In addition, recent studies have shown novel nondegradative functions of the autophagy-related genes in the host defense against intracellular pathogens. Thus, autophagy can play crucial roles in the host defense as a degradative pathway and nondegradative effectors. Recent reports have also demonstrated, however, that many pathogens have evolved their unique strategies to escape from both degradative and nondegradative autophagic elimination and to usurp the autophagy pathway/genes for their own benefits [3, 5, 18, 20, 28, 78]. During the infection of intracellular pathogens, the cell-autonomous immune system senses the specific molecular patterns present in the microbes or their components. The recognition of distinct microbial patterns by innate receptors not only leads to the activation of innate immunity, but also promotes the induction of acquired immune responses [84]. Numerous data have revealed the connections between the immune system and the autophagy pathway at the multiple levels [72]. Importantly, toll-like receptors (TLRs) are linked to the activation of the autophagy pathways, thus functioning as a cell-autonomous antimicrobial defense [2].

In this chapter, we focus on the key aspects of xenophagy (direct elimination of pathogens through autophagy) process at the molecular and cellular levels. We detail the recent discoveries that provide new insights into the role of xenophagy in eliminating intracellular pathogens and activating the cell-autonomous defense system. We also describe how selective autophagy is activated and regulated during intracellular microbial infection and discuss the association of innate immunity molecules and pathways with the activation of xenophagy. Numerous strategies of bacterial pathogens to evade the xenophagic elimination have been recently reviewed in detail [28, 60, 64], so will not be discussed here.

2 Role of Autophagy in Host Defense Against Intracellular Microbes

The generation of the double-membrane bound autophagosome is the hallmark of the autophagy pathway and 38 autophagy-related genes (ATGs) have been identified so far to play essential roles in multiple steps of the autophagy pathway [1, 45, 53]. Sequestration via autophagosome and subsequent lysosomal degradation of invading pathogens has been the major contribution of the autophagy pathway to the cell-autonomous immune system. However, emerging studies have shown novel and important functions of those ATGs in the host defense against intracellular pathogens, some of which might be completely independent of their role in the degradative autophagy pathway. Further, it has become increasingly evident that the autophagy pathway is essentially selective degradation of specific substrates rather than nonselective bulk degradation process. Thus, we focus on the canonical and noncanonical roles of the ATGs and the function of selective autophagy receptors in controlling intracellular pathogens.

2.1 Overview of the Autophagy Pathway and Autophagy-Related Genes

During the last 25 years our understanding of the cellular and molecular mechanisms of the autophagy pathway has been expanded enormously, using diverse model systems like yeast, *C. elegans*, and mammalian cells. Autophagy dynamics can be divided into three steps; vesicle induction and initiation, formation (elongation) of double-membrane vesicle structure and cargo packaging, and the maturation of autophagosome through fusion with lysosomes. This process is regulated by a defined set of ATGs, but how they do so is still not completely understood [8]. In a simple model, environmental cues like starvation or depletion of growth factors activate an initiation complex of UNC-51-like kinase1 (ULK1), Atg13, Focal Adhesion Kinase interacting protein of 200 kDa (FIP200), and Atg101. This activation further stimulates a nucleation complex of Beclin1, Atg14L, and the class III phosphatidylinositol 3 Kinase (Vps34 and Vps15) to generate phosphatidylinositol-3-phsphate, promoting the nucleation of the autophagosomal membrane. The initiated autophagosomal membrane elongates further while capturing its cytosolic cargos and eventually closes to form an autophagosome. Lastly, the newly formed autophagosome matures through its fusion with endosomes and then finally with lysosomes to form an autolysosome, and the inner membrane and enclosed cytosolic cargoes are degraded by lysosomal enzymes [45].

The elongation and closure of autophagosomal membrane requires two ubiquitin-like conjugation systems of Atg12 and microtubule-associated protein 1 light chain 3 (LC3, a mammalian orthologue of yeast Atg8). Atg12 is conjugated to Atg5 through E1 activating enzyme, Atg7, and E2 conjugating enzyme, Atg10 [52]. The conjugate further binds to Atg16L1 to form an elongation complex. This Atg12-Atg5-Atg16L1 complex functions as E3 ligase for the conjugation of cytosolic LC3 to the phosphatidylethanolamine in growing autophagosomal membrane [35]. LC3 itself is processed for the conjugation by the sequential action of Atg4 homologs (proteases to expose the terminal glycine of LC3 for conjugation), E1 Atg7, and E2 Atg3 [74]. Further, there exist multiple homologs of LC3 in mammalian system, which include LC3A, LC3B, LC3C, GABARAP (γ-aminobutyric acid receptor-associated protein), GABARAP-like 1 (GABARAPL1), and GABARAPL2 (a.k.a. Golgi-associated ATPase enhancer of 16 kDa, GATE-16). The conjugation of these LC3 homologs, especially the LC3B, to the autophagosomal membranes has been an important read-out to monitor the formation of autophagosome and consequent autophagic activity [75]. In addition, the conjugation/localization of the LC3 homologs on different types of membranes, other than the autophagosome, has been reported [10, 16, 21, 47, 65]. Thus, the ubiquitin-like proteins and potentially the conjugation system itself, which is utilized for the generation of the autophagosome in the canonical autophagy pathway, can be used for other cellular pathways dealing with reorganization of intracellular membranes.

2.2 Degradative Role of the Autophagy Pathway Against Intracellular Pathogens

The phenomenon of autophagy was first described in mammalian cells by Dr. de Duve about 40 years ago [17], and infection-triggered autophagy was observed by Dr. Rikihisa nearly 30 years ago [67]. Indeed, the degradative autophagy pathway can be induced in response to pathogen infection and function as cell-autonomous defense against the infection [7, 18, 28]. The intracellular bacteria are classified as vacuolar if they reside within a membrane-bound compartment or cytosolic if they survive in the cytosol [36]. Certain bacteria either modify and stay inside phagosomal compartments or directly lyse and escape from the phagosome/vacuole to cytosol. In mycobacterial infection, activation of the autophagy pathway led to the colocalization of autophagy markers with nonfusogenic phagosomes, modified by *Mycobacterium tuberculosis*, and subsequent reduction in the intracellular survival of the mycobacteria [24]. Further, the autophagic machinery is essential in innate defense against Group A *Streptococcus* (GAS) infection by capturing the bacteria that escape from endosomes to cytosol [56]. So far, many studies revealed several mechanisms by which autophagy mediates the fusion of bacterial phagosomes with autophagosomes or the autophagic capture of bacteria after escaping into the cytoplasm in a variety of infection [44] (Fig. 1). However, intracellular pathogens have also developed various tactics to evade autophagic recognition and elimination as well as to manipulate the autophagy pathway for their own benefits [28, 36, 64].

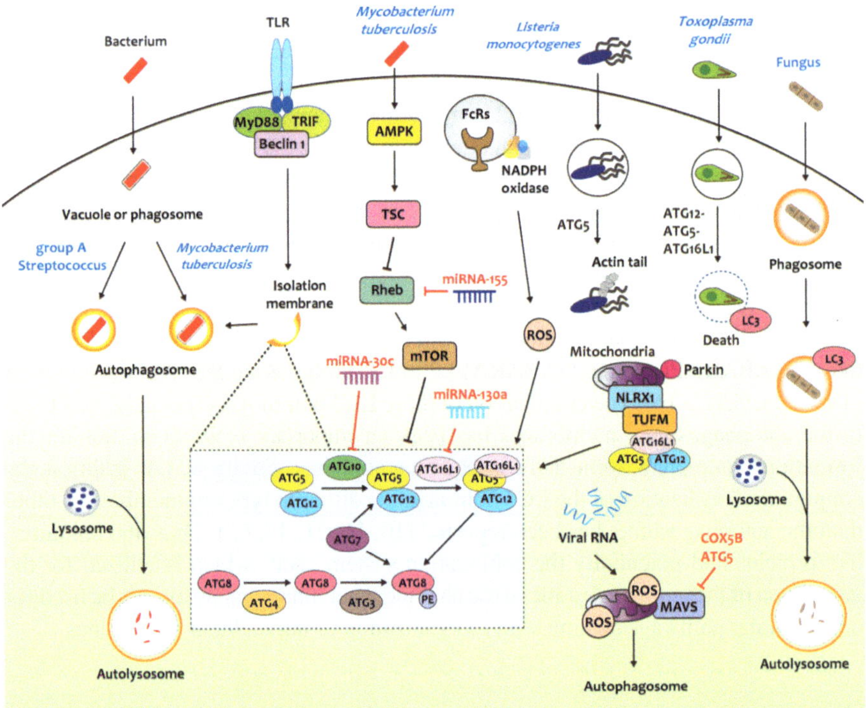

Fig. 1 Regulation of autophagy against intracellular microbes. After infection with the bacterium, *Mycobacterium tuberculosis* escapes from the vacuole (or phagosome) or group A Streptococcus resides in the cytosol. Autophagy eliminates the bacterium by generating the autolysosome. To elongate the autophagophore, many ATGs require to each step. Adaptor molecules of TLR (for example, MyD88 and TRIF) bind to Beclin1 and regulate autophagy. Activation of AMPK restricts *Mycobacterium tuberculosis* by inhibiting MTOR, negative regulator of autophagy. FcγR and NADPH oxidase-dependent ROS generation recruit to phagosomes. In addition, ATG5 is essential in elimination of *Listeria monocytogenes* and *Toxoplasma gondii* independent to autophagosome. Autophagy can target fungus-contained phagosome by recruiting LC3. NLRX1 interact with TUFM and ATG5-ATG12-ATG16L1 complex and subsequently promote virus-induced autophagy. By contrast, COX5B and ATG5 negatively regulate MAVS through the ROS formation during virus infection. Specific microRNAs (for example, miR-155, miR-30c and miR-130a) target each molecule, respectively, and repress the activation of autophagy

2.3 *Nondegradative Role of the ATGs Against Intracellular Pathogens*

The essential component(s) involved in the autophagy pathway have been shown to play key roles in host defense during various bacterial, viral, or parasitic infections. It was previously shown that an essential autophagy gene Atg5 is required for the host immune resistance against *Listeria monocytogenes* and *Toxoplasma gondii* but autophagosome *per se* is not involved in this resistance [94]. Recent studies further showed that the whole or part of the ubiquitin-like conjugation systems (E1 Atg7, E2 Atg3, E3 Atg12-Atg5-Atg16L1), but not the degradative activity, of the autophagy pathway is required to control the infection of *T. gondii* and *Chlamydia trachomatis* [10, 25, 61] (Fig. 1). Similarly, the E3 Atg12-Atg5-Atg16L1 complex is required for the interferon (IFN)-γ-mediated host defense against murine norovirus infection, but the induction of the degradative autophagy pathway or fusion between autophagosomes and lysosomes is not required [29]. Furthermore, the autophagic marker protein LC3 is recruited to the parasitophorus vacuole containing *T. gondii* and phagosomes containing fungal pathogen/signature and is associated with the disruption of the vacuole membrane and the fungicidal activity and subsequent regulation of cytokine production [10, 70, 81] (Fig. 1). These data suggest that the evolutionarily conserved autophagy proteins have evolved and acquired unique functions in the immune system, independent of their roles in the degradation pathway [3, 78].

2.4 *Positive Regulation of Autophagy Against Intracellular Pathogens*

Multiple mechanisms and molecules (other than ATGs) have been identified to activate the autophagy pathway against intracellular pathogens. Reactive oxygen species (ROS) is one of those inducers of the autophagy pathway against bacterial infection (Fig. 1). In macrophages, activation of either TLR or Fcγ receptor (FcγR) signaling during phagocytosis induces recruitment of the autophagy protein LC3 to phagosomes through NOX2 NADPH oxidase-dependent ROS generation [27]. Interestingly, in human epithelial cells, although these cells do not express NOX2, antibacterial autophagy is also dependent on ROS generation [27]. Mitochondrial proteins also play important roles in the activation of the autophagy pathway to promote innate antimicrobial activity. Parkin is a well-known ubiquitin ligase functioning in autophagic degradation of damaged mitochondria (mitophagy). The parkin plays essential roles not only in the ubiquitin-mediated autophagic degradation of *M. tuberculosis* but also in the host defenses against various intracellular bacterial infections, suggesting a link between mitophagy and the clearance of intracellular bacteria [46]. Similarly, NLRX1, a mitochondrial nucleotide-binding leucine-rich repeats (NLR)-containing protein, promoted autophagy during viral infection through its interaction with another mitochondrial protein TUFM (Tu translation elongation factor, mitochondrial, also known as EF-TuMT, COXPD4, and P43) [42] (Fig. 1). TUFM also plays a similar function to NLRX1 in promoting autophagy and inhibiting type I interferon production via the MAVS-DDX58 (RIG-I) pathway [42]. In addition, the activation of AMP-activated protein kinase (AMPK) is required for the restriction of intracellular *M. tuberculosis* by suppressing the mechanistic target of rapamycin (MTOR) and subsequently inducing the autophagy pathway in macrophages [91] (Fig. 1). Peroxisome proliferator-activated receptor-gamma, coactivator 1α (PPARGC1A) was found to be essential in this AMPK-mediated activation of the autophagy pathway through the induction of multiple ATG expression as well as antimicrobial activity [91]. In *Listeria* infection, the fascin1, an actin-bundling protein, was found to be important in autophagy activation and phagolysosomal fusion in dendritic cells for enhanced killing of intracellular *Listeria* through interaction between fascin1 and LC3 [49]. Moreover, the KEAP1 (kelch-like ECH-associated protein 1)-NRF2 (nuclear factor erythroid 2-related factor) signaling pathway, which is critical for the maintenance of the cellular redox balance in response to oxidative stress, is required for the expression of selective autophagy marker p62/SQSTM1 and the expression of the autophagy marker protein LC3B in HeLa cells [13]. However, the p62/SQSTM1 has multiple roles in diverse physiological and pathological responses [57]. Thus the results of modulating the p62/SQSTM1 should be carefully considered, because this molecule is required for the detection of autophagic flux through degradation and also is important for the activation of selective autophagy [34]. The role of p62/SQSTM1 in selective autophagy will be discussed in detail in the later part of this chapter.

2.5 Negative Regulation of Autophagy Against Intracellular Pathogens

Novel inhibitory factors were also identified in the regulation of antimicrobial responses through modulation of autophagy and ROS production. The mitochondrial adaptor protein MAVS, an essential molecule for antiviral response, is negatively regulated by the cytochrome c oxidase complex subunit COX5B and Atg5 through the repression of ROS production [93] (Fig. 1). Axin, a negative regulator of the Wnt signaling pathway, was also shown to inhibit the autophagy-mediated suppression of herpes simplex virus (HSV) replication in L929 cells [9]. Induction of the autophagy pathway by rapamycin suppresses HSV replication, while inhibiting the autophagy pathway using 3-MA and Beclin-1 knockdown facilitates viral replication [9]. MicroRNA (miRNA) also got involved in the regulation of autophagy to control intracellular bacterial infection. Induced miRNA-155 in macrophages upon mycobacteria infection activates the autophagy pathway through the inhibition of Ras homologue enriched in brain (Rheb), a negative regulator of autophagy (Fig. 1). This leads to the maturation of the phagosome containing mycobacteria and consequent killing of the intracellular mycobacteria [87]. In contrast, adherent-invasive *Escherichia coli*, which colonizes the ileal mucosa and invades intestinal epithelial cells, upregulates miRNA-30c and miRNA-130a to modulate the levels of ATG5 and ATG16L1. The modulation of the autophagy proteins by the miRNAs inhibits the autophagy pathway, promoting the survival of the bacteria and aggravating inflammatory response [58].

3 Selective Autophagy Receptors Against Intracellular Microbes

Earlier studies proposed that autophagy is a bulk process without selectivity. However, recent studies have highlighted the selective function of the autophagy pathway to exclusively eliminate specific targets, especially intracellular invaders. The protective role of the selective autophagy pathway is now well appreciated in cell-autonomous immune defense against intracellular pathogens, and we just began to understand the regulation and mechanism of selective autophagy activation during numerous pathogen infections [86]. The ubiquitination machinery and the 26S proteasome are required for the regulated degradation of short-lived cellular proteins and the maintenance of protein quality control [26]. The ubiquitination is also a selective degradation signal for the targeting of intracellular microbes to the degradative autophagy pathway. Now it becomes clear that these two degradation systems are not separated but rather there is an active crosstalk between the proteasome-mediated degradation and the selective autophagy pathway [41]. Selective autophagy can target a wide range of cargos including long-lived proteins, damaged organelles, and intracellular microbes [86]. Keys to understand the immune function of the selective autophagy pathway are the autophagy receptors, which are critical for directing autophagy machinery toward ubiquitinated cargos in response to 'eat-me' signals [86]. The autophagy receptors have the unique structures containing both the ubiquitin-binding domain and LC3-interacting region (LIR) motif. The ubiquitinated proteins or bacteria can be recognized by the autophagy receptors that can bind to both ubiquitin via ubiquitin-binding domains and LC3 via LIR motif [45]. The LIR motif ensures the targeting of the cargo via the receptors to LC3 (or other ATG8 family proteins) anchored in the autophagosomal membrane. Numerous efforts have been made to identify various autophagy receptors that function in the recognition and targeting of pathogens to autophagic machinery [89]. In this session, we will briefly describe the recent findings of the autophagy receptors that target invading bacteria, mainly four selective autophagy receptors, p62/SQSTM1, NDP52, NBR1, and optineurin, and the mechanisms by which selective autophagy activation destroy the intracellular microbes (Fig. 2).

3.1 p62/SQSTM1

p62 [also known as sequestosome-1 (SQSTM1)] is the first identified autophagy receptor [4, 37, 63] and is composed of a C-terminal ubiquitin-associated domain (UBA) that binds to poly-ubiquitinated proteins and a LIR motif required for LC3 interaction [57] (Fig. 2). p62 is also a signaling adaptor with a multidomain structure activating the nuclear factor (NF)-κB to control cell death and survival as well as inflammation [48, 54]. Interestingly, p62 plays a dual role as both autophagy receptor and autophagic substrate for the degradation by selective autophagy [4, 63].

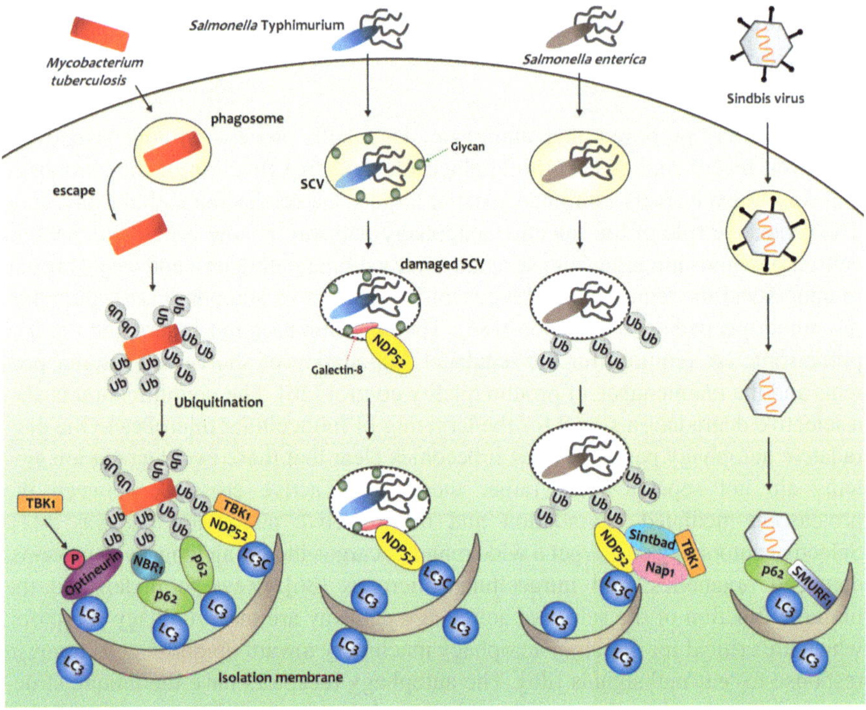

Fig. 2 Selective autophagy receptors are required for elimination microbes. Some bacteria escape from phagosome and are ubiquitinated. Ubiquitination recruits adaptor proteins, including p62, NDP52, NBR1 and optineurin by interacting with ubiquitin, and they bind to LC3. *Salmonella Typhimurium* resides in a Salmonella-containing vacuole (SCV) and glycans are exposed to the cytoplasm by damaged SCV. Galectin-8 detects glycans and activates autophagy by recruiting NDP52. *Salmonella enterica* also fails to maintain the membrane of SCV and is targeted for ubiquitination. NDP52 contributes to remove ubiquitinated-SCV through an interaction with Sintbad-Nap1-TBK1 complex. After Sindbis virus infection, p62 recognizes virus capsid protein and links with SMURF1, autophagy receptor, resulting in phagopore formation

Earlier studies showed that p62 is recruited to *Salmonella Typhimurium*, which are associated with ubiquitinated proteins, targeting the bacteria to the autophagy system [95]. Similar function of p62 was also reported in targeting *Shigella* to the autophagy pathway, and the p62-mediated selective autophagy was enhanced by TNF-α, a host proinflammatory cytokine, to restrict the survival of *Shigella* [55]. Intriguingly, a previous study showed that p62 is crucial for the delivery of specific ribosomal and bulk ubiquitinated cytosolic proteins to autolysosomes in order to produce neo-antimicrobial peptides that kill intracellular *M. tuberculosis* [66]. In addition, together with another autophagy receptor NDP52, p62 have been found to mediate the selective clearance of protein aggregation (aggrephagy).

In Sindbis virus infection, p62 is also essential for the control of viral infection via selective autophagy by linking the Sindbis virus capsid protein to the selective autophagy machinery [80] (Fig. 2). Another selective autophagy receptor SMURF1, a HECT-domain containing E3 ligase, was identified by genome-wide small interfering RNA screen and found to interact with p62. SMURF1 is required for the autophagosomal targeting of Sindbis and herpes simplex viruses and for the clearance of damaged mitochondria, thus functioning in both virophagy and mitophagy [62].

3.2 NDP52 (Nuclear Dot Protein 52 kDa)

NDP52 is responsible for the recognition of ubiquitin-coated *Salmonella enterica* in human cells, through its interaction with the adaptors Nap1 and Sintbad, and it activates the autophagy system to restrict the infection [82] (Fig. 2). The NDP52 is phosphorylated and activated by the TANK-binding kinase (TBK1), which is importantly involved in innate immunity, and play crucial roles in the autophagic clearance of *Salmonella enterica* [82]. TBK1 is also able to phosphorylate the other autophagy receptor optineurin (OPTN), leading to enhanced interaction of OPTN with the LC3 homologs [88]. NDP52 was also reported to be specifically associated with LC3C by a noncanonical LIR domain for efficient antibacterial autophagy [85]. The sugar receptor galectin-8 (a.k.a. LGALS8) is involved in the activation of the antibacterial autophagy pathway and contributes to the restriction of *Salmonella* proliferation by recruiting the NDP52 [83]. A recent study has revealed the key structural determinants of the interaction between NDP52 and galectin-8 and has defined a spatial complex formation of dimeric NDP52 with two monomeric galectin-8 molecules as well as two LC3C molecules [33]. The earliest findings of the defensive role of autophagy was described in nonphagocytic cells against pathogenic GAS that escape from endosomes into the host cell cytoplasm [56]. The ubiquitin (Ub)-NDP52-LC3 pathway is essential for targeting GAS to the autophagy system [60]. Moreover, the pore-forming toxin streptolysin O (SLO) from GAS is required for the activation of xenophagy in epithelial cells and in pharyngeal keratinocytes. However, the coordinated action of the SLO and the co-toxin NADase inhibits the maturation step of GAS-containing autophagosomes, thereby blocking autophagic killing of GAS in pharyngeal cells [59]. Further, the function of NDP52 can be modulated by bacterial effectors that target the ubiquitin system. The OTU-like domain containing protein CCA00261 of *Chlamydia caviae* is a type III secretion effector and putative deubiquitinase. It can target both ubiquitin and NDP52 and mediate the rapid clearance of ubiquitinated proteins, thus explaining bacterial pathogenesis during infection [22].

3.3 NBR1

NBR1 has its own PB1 domain to interact with p62 as well as the UBA domain and LIR motif, for the autophagic degradation of ubiquitinated proteins [39, 40] (Fig. 2). The intracellular bacterium *Francisella tularensis* has a unique characteristics of rapid phagosomal escape and proliferation in host cytoplasm while avoiding autophagic elimination. A recent study showed that a replication-deficient, ΔdipA mutant of *F. tularensis* is captured from the cytosols into double-membrane bound autophagosomes and cleared by the selective autophagy system, through the autophagy receptors p62 and NBR1 [11].

3.4 Optineurin

The phosphorylation of several autophagy receptors is crucial for the activation of selective autophagy. The phosphorylated optineurin was found to promote antibacterial autophagy of ubiquitin-coated *Salmonella enterica*. For this autophagic recognition, TBK1-mediated phosphorylation of optineurin on serine-177 is essential in the enhancement of its LC3-binding affinity and autophagic elimination of cytosolic *Salmonella* [88] (Fig. 2). The structural basis for this increased interaction of the phosphorylated optineurin with LC3 homologs has been elucidated [68]. In addition, the clinical relevance of optineurin has been suggested in the transcriptome study of macrophages from Crohn's disease patients and control groups. This study has shown that the expression of optineurin is significantly reduced in approximately 10 % of the Crohn's disease patients and that the reduced optineurin expression is accompanied with diminished cytokine secretion by macrophages from these patients [76].

4 The Association of Innate Immunity and Xenophagy Activation

In response to pathogen invasion, the host innate immune sensors alarm signals that activate a variety of immune defense system. Toll-like receptors (TLRs) are one of the best characterized sensors and play major roles in detecting specific molecular patterns of various pathogens. TLRs trigger intracellular signaling events involving the ubiquitination of various signaling molecules and recent studies have revealed a molecular link between TLR-mediated innate immunity and the induction of autophagy during infection [30, 79, 92]. In addition, autophagy activation contributes to suppression and subsequent "fine-tuning" of inflammatory responses [32, 79]. This function of autophagy in the control of excessive inflammation might be due to the fundamental ability of the autophagy pathway in maintaining the mitochondrial integrity or in removing aggregated signaling proteins and inflammasome activators [50]. In this session, we briefly discuss the current knowledge regarding the role of xenophagy in TLR-induced innate immunity and the crosstalk between TLR- and autophagic signaling. Instead of providing a comprehensive review, we just highlight a couple of new important issues, because the link between autophagy and TLR signaling has been addressed already in other literatures extensively.

4.1 The Role of Xenophagy in TLR-Mediated Innate Immune Responses During Infection

During the infection of intracellular bacteria or viruses, the innate immune cells recognize pathogen-associated molecular patterns through specialized innate immune receptors like TLRs. The activation of TLR-dependent signaling in the infected cells leads to the production of NF-κB-dependent inflammatory cytokines and antimicrobial proteins [15, 19]. The TLR signaling simultaneously leads to the activation of antibacterial autophagy. Earlier studies showed that a variety of TLR stimulation induces autophagy activation, resulting in anti-microbial responses in phagocytic cells [14, 70, 90]. Moreover, TLR8 activation was found to induce the expression of the human cathelicidin and to activate the vitamin D receptor (VDR) signaling, leading to the conversion of the inactive form of vitamin D, 25-hydroxycholecalciferol, into its active metabolite [6]. It is recently shown that TLR8 agonists control HIV infection through a vitamin D- and cathelicidin-dependent autophagy activation [6]. The innate immune system should be tightly controlled by multiple regulators and/or mechanisms, because the over-activation of signaling cascade can lead to excessive and harmful responses to the host. Although autophagy play essential effector functions during innate immune defense against pathogen infection [44], autophagy is also crucial for the proper control of inflammatory responses. For example, autophagy is enhanced by the stabilization of Beclin-1, mediated by the induced expression of plasminogen activator inhibitor type 2 (PAI-2) upon TLR activation. This further leads to the degradation of NOD-like receptor family, pyrin domain containing 3 (NLRP3) and subsequent reduction in the activation of inflammasome and IL-1β-driven inflammation [12]. Thus, autophagy plays an essential role in the PAI-2-mediated negative feedback loop of TLR signaling.

4.2 The Crosstalk Between TLR Signaling and the Autophagy Pathway

Critical components of the TLR signaling have been identified to interact directly with key players in the autophagy pathway. For example, TLR signaling adaptors MyD88 (myeloid differentiation primary response 88) and TRIF (TIR-domain-containing adapter-inducing interferon-β) associate with Beclin-1, a key molecule in autophagy activation. TLR signaling leads to the interaction of MyD88 and TRIF with Beclin-1 and this interaction dissociates the Beclin-1 from a negative regulator Bcl-2 and thus induce the autophagy pathway [71]. In mycobacterial infection, various mycobacterial ligands stimulate TLRs to induce intracellular signaling cascades involved in the activation of NF-κB and mitogen-activated protein kinase pathways [2]. Mycobacterial antigen-induced TLR signaling leads to the activation of functional vitamin D receptor (VDR) signaling, which eventually enhances antibacterial autophagy against *M. tuberculosis* [2, 73]. Recent studies have also shown that *Pseudomonas aeruginosa*-induced autophagy is mediated through TLR4 and its adaptor TRIF in macrophages. The cleavage of TRIF by caspase-1 activated upon NLRC4 inflammasome formation leads to the inhibition of autophagy and type I interferon secretion. These data suggest that an essential TLR adaptor TRIF plays a key role in the activation of autophagy and the production of type I interferon during *P. aeruginosa* infection of macrophages [31].

5 Concluding Remarks and Future Directions

Considerable progress has been made in elucidating the roles of autophagy and their functional mechanisms in the cell-autonomous immune defense system against pathogen invasion. Recent reports suggest that both selective and nonselective autophagy pathways are important in the host defense against intracellular pathogens. Now it is also clear that both degradative activity of the autophagy pathway and nondegradative functions of the autophagy-related genes are required for the effective control of intracellular pathogens. However, numerous studies have also highlighted diverse mechanisms of pathogens to evade such anti-pathogen functions of the autophagy pathway/genes. Understanding the mechanistic basis for the role of the autophagy pathway/genes during pathogen infection is a challenge with substantial clinical impact, since restoration or enhancement of those autophagic functions is likely to be of great importance with gaining insights of new therapeutic modalities. Knowledge from the recent studies on the negative regulation of autophagy also offers the potential possibility of developing new strategies against infection and inflammation. Further, emerging data on the mechanisms of linking innate immune function to autophagy during microbial infection provides insights into xenophagy regulation. In conclusion, ever-expanding knowledge of the autophagic regulators and regulatory mechanisms during infection promises new avenues for harnessing and shaping innate host responses against pathogen infections.

References

1. Abada A, Elazar Z (2014) Getting ready for building: signaling and autophagosome biogenesis. EMBO Rep 15:839–852
2. Basu J, Shin DM, Jo EK (2012) Mycobacterial signaling through toll-like receptors. Front Cell Infect Microbiol 2:145. doi:10.3389/fcimb.2012.00145
3. Bestebroer J, V'kovski P, Mauthe M, Reggiori F (2013) Hidden behind autophagy: the unconventional roles of ATG proteins. Traffic 14:1029–1041
4. Bjørkøy G, Lamark T, Brech A, Outzen H, Perander M, Øvervatn A, Stenmark H, Johansen T (2005) p62/SQSTM1 forms protein aggregates degraded by autophagy and has a protective effect on huntingtin-induced cell death. J Cell Biol 171:603–614
5. Boyle KB, Randow F (2013) The role of 'eat-me' signals and autophagy cargo receptors in innate immunity. Curr Opin Microbiol 16:339–348

6. Campbell GR, Spector SA (2012) Vitamin D inhibits human immunodeficiency virus type 1 and *Mycobacterium tuberculosis* infection in macrophages through the induction of autophagy. PLoS Pathog 8:e1002689. doi:10.1371/journal.ppat.1002689
7. Cemma M, Brumell JH (2012) Interactions of pathogenic bacteria with autophagy systems. Curr Biol 22:R540–R545
8. Chen Y, Klionsky DJ (2010) The regulation of autophagy – unanswered questions. J Cell Sci 124:161–170
9. Choi EJ, Kee SH (2014) Axin expression delays herpes simplex virus-induced autophagy and enhances viral replication in L929 cells. Microbiol Immunol 58:103–111
10. Choi J, Park S, Biering SB, Selleck E, Liu CY, Zhang X, Fujita N, Saitoh T, Akira S, Yoshimori T et al (2014) The parasitophorous vacuole membrane of Toxoplasma gondii is targeted for disruption by ubiquitin-like conjugation systems of autophagy. Immunity 40:924–935
11. Chong A, Wehrly TD, Child R, Hansen B, Hwang S, Virgin HW, Celli J (2012) Cytosolic clearance of replication-deficient mutants reveals Francisella tularensis interactions with the autophagic pathway. Autophagy 8:1342–1356
12. Chuang SY, Yang CH, Chou CC, Chiang YP, Chuang TH, Hsu LC (2013) TLR-induced PAI-2 expression suppresses IL-1β processing via increasing autophagy and NLRP3 degradation. Proc Natl Acad Sci U S A 110:16079–16084
13. Darvekar SR, Elvenes J, Brenne HB, Johansen T, Sjøttem E (2014) SPBP is a sulforaphane induced transcriptional coactivator of NRF2 regulating expression of theautophagy receptor p62/SQSTM1. PLoS One 9:e85262. doi:10.1371/journal.pone.0085262
14. Delgado MA, Elmaoued RA, Davis AS, Kyei G, Deretic V (2008) Toll-like receptors control autophagy. EMBO J 27:1110–1121
15. Deretic V, Levine B (2009) Autophagy, immunity, and microbial adaptations. Cell Host Microbe 5:527–549
16. DeSelm CJ, Miller BC, Zou W, Beatty WL, van Meel E, Takahata Y, Klumperman J, Tooze SA, Teitelbaum SL, Virgin HW (2011) Autophagy proteins regulate the secretory component of osteoclastic bone resorption. Dev Cell 21:966–974
17. Deter RL, Duve CD (1967) Influence of glucagon, an inducer of cellular autophagy, on some physical properties of rat liver lysosomes. J Cell Biol 33:437–449
18. Dong X, Levine B (2013) Autophagy and viruses: adversaries or allies? J Innate Immun 5:480–493
19. Faure M, Lafont F (2013) Pathogen-induced autophagy signaling in innate immunity. J Innate Immun 5:456–470
20. Fentress SJ, Behnke MS, Dunay IR, Mashayekhi M, Rommereim LM, Fox BA, Bzik DJ, Taylor GA, Turk BE, Lichti CF et al (2010) Phosphorylation of immunity-related GTPases by a *Toxoplasma gondii*-secreted kinase promotes macrophage survival and virulence. Cell Host Microbe 8:484–495
21. Florey O, Kim SE, Sandoval CP, Haynes CM, Overholtzer M (2011) Autophagy machinery mediates macroendocytic processing and entotic cell death by targeting single membranes. Nat Cell Biol 13:1335–1343
22. Furtado AR, Essid M, Perrinet S, Balañá ME, Yoder N, Dehoux P, Subtil A (2013) The chlamydial OTU domain-containing protein *Chla*OTU is an early type III secretion effector targeting ubiquitin and NDP52. Cell Microbiol 15:2064–2079
23. Green DR, Levine B (2014) To be or not to be? How selective autophagy and cell death govern cell fate. Cell 157:65–75
24. Gutierrez MG, Master SS, Singh SB, Taylor GA, Colombo MI, Deretic V (2004) Autophagy is a defense mechanism inhibiting BCG and *Mycobacterium tuberculosis* survival in infected macrophages. Cell 119:753–766
25. Haldar AK, Piro AS, Pilla DM, Yamamoto M, Coers J (2014) The E2-like conjugation enzyme Atg3 promotes binding of IRG and Gbp proteins to *Chlamydia*- and *Toxoplasma*-containing vacuoles and host resistance. PLoS One 9:e86684. doi:10.1371/journal.pone.0086684
26. Hershko A, Ciechanover A (1998) The ubiquitin system. Annu Rev Biochem 67:425–479

27. Huang J, Canadien V, Lam GY, Steinberg BE, Dinauer MC, Magalhaes MA, Glogauer M, Grinstein S, Brumell JH (2009) Activation of antibacterial autophagy by NADPH oxidases. Proc Natl Acad Sci U S A 106:6226–6231
28. Huang J, Brumell JH (2014) Bacteria–autophagy interplay: a battle for survival. Nat Rev Microbiol 12:101–114
29. Hwang S, Maloney NS, Bruinsma MW, Goel G, Duan E, Zhang L, Shrestha B, Diamond MS, Dani A, Sosnovtsev SV et al (2012) Nondegradative role of Atg5-Atg12/Atg16L1 autophagy protein complex in antiviral activity of interferon gamma. Cell Host Microbe 11:397–409
30. Into T, Inomata M, Takayama E, Takigawa T (2012) Autophagy in regulation of Toll-like receptor signaling. Cell Signal 24:1150–1162
31. Jabir MS, Ritchie ND, Li D, Bayes HK, Tourlomousis P, Puleston D, Lupton A, Hopkins L, Simon AK, Bryant C et al (2014) Caspase-1 cleavage of the TLR adaptor TRIF inhibits autophagy and β-interferon production during Pseudomonas aeruginosa infection. Cell Host Microbe 15:214–227
32. Jo EK, Shin DM, Choi AM (2012) Autophagy: cellular defense to excessive inflammation. Microbes Infect 14:119–125
33. Kim BW, Hong SB, Kim JH, Kwon DH, Song HK (2013) Structural basis for recognition of autophagic receptor NDP52 by the sugar receptor galectin-8. Nat Commun 4:1613. doi:10.1038/ncomms2606
34. Klinosky DJ, Abdalla FC, Abeliovich H, Abraham RT, Acevedo-Arozena A, Adeli K, Agholme L, Agnello M, Agostinis P, Aguirre-Ghiso JA et al (2012) Guidelines for the use and interpretation of assays for monitoring autophagy. Autophagy 8:445–544
35. Klionsky DJ, Schulman BA (2014) Dynamic regulation of macroautophagy by distinctive ubiquitin-like proteins. Nat Struct Mol Biol 21:336–345
36. Knodler LA, Celli J (2011) Eating the strangers within: host control of intracellular bacteria via xenophagy. Cell Microbiol 13:1319–1327
37. Komatsu M, Wagun S, Koike M, Sou YS, Ueno T, Hara T, Mizushima N, Iwata J, Ezaki J, Murata S et al (2007) Homeostatic levels of p62 control cytoplasmic inclusion body formation in autophagy-deficient mice. Cell 131:1149–1163
38. Kraft C, Peter M, Hofmann K (2010) Selective autophagy: ubiquitin-mediated recognition and beyond. Nat Cell Biol 12:836–841
39. Lamark T, Perander M, Outzen H, Kristiansen K, Øvervatn A, Michaelsen E, Bjørkøy G, Johansen T (2003) Interaction codes within the family of mammalian Phox and Bem1p domain-containing proteins. J Biol Chem 278:34568–34581
40. Lamark T, Kirkin V, Dikic I, Johansen T (2009) NBR1 and p62 as cargo receptors for selective autophagy of ubiquitinated targets. Cell Cycle 8:1986–1990
41. Lamark T, Johansen T (2010) Autophagy: links with the proteasome. Curr Opin Cell Biol 22:192–198
42. Lei Y, Wen H, Yu Y, Taxman DJ, Zhang L, Widman DG, Swanson KV, Wen KW, Damania B, Moore CB et al (2012) The mitochondrial proteins NLRX1 and TUFM form a complex that regulates type I interferon and autophagy. Immunity 36:933–946
43. Levine B, Deretic V (2007) Unveiling the roles of autophagy in innate and adaptive immunity. Nat Rev Immunol 7:767–777
44. Levine B, Mizushima N, Virgin HW (2011) Autophagy in immunity and inflammation. Nature 469:323–335
45. Lippai M, Lőw P (2014) The role of the selective adaptor p62 and ubiquitin-like proteins in autophagy. Biomed Res Int 2014:832704. doi:10.1155/2014/832704
46. Manzanillo PS, Ayres JS, Watson RO, Collins AC, Souza G, Rae CS, Schneider DS, Nakamura K, Shiloh MU, Cox JS (2013) The ubiquitin ligase parkin mediates resistance to intracellular pathogens. Nature 501:512–516
47. Martinez J, Almendinger J, Oberst A, Ness R, Dillon CP, Fitzgerald P, Hengartner MO, Green DR (2011) Microtubule-associated protein 1 light chain 3 alpha (LC3)-associated phagocytosis is required for the efficient clearance of dead cells. Proc Natl Acad Sci U S A 108:17396–173401

48. Mathew R, Karp CM, Beaudoin B, Vuong N, Chen G, Chen HY, Bray K, Reddy A, Bhanot G, Gelinas C et al (2009) Autophagy suppresses tumorigenesis through elimination of p62. Cell 137:1062–1075
49. Matsumura F, Yamakita Y, Starovoytov V, Yamashiro S (2013) Fascin confers resistance to *Listeria* infection in dendritic cells. J Immunol 191:6156–6164
50. Meunier E, Dick MS, Dreier RF, Schürmann N, Broz DK, Warming S, Roose-Girma M, Bumann D, Kayagaki N, Takeda K et al (2014) Caspase-11 activation requires lysis of pathogen-containing vacuoles by IFN-induced GTPases. Nature 509:366–370
51. Mihalache CC, Simon HU (2012) Autophagy regulation in macrophages and neutrophils. Exp Cell Res 318:1187–1192
52. Mizushima N, Noda T, Yoshimori T, Tanaka Y, Ishii T, George MD, Klionsky DJ, Ohsumi M, Ohsumi Y (1998) A protein conjugation system essential for autophagy. Nature 395:395–398
53. Mizushima N, Yoshimori T, Ohsumi Y (2011) The role of Atg proteins in autophagosome formation. Annu Rev Cell Dev Biol 27:107–132
54. Moscat J, Diaz-Meco MT (2009) P62 at the crossroads of autophagy, apoptosis, and cancer. Cell 137:1001–1004
55. Mostowy S, Sancho-Shimizu V, Hamon MA, Simeone R, Brosch R, Johansen T, Cossart P (2011) p62 and NDP52 proteins target intracytosolic *Shigella* and *Listeria* to different autophagy pathways. J Biol Chem 286:26987–26995
56. Nakagawa I, Amano A, Mizushima N, Yamamoto A, Yamaguchi H, Kamimoto T, Nara A, Funao J, Nakata M, Tsuda K et al (2004) Autophagy defends cells against invading group A *Streptococcus*. Science 306:1037–1040
57. Nezis IP, Stenmark H (2012) p62 at the interface of autophagy, oxidative stresssignaling, and cancer. Antioxid Redox Signal 17:786–793
58. Nguyen HT, Dalmasso G, Müller S, Carrière J, Seibold F, Darfeuille-Michaud A (2014) Crohn's disease-associated adherent invasive Escherichia coli modulate levels of microRNAs in intestinal epithelial cells to reduce autophagy. Gastroenterology 146:508–519
59. O'Seaghdha M, Wessels MR (2013) Streptolysin O and its co-toxin NAD-glycohydrolase protect group A *Streptococcus* from Xenophagic killing. PLoS Pathog 9:e1003394. doi:10.1371/journal.ppat.1003394
60. Ogawa M, Yoshikawa Y, Mimuro H, Hain T, Chakraborty T, Sasakawa C (2011) Autophagy targeting of *Listeria monocytogenes* and the bacterial countermeasure. Autophagy 7:310–314
61. Ohshima J, Lee Y, Sasai M, Saitoh T, Ma JS, Kamiyama N, Matsuura Y, Pann-Ghill S, Hayashi M, Ebisu S et al (2014) Role of mouse and human autophagy proteins in IFN-γ-induced cell-autonomous responses against *Toxoplasma gondii.* J Immunol 192:3328–3335
62. Orvedahl A, Sumpter R Jr, Xiao G, Ng A, Zou Z, Tang Y, Narimatsu M, Gilpin C, Sun Q, Roth M et al (2011) Image-based genome-wide siRNA screen identifies selective autophagy factors. Nature 480:113–117
63. Pankiv S, Clausen TH, Lamark T, Brech A, Bruun JA, Outzen H, Øvervatn A, Bjørkøy G, Johansen T (2007) p62/SQSTM1 binds directly to Atg8/LC3 to facilitate degradation of ubiquitinated protein aggregates by autophagy. J Biol Chem 282:24131–24145
64. Pareja ME, Colombo MI (2013) Autophagic clearance of bacterial pathogens: molecular recognition of intracellular microorganisms. Front Cell Infect Microbiol 3:54. doi:10.3389/fcimb.2013.00054
65. Patel KK, Miyoshi H, Beatty WL, Head RD, Malvin NP, Cadwell K, Guan JL, Saitoh T, Akira S, Seglen PO et al (2013) Autophagy proteins control goblet cell function by potentiating reactive oxygen species production. EMBO J 32:3130–3144
66. Ponpuak M, Davis AS, Roberts EA, Delgado MA, Dinkins C, Zhao Z, Virgin HW 4th, Kyei GB, Johansen T, Vergne I et al (2010) Delivery of cytosolic components by autophagic adaptor protein p62 endows autophagosomes with unique antimicrobial properties. Immunity 32:329–341
67. Rikihisa Y (1984) Glycogen autophagosomes in polymorphonuclear leukocytes induced by rickettsiae. Anat Rec 208:319–327

68. Rogov VV, Suzuki H, Fiskin E, Wild P, Kniss A, Rozenknop A, Kato R, Kawasaki M, McEwan DG, Löhr F et al (2013) Structural basis for phosphorylation-triggered autophagic clearance of *Salmonella*. Biochem J 454:459–466
69. Rogov V, Dötsch V, Johansen T, Kirkin V (2014) Interactions between autophagy receptors and ubiquitin-like proteins form the molecular basis for selective autophagy. Mol Cell 53:167–178
70. Sanjuan MA, Dillon CP, Tait SW, Moshiach S, Dorsey F, Connell S, Komatsu M, Tanaka K, Cleveland JL, Withoff S et al (2007) Toll-like receptor signalling in macrophages links the autophagy pathway to phagocytosis. Nature 450:1253–1257
71. Shi CS, Kehrl JH (2008) MyD88 and Trif Target Beclin 1 to Trigger Autophagy in Macrophages. J Biol Chem 48:33175–33182
72. Shibutani ST, Yoshimori T (2014) Autophagosome formation in response to intracellular bacterial invasion. Cell Microbiol 16:1619–1626
73. Shin DM, Yuk JM, Lee HM, Lee SH, Son JW, Harding CV, Kim JM, Modlin RL, Jo EK (2010) Mycobacterial lipoprotein activates autophagy via TLR2/1/CD14 and a functional vitamin D receptor signaling. Cell Microbiol 12:1648–1665
74. Shpilka T, Weidberg H, Pietrokovski S, Elazar Z (2011) Atg8: an autophagy-related ubiquitin-like protein family. Genome Biol 12:226. doi:10.1186/gb-2011-12-7-226
75. Slobodkin MR, Elazar Z (2013) The Atg8 family: multifunctional ubiquitin-like key regulators of autophagy. Essays Biochem 55:51–64
76. Smith A, Sewell G, Levine A, Chew T, Dunne J, O'Shea N, Smith P, Harrison P, Macdonald C, Bloom S et al (2014) Disruption of macrophage pro-inflammatory cytokine release in Crohn's disease is associated with reduced optineurin expression in a subset of patients. Immunology 144:45–55. doi:10.1111/imm.12338
77. Sosa MS, Bragado P, Aguirre-Ghiso JA (2014) Mechanisms of disseminated cancer cell dormancy: an awakening field. Nat Rev Cancer 14:611–622
78. Subramani S, Malhotra V (2013) Non-autophagic roles of autophagy-related proteins. EMBO Rep 14:143–151
79. Sumpter R Jr, Levine B (2010) Autophagy and innate immunity: triggering, targeting and tuning. Semin Cell Dev Biol 21:699–711
80. Sumpter R Jr, Levine B (2011) Selective autophagy and viruses. Autophagy 7:260–265
81. Tam JM, Mansour MK, Khan NS, Seward M, Puranam S, Tanne A, Sokolovska A, Becker CE, Acharya M, Baird MA et al (2014) Dectin-1-dependent LC3 recruitment to phagosomes enhances fungicidal activity in macrophages. J Infect Dis 210:1844–1854
82. Thurston TL, Ryzhakov G, Bloor S, von Muhlinen N, Randow F (2009) The TBK1 adaptor and autophagy receptor NDP52 restricts the proliferation of ubiquitin-coated bacteria. Nat Immunol 10:1215–1221
83. Thurston TL, Wandel MP, von Muhlinen N, Foeglein A, Randow F (2012) Galectin 8 targets damaged vesicles for autophagy to defend cells against bacterial invasion. Nature 482:414–418
84. Trinchieri G, Sher A (2007) Cooperation of toll-like receptor signals in innate immune defence. Nat Rev Immunol 7:179–190
85. von Muhlinen N, Akutsu M, Ravenhill BJ, Foeglein Á, Bloor S, Rutherford TJ, Freund SM, Komander D, Randow F (2012) LC3C, bound selectively by a noncanonical LIR motif in NDP52, is required for antibacterial autophagy. Mol Cell 48:329–342
86. von Muhlinen N, Akutsu M, Ravenhill BJ, Foeglein Á, Bloor S, Rutherford TJ, Freund SM, Komander D, Randow F (2013) An essential role for the ATG8 ortholog LC3C in antibacterial autophagy. Autophagy 9:784–786
87. Wang J, Yang K, Zhou L, Minhaowu, Wu Y, Zhu M, Lai X, Chen T, Feng L, Li M et al (2013) MicroRNA-155 promotes autophagy to eliminate intracellular mycobacteria by targeting Rheb. PLoS Pathog 9:e1003697. doi:10.1371/journal.ppat.1003697
88. Wild P, Farhan H, McEwan DG, Wagner S, Rogov VV, Brady NR, Richter B, Korac J, Waidmann O, Choudhary C et al (2011) Phosphorylation of the autophagy receptor optineurin restricts *Salmonella* growth. Science 333:228–233
89. Wild P, McEwan DG, Dikic I (2014) The LC3 interactome at a glance. J Cell Sci 127:3–9

90. Xu Y, Jagannath C, Liu XD, Sharafkhaneh A, Kolodziejska KE, Eissa NT (2007) Toll-like receptor 4 is a sensor for autophagy associated with innate immunity. Immunity 27:135–144
91. Yang CS, Kim JJ, Lee HM, Jin HS, Lee SH, Park JH, Kim SJ, Kim JM, Han YM, Lee MS et al (2014) The AMPK-PPARGC1A pathway is required for antimicrobial host defense through activation of autophagy. Autophagy 10:785–802
92. Yuk JM, Yoshimori T, Jo EK (2012) Autophagy and bacterial infectious diseases. Exp Mol Med 44:99–108
93. Zhao Y, Sun X, Nie X, Sun L, Tang TS, Chen D, Sun Q (2012) COX5B regulates MAVS-mediated antiviral signaling through interaction with ATG5 and repressing ROS production. PLoS Pathog 8:e1003086. doi:10.1371/journal.ppat.1003086
94. Zhao Z, Fux B, Goodwin M, Dunay IR, Strong D, Miller BC, Cadwell K, Delgado MA, Ponpuak M, Green KG et al (2008) Autophagosome-independent essential function for the autophagy protein Atg5 in cellular immunity to intracellular pathogens. Cell Host Microbe 4:458–469
95. Zheng YT, Shahnazari S, Brech A, Lamark T, Johansen T, Brumell JH (2009) The adaptor protein p62/SQSTM1 targets invading bacteria to the autophagy pathway. J Immunol 183:5909–5916

Autophagy and Antigen Presentation

Christophe Macri and Justine D. Mintern

Abstract Macroautophagy (hereafter referred to as autophagy) is a eukaryotic catabolic pathway that eliminates intracellular contents and thereby provides cells with a source of nutrients. This fundamental pathway of cellular degradation also participates in immunity and inflammation. The capacity of autophagy to traffic, and to degrade, cytosolic protein is exploited by immune cells as a means to elicit antigen presentation. Here, we will summarize the current knowledge about the intersection of both classical and non-canonical autophagy with pathways of antigen presentation. While the role of autophagy in major histocompatibility complex class II (MHC II) presentation is well documented, autophagy is also implicated in endogenous MHC I presentation and MHC I cross-presentation. Thus, autophagy is a major effector of antigen presentation and has the capacity to impact host immunity and tolerance.

Abbreviations

3-MA	3-methyladenine
Ag85B	Antigen 85B
AMPKα1	Adenosine monophosphate activating kinase-α1
APC	Antigen presenting cell
Atg	Autophagy-related gene
BMDC	Bone marrow-derived dendritic cell
BMDM	Bone marrow-derived macrophage
CLIP	Class-II-associated li peptide
CMA	Chaperone-mediated autophagy
CRP	C-reactive protein
DALIS	Dendritic cell aggresome-like induce structure
DC	Dendritic cell

C. Macri • J.D. Mintern (✉)
Department of Biochemistry and Molecular Biology, Bio21 Molecular Science and Biotechnology Institute, The University of Melbourne,
30 Flemington Rd, Parkville, VIC 3010, Australia
e-mail: jmintern@unimelb.edu.au

M.C. Maiuri, D. De Stefano (eds.), *Autophagy Networks in Inflammation*,
Progress in Inflammation Research, DOI 10.1007/978-3-319-30079-5_8

DRibbles	Defective ribosomal products-containing blebs
EBNA1	Epstein-Barr virus nuclear antigen 1
EBV	Epstein-Barr virus
ENMA	Endosome-mediated autophagy
ER	Endoplasmic reticulum
gB	Glycoprotein B
GCN2	General control nonderepressible 2
GFP	Green fluorescent protein
HEL	Hen egg lysozyme
HIV	Human immunodeficiency virus
Hsc70	Heat shock cognate protein of 70 kDa
HSV	Herpes simplex virus
li	Invariant chain
Lamp-2a	Lysosomal-associated membrane protein 2a
LAP	LC3-associated phagocytosis
LC3	Microtubule-associated protein 1 light chain 3
MEF	Mouse embryonic fibroblasts
MHC	Major histocompatibility complex
MIIC	Major histocompatibility complex class II compartment
moDC	Monocyte-derived dendritic cell
MP1	Matrix protein 1
OVA	Ovalbumin
TAP	Transporter-associated with antigen processing
TCR	T cell receptor
TEC	Thymic epithelial cell
TIM-4	T cell immunoglobulin and mucin domain-containing molecule 4
TLR	Toll-like receptor
αAl2O3-OVA	Ovalbumin-conjugated αAl2O3

1 Introduction

Macroautophagy (referred to as autophagy below) is intracellular catabolic mechanism that is highly conserved in eukaryotes. This process starts with the formation of an isolation membrane that elongates to sequester cytosolic components within a double-membrane vesicle termed "autophagosome". Autophagosomes ultimately fuse with lysosomes to form autolysosomes where the luminal content is degraded. This ancient response is governed by more than thirty autophagy-related genes (*Atg*) that were initially discovered in unicellular eukaryotic cells, notably in yeast. Subsequent studies have identified orthologs genes in mammals [12, 49].

Initial reports described autophagy as an intracellular pathway critical for the preservation of cellular homeostasis. This pathway is constitutively active at basal levels and enables the removal of long-lived proteins and damaged organelles. Furthermore, autophagy constitutes a cellular stress response to numerous endogenous and exogenous stressors including starvation and chemical or metabolic dysfunction. In these settings, the cell upregulates autophagy to reduce potential damage and to promote its survival via the recycling of nutrients liberated by lysosomal proteolysis [40]. Over the last decade, the concept of autophagy has evolved to include a broader range of biological functions, one of which is the generation of an effective immune response [9, 34]. In particular, autophagy machinery is implicated in the process of antigen presentation [38]. Antigen presentation refers to the critical immune process that initiates an adaptive immune response that will ultimately protect mammalian organisms against pathogens. Antigen presentation encompasses the intracellular pathways that are implicated in the trafficking and processing of antigen that is required for its delivery and presentation by major histocompatibility complex (MHC) molecules. These pathways precede the recognition of antigenic peptide, presented in the context of MHC molecules, by antigen-specific T cells. An increasing number of studies have highlighted the intersection of autophagy with different antigen-presenting routes. In this case, the capacity of autophagy to transport and degrade intracellular constituents is utilized to effectively traffic and deliver antigens to compartments for MHC loading.

In this Chapter, the current knowledge of autophagy and its contribution to antigen presentation will be discussed. First, we will focus on MHC II that presents antigenic peptides to $CD4^+$ T lymphocytes. Second, we will describe the role of autophagy and its impact on MHC I presentation of antigen to $CD8^+$ T lymphocytes. We will highlight the multifactorial roles of autophagy in antigen presentation that contribute to generating effective immunity.

2 Autophagy in MHC II Presentation

2.1 Overview of the Classical MHC II Presentation Pathway

Antigen presentation is pivotal to initiating T cell immunity that is responsible for the active clearance of infectious pathogen from the host. Presentation of antigen by MHC II molecules is recognized by the T cell receptor (TCR) expressed by $CD4^+$ T lymphocytes. This subpopulation of T cells orchestrates the humoral and cellular immune response, largely via the secretion of a broad panel of cytokines. MHC II molecules are mostly restricted to a subset of cells, named antigen presenting cells (APCs), that comprise of dendritic cells (DCs), macrophages, B cells and epithelial cells [36].

Traditionally, the classical MHC II presentation pathway has been defined as a route that exclusively presents antigens from an extracellular source. In humans, MHC II molecules are encoded by three polymorphic genes, *HLA-DP*, *–DQ*, and *-DR*. During their synthesis, the α and β chains are assembled in the endoplasmic reticulum (ER), together with the invariant chain (li). Binding of li to newly synthesised MHC II molecules ensures the peptide-binding cleft is blocked preventing it from being loaded with peptides in the ER. Li also promotes the export of MHC II molecules to late endosomes and the MHC II loading compartment (MIIC). Proteases, including cathepsins, degrade li, leaving the remnant class-II-associated li peptide (CLIP) bound to MHC II molecules. The chaperone HLA-DM associates with this complex, facilitates the release of

CLIP and promotes binding of high affinity antigen-derived peptides. Antigens from the extracellular environment are trafficked to MIIC after being endocytosed and degraded in late endosomes-lysosomes. Resulting peptides are loaded on MHC II molecules and the MHC II-peptide complex is exported to the cell surface [6, 36]. The paradigm that MHC II molecules largely present exogenous antigens has evolved. Pioneering works have reported that several endogenous viral antigens gain access to the MHC II presentation pathways. For instance, HLA-DR1$^+$ or HLA-DR4$^+$ murine fibroblasts that express measles virus matrix or measles virus nucleocapsid in their cytoplasm are competent in priming CD4$^+$ T cells specific for these two antigens [17]. In another study, human lymphoblastoid cells infected with vaccinia virus that express the influenza A matrix protein 1 (MP1) gain the capacity to prime MP1-specific CD4$^+$ T cells [19]. For both studies, evidence was provided arguing that MHC II antigen presentation occurs independently of the exogenous pathway. Therefore, it is now clear that MHC II molecules present both intracellular and extracellular antigens, with autophagy acting as an active effector of endogenous MHC II presentation (Fig. 1).

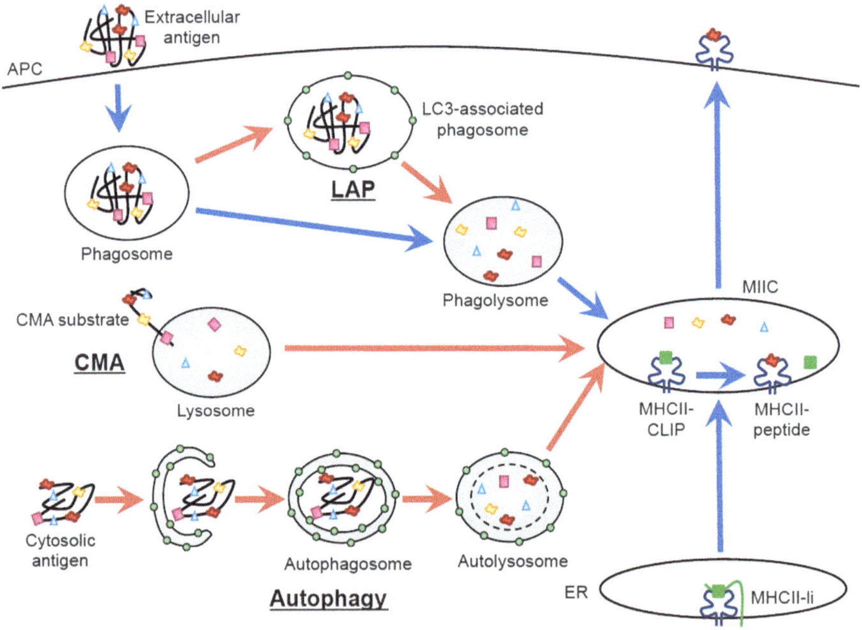

Fig. 1 Autophagy-independent and dependent MHC II presentation pathways. During classical MHC II antigen presentation (*blue arrows*), MHC II-Ii complexes are exported to MIIC. After degradation of Ii, the CLIP peptide is replaced by peptides generated by proteolysis of phagocytosed extracellular antigen. MHC II-peptide complexes are transported to the cell surface. In autophagy-dependent MHC II antigen presentation (*red arrows*), cytosolic antigen is degraded and trafficked to MIIC by autophagy or by chaperone-mediated autophagy (CMA). The proteolysis of extracellular antigen in the endosomal compartment can also be assisted by LC3-associated phagocytosis (LAP)

2.2 Intersection of Autophagy and MHC II Presentation Pathways

2.2.1 Autophagy in the Presentation of Endogenous Antigens

Autophagy plays a major role in routing intracellular antigen to MHC II molecules. Proteomic analysis of the peptidome that is presented by HLA-DR molecules from B lymphoblastoid cells reveals that 35 % of antigens correspond to intracellular proteins. Interestingly, this proportion is augmented when autophagy is induced by nutrient deprivation, raising the possibility that autophagy contributes to MHC II presentation [8]. Schmid et al. have reported that autophagy flux is elevated in different APC populations, including human epithelial cells, B lymphoblastoid cells, monocytes, immature and mature monocyte-derived DCs (moDCs). Interestingly, the authors have also shown that the autophagosome marker microtubule-associated protein 1 light chain 3 (LC3) co-stains with a high proportion of MIIC. This strongly suggested the intersection of autophagy and MHC II presentation. To further investigate this, the cytosolic viral antigen MP1 was fused to LC3. In this case, the fusion protein was efficiently targeted to autophagosomes and enhanced the activation of MP1-specific CD4+ T cells [44]. Therefore, the trafficking of autophagosomes to MIIC in APCs provides a pathway whereby cytosolic antigen can access the MHC II antigen presentation pathway.

Autophagy-dependent presentation of endogenous antigens, and more specifically self-antigens, by MHC II molecules is critical for immune tolerance. In the thymus, immature T cells undergo positive and negative selection by recognizing self-antigen presented by MHC molecules on thymic epithelial cells (TECs). This central tolerance mechanism induces the deletion of potentially harmful self-reactive T cells [14]. Interestingly, autophagy is constitutively elevated in TECs, as evidenced by the presence of numerous intracellular GFP-LC3 aggregates [35]. A high proportion of these autophagosomes co-localize with MIIC, both in cortical TECs, implicated in positive selection, and in medullary TECs that are required for negative selection [23]. To investigate the impact of autophagy in TECs on T cell selection, Aichinger et al. have cloned a neo-antigen encompassing an epitope derived from the C-reactive protein (CRP) fused to the carboxyterminal domain of LC3 and to green fluorescent protein (GFP). This fusion protein is efficiently targeted to autophagosomes in transfected cells. Interestingly, endogenous expression of this antigen by medullary TECs triggers the deletion of CRP-specific $CD4^+$ T cells in mice while *Atg5* gene deletion abrogates this process [3]. In addition, when $Atg5^{-/-}$ thymi are grafted to wild type or TCR transgenic mice, the repertoire of mature $CD4^+$ T cells, but not $CD8^+$ T cells, is modified. Transplantation of $Atg5^{-/-}$ thymi into athymic mice induces the development of colitis and systemic inflammation, mostly triggered by mature $CD4^+$ T cells. Collectively, these data suggest that autophagy in TECs is required for both negative and positive selection of immature $CD4^+$ T cells in the thymus and that the absence of an intact autophagy pathway in these cells leads to a breach of tolerance [35].

Endogenous MHC II antigens that undergo citrullination are particularly dependent on autophagy for their processing [16]. Citrullination is a post-translational modification associated with autoimmune pathologies such as rheumatoid arthritis. In mice expressing hen egg lysozyme (HEL), DCs and macrophages constitutively present HEL-derived citrullinated epitopes by MHC II. Strikingly, pharmacological inhibition of autophagy with 3-methyladenine (3-MA) impairs this mechanism whereas it does not affect MHC II presentation of non-citrullinated epitopes. At steady-state, B cells are not competent in presenting citrullinated epitopes. However, autophagy induction with serum starvation of B lymphoblastoid cells restores MHC II presentation of citrullinated HEL, and this process is reduced in the presence of 3-MA or with *Atg5* shRNA knockdown. Supporting the importance of these findings, this mechanism is also observed in primary B cells following BCR stimulation that activates autophagosome formation. Foreign antigens that are endogenously expressed by APCs are also processed via autophagy for MHC II loading. Immuno-electron microscopy sections of Epstein-Barr virus (EBV)-transformed lymphoblastoid cells demonstrates the presence of EBV nuclear antigen 1 (EBNA1) in double membrane autophagosomes. The use of 3-MA or *Atg12* siRNA knockdown to inhibit autophagy in these cells impairs the priming of EBNA1-specific $CD4^+$ T cells. Therefore, autophagy facilitates MHC II presentation of EBNA1 [37]. In murine macrophages and DCs infected with *Mycobacterium Tuberculosis*, autophagy induction by rapamycin, starvation or interferon-γ relocates the mycobacterial antigen 85B (Ag85B) into autophagosomes and enhances priming of Ag85-specific $CD4^+$ T cells. This process is repressed by *Atg6/Beclin-1* knockdown, showing that autophagy acts as an effector of Ag85B presentation by MHC II molecules. Importantly, the engulfment of Ag85B into autophagosomes was exclusively observed in *Mycobacterium Tuberculosis*-infected cells, thus excluding a scenario where the antigen was taken up from an extracellular source. The stimulatory property of rapamycin on MHC II presentation of mycobacterial antigens can be exploited to improve the efficiency of vaccination. Indeed, *Mycobacterium Tuberculosis*-infected mice that have been vaccinated with rapamycin-treated DCs elicit a higher $CD4^+$ T cell immune response and improved bacterial clearance than mice injected with untreated DCs [18]. Some tumor antigens also depend on autophagy for MHC II presentation. Transfected human moDCs that express the tumor antigen Mucin-1 are able to induce the proliferation of Mucin-1-specific $CD4^+$ T cells, and this capacity is abolished in the presence of 3-MA and wortmannin that block autophagosome formation. Therefore, autophagy activity in moDCs is required for MHC II presentation of the tumor antigen Mucin-1 [10].

One study reported that cytosolic MHC II antigens comprising the amino acid motif KFERQ are processed via chaperone-mediated autophagy (CMA), a selective type of autophagy. During CMA, KFERQ-containing substrates are recognized in the cytosol by a complex of chaperones, including heat shock cognate protein of 70 kDa (Hsc70), and transported to a multimer of lysosomal-associated membrane protein 2a (Lamp-2a) at the lysosome membrane. The substrate is subsequently translocated across the lysosomal membrane and degraded by lysosomal proteases [24]. *LAMP-2* and *Hsc70* knockdown in human B lymphoblasts reduces MHC II presentation of the cytosolic glutamate decarboxylase while Lamp-2a and Hsc70 overexpression enhances this process. Therefore, these findings suggest that the degradation of CMA substrates elicits epitopes for MHC II presentation [51]. Altogether, these studies highlight a unique and critical role of autophagy to transport and process intracellular antigens for the MHC II presentation pathway.

2.2.2 Autophagy in the Presentation of Exogenous Antigens

Besides its contribution to MHC II presentation of endogenous antigen, evidence also exists to support a role for autophagy in the routing and processing of exogenous antigens for MHC II loading. Several studies have highlighted this mechanism during pathogen infection. For example, when human moDCs loaded with inactivated human immunodeficiency virus (HIV) are used as APCs, inhibitors of autophagosome formation (3-MA, *LC3* and *Atg5* siRNA knockdown) impair their capacity to prime $CD4^+$ T cells specific for the HIV antigen Gag. In contrast, autophagy stimulation with rapamycin enhances MHC II presentation of Gag. Importantly, the use of the Gag-derived peptide to pulse DCs does not elicit any difference in MHC II presentation outcomes, suggesting that autophagy is involved in the processing of the exogenously acquired Gag protein [5]. In another study, Lee et al. have reported that, upon infection with herpes simplex virus (HSV)-1, chimeric mice reconstituted with a hematopoietic $Atg5^{-/-}$ immune system show weaker $CD4^+$ T cell immunity compared to mice reconstituted with wild-type cells. Furthermore, infection of $Atg5^{-/-}$ chimeric mice with ovalbumin (OVA)-expressing HSV-1 or OVA-expressing *Listeria monocytogenes* leads to a defect in the proliferation of adoptively transferred OVA-specific $CD4^+$ T cells. Mice with a conditional knockout of *Atg5* in DCs and infected with HSV-2 also demonstrate a defect in $CD4^+$ T cell priming and reduced survival due to higher disease severity. Of note, DCs that lack Atg5 undergo normal maturation, are competent in cell migration and are not defective in endocytosis and phagocytosis. Therefore, the impaired $CD4^+$ T cell immune response is not the result of a global reduction in DC function, but rather occurs due to impaired processing of the phagocytosed antigens. Indeed, using lipopolysaccharide and OVA-coated beads, $Atg5^{-/-}$ DCs exhibited a delay in phagosome-lysosome fusion and a defect in the delivery of lysosomal proteases to antigen-containing phagosomes. Surprisingly, electron microscopy images did not reveal double-membrane autophagosomes engulfing phagosomes. The authors raise the possibility that a non-canonical form of autophagy is occurring that may facilitate the proteolysis of extracellular antigens in DCs [26]. These results are in line with the recent description of a novel form of autophagy in DCs that promotes the degradation of phagocytosed material, known as LC3-associated phagocytosis (LAP). In this pathway, LC3-II is directly recruited to the single membrane of phagosomes, a process that requires the autophagy machinery components Atg5, Atg7 and Beclin-1. Importantly, this non-canonical form of autophagy is activated downstream of the stimulation of several innate immune receptors by microbial components, including Toll-like receptors (TLRs) 1/2, TLRs2/6, TLR4 and Dectin-1 [33, 43]. Other evidence demonstrates the involvement of LAP in the proteolysis of phagocytosed antigens for MHC II presentation. Pulsing of bone marrow-derived DCs (BMDCs) with yeast stimulates Dectin-1 and activates LAP downstream of the receptor. Interestingly, when OVA-expressing yeast are used,

$LC3^{-/-}$ DCs are defective in activating OVA-specific $CD4^+$ T cells whereas the presentation of exogenously supplied OVA peptide is not affected [31]. Using live cell imaging of human macrophages, Romao et al. have reported that LC3-conjugated phagosomes formed during LAP are delayed in acquiring lysosomal markers compared to phagosomes that do not harbor LC3. Furthermore, inhibition of LAP with *Atg5* or *Atg16L1* knockdown abrogates the presentation of fungal antigens by MHC II molecules. These findings conflict with the data of Lee et al. that show enhanced MHC II presentation through LAP is a consequence of improved phagosomal maturation [26]. Romao et al. rather propose a mechanism where the conjugation of LC3 to phagosomes prolongs antigen storage [42]. Together, these results demonstrate that autophagy, and more particularly non-canonical LAP, assists the processing of phagocytosed antigens for MHC II presentation.

3 Autophagy in MHC I Presentation

3.1 Overview of the MHC I Presentation Pathways

3.1.1 The Classical MHC I Presentation Pathway

MHC I presentation is critical for the priming of $CD8^+$ T cell immune responses. The complex formed by MHC I bound to an antigenic peptide is directly recognized by the TCR of antigen-specific $CD8^+$ T cells. Upon activation, $CD8^+$ T cells differentiate into cytotoxic T cells that detect and kill pathogen-infected cells or tumor cells. All nucleated cells express MHC I molecules, however only DCs have the ability to prime $CD8^+$ T cell immune responses [22, 36].

A major source of antigens for MHC I molecules are proteins expressed by the cell itself. Antigenic peptides are produced following proteasomal degradation of endogenously expressed proteins. Peptides are imported into the ER through the transporter-associated with antigen processing (TAP) expressed at the ER membrane. These peptides can be further processed by ER aminopeptidases. Genes encoding MHC I molecules are *HLA-A*, *–B*, *–C* in humans. After synthesis, the α chain interacts with β2-microglobulin in the ER and this dimer associates with peptides that fit in the binding groove of the complex. This step is facilitated by the action of ER chaperones such as tapasin, calnexin, calreticulin. After leaving the ER, the MHC I-peptide complex transverses the Golgi and is displayed at the cell surface [6, 36].

3.1.2 The MHC I Cross-Presentation Pathway

The function of MHC I is not confined to the presentation of endogenous antigen. Certain types of APCs, like DCs, are able to present MHC I peptides that are derived from extracellular antigens via a mechanism termed “cross-presentation”. This immune process is particularly significant to elicit CD8+ T cell immunity against pathogens or cancer where APCs are not directly targeted by pathogens or when they do not express tumor antigens [20, 45].

The intracellular mechanisms underlying cross-presentation are not well defined. Two main models of cross-presentation have been proposed according to the intracellular trafficking route of the antigen. In the "vacuolar pathway", the phagocytosed antigen is processed by lysosomal proteases. Resulting peptides are loaded on MHC I molecules that recycle from the cell surface, and the MHC I peptide complex is re-exposed on the cell surface. In the "cytosolic pathway", the phagocytosed antigen is exported from the endosomal compartment to the cytosol. Although the identity of the transporter is unknown, evidence suggests that the translocon Sec61 is an attractive candidate to undertake this mechanism [1]. Once in the cytosol, antigen is degraded by the proteasome into peptides that are transported into the ER via TAP for loading on MHC I molecules. An alternative mechanism has also been proposed where antigen-containing phagosomes directly fuse with the ER or with vesicles containing ER-membrane. This fusion leads to the formation of a cross-presenting compartment comprising Sec61, TAP, MHC I molecules and the ER loading chaperones. In this scenario, antigen is transferred from the phagosomal lumen to the cytosol for proteasomal degradation, however the liberated peptides are directly reimported into the cross-presenting compartment and loaded on MHC I molecules [2, 13, 15]. The exact mechanisms that facilitate cross-presentation remain a matter of debate [20, 45, 47].

3.2 Intersection of Autophagy with MHC I Presentation Pathways

3.2.1 Autophagy in the Classical MHC I Presentation Pathway

Although MHC I presentation largely relies on the proteasome for antigen degradation, several groups have shown that autophagy can assist MHC I antigen processing (Fig. 2). For MHC I presentation of endogenous antigen, evidence illustrates that the human cytomegalovirus protein pUL138 is presented to CD8+ T cells via an autophagy-dependent mechanism rather than via the classical proteasomal-dependent route. Indeed, $TAP^{-/-}$ cells are a potent activator of pUL138–specific CD8+ T cells. Furthermore, incubation of cells with lactocystin or epoxomicin, two proteasome inhibitors, or with an inhibitor of ER aminopeptidases did not impair this immune mechanism. In contrast, blocking lysosome proteolysis with chloroquine strongly reduces pUL138 presentation by MHC I molecules. Autophagy downregulation with 3-MA or *Atg12* knockdown also abrogates MHC I presentation. The authors speculate that pUL138 is targeted to the lysosome by autophagy for breakdown, and peptides are directly loaded on MHC I molecules in the endosomal compartment [46]. Another example of autophagy-dependent endogenous MHC I presentation is the presentation of an epitope issued by respiratory syncytial

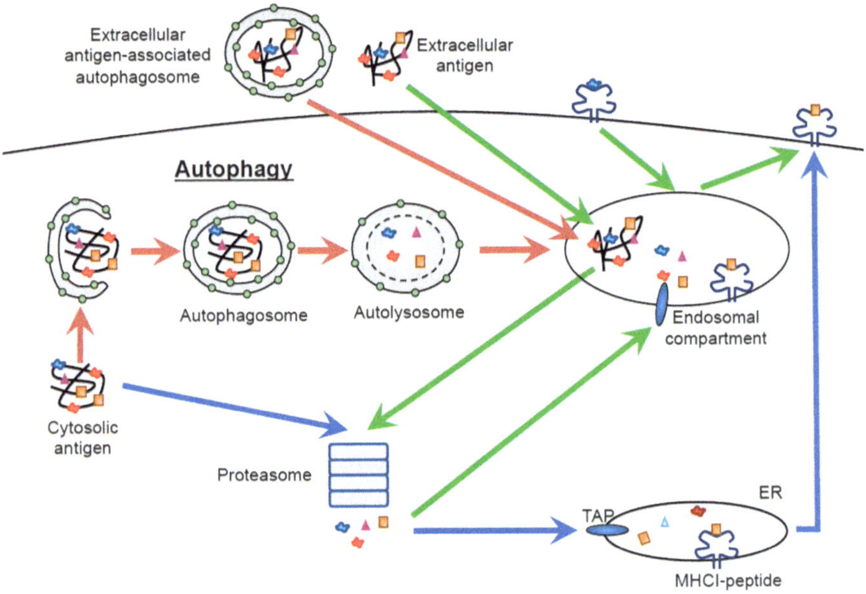

Fig. 2 Autophagy-independent and dependent MHC I presentation pathways. Classical MHC I antigen presentation (*blue arrows*) relies on degradation of endogenous protein by the proteasome and import of peptides into the endoplasmic reticulum (ER) via the transported associated with antigen processing (TAP). Peptides are loaded into the peptide-binding groove of newly synthesized MHC I molecules and trafficked to the cell surface. During cross-presentation (*green arrows*), antigens are transferred into the cytosol for proteasomal degradation. Peptides can be imported into the ER or into endosomal compartments for MHC I loading. Alternatively, phagocytosed antigens can be degraded by lysosomal proteases, eliciting peptides that are loaded on recycling MHC I molecules in the endosomal compartment. Autophagy can intersect with MHC I antigen presentation (*red arrows*) by trafficking cytosolic antigens to endosomal MHC I molecules or by modulating the efficiency of cross presentation by unknown mechanisms. Extracellular antigens associated to autophagosomes readily access the cross-presentation pathway

virus that proceeds in a TAP and proteasome-independent manner. The use of 3-MA reduces the priming of epitope-specific $CD8^+$ T cells, suggesting that access of this epitope to the MHC I machinery relies on autophagy [21]. In HSV-1-infected macrophages, English et al. reported that MHC I presentation of viral antigens depends on autophagy during the late stage of infection. In this case, bafilomycin A1, an inhibitor of lysosomal acidification, abrogates presentation of HSV-1 glycoprotein B (gB) to $CD8^+$ T cells from 8–10 h after the beginning of infection. Similarly, autophagy inhibition with 3-MA or *Atg5* siRNAs knockdown impairs this mode of MHC I presentation. Interestingly, electron microscopy images of HSV-1-infected macrophages reveal the formation of double-membrane autophagosomes emerging from the nuclear envelope containing viral particles in their lumen. Although the autophagy-mediated mechanism underlying MHC I presentation of gB remains elusive, the authors suggest that the viral antigen undergoes proteolysis by autophagy, followed by the export of proteolysis products to the cytosol for further processing by the proteasome [11]. The capture of cytosolic antigens into autophagosomes for

MHC I presentation is supported by a study of polyubiquitinated aggregates that form in mature DCs. These aggregates are termed DC aggresome-like induced structures (DALIS) and are thought to contain misfolded ribosomal products that are a major source of MHC I antigen in DCs [39]. Fluorescence microscopy and electron microscopy images demonstrate that DALIS co-localise with LC3 and that these aggregates are frequently surrounded by autophagosome-like structures. Evidence suggests that these double-membrane vesicles emerge from the MIIC and correspond to an unconventional type of autophagy, termed endosome-mediated autophagy (ENMA) [25]. The impact of ENMA on MHC I presentation is unknown. There is some evidence to suggest that autophagy can modulate the cell surface expression of MHC I molecules. Li et al. reported that the use of pharmacological autophagy inhibitors (3-MA, chloroquin, wortmannin) and siRNA knockdown (*Atg6/Beclin1, Atg5, Atg7*) upregulates MHC I expression at the surface of mouse macrophages. This observation may reflect reduced degradation of surface MHC I molecules by autophagy. In line with this data, rapamycin treatment of B16 melanoma cells induces the re-localization of a population of MHC I molecules into autophagosomes, thereby decreasing their expression at the cell surface. Intriguingly, rapamycin used in combination with interferon-γ, a cytokine that stimulates MHC I expression, has the opposite effect with the expression of MHCI molecules on cell surface being further increased. This synergy between rapamycin and interferon-γ strengthens the killing of B16 tumor cells by antigen-specific cytotoxic T cells and this effect is suppressed when autophagy is inhibited. In this condition, the presence of MHC I molecules in the lumen of autophagosomes is diminished, however the mechanism explaining the synergy between rapamycin and interferon-γ remains undefined. Taken together, these findings highlight the potential for autophagy to impact the display of MHC I molecules at the cell surface [27].

3.2.2 Autophagy in Cross-Presentation: Role in Cross-Presenting Cells

In the cross-presentation pathway, antigen is captured from an extracellular source by APCs and processed for MHC I loading. The potential participation of autophagy in cross-presentation is unclear. $LC3^{-/-}$ BMDCs pulsed with OVA-expressing yeast do not demonstrate any reduced capacity to activate OVA-specific $CD8^{+}$ T cells [31]. Similarly, Lee et al. show that wild-type and $Atg5^{-/-}$ DCs are equally competent in cross-presenting OVA. This was observed when splenic DCs were co-incubated with either soluble OVA taken up by pinocytosis, or with OVA-coated MHC $I^{-/-}$ splenocytes whose uptake depends on phagocytosis [26]. In contrast with these observations, several studies show that autophagy assists cross-presentation. For instance, Ravindran et al. have highlighted that cross presentation of antigen from the yellow fever vaccine YF-17D depends on autophagy. Infection of human moDCs or mouse BMDCs with YF-17D induces autophagy through a mechanism that requires the general control nonderepressible 2 (GCN2) kinase, a sensor of amino acid starvation. To examine the role of vaccine-induced autophagy on cross-presentation, *Atg5*, *Atg7* or *Atg6/Beclin-1* knockout BMDCs were pulsed with YF-17D-OVA-infected cells. In this case, OVA cross-presentation was significantly impaired in autophagy-deficient BMDCs [41].

Autophagy also plays a critical role in the cross-presentation of antigens conjugated to αAl2O3 nanoparticles. OVA-conjugated αAl2O3 (αAl2O3-OVA) exhibits superior capacity to be cross-presented by mouse BMDCs *in vitro* and induce an OVA-specific $CD8^+$ T cell immune response *in vivo* compared to OVA alone. Using immunofluorescence and electron microscopy, the authors show that cross-presentation of the conjugated antigen relies on the localisation of αAl2O3-OVA, but not OVA, in autophagosomes. Blockade of autophagy in BMDCs with pharmacological inhibitors (3-MA, wortmannin) and siRNA knockdown (*Atg6/Beclin-1* and *Atg12*) reduces cross-presentation of αAl2O3-OVA whereas no difference is observed with OVA alone. How the localization of αAl2O3-OVA in the lumen of autophagosomes facilitates OVA cross-presentation remains unknown [28]. Autophagy has been suggested to assist cross-presentation of fungal antigens from *Aspergillus Conidia*. Treatment of lung DCs with 3-MA prevents the routing of *Aspergillus Conidia* to $Rab14^+$ compartments where cross-presentation occurs [7].

It is worth mentioning that one study argues that autophagy has an inhibitory role in cross-presentation. Baghdadi et al. have reported that engulfment of dying cells by bone-marrow-derived macrophages (BMDMs) induces autophagy. This process depends on the stimulation at the cell surface of the T cell immunoglobulin and mucin domain-containing molecule 4 (TIM-4) that recruits the adenosine monophosphate activating kinase-α1 (AMPKα1), an intracellular energy sensor that regulates autophagosome formation. Interestingly $TIM\text{-}4^{-/-}$ BMDMs co-incubated with apoptotic debris from OVA-expressing cells display increased capacity to activate OVA-specific $CD8^+$ T cells. This enhanced cross-presentation is repressed when autophagy is induced by rapamycin. Conversely, pharmacological inhibitors of autophagy (3-MA, bafilomycin A1) or *Atg5* knockout in BMDMs induce a stronger cross-presentation of dying cell-associated OVA. The authors suggest that TIM-4-induced autophagy facilitates the routing of phagocytosed cellular debris to lysosomes for breakdown and prevents the production of epitopes for MHC I molecules [4]. Although TIM-4 engagement has been shown by others to activate LAP [32], electron microscopy images in this study clearly identify the presence of double-membrane autophagosomes containing cellular debris, therefore this immune tolerance mechanism involves classical autophagy. Interestingly, in tumor-bearing mice injected with chemotherapeutic agents, knockout of *Prkaa1* (encoding AMPKα1) or *Atg5* in myeloid cells enhances anti-tumor $CD8^+$ T cell immunity and reduces tumor growth. Therefore, inhibiting autophagosome formation downstream TIM-4 is of interest in enhancing anti-tumor immunity triggered by chemotherapy-induced cell death [4]. In conclusion, the role of autophagy in cross-presenting APCs remains controversial. Several studies exist that either support or reject such a role with no clear consensus as to why this is the case.

3.2.3 Autophagy in Cross-Presentation: Role in Antigen-Donor Cells

Autophagosomes in antigen-donor cells can influence the capacity of antigen to be efficiently cross-presented by APCs. For instance, $Bax/Bak^{-/-}$ mouse embryonic fibroblasts (MEFs) undergo cell death with high autophagy activity. Interestingly, immunization of mice with cellular debris from influenza A-infected Bax/Bak$^{-/-}$ MEFs primes a stronger CD8^{+} T cell immune response than the use of wild-type MEFs. *Atg5* knockdown in Bax/Bak$^{-/-}$ MEFs represses this immune response [48]. Similarly, autophagy flux in human melanoma cells, that express the tumor antigen gp100, impacts the efficiency of cross-presentation of this antigen *in vivo*. Treatment of these cells with rapamycin boosts the CD8^{+} T cell immune response while inhibitors of autophagosome formation, including 3-MA, wortmannin, *Atg6/Beclin-1* shRNAs or *Atg12* siRNAs, impairs this mechanism. Data also exists to support a role of autophagosomes themselves as antigen carriers that promote MHC I cross-presentation. For instance, co-incubation of DCs with autophagosomes purified from OVA-expressing fibroblasts efficiently induces the proliferation of OVA-specific CD8^{+} T cells [30]. Defective ribosomal products-containing blebs (DRibbles) are autophagosome-enriched fractions that can be purified from cells whose proteasome activity and autophagosome maturation are inhibited. DRibbles-associated antigens readily access the cytosolic pathway of cross-presentation. Strikingly, vaccination of lung tumor-bearing mice with DCs pulsed with DRibbles elicits higher protection against tumor growth than vaccination with DCs loaded with dying cells or cell lysates [29]. DRibbles are also effective in human APCs. Pulsing of peripheral blood mononuclear cells or monocytes with DRibbles containing viral antigens elicits activation of memory CD8^{+} T cells [50]. In conclusion, autophagosomes associate with antigen in donor cells and can enhance MHC I cross presentation outcomes following their phagocytosis by APCs.

4 Conclusions

It is now evident that autophagy regulates different immune functions. Specifically, autophagy plays a major role in antigen presentation that is critical to elicit adaptive immunity and priming of T cell-mediated immune responses. The intersection of autophagy and MHC II antigen presentation is well established. Self and foreign antigens expressed in the cytosol of APCs are taken up into autophagosomes and the continuous fusion of these double-membrane structures with MIIC ensures the loading of epitopes into MHC II molecules. This intracellular process regulates central tolerance and promotes effective immune responses against infectious pathogens. More recent studies have also provided evidence that autophagy assists the processing of extracellular MHC II antigens through non-canonical LAP. The relevance of LAP-mediated MHC II presentation in settings of live immunity remains to be further examined. In contrast, little is known about intracellular connections between autophagy and MHC I presentation pathways. A handful of studies have suggested that autophagy may participate to endogenous antigen processing for loading on MHC I molecules. Controversial data also demonstrate that APCs exploit autophagy to enhance antigen-cross-presentation. Whether only certain type antigens are processed in an autophagy-dependent manner for MHC I presentation and cross-presentation needs to be clarified. More work is also required to determine whether these findings correspond to classical autophagy or whether they involve non-canonical types of autophagy. Finally, extracellular antigens packaged into autophagosomes are efficiently targeted to cross-presentation in APCs. Therefore, autophagosomes are candidates of major interest in vaccination to be used as adjuvants and to enhance cross-presentation.

In summary, the intersection of autophagy and antigen presentation may increase the severity of pathologies, including autoimmune diseases, by augmenting detrimental antigen presentation. Conversely, the communication between autophagy and antigen presentation may act synergistically to benefit the host by efficiently priming T cell immune response against pathogens. Therefore, pharmacological modulators of autophagy offer a promising therapeutic approach to manipulate antigen presentation and to elicit the desired immune outcome.

Acknowledgments C. M. is supported by La Fondation pour la Recherche Médicale.

References

1. Ackerman AL, Giodini A, Cresswell P (2006) A role for the endoplasmic reticulum protein retrotranslocation machinery during crosspresentation by dendritic cells. Immunity 25:607–617
2. Ackerman AL, Kyritsis C, Tampé R, Cresswell P (2003) Early phagosomes in dendritic cells form a cellular compartment sufficient for cross presentation of exogenous antigens. Proc Natl Acad Sci U S A 100:12889–12894
3. Aichinger M, Wu C, Nedjic J, Klein L (2013) Macroautophagy substrates are loaded onto MHC class II of medullary thymic epithelial cells for central tolerance. J Exp Med 210:287–300
4. Baghdadi M, Yoneda A, Yamashina T et al (2013) TIM-4 glycoprotein-mediated degradation of dying tumor cells by autophagy leads to reduced antigen presentation and increased immune tolerance. Immunity 39:1070–1081
5. Blanchet FP, Moris A, Nikolic DS et al (2010) Human immunodeficiency virus-1 inhibition of immunoamphisomes in dendritic cells impairs early innate and adaptive immune responses. Immunity 32:654–669
6. Blum JS, Wearsch PA, Cresswell P (2013) Pathways of antigen processing. Annu Rev Immunol 31:443–473
7. De Luca A, Iannitti RG, Bozza S et al (2012) CD4(+) T cell vaccination overcomes defective cross-presentation of fungal antigens in a mouse model of chronic granulomatous disease. J Clin Invest 122:1816–1831
8. Dengjel J, Schoor O, Fischer R et al (2005) Autophagy promotes MHC class II presentation of peptides from intracellular source proteins. Proc Natl Acad Sci U S A 102:7922–7927
9. Deretic V, Saitoh T, Akira S (2013) Autophagy in infection, inflammation and immunity. Nat Rev Immunol 13:722–737
10. Dörfel D, Appel S, Grünebach F et al (2005) Processing and presentation of HLA class I and II epitopes by dendritic cells after transfection with in vitro-transcribed MUC1 RNA. Blood 105:3199–3205
11. English L, Chemali M, Duron J et al (2009) Autophagy enhances the presentation of endogenous viral antigens on MHC class I molecules during HSV-1 infection. Nat Immunol 10: 480–487
12. Feng Y, He D, Yao Z, Klionsky DJ (2014) The machinery of macroautophagy. Cell Res 24:24–41

13. Guermonprez P, Saveanu L, Kleijmeer M et al (2003) ER-phagosome fusion defines an MHC class I cross-presentation compartment in dendritic cells. Nature 425:397–3402
14. Hogquist KA, Baldwin TA, Jameson SC (2005) Central tolerance: learning self-control in the thymus. Nat Rev Immunol 5:772–782
15. Houde M, Bertholet S, Gagnon E et al (2003) Phagosomes are competent organelles for antigen cross-presentation. Nature 425:402–406
16. Ireland JM, Unanue ER (2011) Autophagy in antigen-presenting cells results in presentation of citrullinated peptides to CD4 T cells. J Exp Med 208:2625–2632
17. Jacobson S, Sekaly RP, Jacobson CL et al (1989) HLA class II-restricted presentation of cytoplasmic measles virus antigens to cytotoxic T cells. J Virol 63:1756–1762
18. Jagannath C, Lindsey DR, Dhandayuthapani S et al (2009) Autophagy enhances the efficacy of BCG vaccine by increasing peptide presentation in mouse dendritic cells. Nat Med 15:267–276
19. Jaraquemada D, Marti M, Long EO (1990) An endogenous processing pathway in vaccinia virus-infected cells for presentation of cytoplasmic antigens to class II-restricted T cells. J Exp Med 172:947–954
20. Joffre OP, Segura E, Savina A, Amigorena S (2012) Cross-presentation by dendritic cells. Nat Rev Immunol 12:557–569
21. Johnstone C, Ramos M, García-Barreno B et al (2012) Exogenous, TAP-independent lysosomal presentation of a respiratory syncytial virus CTL epitope. Immunol Cell Biol 90: 978–982
22. Jung S, Unutmaz D, Wong P et al (2002) In vivo depletion of CD11c+ dendritic cells abrogates priming of CD8+ T cells by exogenous cell-associated antigens. Immunity 17:211–220
23. Kasai M, Tanida I, Ueno T et al (2009) Autophagic compartments gain access to the MHC class II compartments in thymic epithelium. J Immunol 183:7278–7285
24. Kaushik S, Cuervo AM (2012) Chaperone-mediated autophagy: a unique way to enter the lysosome world. Trends Cell Biol 22:407–417
25. Kondylis V, van Nispen Tot Pannerden HE, van Dijk S et al (2013) Endosome-mediated autophagy: an unconventional MIIC-driven autophagic pathway operational in dendritic cells. Autophagy 9:861–880
26. Lee HK, Mattei LM, Steinberg BE et al (2010) In vivo requirement for Atg5 in antigen presentation by dendritic cells. Immunity 32:227–239
27. Li B, Lei Z, Lichty BD et al (2010) Autophagy facilitates major histocompatibility complex class I expression induced by IFN-γ in B16 melanoma cells. Cancer Immunol Immunother 59:313–321
28. Li H, Li Y, Jiao J, Hu H-M (2011) Alpha-alumina nanoparticles induce efficient autophagy-dependent cross-presentation and potent antitumour response. Nat Nanotechnol 6:645–650
29. Li Y, Wang L-X, Pang P et al (2011) Tumor-derived autophagosome vaccine: mechanism of cross-presentation and therapeutic efficacy. Clin Cancer Res 17:7047–7057
30. Li Y, Wang L-X, Yang G et al (2008) Efficient cross-presentation depends on autophagy in tumor cells. Cancer Res 68:6889–6895
31. Ma J, Becker C, Lowell CA, Underhill DM (2012) Dectin-1-triggered recruitment of light chain 3 protein to phagosomes facilitates major histocompatibility complex class II presentation of fungal-derived antigens. J Biol Chem 287:34149–34156
32. Martinez J, Almendinger J, Oberst A et al (2011) Microtubule-associated protein 1 light chain 3 alpha (LC3)-associated phagocytosis is required for the efficient clearance of dead cells. Proc Natl Acad Sci U S A 108:17396–17401
33. Mehta P, Henault J, Kolbeck R, Sanjuan MA (2014) Noncanonical autophagy: one small step for LC3, one giant leap for immunity. Curr Opin Immunol 26:69–75
34. Mintern JD, Villadangos JA (2012) Autophagy and mechanisms of effective immunity. Front Immunol 3:60
35. Nedjic J, Aichinger M, Emmerich J et al (2008) Autophagy in thymic epithelium shapes the T-cell repertoire and is essential for tolerance. Nature 455:396–400
36. Neefjes J, Jongsma ML, Paul P, Bakke O (2011) Towards a systems understanding of MHC class I and MHC class II antigen presentation. Nat Rev Immunol 11:823–836

37. Paludan C, Schmid D, Landthaler M et al (2005) Endogenous MHC class II processing of a viral nuclear antigen after autophagy. Science 307:593–596
38. Patterson NL, Mintern JD (2012) Intersection of autophagy with pathways of antigen presentation. Protein Cell 3:911–920
39. Pierre P (2005) Dendritic cells, DRiPs, and DALIS in the control of antigen processing. Immunol Rev 207:184–190
40. Ravikumar B, Sarkar S, Davies JE et al (2010) Regulation of mammalian autophagy in physiology and pathophysiology. Physiol Rev 90:1383–1435
41. Ravindran R, Khan N, Nakaya HI et al (2014) Vaccine activation of the nutrient sensor GCN2 in dendritic cells enhances antigen presentation. Science 343:313–317
42. Romao S, Gasser N, Becker AC et al (2013) Autophagy proteins stabilize pathogen-containing phagosomes for prolonged MHC II antigen processing. J Cell Biol 203:757–766
43. Sanjuan MA, Dillon CP, Tait SW et al (2007) Toll-like receptor signalling in macrophages links the autophagy pathway to phagocytosis. Nature 450:1253–1257
44. Schmid D, Pypaert M, Münz C (2007) Antigen-loading compartments for major histocompatibility complex class II molecules continuously receive input from autophagosomes. Immunity 26:79–92
45. Segura E, Villadangos JA (2011) A modular and combinatorial view of the antigen cross-presentation pathway in dendritic cells. Traffic 12:1677–1685
46. Tey S-K, Khanna R (2012) Autophagy mediates transporter associated with antigen processing-independent presentation of viral epitopes through MHC class I pathway. Blood 120:994–1004
47. Touret N, Paroutis P, Terebiznik M et al (2005) Quantitative and dynamic assessment of the contribution of the ER to phagosome formation. Cell 123:157–170
48. Uhl M, Kepp O, Jusforgues-Saklani H et al (2009) Autophagy within the antigen donor cell facilitates efficient antigen cross-priming of virus-specific CD8+ T cells. Cell Death Differ 16:991–1005
49. Yang Z, Klionsky DJ (2010) Eaten alive: a history of macroautophagy. Nat Cell Biol 12:814–822
50. Ye W, Xing Y, Paustian C et al (2014) Cross-presentation of viral antigens in dribbles leads to efficient activation of virus-specific human memory T cells. J Transl Med 12:100
51. Zhou D, Li P, Lin Y et al (2005) Lamp-2a facilitates MHC class II presentation of cytoplasmic antigens. Immunity 22:571–581

Autophagy in T and B Lymphocytes

Alexander J. Clarke and A. Katharina Simon

Abstract Autophagy is now understood to have an active role in the adaptive immune system, participating in lymphocyte development, differentiation, and memory. In T cells, it is activated during thymocyte development, and following receptor ligation, and is also critical for T cell memory. In B cells, there is again activation of autophagy during development, and whilst it is dispensable for maintenance of the peripheral B2 cell pool, it is required for normal formation of B1 B cells. Autophagy is strongly active in plasma cells, which appear to use it as a mechanism for survival of the metabolic stress associated with immunoglobulin production. Finally, as with T cells, autophagy is necessary for memory B cell maintenance.

Abbreviations

AMPK	AMP-activated protein kinase
ASCs	Antibody secreting cells
Atg	Autophagy-related
ATP	Adenosine triphosphate
Bax	Bcl-2-like protein 4
Bcl10	B-cell lymphoma/leukemia 10
Bcl-2	B-cell lymphoma 2
Bcl_{XL}	B-cell lymphoma-extra large
BCR	B cell receptor
BH	Bcl-2 homology
Blimp-1	B lymphocyte-induced maturation protein-1

A.J. Clarke (✉) • A.K. Simon
Kennedy Institute of Rheumatology,
University of Oxford, Headington, Oxford OX3 7FY, UK
e-mail: alexander.clarke@kennedy.ox.ac.uk; katja.simon@imm.ox.ac.uk

M.C. Maiuri, D. De Stefano (eds.), *Autophagy Networks in Inflammation*,
Progress in Inflammation Research, DOI 10.1007/978-3-319-30079-5_9

CLP	Common lymphoid progenitor
cTEC	Cortical thymic epithelium
CXCR4	C-X-C chemokine receptor type 4
DN	Double negative
EAE	Experimental autoimmune encephalomyelitis
ER	Endoplasmic reticulum
GCN2	General control non-repressible-2
GFP	Green fluorescent protein
HIF1α	Hypoxia inducible factor 1α
HIV-1	Human immunodeficiency virus-1
HSCs	Haematopoietic stem cells (HSCs)
Ig	Immunoglobulin
IL	Interleukin
JNK	c-Jun N-terminal kinase
LC3	Microtubule-associated protein 1A/1B-light chain 3
Lck	Lymphocyte protein tyrosine kinase
LCMV	Lymphocytic choriomeningitis virus
LPS	Lipopolysaccharide
MAPK	Mitogen-activated protein kinase
MCMV	Murine cytomegalovirus
MHC	Major histocompatibility complex
mTEC	Medullary thymic epithelial cells
mTOR	Mechanistic target of rapamycin
NF-κB	Nuclear factor-kappaB
NP-KLH	4-hydroxy-3-nitrophenylacetyl conjugated to keyhole limpet hemocyanin
Rag	Recombination-activating genes
ROS	Reactive oxygen species
S6K1	p70 ribosomal protein S6 kinase 1
SLE	Systemic lupus erythematosus
Tat	Trans-Activator of Transcription
TCR	T cell receptor
TD	T cell help
TI	T-independent stimulation
TLR	TLR, Toll-like receptor
T_{reg}	Regulatory T
ULK1	unc-51 like autophagy activating kinase 1
UPR	Unfolded protein response (UPR)
Vps34	Vacuolar protein sorting
XBP-1	X-box binding protein 1.

1 Introduction

The last few years have seen the recognition that autophagy is fundamentally important in the development and homeostasis of both T and B lymphocytes, where it supports cell survival but may also act selectively to regulate their differentiation fate. This chapter will consider the many and varied roles that autophagy has been identified to play in the life cycle of the lymphocyte.

2 Maintenance of Lymphocyte Progenitors

Autophagy plays a critical role in the earliest stage of lymphocyte development: the maintenance of haematopoietic stem cells (HSCs), which self-renew and supply the common lymphoid progenitor (CLP) that progresses to differentiate into T or B cells [24]. HSCs combine a long lifespan with existence in a metabolically challenging environment, and it may therefore be expected that autophagy is important in their homeostasis – and this is indeed the case. The hypoxic nature of the HSC niche leads to glycolysis rather than oxidative phosphorylation, and HSCs have accordingly low mitochondrial density [53]. Autophagy is constitutively active in aged HSCs [62], and deficiency of key autophagy genes leads to HSC failure, manifest as pancytopenia with an associated reduction in CLP and T and B lymphocyte numbers [34]. Whilst the specific role for autophagy in the maintenance of HSCs is unclear, it may at the least be important in the control of mitochondria via mitophagy, which could regulate reactive oxygen species (ROS) production and thereby maintain HSC quiescence [34].

3 T Cells

3.1 Thymocyte Development

Following egress from the bone marrow, T cell precursors migrate to the thymus where further development, including T cell receptor (TCR) formation and testing occurs. T cells progress through a 'double negative' (DN) phase, without expression of the CD3: TCR complex or the co-stimulatory molecules CD4 or CD8, a 'double positive' phase associated with expression of the newly rearranged TCR and both CD4 and CD8, and finally cells become 'single positive' $CD4^+$ or $CD8^+$ T cells and are exported to the periphery. Autophagy within the thymic epithelial cells themselves has been found to have a vital function in T cell maturation. During their development, thymocytes undergo positive selection for the presence of a self-MHC restricted TCR during contact with the cortical thymic epithelium (cTEC), and then negative selection to ensure that self-antigens are not recognised, which are presented to the thymocyte by medullary thymic epithelial cells (mTEC) or thymic dendritic cells [54]. mTECs express ubiquitous self-antigens, under the control of the *Aire* transcription factor, in the context of MHC class II, but have limited ability to present exogenous proteins. There is considerable evidence that mTECs utilize autophagy to allow entry of endogenous antigens into the MHC class II loading pathway, a phenomenon known as non-classical cross-presentation. TECs have high levels of constitutive autophagy, which is starvation-independent [35, 57]. Experiments in which *Atg5*$^{-/-}$ thymi are grafted into recipients demonstrate that autophagy is necessary for negative selection of thymocytes. Endogenous antigens are presented on MHC-II by mTECs by fusion of autophagosomes with MHC-II loading compartments, leading to negative selection of auto-reactive CD4^{+} T cells. Interestingly, the requirement for autophagy as a means of antigen presentation is variable; whilst it is essential for the presentation of mitochondrial antigens, it is however dispensable for the presentation of membrane bound antigens [1]. Autophagy as a means of non-classical cross-presentation also appears to be more important at lower levels of antigen expression. The importance of autophagy in negative selection of T cells is demonstrated by severe autoimmunity in athymic *nu/nu* mouse recipients of an *Atg5*$^{-/-}$ thymic transplant, characterised by an activated T cell phenotype, weight loss, and multiorgan lymphocyte infiltration [35]. It is note-worthy however, that these results have not been reproduced in an alternative model of thymic epithelial autophagy deficiency, which used conditional deletion of *Atg7* by keratin 14-cre [57]. The explanation for this discrepancy is not clear. Experimental work also points to a cell intrinsic function for autophagy in T cell development. There is however, phenotypic variability between models of autophagy deficiency. Within the thymus, autophagy is maximally activated in DN thymocytes compared with the DP stage, both in terms of LC3 turnover and autophagy gene expression [45, 55]. Chimeric models suitable for the study of early thymocyte development consistently show that the total numbers of thymocytes are decreased. Similar findings are observed in a Lck-Cre × Atg7$^{F/F}$ model, in which Atg7 is deleted at an early developmental stage. However, whilst in *Atg5*$^{-/-}$ and *Atg7*$^{-/-}$ chimeras or the Lck-Cre × Atg7$^{F/F}$ model there is a global reduction in thymocyte numbers, in *Beclin1*$^{-/-}$ -> *Rag*$^{-/-}$ transplant recipients there is a block at the DN stage [3, 55]. Given that autophagy is activated at this point in development, it would be expected that genetic deficiency might inhibit further differentiation. The molecular mechanisms responsible for autophagy activation at the DN stage of thymocyte development are unknown, but it is conceivable that DNA damage associated with receptor gene rearrangement may be the inducing stimulus.

3.2 T Cell Activation and Selective Autophagy

Autophagy is active at a low level in naïve CD4^{+} T cells, and is substantially up-regulated following TCR ligation, with an additive effect from co-stimulation through CD28 [18, 26, 40]. The signalling pathways responsible for autophagy induction in T cells have not been fully elucidated, but inhibition of the c-Jun N-terminal kinase (JNK) pathway, either pharmacologically or by gene deletion prevents this process [26]. JNK is activated by combined ligation of CD3 and CD28 [56], and is associated with the cellular response to stress [11]. JNK induces autophagy by multisite phosphorylation of Bcl-2, thereby releasing Beclin-1 to initiate autophagosome formation [64]. How other signalling pathways downstream of the TCR might influence autophagy is largely unknown. However, a number of potential functions for autophagy in modulating the response of the T cell are emerging. Autophagy is important for homeostatic regulation of organelles such as mitochondria and the endoplasmic reticulum (ER), which indirectly influence the response to T cell stimulation [21]. T cells deficient in *Atg7* or *Atg3* have an abnormally expanded ER once development is completed and they have entered the periphery [21, 22]. The resulting consequence of the increase in ER is failure of normal calcium influx following TCR stimulation, due to impaired calcium mobilisation. Similarly, the accumulation of mitochondria in T cells seen when autophagy is suppressed, leads to an increase in ROS, which are important signalling intermediaries [21, 43].

Not only does autophagy contribute to T cell homeostasis by maintenance of organelles, but it may also lead to protein degradation [18], which can selectively affect regulatory proteins, thereby influencing the fate of the cell. The adapter protein Bcl10 connects TCR signals to NF-κB, the activation of which is essential for T cell proliferation and differentiation [60]. Bcl10 is rapidly degraded following TCR ligation, thereby acting as a brake to ongoing activation of NF-κB, which may lead to induction of apoptosis. There is evidence that the mechanism for Bcl10 degradation is by selective autophagy [40]. TCR signalling leads to K63 ubiquitination of Bcl10, which in turn recruits the autophagosome adapter protein p62, through its ubiquitin-binding domain. p62 is then directed to the autophagosome by its LC3-interacting region, and the proteins are degraded. A similar phenomenon is observed in the control of cell death, albeit with variation between knockout models. CD4-Cre × Beclin-$1^{F/F}$ T cells have higher levels of the pro-apoptotic proteins caspase-3, −8, and Bim, but anti-apoptotic Bcl-2 is also increased to a more modest degree [25]. Comparable findings are also present in CD4-Cre × $Vps34^{F/F}$ T cells; increased pro-apoptotic Bax, but also Bcl-2 and Bcl_{XL} [37]. Lck-Cre × Atg7 T cells also have increased Bcl-2, but no increase in Bax, and diminished levels of Bcl_{XL} [46]. Whilst both pro- and anti-apoptotic proteins accumulate, the balance appears to favour apoptosis, as the majority of reports suggest that autophagy deficiency leads to enhanced cell death [37, 45]. Finally, selective autophagy in T cells may be important in resistance to infection. HIV-1 viral infection induces autophagy in $CD4^+$ T cells, triggered by the interaction between envelope glycoproteins and CXCR4 [15]. Autophagy activated in this manner is a viral restriction mechanism, demonstrated by its ability to reduce viral production. The viral transactivator Tat is bound by p62 and targeted for autophagic degradation in an ubiquitin-independent manner [49]. The importance of autophagy on the outcome of TCR signalling is revealed by a series of studies using either gene deletion or pharmacological inhibition, but which again also reveal some disparities between phenotypes. Following CD3 and CD28 co-ligation, that deficiency of autophagy impairs lymphocyte proliferation is a consistent finding in most models [18, 21, 25, 45, 55], but not all [3, 40]. Production of IL-2 is generally reported as suppressed [18, 21], but interestingly, Paul et al. noted the opposite finding in a tamoxifen inducible-*Atg3* knockout model, and also with the class III phosphoinositide 3-kinase inhibitor 3-methyladenine, which was hypothesised to be due to decreased Bcl10 degradation leading to NF-κB activation. However, expression of the activation markers CD69 and CD25 is reported as normal [45, 46, 55], and TCR signalling appears to be intact. Given that the proliferation defects seen in autophagy deficient T cells do not appear to be due to failure of TCR signalling, how else might they be explained? Autophagy inhibition limits ATP levels following T cell stimulation, and reduces lactate production [18]. Associated with these metabolic defects are decreased phosphorylation of the mechanistic target of rapamycin (mTOR) target S6K1, and AMP-activated protein kinase (AMPK). However, TCR stimulation and effector differentiation strongly induces mTOR, which inhibits autophagy via phosphorylation of unc-51 like autophagy activating kinase 1 (ULK1) at Ser 757 [23]. Whilst AMPK, which senses energy deprivation and induces autophagy may be active at an early stage following stimulation, autophagy is seen to increase following effector differentiation, and so the regulatory mechanisms in play are unclear. Given the demonstration that JNK signalling is required for autophagy activation in lymphocytes and other cellular systems [26, 64], it is possible that this pathway provides an mTOR independent route to autophagy induction. Treatment of autophagy deficient cells with methyl-pyruvate in an attempt to correct the bioenergetic defect has been demonstrated to restore IL-2 secretion [18].

3.3 Autophagy in T Cell Differentiation

Following from defects in signalling events and disrupted metabolism, the formation and function of T cell subsets is impaired. Most models demonstrate that when autophagy genes are deleted, T cells assume a $CD44^{+}CD62L^{-}$ memory-like phenotype [21, 29, 37, 46, 47, 55, 65]. This is associated with T cell lymphopenia, and it is therefore likely that the assumption of this phenotype is due to homeostatic proliferation [8, 16], as suggested by increased expression of the proliferation markers Ki67 and CD24 in Atg7 deficient $CD8^{+}$ T cells [47]. Chimeras generated by reconstitution of lethally irradiated hosts with equal ratios of $Atg7^{-/-}$ and control bone marrow from an alternative CD45 allotype (CD45.1) demonstrate that it is indeed probable that the memory-like phenotype is due to homeostatic proliferation [47]. Following stimulation in polarising conditions, Th17 differentiation is notably spared in *Beclin1*$^{-/-}$ chimeras, compared with other effector subsets [25]. Th17 cells have a glycolytic metabolism governed by the expression of hypoxia inducible factor 1α (HIF1α), although aerobic glycolysis is common to all T_{eff} cells. Interestingly, HIF1α is degraded by chaperone-mediated autophagy [19], and the drug halofuginone, which activates the nutrient sensor general control non-repressible-2 (GCN2), associated with autophagy induction, inhibits Th17 polarisation [58, 59]. The possibility exists therefore that autophagy negatively regulates Th17 formation. However, *in vivo* findings in the CD4-Cre × Beclin1$^{F/F}$ mouse are against this, as T cell specific deficiency of Beclin1 completely protects against the development of experimental autoimmune encephalomyelitis (EAE), a disease dominated by the Th17 effector response [25, 38]. Autophagy also plays an important role in regulatory T cell function. When Vps34 is deleted in CD4$^+$ T cells, there is an approximately 30% reduction in T_{reg} numbers. Moreover, these T_{regs} are functionally impaired, with poor suppressive ability *in vitro* [37]. The CD4-Cre × Vps34$^{F/F}$ model develops a progressive wasting syndrome characterised by colitis and anaemia from 18 weeks, which appears to be due to T_{reg} failure. Adoptive transfer of *Vps34*$^{-/-}$ T_{regs} fails to abrogate a colitis model induced by allotransplantation of CD4$^+$CD25$^-$ T cells into a *Rag2*$^{-/-}$ host. In contrast to Th17 cells, there is reason to suppose that autophagy may have a positive effect on T_{reg} function. T_{regs}, in contrast to T_{eff} cells, rely on oxidative phosphorylation and fatty acid metabolism [31]. In fact, blocking glycolysis promotes T_{reg} differentiation [52]. Consistent with this is the finding that mTOR inhibition, either pharmacologically with rapamycin or by genetic deletion limits T_{eff} but enhances T_{reg} formation [12]. Although autophagy has not yet been directly quantified in T_{regs}, there is a high level of AMPK activation in this subset that could be an important regulatory pathway [31].

Following the primary immune response, the generation and maintenance of T cell memory is again associated with a switch to a distinct metabolism, which has much in common with that of T_{regs}. Memory T cells also predominantly utilise oxidative phosphorylation, fuelled by fatty acids [41]. The signalling pathways that promote memory cell formation are also AMPK activation, and mTOR inhibition, both of are expected to induce autophagy. Directly quantification of autophagy in memory CD8$^+$ cells confirms this prediction [47, 66]. There is a greatly impaired CD8$^+$ memory response when autophagy is conditionally disrupted in T cells in CD4-Cre × Atg7$^{F/F}$ or granzyme B-Cre × Atg7$^{F/F}$ mouse models [47, 66]. Interestingly, and perhaps somewhat surprisingly given the evidence that autophagy deficiency affects T cell proliferation, the primary CD8$^+$ response is comparable between Atg7$^{-/-}$ and control mice. However, there is profound impairment of the subsequent generation of influenza, murine cytomegalovirus (MCMV), or lymphocytic choriomeningitis virus (LCMV) specific CD8$^+$ memory T cells [47, 66]. There are many potential roles for autophagy in the maintenance of memory T cells. Mitophagy is impaired in *Atg7*$^{-/-}$ CD8$^+$ memory T cells, and there is mitochondrial membrane hyperpolarization [47]. This finding, in common with other reports of T cell specific deletion of autophagy, may predispose to apoptosis, which is indeed observed. An alternative function of autophagy may be the supply of fatty acids for mitochondrial β-oxidation via the process of lipophagy [27, 47]. Of considerable clinical relevance is that the impairment of CD8$^+$ T cell memory seen in aging can be ameliorated by treatment of mice with the mTOR-independent autophagy inducing agent spermidine [13, 47]. Immune senescence occurs with aging, and is a process associated with a reduction in levels of autophagy [44]. There is therefore a rationale for attempting to restore autophagy and thereby rejuvenate the immune system.

4 B Cells

4.1 Autophagy in B Cell Development

The role of autophagy in B cell development has been less extensively studied than in T cells, and both similarities and substantial differences exist. As has been discussed above, B cells develop from the HSC derived CLP, and autophagy is therefore important at the earliest stage in their life cycle. Following commitment to the B cell lineage, there is progression through the pro-B, pre-B, immature and then mature B cell stages, which is associated with rearrangement and expression of immunoglobulin genes. Both *Beclin1* expression, quantified through a GFP fusion protein, and the number of autophagosomes are increased at the earliest stages of B cell development, and progressively decline as the cell matures [4, 9]. Commensurate with these findings, chimeric mice generated by transplantation of $Atg5^{-/-}$ or $Beclin1^{-/-}$ fetal liver cells into irradiated $Rag1^{-/-}$ hosts demonstrate significant abnormalities in B cell development. There is normal formation of Hardy stages A-C, which includes pre-pro-B to early pre-B cells, *i.e.* before B cell receptor (BCR) expression. However, subsequent development into stages D-F (late pre-B to mature B cell) is significantly impaired [3, 32]. Interestingly, this defect is not seen in CD19-Cre × $Atg5^{F/F}$ or $Atg7^{F/F}$ mice [7, 32], suggesting therefore a requirement for autophagy before expression of CD19; *i.e.* the early pro-B cell stage. Once B cells have entered the periphery however, the B2 pool is essentially normal in the resting state [7, 32, 42]. Whilst B2 subpopulations are normal, a marked reduction in the number of peritoneal B1a and B1b B cells is reported in the $Atg5^{-/-}$ chimera, and in CD19-Cre × $Atg7^{F/F}$ mice, but in a $Beclin1^{-/-}$ chimera [3, 7, 32]. There are differing reports in CD19-Cre × $Atg5^{F/F}$ mice, with two groups reporting a decrease in B1a cells [10, 42], but another finding no change [32]. Why these discrepancies might exist is not clear, given the similarities between the experimental systems. B1 cells are innate-like lymphocytes that are the major source of natural immunoglobulin, producing a limited repertoire of antibodies that tend to cross-react with self and microbial antigens [5]. Notably, in the adult mouse B1 cells have long-lived, self-renewing properties akin to stem cells, and may therefore also be reliant on autophagy for homeostasis. The requirement for autophagy at the early stages of B cell development, before surface expression of immunoglobulin, is paralleled with maximal autophagy in DN thymocytes. The role of autophagy in this circumstance is unknown, but Bcl-2 is up-regulated in a similar pattern [30]. It seems likely that autophagy is activated as a survival mechanism when the B cell is threatened, which may occur as a consequence of failure to produce a functional BCR, or through excessive self-reactivity. The BH3-only pro-apoptotic protein Bim is of great importance in the death of auto-reactive B cells, and acts by inhibition of Bcl-2 [14]. Bim may modulate autophagy, dependent on its subcellular location [28]. Alternatively, reductions in Bcl-2 may liberate Beclin-1, and initiate autophagosome formation [39].

4.2 Autophagy in B Cell Activation

Compared with T cells, the effects of B cell stimulation on autophagy have received relatively little attention. Stimulation of murine B cells through TLR4 by LPS leads to a progressive increase in LC3-GFP puncta formation, in association with increased LC3-II conversion [42]. After 4 days however, autophagy levels diminish. Engagement of the BCR also leads to activation of autophagy in murine primary B cells cultured for 24–36 h [20, 63], with attenuation of this effect if CD40 co-stimulation is provided [63]. As with T cells, the signalling pathways leading to, and functional consequences of early autophagy activation are largely unclear. However, it has been reported that BCR engagement leads to TLR9 recruitment from endosomes to autophagosome-like compartments, and it is here that hyperactivation of mitogen-activated protein kinase (MAPK) occurs [6]. Largely similar results are seen in human primary B cells, although autophagy is more potently induced if the cells are untreated, which is in keeping with the function of autophagy as a cell survival mechanism triggered by growth stimulus withdrawal [9]. There is normal early cell proliferation following TLR stimulation in autophagy deficient B cells, and the proportion of apoptotic naïve cells is similar [3, 7, 42]. However, terminal differentiation into antibody secreting plasma cells or memory B cells is markedly impaired when autophagy is deficient [7, 9, 10, 42], although as with T cells, there are significant differences in the details of the phenotypic descriptions. Following encounter with their cognate antigen, B cells may differentiate into short-lived plasmablasts through T-independent stimulation (TI), or undergo the germinal centre reaction to produce an affinity matured long-lived plasma cell or memory B cell if the antigen is T-dependent and the B cell receives T cell help (TD). The extremely high rates of protein synthesis demanded by the plasma cell are supported by extinction of B cell differentiation transcription factors, and the reciprocal expression of a secretory genetic program driven by, notably, Blimp-1 [50, 51]. The transcription factor XBP-1, downstream of Blimp-1, is important for the activation of the unfolded protein response (UPR), which helps adapt the cell to the metabolic stress produced by the accumulation of defective proteins during immunoglobulin synthesis. Following TLR stimulation *in vitro*, in a Vav-Cre × Atg7$^{F/F}$ mouse model, which has conditional deletion of Atg7 in all haematopoietic cells, and in one description of the CD19-Cre × Atg5$^{F/F}$ model, there is inhibition of the differentiation of resting B cells into antigen-secreting plasmablasts following a culture period of 48 h [9, 10]. In the report by Pengo et al. of this latter phenotype however, there was no difference in the formation of CD138^{+} antibody secreting cells (ASCs), but nonetheless double the number of these cells were apoptotic [42]. *In vivo*, there is clear evidence that the antibody response following immunization is impaired. However, as before there are differences in the details of the descriptions. Both reports of CD19-Cre × Atg5$^{F/F}$ mice describe a significant impairment of the primary antibody response, both for TI and TD antigens [10, 42]. However, the equivalent Atg7 conditional knockout mouse has a normal primary antibody response to NP-KLH, a TD antigen [7], although the secondary antibody response is severely limited. Murine memory B cells have a high level of expression of autophagy genes, and increased numbers of $LC3^+$ punctae relative to other B cell subsets [7]. They have intrinsically lower levels of apoptosis than *e.g.* germinal centre B cells, and following deletion of Atg7 there is a highly significant loss of memory B cell survival, accounting for the limitation of the secondary antibody response [7]. *Atg7*$^{-/-}$ memory B cells have evidence of mitochondrial dysfunction, *e.g.* increased ROS and decreased mitochondrial membrane potential, findings in common with *Atg7*$^{-/-}$ $CD8^+$ memory T cells [47]. Treatment of *Atg7*$^{-/-}$ memory B cells with the anti-oxidant N-acetylcysteine is able to improve cell survival *in vitro*, and *in vivo* there is partial restoration of the secondary antibody response [7]. The biological relevance of autophagy in memory B cell survival is illustrated by the high mortality in CD19-Cre × Atg7$^{F/F}$ mice after a re-challenge with influenza following a sub-lethal dose 2 months before, compared with complete protection elicited in wild type controls [7]. The functions of autophagy in plasma cell formation have been explored in the CD19-Cre × Atg5$^{F/F}$ mouse model. As has been reported in T cell autophagy deficiency, there is expansion of the endoplasmic reticulum in *Atg5*$^{-/-}$ ASCs generated following LPS stimulation [22, 42]. Interestingly, it has been demonstrated that there is in fact an increase in immunoglobulin secretion in *Atg5*$^{-/-}$ ASCs generated following LPS stimulation [42]. The mechanism behind this finding is unclear, as whilst it has been attributed to increased *Blimp-1* expression due to enhanced ER stress, it has also been reported that *Blimp-1* mRNA levels are reduced in LPS treated *Atg5*$^{-/-}$ B cells, albeit following a slightly shorter period of stimulation [10]. This enhancement of immunoglobulin synthesis does not appear to be present in a Vav-Atg7$^{-/-}$ *in vitro* differentiation model, and there is also not an apparent increase in basal immunoglobulin levels [9, 10]. Impairment of *in vivo* immune function in the CD19-Cre × Atg5$^{F/F}$ mouse model is demonstrated by reduction in parasite specific IgE and IgG1 levels compared with wild type controls following infection with the intestinal parasite *H. polygyrus*, and a reduction in severity of the dextran-induced colitis model [10].

5 Clinical Perspectives

The importance of autophagy in the adaptive immune system has been demonstrated in experimental systems, but what is its potential as a treatment target in human disease?

Activation of autophagy is an attractive goal for potentiation of the immune response following vaccination. The phenomenon of immune senescence, that is, impairment of immunity with advanced age, represents a great cause of morbidity and mortality in the increasingly elderly population of the developed world [33]. It has been demonstrated that aging is associated with decreased autophagy in $CD8^+$ T cells [44], and that in mice, activation of autophagy with spermidine may offset the normal age-associated reduction in vaccination efficacy [47].

Alternatively, in autoimmune disease such as systemic lupus erythematosus (SLE), reduction of autophagy may be the desired outcome. SLE is a systemic autoimmune disease characterised by activation of multiple arms of the immune system, resulting in the production of pathogenic auto-antibodies against components of the nucleus [48]. SLE is associated with enhanced autophagy in T and B cells, where it may be acting as a survival mechanism to prevent the deletion of auto-reactive cells [2, 9, 17]. One of the most commonly used medications in the treatment of SLE, hydroxychloroquine, inhibits autophagy, and this function may be one of its therapeutic mechanisms [61]. Other experimental agents, such as the therapeutic peptide Lupuzor (P140), may also act by inhibition of autophagy [36].

6 Conclusions

The realisation that autophagy is important in the function of the adaptive immune system has only occurred relatively recently, but already there are many studies exploring its role, both in lymphocyte development and homeostasis. However, it is also clear that whilst there is agreement in the literature in broad terms, there are also significant discrepancies between phenotypic descriptions, often in important details. Whether this lies in experimental variability, or is due to fundamental differences between the phenotypes generated by the choice of autophagy gene deletion in mouse models is unknown. There remain a number of unanswered questions about the role that autophagy plays in lymphocytes. Firstly, what function does autophagy have in early lymphocyte development? In both T and B cells there is maximal activation of autophagy before surface expression of their receptors. Autophagy could potentially be acting as a pro-survival mechanism activated when apoptosis is threatened, *e.g.* when there is failure to productively rearrange receptor genes. Secondly, how is autophagy regulated following receptor engagement? Finally, how can the metabolism associated with autophagy, *i.e.* predominantly oxidative phosphorylation, be reconciled with the glycolytic metabolism driven by mTOR signalling seen in activated T cells? There is therefore scope for much further work before our understanding of the multiple roles for autophagy in lymphocytes is complete.

References

1. Aichinger M, Wu C, Nedjic J, Klein L (2013) Macroautophagy substrates are loaded onto MHC class II of medullary thymic epithelial cells for central tolerance. J Exp Med 210:287–300
2. Alessandri C, Barbati C, Vacirca D, Piscopo P, Confaloni A, Sanchez M, Maselli A, Colasanti T, Conti F, Truglia S et al (2012) T lymphocytes from patients with systemic lupus erythematosus are resistant to induction of autophagy. FASEB J 26:4722–4732
3. Arsov I, Adebayo A, Kucerova-Levisohn M, Haye J, MacNeil M, Papavasiliou FN, Yue Z, Ortiz BD (2011) A role for autophagic protein Beclin 1 early in lymphocyte development. J Immunol 186:2201–2209

4. Arsov I, Li X, Matthews G, Coradin J, Hartmann B, Simon AK, Sealfon SC, Yue Z (2008) BAC-mediated transgenic expression of fluorescent autophagic protein Beclin 1 reveals a role for Beclin 1 in lymphocyte development. Cell Death Differ 15:1385–1395
5. Baumgarth N (2010) The double life of a B-1 cell: self-reactivity selects for protective effector functions. Nat Rev Immunol 11:34–46
6. Chaturvedi A, Dorward D, Pierce SK (2008) The B cell receptor governs the subcellular location of toll-like receptor 9 leading to hyperresponses to DNA-containing antigens. Immunity 28:799–809
7. Chen M, Hong MJ, Sun H, Wang L, Shi X, Gilbert BE, Corry DB, Kheradmand F, Wang J (2014) Essential role for autophagy in the maintenance of immunological memory against influenza infection. Nat Med 20:503–510
8. Cho BK, Rao VP, Ge Q, Eisen HN, Chen J (2000) Homeostasis-stimulated proliferation drives naive T cells to differentiate directly into memory T cells. J Exp Med 192:549–556
9. Clarke AJ, Ellinghaus U, Cortini A, Stranks A, Simon AK, Botto M, Vyse TJ (2014) Autophagy is activated in systemic lupus erythematosus and required for plasmablast development. Ann Rheum Dis 74:912–920
10. Conway KL, Kuballa P, Khor B, Zhang M, Shi HN, Virgin HW, Xavier RJ (2013) ATG5 regulates plasma cell differentiation. Autophagy 9:528–537
11. Davis RJ (2000) Signal transduction by the JNK group of MAP kinases. Cell 103:239–252
12. Delgoffe GM, Kole TP, Zheng Y, Zarek PE, Matthews KL, Xiao B, Worley PF, Kozma SC, Powell JD (2009) The mTOR kinase differentially regulates effector and regulatory T cell lineage commitment. Immunity 30:832–844
13. Eisenberg T, Knauer H, Schauer A, Büttner S, Ruckenstuhl C, Carmona-Gutierrez D, Ring J, Schroeder S, Magnes C, Antonacci L et al (2009) Induction of autophagy by spermidine promotes longevity. Nat Cell Biol 11:1305–1314
14. Enders A, Bouillet P, Puthalakath H, Xu Y, Tarlinton DM, Strasser A (2003) Loss of the Pro-apoptotic BH3-only Bcl-2 family member Bim inhibits BCR stimulation-induced apoptosis and deletion of autoreactive B cells. J Exp Med 198:1119–1126
15. Espert L (2006) Autophagy is involved in T cell death after binding of HIV-1 envelope proteins to CXCR4. J Clin Invest 116:2161–2172
16. Goldrath AW, Bogatzki LY, Bevan MJ (2000) Naive T cells transiently acquire a memory-like phenotype during homeostasis-driven proliferation. J Exp Med 192:557–564
17. Gros F, Arnold J, Page N, Décossas M, Korganow A-S, Martin T, Muller S (2012) Macroautophagy is deregulated in murine and human lupus T lymphocytes. Autophagy 8:1054–1053
18. Hubbard VM, Valdor R, Patel B, Singh R, Cuervo AM, Macian F (2010) Macroautophagy regulates energy metabolism during effector T cell activation. J Immunol 185:7349–7357
19. Hubbi ME, Hu H, Kshitiz, Ahmed I, Levchenko A, Semenza GL (2013) Chaperone-mediated autophagy targets hypoxia-inducible factor-1 (HIF-1) for lysosomal degradation. J Biol Chem 288:10703–10714
20. Ireland JM, Unanue ER (2011) Autophagy in antigen-presenting cells results in presentation of citrullinated peptides to CD4 T cells. J Exp Med 208:2625–2632
21. Jia W, He YW (2011) Temporal regulation of intracellular organelle homeostasis in T lymphocytes by autophagy. J Immunol 186:5313–5322
22. Jia W, Pua HH, Li QJ, He YW (2011) Autophagy regulates endoplasmic reticulum homeostasis and calcium mobilization in T lymphocytes. J Immunol 186:1564–1574
23. Kim J, Kundu M, Viollet B, Guan KL (2011) AMPK and mTOR regulate autophagy through direct phosphorylation of Ulk1. Nat Cell Biol 13:132–141
24. Kondo M, Weissman IL, Akashi K (1997) Identification of clonogenic common lymphoid progenitors in mouse bone marrow. Cell 91:661–672
25. Kovacs JR, Li C, Yang Q, Li G, Garcia IG, Ju S, Roodman DG, Windle JJ, Zhang X, Lu B (2011) Autophagy promotes T-cell survival through degradation of proteins of the cell death machinery. Cell Death Differ 9:144–152

26. Li C, Capan E, Zhao Y, Zhao J, Stolz D, Watkins SC, Jin S, Lu B (2006) Autophagy is induced in CD4+ T cells and important for the growth factor-withdrawal cell death. J Immunol 177:5163–5168
27. Liu K, Czaja MJ (2012) Regulation of lipid stores and metabolism by lipophagy. Cell Death Differ 20:3–11
28. Luo S, Garcia-Arencibia M, Zhao R, Puri C, Toh PP, Sadiq O, Rubinsztein DC (2012) Bim inhibits autophagy by recruiting Beclin 1 to microtubules. Mol Cell 47:359–370
29. McLeod IX, Zhou X, Li QJ, Wang F, He YW (2011) The class III kinase Vps34 promotes T lymphocyte survival through regulating IL-7R surface expression. J Immunol 187:5051–5061
30. Merino R, Ding L, Veis DJ, Korsmeyer SJ, Nuñez G (1994) Developmental regulation of the Bcl-2 protein and susceptibility to cell death in B lymphocytes. EMBO J 13:683–691
31. Michalek RD, Gerriets VA, Jacobs SR, Macintyre AN, MacIver NJ, Mason EF, Sullivan SA, Nichols AG, Rathmell JC (2011) Cutting edge: distinct glycolytic and lipid oxidative metabolic programs are essential for effector and regulatory CD4+ T cell subsets. J Immunol 186:3299–3303
32. Miller BC, Zhao Z, Stephenson LM, Cadwell K, Pua HH, Lee HK, Mizushima NN, Iwasaki A, He YW, Swat W et al (2008) The autophagy gene ATG5 plays an essential role in B lymphocyte development. Autophagy 4:309–314
33. Montoya-Ortiz G (2013) Immunosenescence, aging, and systemic lupus erythematous. Autoimmune Dis 2013:267078. doi:10.1155/2013/267078
34. Mortensen M, Soilleux EJ, Djordjevic G, Tripp R, Lutteropp M, Sadighi-Akha E, Stranks AJ, Glanville J, Knight S et al (2011) The autophagy protein Atg7 is essential for hematopoietic stem cell maintenance. J Exp Med 208:455–467
35. Nedjic J, Aichinger M, Emmerich J, Mizushima N, Klein L (2008) Autophagy in thymic epithelium shapes the T-cell repertoire and is essential for tolerance. Nature 455:396–400
36. Page N, Gros F, Schall N, Décossas M, Bagnard D, Briand JP, Muller S (2010) HSC70 blockade by the therapeutic peptide P140 affects autophagic processes and endogenous MHCII presentation in murine lupus. Ann Rheum Dis 70:837–843
37. Parekh VV, Wu L, Boyd KL, Williams JA, Gaddy JA, Olivares-Villagomez D, Cover TL, Zong WX, Zhang J, Van Kaer L (2013) Impaired autophagy, defective T cell homeostasis, and a wasting syndrome in mice with a T cell-specific deletion of Vps34. J Immunol 190:5086–5101
38. Park H, Li Z, Yang XO, Chang SH, Nurieva R, Wang YH, Wang Y, Hood L, Zhu Z, Tian Q et al (2005) A distinct lineage of CD4 T cells regulates tissue inflammation by producing interleukin 17. Nat Immunol 6:1133–1141
39. Pattingre S, Tassa A, Qu X, Garuti R, Liang XH, Mizushima N, Packer M, Schneider MD, Levine B (2005) Bcl-2 antiapoptotic proteins inhibit Beclin 1-dependent autophagy. Cell 122:927–939
40. Paul S, Kashyap AK, Jia W, He YW, Schaefer BC (2012) Selective autophagy of the adaptor protein Bcl10 modulates T cell receptor activation of NF-kB. Immunity 36:947–958
41. Pearce EL, Walsh MC, Cejas PJ, Harms GM, Shen H, Wang LS, Jones RG, Choi Y (2009) Enhancing CD8 T-cell memory by modulating fatty acid metabolism. Nature 460:103–107
42. Pengo N, Scolari M, Oliva L, Milan E, Mainoldi F, Raimondi A, Fagioli C, Merlini A, Mariani E, Pasqualetto E et al (2013) Plasma cells require autophagy for sustainable immunoglobulin production. Nat Immunol 14:298–305
43. Perl A, Gergely P Jr, Nagy G, Koncz A, Banki K (2004) Mitochondrial hyperpolarization: a checkpoint of T-cell life, death and autoimmunity. Trends Immunol 25:360–367
44. Phadwal K, Alegre-Abarrategui J, Watson AS, Pike L, Anbalagan S, Hammond EM, Wade-Martins R, McMichael A, Klenerman P, Simon AK (2012) A novel method for autophagy detection in primary cells: impaired levels of macroautophagy in immunosenescent T cells. Autophagy 8:677–689
45. Pua HH, Dzhagalov I, Chuck M, Mizushima N, He YW (2007) A critical role for the autophagy gene Atg5 in T cell survival and proliferation. J Exp Med 204:25–31

46. Pua HH, Guo J, Komatsu M, He YW (2009) Autophagy is essential for mitochondrial clearance in mature T lymphocytes. J Immunol 182:4046–4055
47. Puleston DJ, Zhang H, Powell TJ, Lipina E, Sims S, Panse I, Watson AS, Cerundolo V, Townsend AR, Klenerman P et al (2014) Autophagy is a critical regulator of memory CD8(+) T cell formation. ELife. 3. doi:10.7554/eLife.03706
48. Rahman A, Isenberg DA (2008) Systemic lupus erythematosus. N Engl J Med 358:929–939
49. Sagnier S, Daussy CF, Borel S, Robert-Hebmann V, Faure M, Blanchet FP, Beaumelle B, Biard-Piechaczyk M, Espert L (2014) Autophagy restricts HIV-1 infection by selectively degrading Tat in CD4+ T lymphocytes. J Virol 89:615–625
50. Shaffer AL, Lin KI, Kuo TC, Yu X, Hurt EM, Rosenwald A, Giltnane JM, Yang L, Zhao H, Calame K et al (2002) Blimp-1 orchestrates plasma cell differentiation by extinguishing the mature B cell gene expression program. Immunity 17:51–62
51. Shapiro-Shelef M, Calame K (2005) Regulation of plasma-cell development. Nat Rev Immunol 5:230–242
52. Shi LZ, Wang R, Huang G, Vogel P, Neale G, Green DR, Chi H (2011) HIF1-dependent glycolytic pathway orchestrates a metabolic checkpoint for the differentiation of TH17 and treg cells. J Exp Med 208:1367–1376
53. Simsek T, Kocabas F, Zheng J, DeBerardinis RJ, Mahmoud AI, Olson EN, Schneider JW, Zhang CC, Sadek HA (2010) The distinct metabolic profileof hematopoietic stem cells reflects their location in a hypoxic niche. Cell Stem Cell 7:380–390
54. Starr TK, Jameson SC, Hogquist KA (2003) Positive and negative selection of T cells. Annu Rev Immunol 21:139–176
55. Stephenson LM, Miller BC, Ng A, Eisenberg J, Zhao Z, Cadwell K, Graham DB, Mizushima NN, Xavier R, Virgin HW et al (2009) Identification of Atg5-dependent transcriptional changes and increases in mitochondrial mass in Atg5-deficient T lymphocytes. Autophagy 5:625–635
56. Su B, Jacinto E, Hibi M, Kallunki T, Karin M, Ben-Neriah Y (1994) JNK is involved in signal integration during costimulation of T lymphocytes. Cell 77:727–736
57. Sukseree S, Mildner M, Rossiter H, Pammer J, Zhang C-F, Watanapokasin R, Tschachler E, Eckhart L (2012) Autophagy in the thymic epithelium is dispensable for the development of self-tolerance in a novel mouse model. PLoS One 7, e38933. doi:10.1371/journal.pone.0038933
58. Sundrud MS, Koralov SB, Feuerer M, Calado DP et al (2009) Halofuginone inhibits TH17 cell differentiation by activating the amino acid starvation response. Science 324:1334–1338
59. Tallóczy Z, Jiang W, Virgin HW, Leib DA, Scheuner D, Kaufman RJ, Eskelinen E-L, Levine B (2002) Regulation of starvation- and virus-induced autophagy by the eIF2alpha kinase signaling pathway. Proc Natl Acad Sci U S A 99:190–195
60. Thome M (2004) CARMA1, BCL-10 and MALT1 in lymphocyte development and activation. Nat Rev Immunol 4:348–359
61. van Loosdregt J, Spreafico R, Rossetti M (2013) Hydroxychloroquine preferentially induces apoptosis of CD45RO+ effector T cells by inhibiting autophagy: a possible mechanism for therapeutic modulation of T cells. J Allergy Clin Immunol 131:1443–1446. e1. doi:10.1016/j.jaci.2013.02.026
62. Warr MR, Binnewies M, Flach J, Reynaud D, Garg T, Malhotra R, Debnath J, Passegué E (2013) FOXO3A directs a protective autophagy program in haematopoietic stem cells. Nature 494:323–327
63. Watanabe K, Ichinose S, Hayashizaki K, Tsubata T (2008) Induction of autophagy by B cell antigen receptor stimulation and its inhibition by costimulation. Biochem Biophys Res Commun 374:274–281
64. Wei Y, Pattingre S, Sinha S, Bassik M, Levine B (2008) JNK1-mediated phosphorylation of Bcl-2 regulates starvation-induced autophagy. Mol Cell 30:678–688
65. Willinger T, Flavell RA (2012) Canonical autophagy dependent on the class III phosphoinositide-3 kinase Vps34 is required for naive T-cell homeostasis. Proc Natl Acad Sci U S A 109:8670–8675
66. Xu X, Li S, Han J-H, Ye L, Tan WG, Konieczny BT, Bruinsma MW, Martinez J, Pearce EL, Green DR et al (2014) Autophagy is essential for effector CD8. Nat Immunol 15:1152–1161

Autophagy in Host Defense Against Viruses

Jin Wang and Min Chen

Abstract The immune system employs both the adaptive and innate immune responses to combat viral infections. Autophagy, a cytoplasmic lysosomal degradation process, has emerged as an important mechanism in the regulation of multiple aspects of cellular functions in the immune system against viruses. Autophagy may directly contribute to the degradation of viral components. Autophagy can also promote innate immunity and facilitate the processing of viral antigens for the activation of antigen-specific T cells. On the other hand, some viruses have adapted to use autophagy machinery for replication. As an important cellular mechanism to remove damaged or excess protein aggregates and organelles, autophagy plays a critical role in the protection of lymphocytes against stress to prolong their survival, especially the maintenance of long-lived memory cells. Here we discuss the involvement of autopahgy in the regulation of immune responses against viral infections, with emphasis on the roles for autophagy in the protection of immunological memory.

Abbreviations

5'pp	5'-diphosphates
5'ppp	5'-triphosphate
AIF	Apoptosis Inducing Factor
Alox5	Arachidonate 5-lipoxygenase
APC	Antigen-presenting cell
ASCs	Antibody-secreting plasma cells
Atg	Autophagy-related
ATP	Adenosine triphosphate
Bcl-2	B-cell lymphoma 2

J. Wang (✉) • M. Chen
Department of Pathology and Immunology, Baylor College of Medicine,
One Baylor Plaza, Houston, TX 77030, USA
e-mail: jinwang@bcm.edu

M.C. Maiuri, D. De Stefano (eds.), *Autophagy Networks in Inflammation*,
Progress in Inflammation Research, DOI 10.1007/978-3-319-30079-5_10

cGAMP (cGMP-AMP)	Cyclic guanosine monophosphate–adenosine monophosphate
cGAS	cGAMP synthase
CNS	Central nervous system
EBNA1	Epstein–Barr virus
eIF2α	Eukaryotic Initiation Factor 2α
ER	Endoplasmic reticulum
FCCP	Carbonyl cyanide p-trifluoromethoxyphenylhydrazone
FOXO3a	Forkhead box O3a
GABARAP	Gamma-aminobutyric acid receptor-associated protein
GCs	Germinal centers
HA	Hemagglutin
HBV	Hepatitis B virus
HCV	Hepatitis C virus
HIF	Hypoxia Inducible Factors
HIV-1	Human immunodeficiency virus-1
HSV	Herpes simplex virus
IFN	Interferon
IL	Interleukin
IRGM	Immunity-associated GTPase family M
LC3	Microtubule-associated protein 1 light chain 3
LGP2	Laboratory of Genetics and Physiology 2
LPS	Lipopolysaccharide
MAVS	Mitochondrial antiviral signaling protein
MDA5	Melanoma Differentiation-Associated protein 5
MEFs	Mouse embryonic fibroblasts
MeV	Measles virus
MHC	Major histocompatibility complex
miRNA	Micro RNA
mTOR	Mammalian target of rapamycin
NLRs	NOD-like receptors
NLRX1	Nucleotide-binding oligomerization domain, leucine rich repeat containing X1
p62/SQSTM1	Sequestosome 1
RIG-1	RIG-1, retinoic acid-inducible gene 1
RLRs	RIG-1-like receptors
ROS	Reactive oxygen species
SH3GLB1	SH3-Domain GRB2-Like Endophilin B1
STING	Stimulator of interferon genes
TCR	T cell receptor
TLRs	Toll-like receptors
TUFM	Tu translation elongation factor
ULK1	UNC-51-like kinase 1.

1 Autophagy as a Direct Degradation Mechanism for the Suppression of Viruses

Autophagy involves the sequestration of cytoplasmic components into double-membraned autophagosomes for lysosomal degradation. Autophagy can also serve as a direct degradation mechanism for the clearance of pathogens invading the cells. Sensing of pathogen invasion by host cells can induce autophagy. For example, responses of macrophages to Toll-like receptor 4 (TLR4) or TLR7 stimulation can induce autophagy to facilitate the removal of bacteria and viruses cite a difference reference [11, 75]. Ultrastructural studies indicate that intact HSV-1 virions can be targeted to autophagososmes, and that PKR-mediated autophagy can degrade intact HSV-1 virions and viral vesicles [69]. A role for autophagy in the clearance of viruses is further supported by studies using mouse models deficient in essential autophagy genes [13]. It has been shown that inhibition of autophagy by deletion of Atg5 impairs the clearance of Sindbis virus from the central nervous system (CNS) and increased the fatality in mice [50]. Beclin 1 can protect against the fatal encephalitis caused by Sindbis virus infection in the brain [39]. Beclin 1 also contributes to the inhibition of VSV infection in vivo [49]. Additionally, Atg16L-deficient mice are more prone to infection by Chikungunya virus or mouse norovirus [5, 23].

Besides direct effect on virions, autophagy may help to clear the viruses by degrading essential viral components. Indeed, it has been shown that interaction with an adaptor protein p62/SQSTM1 targets the nucleocapsid protein of Sindbis virus to autophagosome for degradation [51]. Although this does not lead to the inhibition of viral replication, such degradation inhibited neuronal cell death and reduced the mortality caused by Sindbis virus [51]. These studies suggest that autophagy can degrade the viruses or their components to control the infections.

Despite an important role for autophagy in antimicrobial host defense against pathogens, autophagy may be a culprit in the pathogenesis of some viruses. Pathogens have evolved to evade or exploit the autophagy machinery. For example, H5N1 influenza-induced lung damage and increased mortality in mice is observed in Atg5-deficient mice [66]. It has been demonstrated that liver-specific knockout of Atg5 inhibited the replication of hepatitis B virus (HBV) in a HBV transgenic mouse model [72]. This sugests that Atg5 is important for HBV replication *in vivo*. Hepatitis C virus also requires autophagy machinery, including Beclin 1, Atg4B, Atg5 and Atg12 for translation of viral RNA genome and the initiation of HCV replication [14]. Once HCV infection is established, the autophagy process become dispensable [14], indicating that some viruses have successfully adapted to use the components of autophagy as a integrated part of their life cycle.

2 Autophagy in Antigen Presenting Cells for the Processing and Presentation of Viral Antigens

Besides direct degradation and clearance of pathogens, autophagy can promote viral antigen processing for antigen presentation by MHC class II molecules. For example, the nuclear antigen 1 of the Epstein–Barr virus (EBNA1) in dendritic cells is targeted to autophagsome for degradation and presentation by class II MHC molecules [52]. Targeting such antigens to autophagosomes is important for the activation of antigen-specific T cells [46, 52]. Although not absolutely required, this process can enhance the activation of virus-specific T cells for viral clearance. The autophagic process interacts with the endosomal pathways, thereby allowing the loading of endogenous antigens, including those from invading pathogens, to MHC class II molecules [46]. This may also explain why autophagy is found to promote antigen processing for the class II, but not class I MHC molecules, which employs the proteosomal pathway.

The genetic evidence supporting a role for autophagy in class II antigen presentation is shown by a study using a mouse model with dendritic cell-specific deletion of Atg5 [30]. Atg5-deficient dendritic cells fail to elicit IFN-γ production by Th1 cells. Consequently, these mice display increased mortality after Herpes simplex virus 2 infection [30]. These studies suggest autophagy promotes the presentation of endogenous antigen to MHC II molecules to stimulate virus-specific T cell responses.

3 Regulation of Autophagy by Viruses

Although autophagy can restrict virus replication in the cells, viruses have evolved mechanisms to counteract autophagy. Kaposi's Sarcoma-Associated Herpesvirus K7 modulates the suppression of autophagy through Rubicon [37]. A mitochondrial nucleotide-binding, leucine-rich repeats (NLR)-containing protein NLRX1, forms a complex with another mitochondrial protein, Tu translation elongation factor (TUFM). The NLRX1-TUFM complex promotes autophagy through interactions with Atg5-Atg12 and inhibits type I IFN production [32]. Several RNA viruses, including Measles virus (MeV), HCV and human immunodeficiency virus-1 (HIV-1), recognize a cellular protein, immunity-associated GTPase family M (IRGM) [19]. IRGM interacts with Atg5, Atg10, LC3 and SH3GLB1 to induce autophagy and increase virus production [19]. HSV-1 suppresses autophagy via its neurovirulence factor ICP34.5, which dephosphorylates eIF2α and inhibits Beclin 1 in permissive cells. However, genomic DNA of HSV-1 can induce autophagy in nonpermissive cells in a STING (stimulator of interferon genes)-dependent manner [61]. Induction of both type I interferon responses and autophagy may inhibit viral replication in the non-permissive cells.

Different pattern recognition receptors, including toll-like receptors (TLRs), NOD-like receptors (NLRs) and RIG-1-like receptors (RLRs), are used to recognize components of infecting viruses or bacteria. These receptors have been shown to induce autophagy. ***TLRs***. Viruses and bacteria can be sensed by pattern recognition receptors on the cell surface and intracellularly. Different bacterial components can interact with different TLRs to induce autophagy. Such recognition has been shown to regulate autophagy. Notably, LPS can stimulate autophagy to enhance phagocytosis of bacteria in macrophages [75]. Single-stranded RNA can also stimulate TLR7 to activate autophagy [11]. Autophagy induced by TLRs can lead to direct pathogen clearance [75]. Additionally, TLRs can stimulate autophagy for better antigen presentation to induce adaptive immune responses against viruses [30]. ***NLRs***. Due to the lack of signal peptides or transmembrane domains, NOD like receptors (NLRs) are likely to be exclusively located inside the cell [7]. NLRs recognize peptideoglycan or other bacterial cell wall components. The interactions between NLRs and autophagy have been well established [62, 73]. As described above, NLRX1 increases autophagy through interaction with the Tu translation elongation factor (TUFM). This complex, in turn, interacts with the Atg5-Atg12 complex topromotes autophagy [31, 32]. ***RLRs***. RLRs are intracellular nucleic acid sensors. Different RLRs, such as RIG-I, MDA5, and LGP2, can recognize RNAs bearing 5'-triphosphate (5'ppp) or 5'-diphosphates (5'pp) associated with viral infections. Such detection triggers a signaling cascade via the adaptor mitochondrial antiviral signaling protein (MAVS) that results in the production of type I interferons [17]. Interestingly, Atg5-deficient mouse embryonic fibroblasts (MEFs) show enhanced RLR signaling, increased IFN production and resistance to infection by vesicular stomatitis virus [68]. This suggests that autophagy counteract RLR signaling in antiviral responses.

Sensing of cytoplasmic DNA Cellular DNAs are found in the nucleus or mitochondria, while the presence of DNA in the cytosol usually indicates infection by viruses or bacteria. In response to cytoplasmic DNA, the cell can generate cyclic GMP-AMP (cGAMP) by cGAMP synthase (cGAS) [6]. cGAMP can bind to Stimulator of Interferon Genes (STING) to induce type I interferon production [10, 36, 76]. Such generation of a secondary messenger in the cytoplasm enables the host cells to mount interferon responses to DNA carried by invading pathogens. Interestingly, binding of cGAMP to STING can also activate Beclin 1 to induce autophagy [38]. During dengue virus infection, STING-dependent induction of autophagy may help to promote innate immune responses to viral infections [18, 53]. Both type I interferon responses and autophagy induced by STING signaling may help to provide protection against pathogens that introduce DNA in the cytoplasm.

4 Autophagy in Lymphocyte Development and Functions

After encountering viral infections, the immune system is capable of mounting innate and adaptive immune responses to combat viral infections. Autophagy has been shown to be important for the development or functions of different cell types involved in innate and adaptive systems. While autophagy can contribute to the inhibition of viral replication by degradation of viral components, autophagy-mediated degradation also facilitate the processing of viral components for presentation by APCs to activate virus-specific T cells. In addition, autophagy is important for the development of T cells. Mice with mutations in key autophagy genes in the T cell compartment show a significant reduction in mature T cells [22, 54, 58, 74]. Due to an important role for autophagy in T cell development, the use of mouse models for studying autophagy deficiency in T cells *in vivo* has been limited. Temporal deletion of autophagy genes in mature T cells without affecting T cell development will be valuable for evaluating the function of autophagy in mature T cells, including primary and memory T cell responses against infections.

In the B cell compartment, autophagy is required for the development of B-1a cells [44], but dispensable for convention B cell development [57]. We found that B-1a cells display increased expression of multiple autophagy genes compared to conventional B-2 cells [8]. The mechanisms for autophagy in the maintenance of B-1a cells are unclear. One possibility is that autophagy is required during a critical step in B-1a development. Another possibility is that B-1a cells may rely on autophagy to maintain their survival. Indeed, we found that $Atg7^{-/-}$ B-1a cells undergo increased cell death [8]. Interestingly, autophagy is not required for the development of conventional B-2 cells. Primary B cells and GC B cells have relatively low levels of autophagy compared to memory B cells [8].

Immunization of $B/Atg7^{-/-}$ mice leads to the production of normal levels of primary antibodies against antigens. The affinities and class switching of antibodies were normal in $B/Atg7^{-/-}$ mice. This suggests that autophagy is not essential for primary B cell responses. Rather, autophagy appears to be most critical for certain steps during B cell differentiation, such as the memory B cell stage. Autophagy is not required for primary B cell responses. Autophagy has also been shown to be important for the survival of antibody secreting cells using B cell-specific Atg5-deficient mice [9, 57]. Autophagy helps to sustain the ability for plasma cells in antibody production. Autophagy may protect plasma cells from ER stress induced by antibody secretion [57]. It is likely that such a function for autophagy in sustainable antibody production is important for the protection against virus infection.

Autophagy is not required for dendritic cell development [30]. However, stimulation of dendritic cells or macrophages can induce autophagy that may lead to antigen processing and pathogen clearance [46, 52, 75]. In T cells, autophagy can be induced by TCR stimulation and help to maintain proper metabolic T cell functions [20, 34, 58]. How autophagy is regulated in B cells is yet to be fully characterized. We have observed that autophagy in constitutively active in antigen-specific memory B cells and protect the survival of these cells [8], suggesting that autophagy is required for the long-term maintenance of memory B cells *in vivo*.

5 Active Autophagy in Memory B Cells

One of the important consequences of primary immune responses after encountering an infection is the establishment of long-lasting protection against the same pathogens. Compared to primary antibody responses, memory B cell responses are more rapid and are also qualitatively improved. The primary antibody response is characterized by initial production of IgM antibodies. This is followed by the class switching and production of other class of antibodies, such as IgG and IgA. Secondary responses usually contain class-switched IgG, IgA or IgE. Moreover, the V_H and V_L segments of antibodies also contain somatic hypermutations that increase the affinity of the antibodies to antigens. The formation and long-lasting persistence of immune memory cells is the basis of immunological memory.

Memory B cells are quiescent antigen-experienced, long-lived B cells generated after the primary antibody responses [27]. Memory B cells are heterozygous can be generated in response to both T cell–dependent and T cell–independent manners [43]. During T cell–dependent antigen stimulation, the interaction of follicular B cells with by follicular T helper cells in the germinal centers (GCs) leads to isotype class switching and somatic hypermutations in the immunoglobulin genes [42]. These antigen-specific GC B cells can give rise to memory B cells. After re-exposure to the same antigens, memory B cells rapidly proliferate and differentiate into antibody-secreting plasma cells (ASCs) to produce high-affinity antibodies that neutralize antigens.

We have found that genes that regulate different steps of autophagy are increased in memory B cells compared to other B cell subtypes [8]. In particular, ULK1, Atg14 and Beclin 1 that are important for the initiation of autophagy, are significantly increased in memory B cells compared to other B cell subsets. In addition, Atg7, Atg5, MAPLC3a, MAPLC3B and GABARAP that are required for autophagosome formation, are also increased in memory B cells. Both mouse and human antigen-specific memory B cells display active autophagy by LC3 punctate formation (Fig. 1). Disruption of mitochondrial membrane potential with carbonyl cyanide p-trifluoromethoxyphenylhydrazone (FCCP) further increases autophagy in memory B cells (Fig. 1a). This suggests that cellular stress, such as mitochondrial disruption, can promote autophagy in memory B cells. Consistently, a microarray study show that Atg7 is increased in memory B cells compared to GC or naive B cells [3]. This suggests that the level of autophagy is increased in memory B cells.

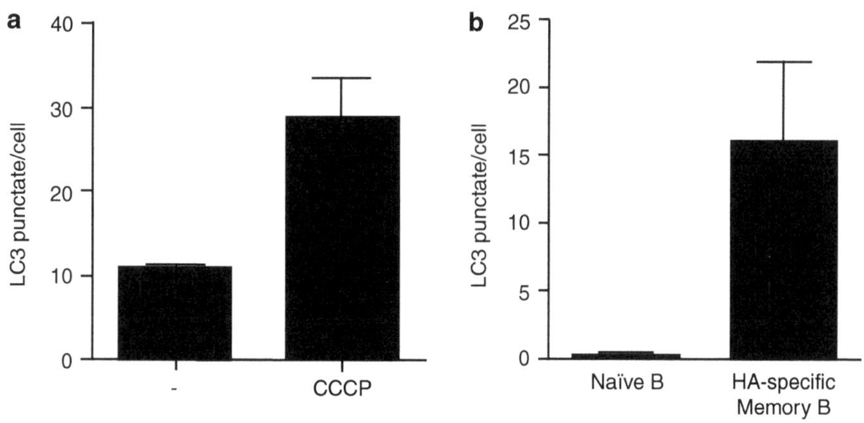

Fig. 1 Autophagy in mouse and human memory B cells. (**a**) Mouse memory B cells specific for 4-hydroxy-3-nitrophenylacetyl (*NP*) were incubated with or without 10 μM carbonyl cyanide p-trifluoromethoxyphenylhydrazone (FCCP) for 1 h. The cells were used for immunocytochemistry staining of LC3. LC3 punctates in the cells were quantitated. (**b**) Human $CD19^{+}CD27^{-}IgG^{-}IgA^{-}$ naïve B cells and H3N2 Influenza A hemagglutinin (HA)-specific $CD19^{+}CD27^{+}IgG^{+}HA^{+}$ memory B cells were sorted and used for immunocytochemistry staining of LC3 as in (**a**)

Currently, the molecular mechanisms for the increased expression of autophagy genes in memory B cells are unclear. Epigenetic regulation has emerged as an important mechanism for the regulation of autophagy gene expression [15]. Investigating different mechanisms for epigenetic regulation, including DNA methylation in the promoter and enhancer regions of autophagy genes, as well as histone methylation and histone acetylation at the autophagy gene loci, will help to uncover the potential molecular mechanisms for autophagy gene expression in memory B cells.

In addition to epigenetic regulation, several transcription factors that active autophagy under stress, including HIF-1α and FOXO3a, have been shown to be up-regulated in memory B cells by microarray [3]. Both the "open" chromatin structure and the availability of transcription factors are likely to be important for increased transcription of autophagy genes in memory B cells. However, the expression profile of different transcription factors for autophagy in memory B cells will need to be systemically investigated.

In addition to the increased gene expression, memory B cells contain autophagosomes that can be detected by immunocytochemistry staining of LC3 (Fig. 1), supporting the conclusions that autophagy is constitutively active in memory B cells. Autophagy can be induced by different stimuli or stress signals. Whether there is a constant trigger to keep the autophagy process active in memory B cells is unknown. We have stimulated IgG memory B cells by surface IgG crosslinking. However, such stimulation did not further increase autophagy in memory B cells [8]. Our study therefore suggests that the activation of autophagy in memory B cells is antigen-independent. Consistently, it has been reported that the persistence of memory B cells is independent of antigens [41], indicating that antigen-receptor stimulation is not required for the maintenance of memory B cells. Other extrinsic factors and intrinsic signaling mechanisms might be responsible for the activation of autophagy to protect memory B cells. The signaling mechanism for constitutive autophagy in memory B cells will be an important area for future investigation. It may provide valuable insights into the design of effective approaches to stimulate autophagy in memory B cells to boost the efficacy of vaccination.

6 Memory B Cells Require Autophagy for Long-Term Survival and Functions

The immune system is able to maintain long-term memory against pathogens critical for rapid induction of immunity upon later infections, a phenomenon called immunological memory [16]. Vaccination has been the most widely used strategy to protect against viral infections by inducing virus-specific immunological memory. After the initial exposure to antigens, antigen-specific lymphocytes are activated and undergo considerable expansion. Most of these expanded lymphocytes are removed by programmed cell death after the infections are cleared [21, 26, 33, 45, 47, 65], while some of the activated antigen-specific lymphocytes can survive and develop into memory cells [24, 28]. The persistence of antigen-specific memory lymphocytes after an immune response is the cellular basis for the maintenance of long-term immunological memory [25, 27, 28, 43, 60, 70].

For memory cells to successfully carry out the function of immunological memory, they must be in a poised state so that they can be readily activated after re-encountering antigens. Antigen-specific memory cells also have to survive for an extended period of time. Memory cells may also undergo self renew to maintain the colonies of antigen-specific memory [77]. For immunological memory to persist, the immune memory cells need to survive for a long period of time. The utmost important task for memory B cells to persist is to inhibit cell death machinery in the cells. We found significant reduction of influenza hemagglutin (HA)-specific memory B cells after immunization (Fig. 2). This is likely caused by the increased cell death of these autophagy-deficient memory B cells [8].

During B cell differentiation, Bcl-2 is expressed in naïve B cells [67]. When developed into GC B cells, Bcl-2 mRNA is dramatically suppressed in GC B cells [29, 67]. The long-lived memory B cells re-express Bcl-2 mRNA that is higher than in naïve B cells and GC B cells [67]. Interestingly, it has been reported that miRNA responsible for the inhibition of bcl-2, including miR-15a and miR-16, are expressed in GC B cells but down-regulated in memory B cells [1, 4, 71]. It is known that the intrinsic cell death pathways involving mitochondrial disruption. We therefore examined whether the expression of anti-apoptotic Bcl-2 family members are changed in memory B cells. We observed that Bcl-2 was highly expressed in memory B cells, but significantly reduced in GC B cells [8]. The repression of miRNAs targeting Bcl-2 in GC B cells likely contributes to increased Bcl-2 expression in memory B cells [35]. Transgenic expression of Bcl-2 in B cells increased the numbers of antigen-specific memory B cells. However, it was found that the increases were predominantly in low-affinity memory B cells [63]. Although Bcl-2 expression is correlated with reduced sensitivity to spontaneous cell death in GC and memory B cells, the precise contribution of Bcl-2 and other Bcl-2 family to the regulation of the survival in GC and memory B cells remained undefined. In addition to inhibiting apoptosis, Bcl-2 can also bind to Beclin 1 to inhibit autophagy [55]. How can the high levels of Bcl-2 and the observation of active autophagy in memory B cells be reconciled? One explanation is that the intracellular location of Bcl-2 might determine its functions.

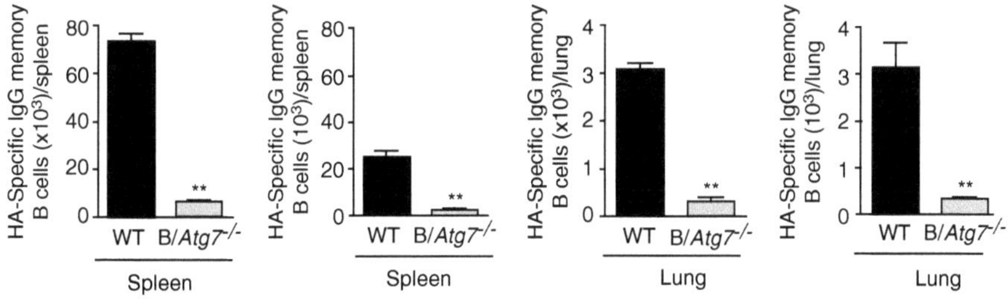

Fig. 2 Defective memory B cell responses to influenza infection in mice with B cell-specific deletion of Atg7 (B/Atg7$^{-/-}$). B/Atg7$^{-/-}$ or wild type mice were immunized with influenza intranasally. Two months later, IgG^{+} or IgA^{+} memory B cells in the spleen and lungs were examined

Bcl-2 localized to the ER binds to Beclin 1 to inhibit autophagy, while Bcl-2 on mitochondria inhibits apoptosis [40]. It will be interesting to determine whether Bcl-2 is indeed predominantly found on the mitochondrion in memory B cells.

Autophagy can help to maintain cell survival by promoting energy production. In particular, memory T cells have been shown to engage a program of fat metabolism to maintain energy production that is critical for the generation or maintenance of memory T cells [48, 56]. Whether autophagy plays a role in the regulation of fat metabolism in memory T cells is unclear. It is also unclear whether autophagy might regulate the metabolic process in memory B cells. Whether memory B cells might require the similar metabolic regulation will be very interesting for further investigation.

Compared to GC B cells that represent recently activated antigen-specific B cells, memory B cells express higher levels of autophagy genes [8]. Because autophagy has been implicated in the protection of long-lived cells, such as neurons, we tested whether memory B cells might be dependent on autophagy for survival. Indeed, we have found that incubation of memory B cells with 3-methyladenine induced rapid cell death in memory B cells. Moreover, Atg7-deficient memory B cells also displayed accelerated cell death. This indicates that autophagy is required for the protection of memory B cell survival.

We determined whether reduced cell death in memory B cells is due to inhibition of caspase activation. As expected, the activation of caspase-3 and caspase-9 can be readily detected in GC B cells after *in vitro* culture for 4 to 6 h [8]. This suggests that caspase-dependent intrinsic apoptosis that involves caspase-9 to caspase-3 signaling cascade, is involved in cell death in GC B cells. However, memory B cells do not display caspase activation during *in vitro* culture [8]. Due to the lack of caspase activation in memory B cells, one interesting possibility is that caspase-dependent apoptosis might be suppressed by autophagy [8]. Surprisingly, however, autophagy-deficient memory B cells do not display detectable caspase activation during cell death. Moreover, cell death in autophagy-deficient memory B cells is not affected by a pan-caspase inhibitor. This indicates that autophagy protects memory B cells against cell death that is distinct from classic caspase-dependent apoptosis.

In the absence of autophagy, memory B cells show increased ROS production and membrane lipid peroxidation. This may lead to increased disruption of cellular membranes, including plasma and nuclear membranes, as well as membranes of different organelles. Alox5-dependent lipid peroxidation is likely to be involved in such caspase-independent cell death because additional deletion of Alox5 partially protected autophagy-deficient memory B cells [8]. Although the precise mechanism for such a mechanism of cell death involving membrane lipid peroxidation is not yet elucidated, it is conceivable that membrane lipid peroxidation leads to the disruption of membrane and the loss of organelle function requiring the intact membranes, such as loss of ATP production on the mitochondrial inner membranes. Additionally, loss of membrane integrity may cause the leak of cell death molecules from the mitochondrion into the cytosol and nucleus to cause cell death. Although not yet tested in memory B cells, it is conceivable that cell death molecules in mitochondria, such as AIF and endonuclease G, will induce cell death if released from mitochondria and entering the nucleus.

7 Autophagy in T Cell-Mediated Immunity Against Viral Infection

Autophagy is important for T cell development [22, 54, 58, 74]. Autophagy is also found to be induced during T cell activation [20]. Autophagy-deficient T cells show defective cell proliferation and cytokine production [20, 58]. While deficiency in autophagy results in the decrease of mature T cells, the $CD8^+$ T cell frequency is reduced more than the $CD4^+$ T cells [54, 59, 64, 66, 74, 77]. Autophagy might play a more dominant in $CD8^+$ T cell function. Because autophagy is inducible by activation from the T cell antigen receptor, it is possible that autophagy plays important roles in the regulation of different T cell subsets during immune responses to viral infections. Due to an important function for autophagy in the protection of long-lived cell types, it is likely that autophagy also play an important role in memory T cell maintenance. The functions for autophagy in the protection of immunological memory by T cells and B cells against infections will be an important topic for future studies.

8 Can Promoting Autophagy Improve Immunological Memory and the Efficacy of Vaccination?

We have found that treatment with rapamycin can promote autophagy in memory B cells. Rapamycin can also improve the survival and function of memory B cells against influenza infection in mice [8]. Rapamycin has also been shown to increase the generation of memory T cells against lymphocytic choriomeningitis virus infection and improve memory T cell responses *in vivo* [2]. These studies support the possibility that promoting autophagy can enhance immune memory against viral infections. However, rapamycin potently inhibits mTOR in TORC1 complex that have diverse biological functions, such as inhibition of T cell proliferation and IL-2 production in addition to promoting autophagy [12]. It might not be clinically feasible to use rapamycin solely for the purpose of promoting immune memory after vaccination. A more specific autophagy inducer that increases autophagy with minimum side effects of immunosuppression will be needed for clinical useful for the purpose of promoting immunological memory.

References

1. Aqeilan RI, Calin GA, Croce CM (2010) miR-15a and miR-16-1 in cancer: discovery, function and future perspectives. Cell Death Differ 17:215–220
2. Araki K, Turner AP, Shaffer VO, Gangappa S, Keller SA, Bachmann MF, Larsen CP, Ahmed R (2009) mTOR regulates memory CD8 T-cell differentiation. Nature 460:108–112
3. Bhattacharya D, Cheah MT, Franco CB, Hosen N, Pin CL, Sha WC, Weissman IL (2007) Transcriptional profiling of antigen-dependent murine B cell differentiation and memory formation. J Immunol 179:6808–6819

4. Bonci D, Coppola V, Musumeci M, Addario A, Giuffrida R et al (2008) The miR-15a-miR-16-1 cluster controls prostate cancer by targeting multiple oncogenic activities. Nat Med 14:1271–1277
5. Cadwell K, Patel KK, Maloney NS, Liu TC, Ng AC, Storer CE, Head RD, Xavier R, Stappenbeck TS, Virgin H (2010) Virus-plus-susceptibility gene interaction determines Crohn's disease gene Atg16L1 phenotypes in intestine. Cell 141:1135–1145
6. Cai X, Chiu YH, Chen ZJ (2014) The cGAS-cGAMP-STING pathway of cytosolic DNA sensing and signaling. Mol Cell 54:289–296
7. Carneiro LA, Travassos LH (2013) The interplay between NLRs and autophagy in immunity and inflammation. Front Immunol 4:361. doi:10.3389/fimmu.2013.00361
8. Chen M, Hong MJ, Sun H, Wang L, Shi X, Gilbert BE, Corry DB, Kheradmand F, Wang J (2014) Essential role for autophagy in the maintenance of immunological memory against influenza infection. Nat Med 20:503–510
9. Conway KL, Kuballa P, Khor B, Zhang M, Shi HN, Virgin HW, Xavier RJ (2013) ATG5 regulates plasma cell differentiation. Autophagy 9:528–537
10. Dai P, Wang W, Cao H, Avogadri F, Dai L, Drexler I, Joyce JA, Li XD, Chen Z, Merghoub T, Shuman S, Deng L (2014) Modified vaccinia virus Ankara triggers type I IFN production in murine conventional dendritic cells via a cGAS/STING-mediated cytosolic DNA-sensing pathway. PLoS Pathog 10(4):e1003989. doi:10.1371/journal.ppat.1003989
11. Delgado MA, Elmaoued RA, Davis AS, Kyei G, Deretic V (2008) Toll-like receptors control autophagy. EMBO J 27:1110–1121
12. Delgoffe GM, Powell JD (2009) mTOR: taking cues from the immune microenvironment. Immunology 127:459–465
13. Dong X, Levine B (2013) Autophagy and viruses: adversaries or allies? J Innate Immun 5:480–493
14. Dreux M, Gastaminza P, Wieland SF, Chisari FV (2009) The autophagy machinery is required to initiate hepatitis C virus replication. Proc Natl Acad Sci U S A 106:14046–14051
15. Fullgrabe J, Klionsky DJ, Joseph B (2014) The return of the nucleus: transcriptional and epigenetic control of autophagy. Nat Rev Mol Cell Biol 15:65–74
16. Gomez Lorenzo MM, Fenton MJ (2013) Immunobiology of influenza vaccines. Chest 143:502–510
17. Goubau D, Schlee M, Deddouche S et al (2014) Antiviral immunity via RIG-I-mediated recognition of RNA bearing 5'-diphosphates. Nature 514:372–375
18. Green AM, Beatty PR, Hadjilaou A, Harris E (2014) Innate immunity to dengue virus infection and subversion of antiviral responses. J Mol Biol 426:1148–1160
19. Gregoire IP, Richetta C, Meyniel-Schicklin L, Borel S et al (2011) IRGM is a common target of RNA viruses that subvert the autophagy network. PLoS Pathog 7:e1002422. doi:10.1371/journal.ppat.1002422
20. Hubbard VM, Valdor R, Patel B, Singh R, Cuervo AM, Macian F (2010) Macroautophagy regulates energy metabolism during effector T cell activation. J Immunol 185:7349–7357
21. Hughes P, Bouillet P, Strasser A (2006) Role of Bim and other Bcl-2 family members in autoimmune and degenerative diseases. Curr Dir Autoimmun 9:74–94
22. Jia W, He YW (2011) Temporal regulation of intracellular organelle homeostasis in T lymphocytes by autophagy. J Immunol 186:5313–5322
23. Joubert PE, Werneke SW, de la Calle C, Guivel-Benhassine F, Giodini A, Peduto L, Levine B, Schwartz O, Lenschow DJ, Albert ML (2012) Chikungunya virus-induced autophagy delays caspase-dependent cell death. J Exp Med 209:1029–1047
24. Kalia V, Sarkar S, Gourley TS, Rouse BT, Ahmed R (2006) Differentiation of memory B and T cells. Curr Opin Immunol 18:255–264
25. Kasturi SP, Skountzou I, Albrecht RA, Koutsonanos D, Hua T, Nakaya HI et al (2011) Programming the magnitude and persistence of antibody responses with innate immunity. Nature 470:543–547
26. Krammer PH, Arnold R, Lavrik IN (2007) Life and death in peripheral T cells. Nat Rev Immunol 7:532–542

27. Kurosaki T, Aiba Y, Kometani K, Moriyama S, Takahashi Y (2010) Unique properties of memory B cells of different isotypes. Immunol Rev 237:104–116
28. Lanzavecchia A, Sallusto F (2002) Progressive differentiation and selection of the fittest in the immune response. Nat Rev Immunol 2:982–987
29. Lebecque S, de Bouteiller O, Arpin C, Banchereau J, Liu YJ (1997) Germinal center founder cells display propensity for apoptosis before onset of somatic mutation. J Exp Med 185: 563–571
30. Lee HK, Mattei LM, Steinberg BE, Alberts P, Lee YH, Chervonsky A, Mizushima N, Grinstein S, Iwasaki A (2010) In vivo requirement for Atg5 in antigen presentation by dendritic cells. Immunity 32:227–239
31. Lei Y, Wen H, Ting JP (2013) The NLR protein, NLRX1, and its partner, TUFM, reduce type I interferon, and enhance autophagy. Autophagy 9:432–433
32. Lei Y, Wen H, Yu Y, Taxman DJ, Zhang L et al (2012) The mitochondrial proteins NLRX1 and TUFM form a complex that regulates type I interferon and autophagy. Immunity 36:933–946
33. Lenardo M, Chan KM, Hornung F, McFarland H, Siegel R, Wang J, Zheng L (1999) Mature T lymphocyte apoptosis--immune regulation in a dynamic and unpredictable antigenic environment. Annu Rev Immunol 17:221–253
34. Li C, Capan E, Zhao Y, Zhao J, Stolz D, Watkins SC, Jin S, Lu B (2006) Autophagy is induced in CD4+ T cells and important for the growth factor-withdrawal cell death. J Immunol 177:5163–5168
35. Li G, Zan H, Xu Z, Casali P (2013) Epigenetics of the antibody response. Trends Immunol 34:460–470
36. Li XD, Wu J, Gao D, Wang H, Sun L, Chen ZJ (2013) Pivotal roles of cGAS-cGAMP signaling in antiviral defense and immune adjuvant effects. Science 341:1390–1394
37. Liang Q, Chang B, Brulois KF, Castro K, Min CK, Rodgers MA, Shi M, Ge J, Feng P, Oh BH, Jung JU (2013) Kaposi's sarcoma-associated herpesvirus K7 modulates Rubicon-mediated inhibition of autophagosome maturation. J Virol 87:12499–12503
38. Liang Q, Seo GJ, Choi YJ, Kwak MJ, Ge J, Rodgers MA, Shi M, Leslie BJ, Hopfner KP, Ha T, Oh BH, Jung JU (2014) Crosstalk between the cGAS DNA sensor and Beclin-1 autophagy protein shapes innate antimicrobial immune responses. Cell Host Microbe 15:228–238
39. Liang XH, Kleeman LK, Jiang HH, Gordon G, Goldman JE, Berry G, Herman B, Levine B (1998) Protection against fatal Sindbis virus encephalitis by beclin, a novel Bcl-2-interacting protein. J Virol 72:8586–8596
40. Marquez RT, Xu L (2012) Bcl-2:Beclin 1 complex: multiple, mechanisms regulating autophagy/apoptosis toggle switch. Am J Cancer Res 2:214–221
41. Maruyama M, Lam KP, Rajewsky K (2000) Memory B-cell persistence is independent of persisting immunizing antigen. Nature 407:636–642
42. McHeyzer-Williams LJ, McHeyzer-Williams MG (2005) Antigen-specific memory B cell development. Annu Rev Immunol 23:487–513
43. McHeyzer-Williams M, Okitsu S, Wang N, McHeyzer-Williams L (2012) Molecular programming of B cell memory. Nat Rev Immunol 12:24–34
44. Miller BC, Zhao Z, Stephenson LM, Cadwell K, Pua HH, Lee HK, Mizushima NN, Iwasaki A, He YW, Swat W, Virgin HW (2008) The autophagy gene ATG5 plays an essential role in B lymphocyte development. Autophagy 4:309–314
45. Mountz JD, Wu J, Cheng J, Zhou T (1994) Autoimmune disease. A problem of defective apoptosis. Arthritis Rheum 37:1415–1420
46. Munz C (2009) Enhancing immunity through autophagy. Annu Rev Immunol 27:423–449
47. Nagata S, Suda T (1995) Fas and Fas ligand: lpr and gld mutations. Immunol Today 16:39–43
48. O'Sullivan D, van der Windt GJ, Huang SC, Curtis JD, Chang CH, Buck MD, Qiu J, Smith AM, Lam WY, DiPlato LM, Hsu FF, Birnbaum MJ, Pearce EJ, Pearce EL (2014) Memory CD8(+) T cells use cell-intrinsic lipolysis to support the metabolic programming necessary for development. Immunity 41:75–88

49. Orvedahl A, Alexander D, Talloczy Z, Sun Q, Wei Y, Zhang W, Burns D, Leib DA, Levine B (2007) HSV-1 ICP34.5 confers neurovirulence by targeting the Beclin 1 autophagy protein. Cell Host Microbe 1:23–35
50. Orvedahl A, MacPherson S, Sumpter R Jr, Talloczy Z, Zou Z, Levine B (2010) Autophagy protects against sindbis virus infection of the central nervous system. Cell Host Microbe 7:115–127
51. Orvedahl A, Sumpter R Jr, Xiao G, Ng A, Zou Z, Tang Y, Narimatsu M, Gilpin C, Sun Q et al (2011) Image-based genome-wide siRNA screen identifies selective autophagy factors. Nature 480:113–117
52. Paludan C, Schmid D, Landthaler M, Vockerodt M, Kube D, Tuschl T, Munz C (2005) Endogenous MHC class II processing of a viral nuclear antigen after autophagy. Science 307:593–596
53. Panyasrivanit M, Khakpoor A, Wikan N, Smith DR (2009) Co-localization of constituents of the dengue virus translation and replication machinery with amphisomes. J Gen Virol 90: 448–456
54. Parekh VV, Wu L, Boyd KL et al (2013) Impaired autophagy, defective T cell homeostasis, and a wasting syndrome in mice with a T cell-specific deletion of Vps34. J Immunol 190: 5086–5101
55. Pattingre S, Tassa A, Qu X, Garuti R, Liang XH, Mizushima N, Packer M, Schneider MD, Levine B (2005) Bcl-2 antiapoptotic proteins inhibit Beclin 1-dependent autophagy. Cell 122:927–939
56. Pearce EL, Walsh MC, Cejas PJ, Harms GM, Shen H, Wang LS, Jones RG, Choi Y (2009) Enhancing CD8 T-cell memory by modulating fatty acid metabolism. Nature 460:103–107
57. Pengo N, Scolari M, Oliva L, Milan E, Mainoldi F, Raimondi A, Fagioli C, Merlini A, Mariani E et al (2013) Plasma cells require autophagy for sustainable immunoglobulin production. Nat Immunol 14:298–305
58. Pua HH, Dzhagalov I, Chuck M, Mizushima N, He YW (2007) A critical role for the autophagy gene Atg5 in T cell survival and proliferation. J Exp Med 204:25–31
59. Pua HH, Guo J, Komatsu M, He YW (2009) Autophagy is essential for mitochondrial clearance in mature T lymphocytes. J Immunol 182:4046–4055
60. Pulendran B, Ahmed R (2011) Immunological mechanisms of vaccination. Nat Immunol 12:509–517
61. Rasmussen SB, Horan KA, Holm CK, Stranks AJ, Mettenleiter TC, Simon AK, Jensen SB, Rixon FJ, He B, Paludan SR (2011) Activation of autophagy by alpha-herpesviruses in myeloid cells is mediated by cytoplasmic viral DNA through a mechanism dependent on stimulator of IFN genes. J Immunol 187:5268–5276
62. Saitoh T, Akira S (2010) Regulation of innate immune responses by autophagy-related proteins. J Cell Biol 189:925–935
63. Smith KG, Light A, O'Reilly LA, Ang SM, Strasser A, Tarlinton D (2000) bcl-2 transgene expression inhibits apoptosis in the germinal center and reveals differences in the selection of memory B cells and bone marrow antibody-forming cells. J Exp Med 191:475–484
64. Stephenson LM, Miller BC, Ng A, Eisenberg J, Zhao Z, Cadwell K, Graham DB, Mizushima NN, Xavier R, Virgin HW, Swat W (2009) Identification of Atg5-dependent transcriptional changes and increases in mitochondrial mass in Atg5-deficient T lymphocytes. Autophagy 5:625–635
65. Strasser A, Jost PJ, Nagata S (2009) The many roles of FAS receptor signaling in the immune system. Immunity 30:180–192
66. Sun Y, Li C, Shu Y, Ju X, Zou Z, Wang H, Rao S, Guo F, Liu H, Nan W et al (2012) Inhibition of autophagy ameliorates acute lung injury caused by avian influenza A H5N1 infection. Sci Signal 5:ra16. doi:10.1126/scisignal.2001931
67. Takahashi Y, Ohta H, Takemori T (2001) Fas is required for clonal selection in germinal centers and the subsequent establishment of the memory B cell repertoire. Immunity 14:181–192

68. Tal MC, Sasai M, Lee HK, Yordy B, Shadel GS, Iwasaki A (2009) Absence of autophagy results in reactive oxygen species-dependent amplification of RLR signaling. Proc Natl Acad Sci U S A 106:2770–2775
69. Talloczy Z, Virgin HW, Levine B (2006) PKR-dependent autophagic degradation of herpes simplex virus type 1. Autophagy 2:24–29
70. Tarlinton D (2006) B-cell memory: are subsets necessary? Nat Rev Immunol 6:785–790
71. Thai TH, Calado DP, Casola S, Ansel KM, Xiao C, Xue Y, Murphy A, Frendewey D, Valenzuela D, Kutok JL, Schmidt-Supprian M, Rajewsky N, Yancopoulos G, Rao A, Rajewsky K (2007) Regulation of the germinal center response by microRNA-155. Science 316:604–608
72. Tian Y, Sir D, Kuo CF, Ann DK, Ou JH (2011) Autophagy required for hepatitis B virus replication in transgenic mice. J Virol 85:13453–13456
73. Travassos LH, Carneiro LA, Ramjeet M, Hussey S, Kim YG, Magalhaes JG, Yuan L, Soares F, Chea E, Le Bourhis L, Boneca IG, Allaoui A, Jones NL, Nunez G, Girardin SE, Philpott DJ (2010) Nod1 and Nod2 direct autophagy by recruiting ATG16L1 to the plasma membrane at the site of bacterial entry. Nat Immunol 11:55–62
74. Willinger T, Flavell RA (2012) Canonical autophagy dependent on the class III phosphoinositide-3 kinase Vps34 is required for naive T-cell homeostasis. Proc Natl Acad Sci U S A 109:8670–8675
75. Xu Y, Jagannath C, Liu XD, Sharafkhaneh A, Kolodziejska KE, Eissa NT (2007) Toll-like receptor 4 is a sensor for autophagy associated with innate immunity. Immunity 27:135–144
76. Zhang X, Wu J, Du F, Xu H, Sun L, Chen Z, Brautigam CA, Chen ZJ (2014) The cytosolic DNA sensor cGAS forms an oligomeric complex with DNA and undergoes switch-like conformational changes in the activation loop. Cell Rep 6:421–430
77. Zuccarino-Catania GV, Sadanand S, Weisel FJ, Tomayko MM, Meng H, Kleinstein SH, Good-Jacobson KL, Shlomchik MJ (2014) CD80 and PD-L2 define functionally distinct memory B cell subsets that are independent of antibody isotype. Nat Immunol 15:631–637

Part II
Targeting Autophagy as a Novel Therapeutic Strategy in Inflammation-Based Pathologies

Role of Autophagy in Brain Sculpture: Physiological and Pathological Implications

Annalisa Nobili, Virve Cavallucci, and Marcello D'Amelio

Abstract The brain has the ability to change during the life rearranging itself by the elimination and the formation of new connections between neurons. This dynamic capacity is known as brain plasticity or neuroplasticity, and is associated with functional changes involving functional recovery after brain damage, learning, memory, and addiction. It is well defined that protein synthesis is required for neuroplasticity and the establishment of long-term memories, but protein degradation plays also a crucial role in neuronal physiology and pathology. Ubiquitin-proteasome system, which degrades short-lived proteins, is important in synaptic plasticity, learning and memory, as well as lysosome system, which involves endocytosis to degrade proteins, plays a role in synaptic plasticity regulating receptor trafficking. The third major degradation pathway is the autophagy which degrades long-lived cytoplasmic proteins or damaged organelles to maintain normal cell homeostasis. Recent evidence suggests the involvement of autophagy in synaptic plasticity, in addition to its crucial role in the quality control of proteins and organelles in neurons. Thus an impairment of the autophagic machinery is closely connected with the alteration of neuronal function and neuron ability to respond to damage. A clear understanding of neuronal autophagy in brain physiology and pathology could help to develop new pharmaceutical approaches for the treatment of neurological disorders. The current Chapter will focus on the key role of autophagy in the development and function of the central nervous system (CNS), and on the emerging evidence of autophagy deregulation in neurodegenerative disease and acute brain damage.

A. Nobili • M. D'Amelio (✉)
Department of Medicine, University Campus-Biomedico, Via Alvaro del Portillo, 21, Rome, Italy

Department of Experimental Neurosciences, IRCCS S. Lucia Foundation, Rome, Italy
e-mail: m.damelio@unicampus.it

V. Cavallucci
Department of Experimental Neurosciences, IRCCS S. Lucia Foundation, Via del Fosso di Fiorano, 64, Rome, Italy

M.C. Maiuri, D. De Stefano (eds.), *Autophagy Networks in Inflammation*,
Progress in Inflammation Research, DOI 10.1007/978-3-319-30079-5_11

Abbreviations

3-MA	3-methyladenine
AD	Alzheimer's Disease
Akt (PKB)	Protein kinase B
Ambra1	Autophagy/Beclin-1 Regulator 1
AMPAR	α-amino-3-hydroxy-5-methyl-4-isoxazolepropionic acid receptor
APP	Amyloid precursor protein
ASD	Autism spectrum disorders
Atg	Autophagy-related
ATP	Adenosine triphosphate
AVs	Autophagic vacuoles
Aβ	β-amyloid peptide
Bcl-2	B-cell lymphoma 2
Bcl-XL	B-cell lymphoma-extra large
Bif-1	BAX-interacting factor-1
Bim	B-cell lymphoma 2 interacting mediator of cell death
CCI	Controlled cortical impact injury
CIII PI3K	Class III phosphatidylinositol 3-kinase
CMA	Chaperone-mediated autophagy
CNS	Central nervous system
COMT	Catechol-o-methyltransferase
Deptor	DEP domain containing MTOR-interacting protein
ER	Endoplasmic reticulum
FAD	Familial AD
Fbxo7	F-box protein 7
FIP200	FAK Family Kinase-Interacting Protein of 200 kDa
FPI	Fluid percussion injury
$GABA_A$,	γ-aminobutyric acid
GCEE	γ-glutamylcysteinyl ethyl ester
H/I	Hypoxia/ischemia
HCb	Hemicerebellectomy
HD	Huntington's disease
HDAC6	Histone deacetylase 6
Htt	Huntingtin
i.c.v.	Intracerebroventricular
IO	Inferior olive
IRGM	Immune-related GTPase M
KO	Knockout
LAMP	Lysosome-associated membrane protein type
LC3	Microtubule-associated protein 1 light chain 3
LRRK2	Leucine-rich repeat kinase 2
LTD	Long-term depression
LTP	Long-term potentiation
MAO-B	Monoamine oxidase B
Mfn1	Mitofusin 1

MPP+	1-methyl-4-phenylpyridinium
mTOR	Mammalian target of rapamycin
mTORC1	mTOR complex 1
NBR1	Neighbor Of BRCA1 Gene 1
Ndp52	Nuclear dot protein 52 kDa
NGF	Nerve growth factor
NMDAR	N-methyl-D-aspartate receptor
NPCs	Neural progenitor cells
NSF	N-ethylmaleimide-sensitive factor
p62/SQSTM1	Sequestosome 1
PD	Parkinson's Disease
PE	Phosphatidylethanolamine
PI3K	Phosphatidylinositol 3-kinase
PI3P	Phosphatidylinositol 3-phosphate
PINK1	PTEN-induced putative kinase 1
pMCAO	Permanent middle cerebral artery occlusion
Pn	Pontine nuclei
polyQ	Polyglutamine
PP1	Protein phosphatase 1
PRAS40	Proline-rich Akt substrate of 40 kDa
PSD95	Postsynaptic density protein 95
PSEN	Presenilin
PTEN	Phosphatase and tensin homolog
Raptor	Regulatory-associated protein of mTOR
Rubicon	RUN and cysteine rich domain containing beclin 1 interacting protein
SCI	Spinal cord injury
SVZ	Subventricular zone
TBI	Traumatic brain injury
tMCAO	Transient middle cerebral artery occlusion
TrkA	Tropomyosin receptor kinase A
Tsc2	Tuberous sclerosis proteins
ULK1	UNC-51-like kinase 1
UVRAG	UV Radiation Resistance-Associated Gene
VDAC1	Voltage-dependent anion channel 1
VMP1	Vacuole Membrane Protein 1
WIPI-1	WD-repeat protein Interacting with PhosphoInositides-1

1 Introduction

Autophagy is a catabolic mechanism that mediates degradation and recycling of cellular constituents, delivering portion of cytoplasm to the lysosomes for degradation. Autophagy is considered to be important to maintain cellular homeostasis, especially under nutrient deprivation or stress conditions, and to guarantee proteins quality control and organelles turnover. Furthermore, autophagy has been implicated in various cellular processes, such as development, differentiation, ageing and immunity [77, 101, 102].

Autophagy is highly conserved from yeast to mammals and it can be classified in three principal types: microautophagy, chaperone-mediated autophagy (CMA) and macroautophagy [115]. Microautophagy leads to degradation of sequestered portions of cytosol by direct invagination of lysosome membrane [99]. CMA delivers cytosolic protein containing a KFERQ-like motif to the lysosomal lumen via chaperone Hsc70 and LAMP-2A complex [66]. Macroautophagy (hereafter referred to as autophagy) is the well-characterised form of autophagy that leads to the formation of a double-membraned vacuole, the autophagosome, containing cytoplasmic material, such as macromolecules and organelles. Autophagy requires several steps: induction and nucleation of phagophore (the isolation membrane), elongation of phagophore to constitute the autophagosome, maturation of autophagosome into amphisome/autolysosome by fusion with endosome/lysosome and, finally, degradation of membrane contents (Fig. 1). Autophagy is constitutively activated at a basal level to maintain cellular homeostasis but it can also be induced by several input signals, such as nutrient deprivation, change of intracellular levels of Ca^{2+}, ATP and cAMP, hormones, protein accumulation and damaged organelles [136]. The main regulator of autophagy is mTOR (mammalian target of rapamycin) complex 1 (mTORC1), a polyprotein complex that contains mTOR, Raptor, mLST8/GßL, Deptor and PRAS40 [22]; however the autophagy is also regulated by mTOR-independent pathways even though the effectors involved in the autophagosome biogenesis are not clear [136].

More than 30 autophagy-related (Atg) proteins have been identified and characterized in yeast [60, 61] and conserved in mammals, where additional Atg proteins have been identified. However, less than half, the "core Atg proteins" [110, 166], are

Fig. 1 Schematic representation of mammalian autophagy pathway. Following inhibition of mTOR, ULK complex, composed of ULK1, Atg13, FIP200 and Atg101, is activated and starts the autophagosome nucleation. ULK1 also phosphorylates Ambra1, leading to the activation of Beclin1, a component of CIII PI3K complex (or Beclin1 complex), which consists of PI3K or hVps34, Beclin1 and p150. In this step, VMP1 recruits Beclin1 to the phagophore where the complex is required for generating a pool of phosphatidylinositol 3-phosphate (PI3P). To expand the autophagosome membrane, Atg12-Atg5-Atg16L complex and LC3 ubiquitin-like conjugation systems are required. The Atg12-Atg5-Atg16L multimeric complex is formed by subsequent steps involving Atg7 and Atg10 and is also required for the efficient function of LC3. The second system mediates the conjugation of LC3 to phosphatidylethanolamine (PE). LC3 is processed by Atg4, forming the cytosolic LC3 I. Atg7 binds LC3 I and transfers it to Atg3, which catalyses the conjunction to the lipid PE and the conversion of LC3 I in LC3 II. The complete autophagosome fuses with the lysosome and cargo molecules engulfed by autophagosomes are degraded by lysosomal hydrolases and recycled back to the cytoplasm. The specificity of cargo degradation is mediated by selective adaptor proteins, such as p62 that binds ubiquitinated residues of the target proteins or organelles. The steps known to be affected in neurodegenerative diseases and the action of drugs modulating autophagy are indicated. *AD* Alzheimer disease, *HD* Huntington disease, *PD* Parkinson disease. Regulators of autophagy are also indicated. 3-MA, 3-methyladenine

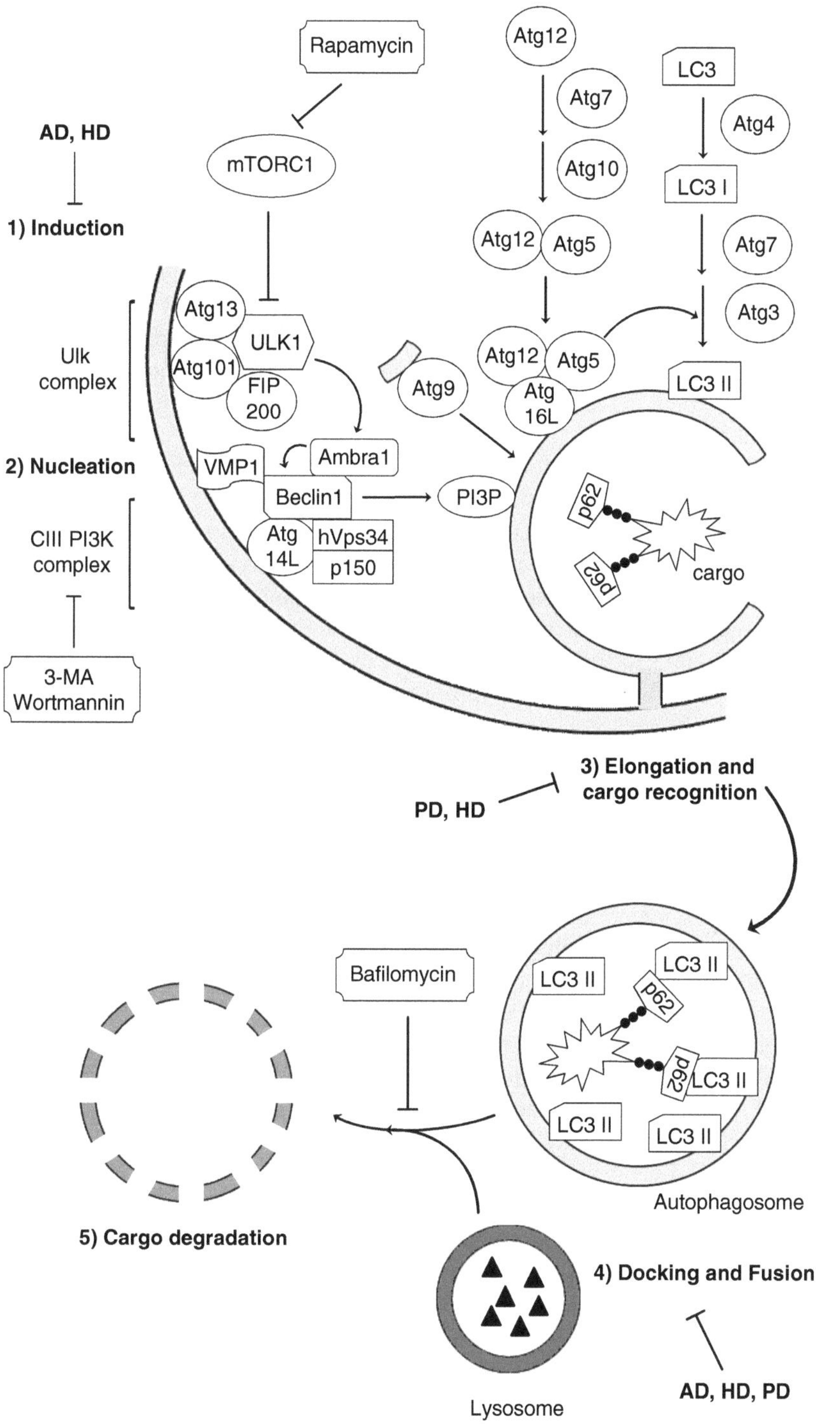
Rapamycin
Atg12
LC3
AD, HD
Atg7
Atg4
mTORC1
Atg10
LC3 I
1) Induction
Atg12
Atg5
Atg7
Atg13
Atg3
Ulk complex
ULK1
Atg101
Atg12
Atg5
FIP 200
Atg9
Atg 16L
LC3 II
2) Nucleation
VMP1
Ambra1
Beclin1
PI3P
p62
CIII PI3K complex
Atg 14L
hVps34
p150
p62
cargo
3-MA Wortmannin
3) Elongation and cargo recognition
PD, HD
Bafilomycin
LC3 II
LC3 II
p62
p62
LC3 II
LC3 II
LC3 II
Autophagosome
5) Cargo degradation
4) Docking and Fusion
Lysosome
AD, HD, PD

involved in autophagosome formation. These proteins can be subdivided in three functional groups: (1) two kinase complex, ULK complex (consisting of ULK1, Atg13, FIP200 and Atg101) and the class III phosphatidylinositol 3-kinase (CIII PI3K) complex (or Beclin1 complex, comprising PI3K or hVps34, Beclin1 and p150); (2) two ubiquitin-like protein conjugation systems, ATG16L1 complex (Atg12-Atg5-Atg16L) and LC3; (3) two transmembrane proteins, Atg9 (and associated proteins involved in its movement such as WIPI-1) and VMP1.

The activity of ULK complex is important in the induction step of autophagy and is negatively regulated by mTORC1. After autophagy induction, mTORC1 is inactivated, leading to dephosphorylation of ULK1 and Atg13. This event causes the activation of these proteins and the consequent phosphorylation of FIP200 by ULK1. Recently it has been identified a novel ULK complex interactor, Atg101, important for both the stability of Atg13 and basal phosphorylation of Atg13 and ULK1 [48, 98].

Beclin1-CIII PI3K complex is required for the isolation membrane nucleation, generating a pool of phosphatidylinositol 3-phosphate (PI3P), which acts as platform for other Atg proteins recruitment. Beclin1 interacts with several proteins, which are able to modulate the complex activity: Atg14L, UVRAG, Ambra1, and Bif-1 enhance Beclin1 complex activity; the transmembrane protein VMP1 is crucial for the nucleation step, recruiting Beclin1 (and other components of the Beclin1 complex) to the phagophore; whereas Rubicon, the anti-apoptotic proteins Bcl-2 and Bcl-XL and the pro-apoptotic protein Bim negatively regulate autophagy [136].

Following the autophagy induction, ULK1 phosphorylates Ambra1, leading to the activation of Beclin1 and the translocation of the Beclin1-CIII PI3K complex from the microtubule network to the endoplasmic reticulum (ER), which is considered the main membrane source for the biogenesis of phagophore [1, 28, 44]. The fundamental contribution to the biogenesis and elongation of the phagophore from other organelles membrane is mediated by Atg9 [169].

The elongation of the phagophore requires two ubiquitin-like protein conjugation systems. Firstly, Atg5 and Atg12 are conjugated by the ubiquitin-activating E1-like enzyme Atg7. In particular, Atg7 activates Atg12, which is transferred to Atg10, an E2-like enzyme, and then covalently linked with Atg5 ([105, 148]). The Atg12-Atg5 heterodimer binds Atg16L forming a large multimeric complex required for elongation of phagophore [103] and for the efficient function of LC3. The second system mediates the conjugation of LC3 to phosphatidylethanolamine (PE) [41, 103]. LC3 is processed by the protease Atg4, forming the cytosolic LC3 I. Atg7 binds LC3 I and transferred it to Atg3 (E2-like), which catalyses the conjunction to the lipid PE and the conversion of LC3 I in LC3 II. During the expansion of autophagosome, LC3 II is attached both to the inner and outer membrane of this structure. However, until the fusion with the lysosome LC3 II is removed from the outer membrane, while the one in the inner membrane is degraded with the autophagosome cargo [52, 59].

Autophagy has been originally identified as a non-selective process, but recent evidence underlines the importance of selective mechanism, specially involved in the quality control of proteins and organelles. Several post-translational modifications have been implicated in the regulation of autophagy, one of which is the ubiquitination. The residues of ubiquitin facilitate the recruitment of autophagic receptors and selective adaptor proteins, such as p62 (also called SQSTM1), NBR1, HDAC6, Nix, Ndp52, which tether the targeted substrate to core of Atg proteins, such LC3 [139].

The autophagy mechanism has been analyzed in several tissues and although it occurs in all cell types and involves the same protein complexes, recent studies suggest that this mechanism is tissue-specific regulated. In particular, autophagy plays a crucial role in the brain, in which it is involved as well as in stress response, in quality control of proteins and organelles, even in specific function, such as neurodevelopment, differentiation, learning and memory.

In this Chapter we will discuss the physiological roles of autophagy in neurons and the pathological implications occurring when this process is dysregulated.

2 Neuronal Autophagy

Neurons are highly specialized cells, composed by specific compartments including neurites (axons and dendrites), synapses and soma, in which the synthesis, transport and degradation processes are finely regulated. Moreover, mature neurons are post-mitotic cells, which require an efficient protein quality control system to avoid accumulation of misfolded or aggregated proteins and damaged organelles that cannot be diluted through cell division. Autophagy is of particular importance in the synaptic compartments, characterized by high-energy demand and where a fine control of proteins and organelles turnover is necessary to ensure the activity.

Compared to other organs (such as heart, liver, pancreas, kidney or muscle), in basal condition the brain shows higher level of LC3 and a low number of autophagosomes [112], which do not increase even under nutrient deprivation [106]. Several studies demonstrated that the low presence of autophagic vesicles, analyzed as a low amount of LC3 II then LC3 I, depends on the fast kinetics of vesicles formation and degradation, that reflects the high efficiency of autophagosome turnover [8, 107].

3 The Role of Autophagy in Neurodevelopment and Neurogenesis

The importance of autophagy in development has long been suggested by several studies in which autophagic structures were analyzed during embryogenesis [20, 83] and by studying mutant mice for several *Atg* genes, which have different role in the regulation of development. In fact, mice lacking *Atg3, Atg5, Atg7, Atg9* and *Atg16L1* complete the embryonic development, but die shortly after birth, suggesting that the proteins encoded by these genes are not essential for embryogenesis but have a crucial role in the regulation of perinatal starvation [64, 68, 131, 132, 145]. Instead, Beclin1, Ambra1 or FIP200 deficient mice are embryonic lethal at the stage E7.5, E10–E14 and E13.5–E16, respectively [35, 37, 174]. The differences among *Atg* genes are not well understood. It seems possible that the role of Beclin1 and FIP200, which interact with several factors, may be related to other function; alternatively, the different lethality may depend on the step in which each factor is involved [104]. In the latter case, Atg9 constitutes an exception because it acts in early phase of autophagy process but its loss causes a less severe phenotype [146].

The short survival time after birth of *Atg* knockout (KO) mice prevents the study of the Atg proteins role, whereas the generation of conditional KO mice allowed to understand their function in specific tissues. Brain-specific deletion of *Atg5* and *Atg7* suggests that these proteins are involved in motor function and that their deficiency results in the development of progressive motor and behavioural deficits. The histological analysis of Atg5 conditional KO mice shows partial loss of Purkinje cells and neuronal inclusion bodies accumulation. Aggregates of ubiquitinated proteins, which accumulate in a time-dependent manner, are also detected starting at embryonic day E15.5. It has been proposed that basal autophagy in Purkinje neurons is necessary to ensure the correct protein turnover, avoiding aggregates formation [42]. Atg5 KO embryos also display defects in apoptotic corpse engulfment in photoreceptor and ganglion cell layers of the retina at E18.5 [123]. A recent work has also proposed that Atg5 plays a central role in developing embryonic cortex to ensure the formation of correct multiple layers architecture. By silencing Atg5 expression in cortical neural progenitor cells (NPCs), it has been observed that the loss of Atg5 function leads to unbalanced cortical NPCs differentiation and proliferation and causes the abnormal morphology of cortical neurons. Atg5 exerts its function in strictly cooperation with β-Catenin and both are required to regulate cortical NPCs differentiation and proliferation [87].

Studies performed on Atg7 conditional KO mice show that the loss of this protein leads to axonal swellings, with accumulation of aberrant membrane structures and progressive dystrophy and degeneration of the axon terminals in Purkinje cells. Interestingly, these events occur much earlier than the neuronal death, suggesting that basal autophagy is required to regulate membrane homeostasis in the axonal terminals in addition to protein quality control [65].

ULK1 protein regulates axon outgrowth in cerebellum granule neurons and it has been demonstrated that ULK1 is expressed in different neuron population during development and is particular abundant in developing cerebellar granular cells. The inactivation of ULK1 by retroviral injection of its dominant negative form demonstrates that this kinase has a pivotal role in neurite extension/parallel fiber formation [151]. Recent studies also suggest that ULK1 regulates endocytotic trafficking of growth factors, that are necessary during polarized axon elongation. However is not well understood whether ULK1 acts influencing the autophagy mechanism or has a different function through the interaction with several proteins, for example in NGF-TrkA endocytosis [152, 180].

Another example of the importance of autophagy during neurodevelopment comes from the studies conducted on Ambra1 protein. Ambra1 is strongly expressed in developing neuronal tissue starting at the embryonic day E8.5, where is detected in neuroepithelium. Subsequently, a massive expression is observable in the ventral part of the spinal cord, in the encephalic vesicles, in the neural retina, in the limbs and in the dorsal root ganglia (E11.5); at later stages, Ambra1 is expressed in the entire developing nervous system; finally, in postnatal brain Ambra1 is particularly abundant in the cortex, hippocampus and striatum. Ambra1-deficient mice display defects in neural tube closure, as well as increased cell proliferation and cell death, that cause exencephaly and/or spina bifida phenotypes [35]. Therefore, Ambra1 appears to be involved in cell proliferation and survival during neurodevelopment, perhaps controlling the degradation of key development regulators, as demonstrated by the increase of ubiquitinated proteins in Ambra1-deficient mice, or directly regulating cell proliferation.

A central role for autophagy has recently been proposed also in adult neurogenesis. Ambra1 and Beclin1 are highly expressed in adult subventricular zone (SVZ) of the lateral ventricles, an area where new neurons are generated and then migrate through the rostral migratory stream to the olfactory bulb to become interneurons [100]. It has been demonstrated that autophagy is involved in two different mechanisms in SVZ: on the one hand it sustains the pool of stem cell, on the other hand it enhances the survival of neuronal precursors. In fact, *Beclin1* heterozygous mice show a significant decrease in cell division and an increase in the number of apoptotic cells in SVZ compared to wild-type [171].

Recent evidences support the idea that autophagy plays a fundamental role also in dendritic spine pruning during postnatal life. It has been reported an increase in spine density with reduced developmental spine pruning in layer V pyramidal neurons in postmortem temporal lobe of patients with autism spectrum disorders (ASD). These spine defects correlate with mTOR hyperactivation and autophagy impairment. In fact, Tang and colleagues, using a mTOR costitutively activated mouse model ($Tsc2^{+/-}$ mouse, with mutated Tsc2, a protein that indirectly inhibit mTOR) observed a postnatal spine pruning defects, blockage of autophagy and ASD-like behaviours. Interestingly, rapamacyn is able to revert these phenotypes in $Tsc2^{+/-}$ mice but not in $Tsc2^{+/-}$: Atg7 conditional KO double mutants, suggesting that autophagy is required for the correct remodelling of dendritic and spine architecture [147].

4 The Synaptic Role of Autophagy

Recent evidences have suggested that autophagy plays an essential role in synaptic plasticity, influencing both pre- and post-synaptic compartments by altering the efficacy of neurotransmitter release or by modifying the post-synaptic density composition (Fig. 2). There are two main forms of synaptic plasticity: long-term potentiation (LTP), where occurs a persistent increase of synaptic efficacies in response to a high-frequency stimulation, and long-term depression (LTD), characterized by a lasting decrease in synaptic effectiveness that follows a pattern of low-frequency stimulation [4, 7]. Both LTP and LTD are considered cellular correlates of learning and memory and for this reason the study of these processes is so intriguing. One of the main receptors involved in synaptic transmission is the

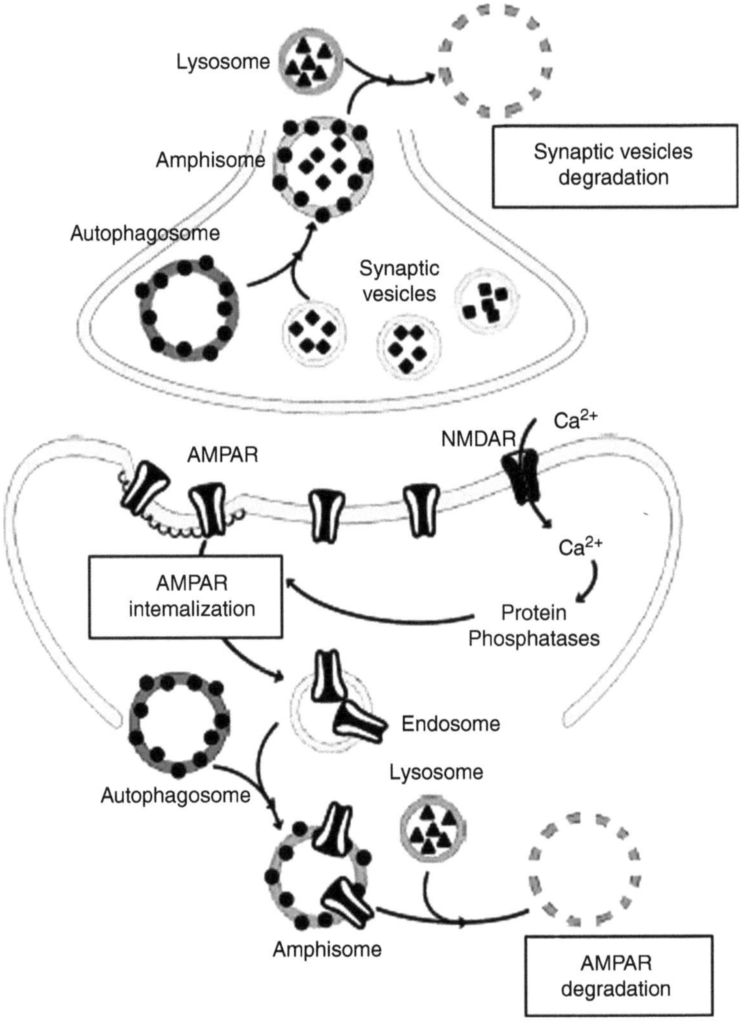

Fig. 2 Representation of autophagy role in pre- and post-synaptic compartment. In the pre-synaptic terminal autophagy alters the efficacy of neurotransmitter release, leading to selective degradation of neurotransmitter-containing vesicles. In the post-synaptic terminal autophagy regulates the degradation of AMPA receptor (AMPAR) after chemical induction of long-term depression (LTD). LTD leads to activation of the calcium ion-permeable NMDA receptor (NMDAR). A series of downstream intermediate signaling steps, including activation of protein phosphatases, cause AMPAR endocytosis and induction of LTD. Endosome vesicles containing AMPAR fuse with autophagosomes, forming amphisomes, that finally fuse with lysosomes leading degradation of the AMPAR

α-amino-3-hydroxy-5-methyl-4-isoxazolepropionic acid receptor (AMPAR), a class of ionotropic glutamate receptors that mediates the excitatory neurotransmission response. The synaptic transmission can be modulated by modifying the exposure of AMPARs in post-synaptic density [144] and by altering the trafficking of AMPARs into and out of synapses [88, 90]. Increased exposure of AMPARs causes a strengthen of synaptic transmission and generally LTP, whereas the removal of AMPARs results in LTD [141, 144]. AMPARs are heterotetrameric complexes, composed of various combination of four subunits (GluA1-4) [47]. In the adult hippocampus GluA subunits are assembled preferably in two major subtypes, GluA1-2, which represent approximately 80 % of synaptic receptors, and GluA2-4, which represent the remaining 20 % [84, 165]. Recent evidences suggest that AMPARs subunit composition and their interactors on synaptic membrane are able to regulate the trafficking and the targeting of AMPARs to synaptic sites [19, 27, 31, 43, 56, 113, 116, 117, 143, 144]. Moreover, several studies proposed that both GluA1 and GluA2 subunits have a central role in LTD [18, 73, 75, 153, 181]. Upon N-methyl-D-aspartate receptor (NMDAR)-dependent chemical LTD induction, the increase of intracellular Ca^{2+} level leads to the activation of several phosphatases, including PTEN, calcineurin and PP1, that beyond other functions are committed to dephosphorylate GluA1 at Ser 845 residue, promoting its internalization [72], and PSD95 at the Ser295, which is responsible to the recruitment of GluA1 subunit to post-synaptic surface and to the stabilization of the synaptic structure [58]. The GluA2 subunit is stabilized at the synapses by the interaction with NSF (N-ethylmaleimide-sensitive factor), an ATPase that is involved in membrane fusion events [56, 74]. During LTD, it is possible that NSF dissociates from GluA2 and that AP2 binds an overlapping region of GluA2, leading to clathrin-dependent receptor internalization [74, 91]. Internalized AMPARs can be recycled back to the synapses, pooled in the endosomes as a reserve [86, 114] or targeted to lysosome for degradation [33, 75]. Recently it has been proposed that autophagy is responsible for the regulation of neuronal activity, leading to the degradation of AMPARs contained in endocytic vesicle. In fact, after chemical LTD induction the number of autophagosome in dendritic shaft and spines of pyramidal neurons increases, together with a reduction of GluA1 subunits [140]. Degradation of AMPARs, and the consequent LTD stabilization, is partially recovered by application of Okadaic acid and dipotassium bisperoxo (5-hydroxypyridine-2-carboxyl) oxovanadate V, which are inhibitors of PP1 and PTEN, respectively. Both PP1 and PTEN regulate the phosphorylation and the activity of Akt and for this reason it has been suggested that chemical LTD leads to autophagy activation by inhibition of PI3K-Akt-mTOR pathway. Moreover, the inhibition of PI3K by wortmannin and the silencing of Atg7 by lentiviral shRNA infection determine a complete block of LTD-induced autophagy or a partial recovery of GluA1 levels, respectively. It is possible that the increase of the number of autophagosomes, observed in dendritic shafts and spines during LTD, increases the probability of fusion between autophagosomes and endosomes containing GluA subunits to form amphisomes, determining lysosomal degradation of cargo and preventing the receptors recycling in the post-synaptic site. Another interesting possibility is that after LTD induction autophagy acts as a selective degradation mechanism. It has been reported that $GABA_A$ (γ-aminobutyric acid) receptors, but not acetylcholine receptors, are internalized and degraded selectively by autophagy in *C. elegans* neurons [128]. Moreover p62 protein, which is a selective adaptor protein between cargo and autophagosomes, has an important role in LTP and spatial memory [124] and interacts directly with AMPAR subunits influencing their trafficking to synaptic membrane [51]. It is probable that autophagosomes merge with endosomes containing AMPARs by a p62-mediated mechanism.

Other evidences suggest that autophagy has also a pre-synaptic role regulating neurotransmission [46]. Conditional Atg7 KO mice, in which autophagy deficiency is restricted to dopaminergic neurons, show that inhibition of autophagy in pre-synaptic terminal alters pre-synaptic structure and neurotransmission. In particular, the absence of autophagy results in increased size of axon terminal, increased evoked dopamine release and more rapid pre-synaptic reuptake of neurotransmitter. The pre-synaptic role of autophagy is mTOR-regulated. In fact, acute rapamycin treatment induces a transient augment in the number of autolysosomes in synaptic terminals and at the same time reduces the number of neurotransmitter vesicles. This effect is absent in Atg7 conditional KO mice, indicating that the induction of autophagosome formation by rapamycin requires Atg7. It has been suggested that this mechanism is not specific to dopaminergic neurons. Indeed, rapamycin induces autophagosome-like structures also in other neurons, such as glutamatergic, GABAergic and cholinergic [46].

5 Role of Autophagy in the CNS Diseases

Considering the complexity of autophagic pathway and its importance in neuronal function is not surprisingly that neurons are particularly sensitive to autophagy defects with several pathological implications. Acute and chronic neurodegenerative diseases – including stroke, brain trauma, spinal cord injury, Alzheimer's disease, Parkinson's disease, Huntington's disease – are characterized by mitochondria dysfunction and extensive neuronal cell death. Moreover, the characterization of the autophagic molecular machinery has been followed by abundant studies supporting the key role of autophagy alterations in several human disorders including the above cited brain diseases (Fig. 1).

6 Acute Brain Damage

6.1 Ischemic Stroke

Stroke is one of the main causes of mortality and long-term disability worldwide and the majority of cases are ischemic. Ischemic stroke occurs as a result of a transient or permanent reduction in cerebral blood flow, with consequent lack of oxygen and nutrients in ischemic areas leading to neuronal death. Cerebral ischemia results in severe intracellular energy stress leading to cell death by a combination of necrotic cell death in infarct core and apoptotic cell death in the ischemic penumbra [32]. In this scenario, the cellular defences against the exhaustion of energy and metabolic stress contribute to neuronal survival. Under stress condition, in neuronal cells exists a subtle balance between life and death and the ultimate fate depends on the cross-talk between different cellular pathways. Since autophagy is induced by cellular stress to enhance cell survival, and many of these stresses (including nutrient deprivation and oxidative stress) occur during cerebral ischemia, in the last years numerous studies have been interested in disentangling the role of autophagy in brain ischemia. Although increased autophagy has been reported in different types of cerebral ischemia (including focal, global and hypoxic/ischemic injury), its role in neuronal death remains controversial. In fact, several works demonstrate that autophagy acts as a protective response after ischemic strokes, whereas other studies show that autophagy inhibition can reduce ischemic damage.

It has been shown that hypoxia/ischemia (H/I) induces Beclin1 expression in the hippocampus and cerebral cortex of neonatal rats (postnatal day 7), and that *in vivo* pharmacological activation of autophagy by rapamycin treatment reduces necrotic cell death and brain injury. Moreover, the inhibition of autophagy by 3-methyladenine (3-MA) accelerates the progression towards necrotic cell death [13]. In addition, both hypoxic preconditioning and prophylactic treatment with simvastatin, which have neuroprotective effect when administered before the onset of H/I in neonatal rats, increase Beclin1 expression after H/I and reduce brain damage, suggesting a protective role of autophagy in the neurodegenerative process that follows neonatal hypoxic-ischemic insult.

Conversely, Koike and Colleagues described that autophagy is strongly induced in the hippocampus following H/I in neonatal mice (postnatal day 7), and they suggested its involvement in neuronal death. In fact Atg7-deficiency protects hippocampal pyramidal neurons from cell death and reduces damaged areas after H/I [63]. In agreement with this work, other studies suggest that autophagy can be implicated in ischemia-induced neuronal death. It has been demonstrated that autophagy activity is increased after transient focal cerebral ischemia in neonatal rats (postnatal day 12) mainly in the border of the lesion, and that post-ischemic intracerebroventricular (i.c.v.) injections of autophagy inhibitor 3-MA reduces the lesion volume. The activation of autophagy in the ischemic penumbra, where delayed cell death occurs, and the neuroprotective effect of post-ischemic autophagy inhibition suggest a detrimental effect of autophagy on neuronal survival [121]. In agreement with these studies, permanent focal ischemia induced in the cortex of adult rats by permanent middle cerebral artery occlusion (pMCAO) causes an increase of autophagosomes, autolysosomes and LC3 conversion (from LC3-I to LC3-II) in cortical neurons. A single i.c.v. injection of bafilomycin or 3-MA after the onset of ischemia reduces the infarct volume [164].

In support of protective role of autophagy, it has been demonstrated that IRGM (immune-related GTPase M, IRGM1 in mouse), a protein that can regulate the survival of immune cells through autophagy, is upregulated in the ischemic side in pMCAO mouse model, concomitantly with a strong autophagic response [45].

Notably, autophagy activation following pMCAO is almost completely lost in IRGM1 KO mice with an increase of infarct volume. Moreover, IRGM1-mediated activation of autophagy in the early phase of ischemia (within 24 h after injury) protects neurons from necrotic cell death in the core of lesion but promotes apoptosis in the penumbra [45].

Furthermore, in adult rats transient middle cerebral artery occlusion (tMCAO) activates autophagy, and in particular mitophagy (a process involved in selective removal of mitochondria by autophagy) [79]. In tMCAO rapamycin attenuates infarct volumes and neurological deficits after cerebral ischemia and 3-MA blocks this protective effect. The observation that rapamycin enhances mitophagy and improves mitochondrial function suggests that the selective removal of damaged mitochondria by autophagy can play an important protective role in ischemic brain injury [79].

Zhang and Colleagues [177] used both pMCAO and tMCAO mouse models of ischemia to understand the role of autophagy in the ischemic and reperfusion phases. Interestingly, they observed that in the pMCAO model, the pre-treatment with 3-MA reduces the ischemia-induced infarct, whereas i.c.v. injection of 3-MA at the onset of reperfusion (tMCAO mice) significantly aggravates brain injury, suggesting that autophagy plays detrimental and protective roles in the ischemia and reperfusion phases, respectively.

Although autophagy is generally considered a cell survival mechanism, massive autophagy can also be associated with cell death through excessive self-digestion and degradation of cellular components. Depending on the context and amount, autophagy may then either protect cell from death or induce cell death. In conclusion, the net effect of ischemic injury-induced autophagy remains controversial and might depend on brain region and maturity, on severity of injury, and on timing of therapeutic interventions. In fact, autophagy could have dissimilar effects depending on stroke model and on the different phases of ischemia (early or late), as well as playing different roles in the core of lesion and in the penumbra area.

6.2 Traumatic Brain Injury

Traumatic brain injury (TBI) refers to brain damage caused by an external mechanical force, which can lead to permanent or temporary impairment of cognitive, physical and psychosocial functions [89]. TBI is the leading cause of mortality and disability in the young aged population under 45 years and represents one of the major causes of hospitalization nowadays: ten million hospitalizations and/or deaths annually are attributable to TBI worldwide [70]. TBI is a complex disease process causing structural damage and motor and cognitive dysfunction produced by both primary and secondary mechanisms. Physical trauma results in the primary early mechanical damage of brain tissue (including hematomas, contusions, ischemia, axonal injury and diffuse swelling). Primary injury can initiate secondary brain damage (from minutes to months after the trauma) including alteration in neurotransmitter release, calcium-mediated damage, mitochondrial dysfunction, oxidative stress, leading to cell death, tissue damage and atrophy [89, 168]. Neuronal death in TBI is widely attributed to the apoptotic process but several studies have also shown an increase of autophagy after TBI.

In light of the heterogeneity of the clinical aspects of TBI, different animal models (principally rodents) have been developed. Among these, fluid percussion injury (FPI), controlled cortical impact injury (CCI), weight drop-impact acceleration injury, and blast injury are widely used [17, 29, 30, 76, 80, 93].

The use of FPI and CCI rat models has allowed to observe that autophagy is persistently activated after TBI [82, 178]. Autophagosomes accumulate early after TBI (1–4 h) and activation of autophagy persists for days (15–32 days). Since oxygen radicals are involved in the pathogenesis of TBI and autophagy can be induced by mitochondrial oxidative stress, Lai and colleagues analyzed the effects of the antioxidant γ-glutamylcysteinyl ethyl ester (GCEE) on autophagy and neurologic outcome in CCI mouse model of TBI [69]. GCEE treatment decreases the oxidative stress in CCI mice and, more interesting, reduces autophagy levels after TBI, with an improvement of cognitive performance in Morris water maze test and a partial reduction of histological damage. These data suggest that oxidative stress is involved in the neuropathology of TBI and can influence autophagy activity after acute brain injury.

In addition to oxidative stress, glutamate excitotoxicity also plays an important role in TBI. Hyperactivation of NMDARs is associated with TBI-induced neuronal death and excitotoxicity has been linked to autopahgy. Interestingly, it has been demonstrated that NMDAR subunit GluN2A (and its signalling intermediates PSD-95, Homer and Shank) interacts with Beclin1 in membrane rafts of rat cerebral cortex neurons. FPI-induced TBI, in addition to increase the levels of GluN2A, causes a rapid redistribution of Beclin1 out of membrane rafts and activates autophagy pathway [6]. These data suggest that the release of Beclin1 (or PSD95/Shank/Homer/Beclin1) from the GluN2B multi-protein signalling complex in response to TBI-induced excessive stimulation of GluN2A could be a key event involved in the activation of neuronal autophagy.

Although several studies demonstrated that autophagy is activated in different TBI models, its role as a protective or detrimental process remains unclear. In order to investigate the function of autophagy after TBI, Luo and Colleagues tested the effects of autophagy inhibitors in weight-drop mouse model. They observed that the administration of autophagy inhibitors (3-MA and bafilomycin; single i.c.v. injection before TBI) blocks TBI-induced autophagy, attenuates TBI-induced cell death and brain lesions, and improves TBI-induced motor and learning deficits [85].

In conclusion, TBI causes pathophysiological responses that lead to oxidative stress, glutamate excitotoxicity, cell death, motor and cognitive outcome deficits. In this context, autophagy has been identified as part of the responses leading to cell injury after TBI and different compounds tested as neuroprotective in TBI have been shown to reduce apoptotic neuronal death as well as autophagic activity [25, 81, 161, 158, 176].

6.3 Spinal Cord Injury

Spinal cord injury (SCI) is a high-cost neurological disability that can leave the individual with severe life-lasting impairment affecting all organ systems, with strong impact on the patient, the family, health care service, and society. SCI refers to any injury, complete or incomplete, to the spinal cord causing different types of motor, sensory and sphincter dysfunction, as well as dystonia [10]. Extreme sports, high-speed transport, and traumatic accidents in general are linked with a particularly high incidence of SCI, which has an annual incidence of 50 individuals per million population with prevalence in young adults [120, 162].

SCI can be studied in mice and rats models, by means the hemisection of spinal cord at different vertebral levels. It has been observed that SCI causes a fast autophagy activation at the lesion site. Specifically, the upregulation of autophagic markers starts 4 h after the lesion, with a peak at 3 days, and lasts for 21 days [49, 53]. The observed persistent increase of key proteins involved in autophagic pathway in degenerating axons [127] suggests that autophagy might be involved in axonal degeneration following traumatic injury.

However, the role of autophagy in SCI needs to be further investigated in order to understand whether this process contributes to neuronal death or represents a neuroprotective response.

6.4 Remote Damage

Acute brain injury is characterized by two events: (1) early primary damage that directly causes cell death and degeneration, and (2) late secondary damage that induces delayed neurodegeneration through other mechanisms that are not limited to the lesion site but can involve remote areas. Remote neurodegeneration is a multifactorial phenomenon that develops days or months after acute damage and strongly affects the clinical outcome in many CNS disorders [156]. Axotomized neurons undergo a series of morphological changes before dying and the severity of remote cell death is associated to several factors, such as the type and extent of the primary lesion, the distance between axonal trauma and the soma, the type of connectivity, and the intrinsic vulnerability of the involved circuits [34].

Remote damage can be studied in animal models by means axotomy and target deprivation in order to analyze the morphological, biochemical, and ultrastructural changes that occur days to months after injury in different brain circuits. The hemicerebellectomy (HCb) is a widely used model to study the mechanisms of remote cell death. HCb consists in the ablation of half of the cerebellum, which leads, because of the crossed input–output cerebellar organization, to the damage of all neuronal axons of the contralateral inferior olive (IO) and pontine nuclei (Pn) and to deprivation of nearly all cerebellar input of the contralateral cerebral cortex [108, 156]. Recently, we analyzed autophagy function and kinetics during apoptotic cell death in HCb-induced remote damage [155]. We demonstrated that acute brain lesions activate autophagy in axotomized neurons and that this event is subsequent to cytochrome c release from the mitochondria. Importantly, we showed that autophagy stimulation by the mTOR inhibitor rapamycin reduces neuronal death and improves functional recovery after HCb. By contrast, autophagy-impaired *Beclin1* heterozygous mice undergo a greater degeneration of axotomized neurons. These data suggest that, in remote damage induced by HCb, activation of autophagy in axotomized neurons acts as a reactive response that protects neurons by engulfing damaged mitochondria and neutralizing pro-apoptotic factors that can cause cell death [155]. In agreement with this hypothesis we recently observed that HCb alters mitochondrial dynamics (fusion/fission) balance in axotomized neurons and that the neuroprotective effect of rapamycin seems to be the result of a dual role: on one hand the stimulation of autophagy leads to damaged mitochondria removal and on the other hand the enhancement of mitochondria fission allows their elimination by mitophagy [14].

Conversely, in remote damage induced by corticovascular focal lesion autophagy activation seems to have detrimental effects. Focal cerebral infarction can be induced by distal MCAO and can cause secondary degeneration of thalamus and delay functional recovery. After the lesion, autophagy is activated in the ipsilateral thalamus and its inhibition (by using *Beclin1* KO and 3-MA treatment) results in a decrease of neuronal loss, gliosis and apoptosis [167]. These data suggest that the inhibition of autophagy can attenuate the secondary thalamic damage after focal cerebral infarction.

Although it is not possible to draw clear and general conclusions on the role of autophagy in remote damage, because of the different responses depending on the type of primary lesion, several lines of evidence implicate autophagy as a pathophysiological mechanism of remote damage and suggest that drugs targeting this process could be useful to reduce remote neurodegeneration.

7 Neurodegenerative Diseases

7.1 Alzheimer's Disease

Alzheimer's Disease (AD) is a progressive and irreversible age-related neurodegenerative disorder leading to cognitive, memory and behavioural impairments and represents the most cause of dementia worldwide. The main histopathological features of AD brains are the presence of extracellular senile plaques principally formed by the β-amyloid peptide (Aβ), intracellular neurofibrillary tangles constituted by hyperphosphorylated aggregates of the microtubule-associated protein tau, altered neuronal connectivity and massive neuronal loss principally in the hippocampus and cerebral cortex [15]. AD is primarily a sporadic pathology, with age as main risk factor; however, autosomal dominant familial forms are known (familial AD,

FAD). The first mutation causing FAD has been recognized in the amyloid precursor protein (APP) encoding-gene on chromosome 21 [40, 149]. Subsequently AD-related mutations in the presenilin 1 (*PSEN1*, on chromosome 14) and presenilin 2 (*PSEN2*, on chromosome 1) genes have also been identified [78, 142]. Pathogenic mutations in these three genes account for the majority of familial cases of early-onset AD and several mutations in the *APP*, *PSEN1* and *PSEN2* genes have been described worldwide. To note, the mutations causing early-onset autosomal dominat AD affect the metabolism and the stability of Aβ peptide, a proteolytic product of APP processing, that plays a crucial role in AD pathogenesis [138].

Aβ is generated in endo-lysosomal pathway and is present in autophagosomes and in lysosomes. The evidence that autophagic vacuoles (AVs) contain APP and are enriched in β- and γ-secretase activity (responsible for amyloidogenic APP processing with production of Aβ) implicates that AVs are active compartments for Aβ generation [172, 173] and a major source of intracellular Aβ in AD brain [172]. The evidence that autophagy is induced in AD comes out from the study of vulnerable neuronal populations before the extracellular deposition of Aβ in PS1/APP transgenic mouse model of AD. Moreover, dystrophic neurites strongly accumulate autophagosomes and other immature AVs [172], involving an impairment in the normal maturation of AVs to lysosomes and suggesting that a change in the autophagy rate or abnormal AVs accumulation in affected AD neurons can contribute to Aβ deposition. However, more recently it has been demonstrated that autophagy stimulation by rapamycin does not alter APP metabolism and Aβ secretion, suggesting that autophagy is not directly involved in APP metabolism [9]. Rather, it seems likely that the accumulation of APP C-terminal fragments (observable when lysosomal flux is impaired) is caused by a defect of endosome-derived APP or APP C-terminal fragments clearance rather than to an increase of newly generated autophagosomes. In addition to Aβ pathology involvement, autophagy-lysosomal system can also degrade both soluble tau and tau aggregates and inhibition of this mechanism leads to enhanced tau aggregation and cytotoxicity [160]. The involvement of autophagy in the pathogenesis of the two hallmarks of AD brain, amyloid plaques and neurofibrillary tangles, is confirmed by the observation that the pharmacological restoring of mTOR signalling with rapamycin rescues cognitive deficits and ameliorates Aβ and tau pathology by increasing autophagy in the 3xTg-AD mouse model of AD [12].

A role of autophagy in AD pathogenesis is also suggested by the evidence that Beclin1 levels are decreased in affected brain regions of AD patients in early stage of the disease [119]. However, in two different lines of APP transgenic mice (J20 and T41 mouse models of AD) the levels of Beclin1 are not reduced at old age, while Beclin1 deficiency (*Beclin1* heterozygous mice) promotes extracellular and intraneuronal Aβ deposition in APP transgenic mice. These data suggest that the overproduction of mutant APP and the development of amyloid pathology is not sufficient to reduce Beclin1 expression in mice and that the reduction of Beclin1 observed in AD brains likely occurs upstream of APP pathology [119]. The enhancement of autophagic protein turnover rate and lysosomal cathepsin activities in TgCRND8 mouse model (by genetic deletion of cystatin B, an inhibitor of lysosomal cysteine proteases) rescues autophagic-lysosomal pathology and the abnormal accumulation of Aβ, ubiquitinated proteins and other autophagic substrates. The improvement of lysosomal function in this model reduces intraneuronal levels of Aβ as well as extracellular amyloid deposition, and prevents learning and memory deficits [170].

In addition to its role in the clearance of Aβ and tau in AD pathology, autophagy is also involved in the removal of damaged organelles. Mitochondrial abnormalities correlate with dystrophic neurites, dendritic branches loss and dendritic spines pathological alteration present in AD brains [2]. Several line of evidence demonstrate that Aβ damages mitochondria and the reduction of lysosomal degradative efficiency can limit mitochondrial recycling. The impairment of autophagy in aging cells and the positive correlation between Aβ content and mitochondrial damage suggest that mitochondrial turnover could progressively decline with age, with consequent increase of oxidative damage, accumulation of dysfunctional mitochondria and finally cell death [16]. The literature about AD and mitophagy is not very abundant but several evidences, including the impairment of mitochondrial fission/fusion events which are involved in mitochondrial elimination by autophagy, imply that an inefficient lysosomal system may compromise the elimination of damaged mitochondria in AD [134, 159].

7.2 *Parkinson's Disease*

Parkinson's Disease (PD) is the second most common neurodegenerative disorder characterized by bradykinesia, resting tremor, rigidity, and postural instability. The movement deficits principally result from the massive and selective degeneration of nigrostriatal dopaminergic neurons with consequent striatal dopamine deficiency. Another pathological feature of PD brains is the abnormal accumulation of fibrillar α-synuclein protein leading to the formation of intracellular insoluble inclusions (Lewy bodies) and degenerating ubiquitin-positive neuronal processes (Lewy neurites) in surviving neurons [36, 109]. Currently there is no convincing therapy to block or slowdown neuronal loss but only symptomatic treatments are available. Indeed, dopaminergic deficit can be temporarily compensated with deep brain stimulation and by treatment with dopamine agonists, dopamine precursor L-dopa, monoamine oxidase B (MAO-B) and catechol-o-methyltransferase (COMT) inhibitors [21]. PD incidence markedly increases with age although a rare young-onset PD occurring before age 40 exists. PD is principally a sporadic disease (90 % of cases) but several PD-related genes have been identified in a subset of familial forms of the disorder. Genetic transmission can be autosomal dominant – mutations in the genes encoding for α-synuclein and LRRK2 (leucine-rich repeat kinase) – or autosomal recessive – mutations in the genes encoding for PINK1 (PTEN-induced putative kinase 1), parkin and DJ-1 [109].

Aging, genetic factors, and environmental exposure to pesticides and heavy metals are implicated in PD pathogenesis, and mitochondrial dysfunction and oxidative stress have long been associated to PD. Mitochondrial quality control ensures the functionality of the mitochondria during cell life and the deterioration of cellular mechanisms involved in mitochondria turnover has been hypothesized to underlie the pathogenesis of several neurodegenerative diseases, in particular PD [150]. Notably, among the genes related to familial forms of PD, there are two genes encoding for proteins involved in mitochondrial quality control. PINK1 is a serine/threonine kinase normally present at very low levels in healthy polarized mitochondria; however, when mitochondria are depolarized, full-length PINK1 accumulates rapidly at damaged organelles, and recruits parkin from the cytosol to the mitochondria. Parkin is an E3 ubiquitin ligase that catalyzes the polyubiquitination of several substrates, including the mitochondrial proteins Mfn1 (mitofusin 1) and VDAC1 (voltage-dependent anion channel 1), and triggers mitochondrial engulfment by autophagosomes and subsequent degradation through mitophagy [38, 39, 96, 111]. Interestingly, another protein linked to recessive juvenile parkinsonism, Fbxo7, is involved in mitochondrial maintenance regulating PINK1/parkin-mediated mitophagy. Fbxo7, in fact, participates in parkin recruitment to damaged mitochondria and Mfn1 ubiquitination [11, 179]. Rare cases of autosomal recessive PD are caused by loss-of-function mutations in the gene encoding for DJ-1, an ubiquitous redox-responsive cytoprotective protein. In addition to antioxidant effects, DJ-1 also regulates autophagy and contributes to the maintenance of mitochondrial function. In fact, the loss of DJ-1, leads to mitochondrial membrane potential reduction, mitochondrial fragmentation, and autophagic markers accumulation [50, 97]. Collectively, these observations implicate the failure of damaged mitochondria removal through mitophagy as a contributing factor in PD pathogenesis.

Aberrant accumulation of α-synuclein, the major protein in Lewy bodies, is associated with PD pathogenesis and mutations in α-synuclein cause early-onset PD. In inducible α-synuclein-overexpressing cell line it has been demonstrated that α-synuclein can be degraded either by ubiquitin/proteasome system and autophagy pathway. In this *in vitro* model, in fact, α-synuclein is presents in structures with the morphological features of autophagic vesicles and stimulation of autophagy by rapamycin increases its clearance [163]. Afterwards, it has been demonstrated that wild-type α-synuclein is internalized and degraded in lysosomes by CMA, and that mutant forms of this protein (both A30P and A53T mutant) bound specific receptors on lysosomes membrane stronger than wild-type but are poorly internalized [24]. Thus, wild-type α-synuclein seems to be efficiently degraded via CMA, whereas the degradation of mutant α-synuclein is impaired. The blockade of CMA then results in a compensatory activation of macroautophagy, which fails to maintain normal rates of protein degradation [24]. Several studies support an involvement of proteasomal, lysosomal and autophagic pathways in PD, although there is no general consensus about the main pathway responsible for α-synuclein degradation. However, it seems that all the three proteolytic pathways are involved in α-synuclein clearance [5, 24, 163]. Numerous studies have suggested a link between LRRK2 (leucine-rich repeat kinase 2), an important genetic contributor to PD, and aberrant autophagy. LRRK2 is associated with late-onset PD displaying variable pathology depending on the type of mutation [92, 95, 175]. Mutated LRRK2 has been shown to induce or inhibit autophagy depending on specific mutation or cell type [130, 133].

In conclusion, mounting evidence implicates dysfunctional autophagy and mitophagy in PD pathogenesis. Many gene mutations causing familial PD have been identified and many of these alter autophagy. In addition, PD toxins (as MPP+, rotenone, 6-hydroxydopamine, and paraquat) also deregulate autophagy [26], highlighting the importance of this process in PD pathogenesis.

7.3 Hungtinton's Disease

Huntington's disease (HD) is an autosomal dominant inherited neurodegenerative disorder characterized by abnormal repetitive movements called chorea, progressive dementia and psychiatric manifestations. The main pathological alteration in HD brains is the selective neuron loss occurring in the striatum and cortex. The onset of symptoms is typically in the middle-age but the disorder can manifest at any time. HD causes death 15–20 years after the onset of neurological manifestations, and although there are treatments that can help to control choreiform movements, successful therapy to block disease progression does not exist [157]. The mutant protein in HD results from an abnormal expansion of the trinucleotide sequence CAG in the first exon of *huntingtin* gene, resulting in an excessive extension of variable length of the polyglutamine (polyQ) tract at the N-terminus of huntingtin (Htt) protein with a toxic gain of function. The normal allele contains a repeated sequence of 10–35 CAG triplets. The expansion of this CAG repeat over 35 is linked with the development of disease, and the length of polyQ tract is inversely correlated with the age-onset of HD [67, 129]. The production and accumulation of misfolded Htt cause the formation of inclusion bodies in HD brain leading to the selective loss of striatal GABAergic neurons.

Different mechanisms have been suggested to contribute to the pathogenesis of HD, including excitotoxic injury, oxidative stress, mitochondrial dysfunction, and apoptosis. As autophagy is primarily responsible for maintaining normal cellular protein homeostasis in the CNS and HD is a neurodegenerative proteinopathy (as other neurodegenerative disease such as AD and PD), it is not surprising that autophagy has attracted the interest of scholars in the field. Several lines of research have reported autophagy dysfunction in HD and have shown that autophagy modulation could represent a potential therapeutic intervention.

Increased number of AVs has been reported in human HD samples and in different experimental models. Initially it has been observed that Htt abnormally accumulates in punctate cytoplasmic structures resembling endosomal-lysosomal organelles in HD brains [135] and that wild-type and mutant Htt associate with endosomes in human-derived primary fibroblasts [154]. Then, in clonal mouse striatal cell line transiently transfected with human Htt it has been demonstrated that both wild-type and mutant Htt accumulate in the cytoplasm forming vacuoles [57], which incorporate the lysosomal enzyme cathepsin D in proportion to polyQ length and have the ultrastructural features of autophagosome [55]. Primary striatal neurons from transgenic HD mice (R6/2 mouse model expressing exon one of the human Htt gene carrying a CAG repeat expansion) exposed to neurotoxic concentration of dopamine not only present increased cell death compared to wild-type neurons, but also exhibit lysosome-associated responses with the induction of autophagic granules and electron-dense lysosomes [118].

Mutant Htt can sequester and inactivate mTOR, thus promoting autophagy in HD brain [126] and accumulation of the autophagic markers p62 and LC3 in the striatum of transgenic mouse models of HD [71]. Therefore, several studies suggest that dysfunctional autophagy can be involved in HD pathogenesis. However, the increase of autophagosomes in cellular and mouse model of HD is not accompanied by an increase of autophagic substrates degradation although autophagic flux does not appear to be affected [94]. In fact, in HD cells AVs form with normal or enhanced rates and are effectively eliminated by lysosomes but there is a defect in autophagic cargo recognition with the formation of "empty" autophagosome. As a result, despite an increase in the initiation of autophagy and the formation of AVs, aggregated proteins (including Htt) and damaged mitochondria are not degraded and accumulate in the cytoplasm contributing to toxicity [94].

Many cell types respond to autophagy blockade by upregulating CMA [54] and CMA results increased in cellular and mouse models of HD [62]. Htt has two putative KFERQ-like CMA-targeting motifs; the pentapeptide motif KFERQ is recognized by cytosolic chaperone Hsc70 that targets the substrates to lysosomes, where they are bound by the lysosome-associated membrane protein type 2A (LAMP-2A), which mediates substrate translocation into the lysosomal lumen [23]. Notably, CMA-dependent degradation of Htt fragments is less efficient for polyQ-expanded Htt fragments, perhaps because polyQ expansion in mutant Htt delays the transport across the lysosomal membrane, with consequent accumulation into the cytosol [62, 122]. CMA targets selectively N-terminal fragments of Htt (while full-length polyQ-Htt is primarily targeted by autophagy), and specific targeting of N-terminal polyQ-Htt fragment to the CMA pathway strongly reduces the formation of cytosolic inclusion and improves HD phenotypes in R6/2 mice [3].

As rapamycin treatment reduces aggregate formation and cell death in cells expressing mutant Htt [125], small molecules which enhance autophagy are able to reduce toxicity in HD models [137], and several mTOR-independent autophagy inducers increase the clearance of mutant Htt and reduce its toxicity [136], the manipulation of autophagy and CMA could represent a promising candidate for therapy development.

8 Conclusions

In this Chapter we have described autophagic mechanism and its fundamental role in neuronal function. We have highlighted that autophagy plays a key role in the physiology of the central nervous system, not only as quality control mechanism but also for brain development and synaptic function. The emerging evidences that autophagy defects are involved in common neurodegenerative diseases and that acute brain injuries are characterized by strong autophagic responses encourage to unravel the role of autophagy in brain diseases and open the possibility of future therapeutic approaches.

References

1. Axe EL, Walker SA, Manifava M, Chandra P, Roderick HL, Habermann A, Griffiths G, Ktistakis NT (2008) Autophagosome formation from membrane compartments enriched in phosphatidylinositol 3-phosphate and dynamically connected to the endoplasmic reticulum. J Cell Biol 182:685–701
2. Baloyannis SJ (2009) Dendritic pathology in Alzheimer's disease. J Neurol Sci 283:153–157
3. Bauer PO, Goswami A, Wong HK, Okuno M, Kurosawa M, Yamada M, Miyazaki H, Matsumoto G, Kino Y, Nagai Y, Nukina N (2010) Harnessing chaperone-mediated autophagy for the selective degradation of mutant huntingtin protein. Nat Biotechnol 28:256–263
4. Bear MF, Abraham WC (1996) Long-term depression in hippocampus. Annu Rev Neurosci 19:437–462
5. Bennett MC, Bishop JF, Leng Y, Chock PB, Chase TN, Mouradian MM (1999) Degradation of α-synuclein by proteasome. J Biol Chem 274:33855–33858
6. Bigford GE, Alonso OF, Dietrich D, Keane RW (2009) A novel protein complex in membrane rafts linking the NR2B glutamate receptor and autophagy is disrupted following traumatic brain injury. J Neurotrauma 2:703–720
7. Bliss TVP, Collingridge GL (1993) A synaptic model of memory: longterm potentiation in the hippocampus. Nature 361:31–39
8. Boland B, Kumar A, Lee S, Platt FM, Wegiel J, Yu WH, Nixon RA (2008) Autophagy induction and autophagosome clearance in neurons: relationship to autophagic pathology in Alzheimer's disease. J Neurosci 28:6926–6937
9. Boland B, Smith DA, Mooney D, Jung SS, Walsh DM, Platt FM (2010) Macroautophagy is not directly involved in the metabolism of amyloid precursor protein. J Biol Chem 285:37415–37426
10. Bregman BS, Kunkel-Bagden E, Schnell L, Dai HN, Gao D, Schwab ME (1995) Recovery from spinal cord injury mediated by antibodies to neurite growth inhibitors. Nature 378:498–501
11. Burchell VS, Nelson DE, Sanchez-Martinez A, Delgado-Camprubi M, Ivatt RM, Pogson JH, Randle SJ, Wray S, Lewis PA, Houlden H, Abramov AY, Hardy J, Wood NW, Whitworth AJ, Laman H, Plun-Favreau H (2013) The Parkinson's disease-linked proteins Fbxo7 and Parkin interact to mediate mitophagy. Nat Neurosci 16:1257–1265
12. Caccamo A, Majumder S, Richardson A, Strong R, Oddo S (2010) Molecular interplay between mammalian target of rapamycin (mTOR), amyloid-beta, and Tau: effects on cognitive impairments. J Biol Chem 285:13107–13120
13. Carloni S, Buonocore G, Balduini W (2008) Protective role of autophagy in neonatal hypoxia-ischemia induced brain injury. Neurobiol Dis 32:329–339
14. Cavallucci V, Bisicchia E, Cencioni MT, Ferri A, Latini L, Nobili A, Biamonte F, Nazio F, Fanelli F, Moreno S, Molinari M, Viscomi MT, D'Amelio M (2014) Acute focal brain damage alters mitochondrial dynamics and autophagy in axotomized neurons. Cell Death Dis 5:e1545. doi:10.1038/cddis.2014.511
15. Cavallucci V, D'Amelio M, Cecconi F (2012) Aβ toxicity in Alzheimer's disease. Mol Neurobiol 45:366–378

16. Cavallucci V, Ferraina C, D'Amelio M (2013) Key role of mitochondria in Alzheimer's disease synaptic dysfunction. Curr Pharm Des 19:6440–6450
17. Cernak I, Savic J, Malicevic Z, Zunic G, Radosevic P, Ivanovic I, Davidovic L (1996) Involvement of the central nervous system in the general response to pulmonary blast injury. J Trauma 40:S100–S104
18. Chung HJ, Steinberg JP, Huganir RL, Linden DJ (2003) Requirement of AMPA receptor GluR2 phosphorylation for cerebellar long-term depression. Science 300:1751–1755
19. Chung HJ, Xia J, Scannevin RH, Zhang X, Huganir RL (2000) Phosphorylation of the AMPA receptor subunit GluR2 differentially regulates its interaction with PDZ domain-containing proteins. J Neurosci 20:7258–7267
20. Clarke PG (1990) Developmental cell death: morphological diversity and multiple mechanisms. Anat Embryol 181:195–213
21. Connolly BS, Lang AE (2014) Pharmacological treatment of Parkinson disease: a review. JAMA 311:1670–1683
22. Efeyan A, Sabatini DM (2011) mTOR and cancer: many loops in one pathway. Curr Opin Cell Biol 22:169–176
23. Cuervo AM (2010) Chaperone-mediated autophagy: selectivity pays off. Trends Endocrinol Metab 21:142–150
24. Cuervo AM, Stefanis L, Fredenburg R, Lansbury PT, Sulzer D (2004) Impaired degradation of mutant α-synuclein by chaperone-mediated autophagy. Science 305:1292–1295
25. Cui C, Cui Y, Gao J, Sun L, Wang Y, Wang K, Li R, Tian Y, Song S, Cui J (2014) Neuroprotective effect of ceftriaxone in a rat model of traumatic brain injury. Neurol Sci 35:695–700
26. Dagda RK, Das Banerjee T, Janda E (2013) How Parkinsonian toxins dysregulate the autophagy machinery. Int J Mol Sci 14:22163–22189
27. Daw MI, Chittajallu R, Bortolotto ZA, Dev KK, Duprat F, Henley JM, Collingridge GL, Isaac JT (2000) PDZ proteins interacting with C-terminal GluR2/3 are involved in a PKC-dependent regulation of AMPA receptors at hippocampal synapses. Neuron 28:873–886
28. Di Bartolomeo S, Corazzari M, Nazio F, Oliverio S, Lisi G, Antonioli M, Pagliarini V, Matteoni S, Fuoco C, Giunta L, D'Amelio M, Nardacci R, Romagnoli A, Piacentini M, Cecconi F, Fimia GM (2010) The dynamic interaction of AMBRA1 with the dynein motor complex regulates mammalian autophagy. J Cell Biol 191:155–168
29. Dixon CE, Clifton GL, Lighthall JW, Yaghmai AA, Hayes RL (1991) A controlled cortical impact model of traumatic brain injury in the rat. J Neurosci Methods 39:253–262
30. Dixon CE, Lyeth BG, Povlishock JT, Findling RL, Hamm RJ, Marmarou A, Young HF, Hayes RL (1987) A fluid percussion model of experimental brain injury in the rat. J Neurosurg 67:110–119
31. Dong H, O'Brien RJ, Fung ET, Lanahan AA, Worley PF, Huganir RL (1997) GRIP: a synaptic PDZ domain-containing protein that interacts with AMPA receptors. Nature 386:279–284
32. Donnan GA, Fisher M, Macleod M, Davis SM (2008) Stroke. Lancet 371:1612–1623
33. Ehlers MD (2000) Reinsertion or degradation of AMPA receptors determined by activity-dependent endocytic sorting. Neuron 28:511–525
34. Faden AI (2002) Neuroprotection and traumatic brain injury: theoretical option or realistic proposition. Curr Opin Neurol 15:707–712
35. Fimia GM, Stoykova A, Romagnoli A, Giunta L, Di Bartolomeo S, Nardacci R, Corazzari M, Fuoco C, Ucar A, Schwartz P, Gruss P, Piacentini M, Chowdhury K, Cecconi F (2007) Ambra1 regulates autophagy and development of the nervous system. Nature 447:1121–1125
36. Forno LS (1996) Neuropathology of Parkinson's disease. J Neuropathol Exp Neurol 55:259–272
37. Gan B, Peng X, Nagy T, Alcaraz A, Gu H, Guan JL (2006) Role of FIP200 in cardiac and liver development and its regulation of TNFalpha and TSC-mTOR signaling pathways. J Cell Biol 175:121–133

38. Geisler S, Holmstrom KM, Skujat D, Fiesel FC, Rothfuss OC, Kahle PJ, Springer W (2010) PINK1/Parkin-mediated mitophagy is dependent on VDAC1 and p62/SQSTM1. Nat Cell Biol 12:119–131
39. Glauser L, Sonnay S, Stafa K, Moore DJ (2011) Parkin promotes the ubiquitination and degradation of the mitochondrial fusion factor mitofusin 1. J Neurochem 118:636–645
40. Goate A, Chartier-Harlin MC, Mullan M, Brown J, Crawford F, Fidani L, Giuffra L, Haynes A, Irving N, James L, Mantparallel R, Newton P, Rooke K, Roques P, Talbot C, Pericak-Vance M, Roses A, Williamson R, Rossor M, Owenparallel M, Hardy J (1991) Segregation of a missense mutation in the amyloid precursor protein gene with familial Alzheimer's disease. Nature 349:704–706
41. Hanada T, Noda NN, Satomi Y, Ichimura Y, Fujioka Y, Takao T, Inagaki F, Ohsumi Y (2007) The Atg12-Atg5 conjugate has a novel E3-like activity for protein lipidation in autophagy. J Biol Chem 282:37298–37302
42. Hara T, Nakamura K, Matsui M, Yamamoto A, Nakahara Y, Suzuki-Migishima R, Yokoyama M, Mishima K, Saito I, Okano H, Mizushima N (2006) Suppression of basal autophagy in neural cells causes neurodegenerative disease in mice. Nature 441:885–889
43. Hayashi Y, Shi SH, Esteban JA, Piccini A, Poncer JC, Malinow R (2000) Driving AMPA receptors into synapses by LTP and CaMKII: requirement for GluR1 and PDZ domain interaction. Science 287:2262–2267
44. Hayashi-Nishino M, Fujita N, Noda T, Yamaguchi A, Yoshimori T, Yamamoto A (2009) A subdomain of the endoplasmic reticulum forms a cradle for autophagosome formation. Nat Cell Biol 11:1433–1437
45. He S, Wang C, Dong H, Xia F, Zhou H, Jiang X, Pei C, Ren H, Li H, Li R, Xu H (2012) Immune-related GTPase M (IRGM1) regulates neuronal autophagy in a mouse model of stroke. Autophagy 8:1621–1627
46. Hernandez D, Torres CA, Setlik W, Cebrián C, Mosharov EV, Tang G, Cheng HC, Kholodilov N, Yarygina O, Burke RE, Gershon M, Sulzer D (2012) Regulation of presynaptic neurotransmission by macroautophagy. Neuron 74:277–284
47. Hollmann M, Heinemann S (1994) Cloned glutamate receptors. Annu Rev Neurosci 17:31–108
48. Hosokawa N, Sasaki T, Iemura S, Natsume T, Hara T, Mizushima N (2009) Atg101, a novel mammalian autophagy protein interacting with Atg13. Autophagy 5:973–979
49. Hou H, Zhang L, Zhang L, Tang P (2014) Acute spinal cord injury in rats should target activated autophagy. J Neurosurg Spine 20:568–577
50. Irrcher I, Aleyasin H, Seifert EL, Hewitt SJ, Chhabra S, Phillips M, Lutz AK, Rousseaux MW, Bevilacqua L, Jahani-Asl A, Callaghan S, MacLaurin JG, Winklhofer KF, Rizzu P, Rippstein P, Kim RH, Chen CX, Fon EA, Slack RS, Harper ME, McBride HM, Mak TW, Park DS (2010) Loss of the Parkinson's disease-linked gene DJ-1 perturbs mitochondrial dynamics. Hum Mol Genet 19:3734–3746
51. Jiang J, Parameshwaran K, Seibenhener ML, Kang MG, Suppiramaniam V, Huganir RL, Diaz-Meco MT, Wooten MW (2009) AMPA receptor trafficking and synaptic plasticity require SQSTM1/p62. Hippocampus 19:392–406
52. Kabeya Y, Mizushima N, Yamamoto A, Oshitani-Okamoto S, Ohsumi Y, Yoshimori T (2004) LC3, GABARAP and GATE16 localize to autophagosomal membrane depending on form-II formation. J Cell Sci 117:2805–2812
53. Kanno H, Ozawa H, Sekiguchi A, Itoi E (2009) Spinal cord injury induces upregulation of Beclin 1 and promotes autophagic cell death. Neurobiol Dis 33:143–148
54. Kaushik S, Massey AC, Mizushima N, Cuervo AM (2008) Constitutive activation of chaperone-mediated autophagy in cells with impaired macroautophagy. Mol Biol Cell 19:2179–2192
55. Kegel KB, Kim M, Sapp E, McIntyre C, Castaño JG, Aronin N, DiFiglia M (2000) Huntingtin expression stimulates endosomal-lysosomal activity, endosome tubulation, and autophagy. J Neurosci 20:7268–7278

56. Kim CH, Chung HJ, Lee HK, Huganir RL (2001) Interaction of the AMPA receptor subunit GluR2/3 with PDZ domains regulates hippocampal long-term depression. Proc Natl Acad Sci U S A 98:11725–11730
57. Kim M, Lee HS, Laforet G, McIntyre C, Martin EJ, Chang P, Kim TW, Williams M, Reddy PH, Tagle D, Boyce FM, Won L, Heller A, Aronin N, DiFiglia M (1999) Mutant huntingtin expression in clonal striatal cells: dissociation of inclusion formation and neuronal survival by caspase inhibition. J Neurosci 19:964–973
58. Kim MJ, Futai K, Jo J, Hayashi Y, Cho K, Sheng M (2007) Synaptic accumulation of PSD-95 and synaptic function regulated by phosphorylation of serine-295 of PSD-95. Neuron 56:488–502
59. Kirisako T, Ichimura Y, Okada H, Kabeya Y, Mizushima N, Yoshimori T, Ohsumi M, Takao T, Noda T, Ohsumi Y (2000) The reversible modification regulates the membrane-binding state of Apg8/Aut7 essential for autophagy and the cytoplasm to vacuole targeting pathway. J Cell Biol 151:263–276
60. Klionsky DJ (2005) The molecular machinery of autophagy: unanswered questions. J Cell Sci 118:7–18
61. Klionsky DJ, Cregg JM, Dunn WA Jr, Emr SD, Sakai Y, Sandoval IV, Sibirny A, Subramani S, Thumm M, Veenhuis M, Ohsumi Y (2003) A unified nomenclature for yeast autophagy-related genes. Dev Cell 5:539–545
62. Koga H, Martinez-Vicente M, Arias E, Kaushik S, Sulzer D, Cuervo AM (2011) Constitutive upregulation of chaperone-mediated autophagy in Huntington's disease. J Neurosci 31:18492–18505
63. Koike M, Shibata M, Tadakoshi M, Gotoh K, Komatsu M, Waguri S, Kawahara N, Kuida K, Nagata S, Kominami E, Tanaka K, Uchiyama Y (2008) Inhibition of autophagy prevents hippocampal pyramidal neuron death after hypoxic-ischemic injury. Am J Pathol 172:454–469
64. Komatsu M, Waguri S, Ueno T, Iwata J, Murata S, Tanida I, Ezaki J, Mizushima N, Ohsumi Y, Uchiyama Y, Kominami E, Tanaka K, Chiba T (2005) Impairment of starvation-induced and constitutive autophagy in Atg7-deficient mice. J Cell Biol 169:425–434
65. Komatsu M, Wang QJ, Holstein GR, Friedrich VL Jr, Iwata J, Kominami E, Chait BT, Tanaka K, Yue Z (2007) Essential role for autophagy protein Atg7 in the maintenance of axonal homeostasis and the prevention of axonal degeneration. Proc Natl Acad Sci U S A 104:14489–14494
66. Kon M, Cuervo AM (2010) Chaperone-mediated autophagy in health and disease. FEBS Lett 584:1399–1404
67. Kremer B, Goldberg P, Andrew SE, Theilmann J, Telenius H, Zeisler J, Squitieri F, Lin B, Bassett A, Almqvist E, Bird TD, Hayden MR (1994) A worldwide study of the Huntington's disease mutation. The sensitivity and specificity of measuring CAG repeats. N Engl J Med 330:1401–1406
68. Kuma A, Hatano M, Matsui M, Yamamoto A, Nakaya H, Yoshimori T, Ohsumi Y, Tokuhisa T, Mizushima N (2004) The role of autophagy during the early neonatal starvation period. Nature 432:1032–1036
69. Lai Y, Hickey RW, Chen Y, Bayir H, Sullivan ML, Chu CT, Kochanek PM, Dixon CE, Jenkins LW, Graham SH, Watkins SC, Clark RS (2008) Autophagy is increased after traumatic brain injury in mice and is partially inhibited by the antioxidant γ-glutamylcysteinyl ethyl ester. J Cereb Blood Flow Metab 28:540–550
70. Langlois JA, Rutland-Brown W, Wald MM (2006) The epidemiology and impact of traumatic brain injury: a brief overview. J Head Trauma Rehabil 21:375–378
71. Lee H, Noh JY, Oh Y, Kim Y, Chang JW, Chung CW, Lee ST, Kim M, Ryu H, Jung YK (2012) IRE1 plays an essential role in ER stress-mediated aggregation of mutant huntingtin via the inhibition of autophagy flux. Hum Mol Genet 21:101–114
72. Lee HK, Kameyama K, Huganir RL, Bear MF (1998) NMDA induces long-term synaptic depression and dephosphorylation of the GluR1 subunit of AMPA receptors in hippocampus. Neuron 21:1151–1162

73. Lee HK, Takamiya K, He K, Song L, Huganir RL (2010) Specific roles of AMPA receptor subunit GluR1 (GluA1) phosphorylation sites in regulating synaptic plasticity in the CA1 region of hippocampus. J Neurophysiol 103:479–489
74. Lee SH, Liu L, Wang YT, Sheng M (2002) Clathrin adaptor AP2 and NSF interact with overlapping sites of GluR2 and play distinct roles in AMPA receptor trafficking and hippocampal LTD. Neuron 36:661–674
75. Lee SH, Simonetta A, Sheng M (2004) Subunit rules governing the sorting of internalized AMPA receptors in hippocampal neurons. Neuron 43:221–236
76. Leung LY, VandeVord PJ, Dal Cengio AL, Bir C, Yang KH, King AI (2008) Blast related neurotrauma: a review of cellular injury. Mol Cell Biomech 5:155–168
77. Levine B, Klionsky DJ (2004) Development by self-digestion: molecular mechanisms and biological functions of autophagy. Dev Cell 6:463–477
78. Levy-Lahad E, Wasco W, Poorkaj P, Romano DM, Oshima J, Pettingell WH, Yu CE, Jondro PD, Schmidt SD, Wang K, Crowley AC, Fu Y-H, Guenette SY, Galas D, Nemens E, Wijsman EM, Bird TD, Schellenberg GD, Tanzi RE (1995) Candidate gene for the chromosome 1 familial Alzheimer's disease locus. Science 269:973–977
79. Li Q, Zhang T, Wang J, Zhang Z, Zhai Y, Yang GY, Sun X (2014) Rapamycin attenuates mitochondrial dysfunction via activation of mitophagy in experimental ischemic stroke. Biochem Biophys Res Commun 444:182–188
80. Lighthall JW (1988) Controlled cortical impact: a new experimental brain injury model. J Neurotrauma 5:1–15
81. Lin CJ, Chen TH, Yang LY, Shih CM (2014) Resveratrol protects astrocytes against traumatic brain injury through inhibiting apoptotic and autophagic cell death. Cell Death Dis 5:e1147. doi:10.1038/cddis.2014.123
82. Liu CL, Chen S, Dietrich D, Hu BR (2008) Changes in autophagy after traumatic brain injury. J Cereb Blood Flow Metab 28:674–683
83. Lockshin RA, Zakeri Z (2004) Apoptosis, autophagy, and more. Int J Biochem Cell Biol 36:2405–2419
84. Lu W, Shi Y, Jackson AC, Bjorgan K, During MJ, Sprengel R, Seeburg PH, Nicoll RA (2009) Subunit composition of synaptic AMPA receptors revealed by a single-cell genetic approach. Neuron 62:254–268
85. Luo CL, Li BX, Li QQ, Chen XP, Sun YX, Bao HJ, Dai DK, Shen YW, Xu HF, Ni H, Wan L, Qin ZH, Tao LY, Zhao ZQ (2011) Autophagy is involved in traumatic brain injury-induced cell death and contributes to functional outcome deficits in mice. Neuroscience 184:54–63
86. Lüscher C, Frerking M (2001) Restless AMPA receptors: implications for synaptic transmission and plasticity. Trends Neurosci 24:665–670
87. Lv X, Jiang H, Li B, Liang Q, Wang S, Zhao Q, Jiao J (2014) The crucial role of Atg5 in cortical neurogenesis during early brain development. Sci Rep 4:6010. doi:10.1038/srep06010
88. Lynch G, Baudry M (1984) The biochemistry of memory: a new and specific hypothesis. Science 224:1057–1063
89. Maas AI, Stocchetti N, Bullock R (2008) Moderate and severe traumatic brain injury in adults. Lancet Neurol 7:728–741
90. Malenka RC, Nicoll RA (1999) Long-term potentiation-A decade of progress. Science 285:1870–1874
91. Man HY, Lin JW, Ju WH, Ahmadian G, Liu L, Becker LE, Sheng M, Wang YT (2000) Regulation of AMPA receptor-mediated synaptic transmission by clathrin-dependent receptor internalization. Neuron 25:649–662
92. Manzoni C, Mamais A, Dihanich S, McGoldrick P, Devine MJ, Zerle J, Kara E, Taanman JW, Healy DG, Marti-Masso JF, Schapira AH, Plun-Favreau H, Tooze S, Hardy J, Bandopadhyay R, Lewis PA (2013) Pathogenic Parkinson's disease mutations across the functional domains of LRRK2 alter the autophagic/lysosomal response to starvation. Biochem Biophys Res Commun 441:862–866

93. Marmarou A, Foda MA, van den Brink W, Campbell J, Kita H, Demetriadou K (1994) A new model of diffuse brain injury in rats. Part I: pathophysiology and biomechanics. J Neurosurg 80:291–300
94. Martinez-Vicente M, Talloczy Z, Wong E, Tang G, Koga H, Kaushik S, de Vries R, Arias E, Harris S, Sulzer D, Cuervo AM (2010) Cargo recognition failure is responsible for inefficient autophagy in Huntington's disease. Nat Neurosci 13:567–576
95. Mata IF, Wedemeyer WJ, Farrer MJ, Taylor JP, Gallo KA (2006) LRRK2 in Parkinson's disease: protein domains and functional insights. Trends Neurosci 29:286–293
96. Matsuda N, Sato S, Shiba K, Okatsu K, Saisho K, Gautier CA, Sou YS, Saiki S, Kawajiri S, Sato F, Kimura M, Komatsu M, Hattori N, Tanaka K (2010) PINK1 stabilized by mitochondrial depolarization recruits Parkin to damaged mitochondria and activates latent Parkin for mitophagy. J Cell Biol 189:211–221
97. Mccoy MK, Cookson MR (2011) DJ-1 regulation of mitochondrial function and autophagy through oxidative stress. Autophagy 7:531–532
98. Mercer CA, Kaliappan A, Dennis PB (2009) A novel, human Atg13 binding protein, Atg101, interacts with ULK1 and is essential for macroautophagy. Autophagy 5:649–662
99. Mijaljica D, Prescott M, Devenish RJ (2011) Microautophagy in mammalian cells Revisiting a 40-year-old conundrum. Autophagy 7:673–682
100. Ming GL, Song H (2011) Adult neurogenesis in the mammalian brain: significant answers and significant questions. Neuron 70:687–702
101. Mizushima N (2005) The pleiotropic role of autophagy: from protein metabolism to bactericide. Cell Death Differ 12:1535–1541
102. Mizushima N, Klionsky DJ (2007) Protein turnover via autophagy: implications for metabolism. Annu Rev Nutr 27:19–40
103. Mizushima N, Kuma A, Kobayashi Y, Yamamoto A, Matsubae M, Takao T, Natsume T, Ohsumi Y, Yoshimori T (2003) Mouse Apg16L, a novel WD-repeat protein, targets to the autophagic isolation membrane with the Apg12–Apg5 conjugate. J Cell Sci 116:1679–1688
104. Mizushima N, Levine B (2010) Autophagy in mammalian development and differentiation. Nat Cell Biol 12:823–830
105. Mizushima N, Sugita H, Yoshimori T, Ohsumi Y (1998) A new protein conjugation system in human. The counterpart of the yeast Apg12p conjugation system essential for autophagy. J Biol Chem 273:33889–33892
106. Mizushima N, Yamamoto A, Matsui M, Yoshimori T, Ohsumi Y (2004) In vivo analysis of autophagy in response to nutrient starvation using transgenic mice expressing a fluorescent autophagosome marker. Mol Biol Cell 15:1101–1111
107. Mizushima N, Yoshimori T, Levine B (2010) Methods in mammalian autophagy research. Cell 140:313–326
108. Molinari M, Viscomi MT, Leggio MG (2013) Hemicerebellectomy. In: Manto M, Schmahmann JD, Rossi F, Gruol DL, Koibuchi N (eds) Handbook of the cerebellum and cerebellar disorders. Springer, Dordrecht, pp 1579–1594
109. Moore DJ, West AB, Dawson VL, Dawson TM (2005) Molecular pathophysiology of Parkinson's disease. Annu Rev Neurosci 28:57–87
110. Nakatogawa H, Suzuki K, Kamada Y, Ohsumi Y (2009) Dynamics and diversity in autophagy mechanisms: lessons from yeast. Nat Rev Mol Cell Biol 10:458–467
111. Narendra DP, Jin SM, Tanaka A, Suen DF, Gautier CA, Shen J, Cookson MR, Youle RJ (2010) PINK1 is selectively stabilized on impaired mitochondria to activate Parkin. PLoS Biol 8:e1000298. doi:10.1371/journal.pbio.1000298
112. Nixon RA, Wegiel J, Kumar A, Yu WH, Peterhoff C, Cataldo A, Cuervo AM (2005) Extensive involvement of autophagy in Alzheimer disease: an immuno-electron microscopy study. J Neuropathol Exp Neurol 64:113–122
113. Osten P, Khatri L, Perez JL, Köhr G, Giese G, Daly C, Schulz TW, Wensky A, Lee LM, Ziff EB (2000) Mutagenesis reveals a role for ABP/GRIP binding to GluR2 in synaptic surface accumulation of the AMPA receptor. Neuron 27:313–325

114. Park M, Penick EC, Edwards JG, Kauer JA, Ehlers MD (2004) Recycling endosomes supply AMPA receptors for LTP. Science 305:1972–1975
115. Parzych KR, Klionsky DJ (2014) An overview of autophagy: morphology, mechanism, and regulation. Antioxid Redox Signal 20:460–473
116. Passafaro M, Piech V, Sheng M (2001) Subunit-specific temporal and spatial patterns of AMPAreceptor exocytosis in hippocampal neurons. Nat Neurosci 4:917–926
117. Perez JL, Khatri L, Chang C, Srivastava S, Osten P, Ziff EB (2001) PICK1 targets activated protein kinase C alpha to AMPA receptor clusters in spines of hipoocampal neurons and reduces surface levels of the AMPA-type glutamate receptor subunit 2. J Neurosci 21:5417–5428
118. Petersén A, Larsen KE, Behr GG, Romero N, Przedborski S, Brundin P, Sulzer D (2001) Expanded CAG repeats in exon 1 of the Huntington's disease gene stimulate dopamine-mediated striatal neuron autophagy and degeneration. Hum Mol Genet 10:1243–1254
119. Pickford F, Masliah E, Britschgi M, Lucin K, Narasimhan R, Jaeger PA, Small S, Spencer B, Rockenstein E, Levine B, Wyss-Coray T (2008) The autophagy-related protein beclin 1 shows reduced expression in early Alzheimer disease and regulates amyloid beta accumulation in mice. J Clin Invest 118:2190–2199
120. Popa C, Popa F, Grigorean VT, Onose G, Sandu AM, Popescu M, Burnei G, Strambu V, Sinescu C (2010) Vascular dysfunctions following spinal cord injury. J Med Life 3:275–285
121. Puyal J, Vaslin A, Mottier V, Clarke PG (2009) Postischemic treatment of neonatal cerebral ischemia should target autophagy. Ann Neurol 66:378–389
122. Qi L, Zhang XD, Wu JC, Lin F, Wang J, DiFiglia M, Qin ZH (2012) The role of chaperone-mediated autophagy in huntingtin degradation. PLoS One 7:e46834. doi:10.1371/journal.pone.0046834
123. Qu X, Zou Z, Sun Q, Luby-Phelps K, Cheng P, Hogan RN, Gilpin C, Levine B (2007) Autophagy gene-dependent clearance of apoptotic cells during embryonic development. Cell 128:931–946
124. Ramesh Babu J, Lamar Seibenhener M, Peng J, Strom AL, Kemppainen R, Cox N, Zhu H, Wooten MC, Diaz-Meco MT, Moscat J, Wooten MW (2008) Genetic inactivation of p62 leads to accumulation of hyperphosphorylated tau and neurodegeneration. J Neurochem 106:107–120
125. Ravikumar B, Duden R, Rubinsztein DC (2002) Aggregate-prone proteins with polyglutamine and polyalanine expansions are degraded by autophagy. Hum Mol Genet 11:1107–1117
126. Ravikumar B, Vacher C, Berger Z, Davies JE, Luo S, Oroz LG, Scaravilli F, Easton DF, Duden R, O'Kane CJ, Rubinsztein DC (2004) Inhibition of mTOR induces autophagy and reduces toxicity of polyglutamine expansions in fly and mouse models of Huntington disease. Nat Genet 36:585–595
127. Ribas VT, Schnepf B, Challagundla M, Koch JC, Bähr M, Lingor P (2014) Early and sustained activation of autophagy in degenerating axons after spinal cord injury. Brain Pathol 25:157–170
128. Rowland AM, Richmond JE, Olsen JG, Hall DH, Bamber BA (2006) Presynaptic terminals independently regulate synaptic clustering and autophagy of GABAA receptors in Caenorhabditis elegans. J Neurosci 26:1711–1720
129. Rubinsztein DC, Leggo J, Coles R, Almqvist E, Biancalana V, Cassiman JJ, Chotai K, Connarty M, Craufurd D, Curtis A, Curtis D, Davidson MJ, Differ AM, Dode C, Dodge A, Frontali M, Ranen NG, Stine OC, Sherr M, Abbott MH, Franz ML, Graham CA, Harper PS, Hedreen JC, Jackson A, Kaplan JC, Losekoot M, MacMillan JC, Morrison P, Trottier Y, Novelletto A, Simpson SA, Theilmann J, Whittaker JL, Folstein SE, Ross CA, Hayden MR (1996) Phenotypic characterization of individuals with 30–40 CAG repeats in the Huntington disease (HD) gene reveals HD cases with 36 repeats and apparently normal elderly individuals with 36–39 repeats. Am J Hum Genet 59:16–22

130. Saha S, Liu-Yesucevitz L, Wolozin B (2014) Regulation of autophagy by LRRK2 in Caenorhabditis elegans. Neurodegener Dis 13:110–113
131. Saitoh T, Fujita N, Hayashi T, Takahara K, Satoh T, Lee H, Matsunaga K, Kageyama S, Omori H, Noda T, Yamamoto N, Kawai T, Ishii K, Takeuchi O, Yoshimori T, Akira S (2009) Atg9a controls dsDNA-driven dynamic translocation of STING and the innate immune response. Proc Natl Acad Sci U S A 106:20842–20846
132. Saitoh T, Fujita N, Jang MH, Uematsu S, Yang BG, Satoh T, Omori H, Noda T, Yamamoto N, Komatsu M, Tanaka K, Kawai T, Tsujimura T, Takeuchi O, Yoshimori T, Akira S (2008) Loss of the autophagy protein Atg16L1 enhances endotoxin-induced IL-1b production. Nature 456:264–268
133. Sánchez-Pérez AM, Claramonte-Clausell B, Sánchez-Andrés JV, Herrero MT (2012) Parkinson's disease and autophagy. Parkinsons Dis 2012:429524. doi:10.1155/2012/429524
134. Santos RX, Correia SC, Wang X, Perry G, Smith MA, Moreira PI, Zhu X (2010) Alzheimer's disease: diverse aspects of mitochondrial malfunctioning. Int J Clin Exp Pathol 3:570–581
135. Sapp E, Schwarz C, Chase K, Bhide PG, Young AB, Penney J, Vonsattel JP, Aronin N, DiFiglia M (1997) Huntingtin localization in brains of normal and Huntington's disease patients. Ann Neurol 42:604–612
136. Sarkar S (2013) Regulation of autophagy by mTOR-dependent and mTOR-independent pathways: autophagy dysfunction in neurodegenerative diseases and therapeutic application of autophagy enhancers. Biochem Soc Trans 41:1103–1130
137. Sarkar S, Perlstein EO, Imarisio S, Pineau S, Cordenier A, Maglathlin RL, Webster JA, Lewis TA, O'Kane CJ, Schreiber SL, Rubinsztein DC (2007) Small molecules enhance autophagy and reduce toxicity in Huntington's disease models. Nat Chem Biol 3:331–338
138. Selkoe DJ (2000) Toward a comprehensive theory for Alzheimer's disease. Hypothesis: Alzheimer's disease is caused by the cerebral accumulation and cytotoxicity of amyloid beta-protein. Ann N Y Acad Sci 924:17–25
139. Shaid S, Brandts CH, Serve H, Dikic I (2013) Ubiquitination and selective autophagy. Cell Death Differ 20:21–30
140. Shehata M, Matsumura H, Okubo-Suzuki R, Ohkawa N, Inokuchi K (2012) Neuronal stimulation induces autophagy in hippocampal neurons that is involved in AMPA receptor degradation after chemical long-term depression. J Neurosci 32:10413–10422
141. Sheng M, Lee SH (2001) AMPA receptor trafficking and the control of synaptic transmission. Cell 105:825–828
142. Sherrington R, Rogaev EI, Liang Y, Rogaeva EA, Levesque G, Ikeda M, Chi H, Lin C, Li G, Holman K, Tsuda T, Mar L, Foncin JF, Bruni AC, Montesi MP, Sorbi S, Rainero I, Pinessi L, Nee L, Chumakov I, Pollen D, Brookes A, Sanseau P, Polinsky RJ, Wasco W, Da Silva HA, Haines JL, Perkicak-Vance MA, Tanzi RE, Roses AD, Fraser PE, Rommens JM, St George-Hyslop PH (1995) Cloning of a gene bearing missense mutations in early-onset familial Alzheimer's disease. Nature 375:754–760
143. Shi S, Hayashi Y, Esteban JA, Malinow R (2001) Subunit-specific rules governing AMPA receptor trafficking to synapses in hippocampal pyramidal neurons. Cell 105:331–343
144. Shi S, Hayashi Y, Petralia RS, Zaman SH, Wenthold RJ, Svoboda K, Malinow R (1999) Rapid spine delivery and redistribution of AMPA receptors after synaptic NMDA receptor activation. Science 284:1811–1815
145. Sou YS, Waguri S, Iwata J, Ueno T, Fujimura T, Hara T, Sawada N, Yamada A, Mizushima N, Uchiyama Y, Kominami E, Tanaka K, Komatsu M (2008) The Atg8 conjugation system is indispensable for proper development of autophagic isolation membranes in mice. Mol Biol Cell 19:4762–4775
146. Suzuki K, Kubota Y, Sekito T, Ohsumi Y (2007) Hierarchy of Atg proteins in pre-autophagosomal structure organization. Genes Cells 12:209–218
147. Tang G, Gudsnuk K, Kuo SH, Cotrina ML, Rosoklija G, Sosunov A, Sonders MS, Kanter E, Castagna C, Yamamoto A, Yue Z, Arancio O, Peterson BS, Champagne F, Dwork AJ,

Goldman J, Sulzer D (2014) Loss of mTOR-dependent macroautophagy causes autistic-like synaptic pruning deficits. Neuron 83:1131–1143

148. Tanida I, Tanida-Miyake E, Ueno T, Kominami E (2001) The human homolog of Saccharomyces cerevisiae Apg7p is a protein-activating enzyme for multiple substrates including human Apg12p, GATE-16, GABARAP, and MAP-LC3. J Biol Chem 276:1701–1706
149. Tanzi RE, Gussella JF, Walkins PC, Bruns GA, St. George-Hyslop P, Van Keuren ML, Patterson D, Pagan S, Kurnit DM, Neve RL (1987) Amyloid beta protein gene: cDNA, mRNA distribution, and genetic linkage near the Alzheimer locus. Science 235:880–884
150. Tatsuta T, Langer T (2008) Quality control of mitochondria: protection against neurodegeneration and ageing. EMBO J 27:306–314
151. Tomoda T, Bhatt RS, Kuroyanagi H, Shirasawa T, Hatten ME (1999) A mouse serine/threonine kinase homologous to C. elegans UNC51 functions in parallel fiber formation of cerebellar granule neurons. Neuron 24:833–846
152. Tomoda T, Kim JH, Zhan C, Hatten ME (2004) Role of Unc51.1 and its binding partners in CNS axon outgrowth. Genes Dev 18:541–558
153. Toyoda H, Wu LJ, Zhao MG, Xu H, Jia Z, Zhuo M (2007) Long-term depression requires postsynaptic AMPA GluR2 receptor in adult mouse cingulate cortex. J Cell Physiol 211:336–343
154. Velier J, Kim M, Schwarz C, Kim TW, Sapp E, Chase K, Aronin N, DiFiglia M (1998) Wild-type and mutant huntingtins function in vesicle trafficking in the secretory and endocytic pathways. Exp Neurol 152:34–40
155. Viscomi MT, D'Amelio M, Cavallucci V, Latini L, Bisicchia E, Nazio F, Fanelli F, Maccarrone M, Moreno S, Cecconi F, Molinari M (2012) Stimulation of autophagy by rapamycin protects neurons from remote degeneration after acute focal brain damage. Autophagy 8:222–235
156. Viscomi MT, Molinari M (2014) Remote neurodegeneration: multiple actors for one play. Mol Neurobiol 50:368–389
157. Walker FO (2007) Huntington's disease. Lancet 369:218–228
158. Wang T, Zhang L, Zhang M, Bao H, Liu W, Wang Y, Wang L, Dai D, Chang P, Dong W, Chen X, Tao L (2013) [Gly14]-Humanin reduces histopathology and improves functional outcome after traumatic brain injury in mice. Neuroscience 231:70–81
159. Wang X, Su B, Zheng L, Perry G, Smith MA, Zhu X (2009) The role of abnormal mitochondrial dynamics in the pathogenesis of Alzheimer's disease. J Neurochem 109:153–159
160. Wang Y, Martinez-Vicente M, Krüger U, Kaushik S, Wong E, Mandelkow EM, Cuervo AM, Mandelkow E (2009) Tau fragmentation, aggregation and clearance: the dual role of lysosomal processing. Hum Mol Genet 18:4153–4170
161. Wang YQ, Wang L, Zhang MY, Wang T, Bao HJ, Liu WL, Dai DK, Zhang L, Chang P, Dong WW, Chen XP, Tao LY (2012) Necrostatin-1 suppresses autophagy and apoptosis in mice traumatic brain injury model. Neurochem Res 37:1849–1858
162. Webb AA, Ngan S, Fowler JD (2010) Spinal cord injury I: a synopsis of the basic science. Can Vet J 51:485–492
163. Webb JL, Ravikumar B, Atkins J, Skepper JN, Rubinsztein DC (2003) α-Synuclein is degraded by both autophagy and the proteasome. J Biol Chem 278:25009–25013
164. Wen YD, Sheng R, Zhang LS, Han R, Zhang X, Zhang XD, Han F, Fukunaga K, Qin ZH (2008) Neuronal injury in rat model of permanent focal cerebral ischemia is associated with activation of autophagic and lysosomal pathways. Autophagy 4:762–769
165. Wenthold RJ, Petralia RS, Blahos J II, Niedzielski AS (1996) Evidence for multiple AMPA receptor complexes in hippo campal CA1/CA2 neurons. J Neurosci 16:1982–1989
166. Xie Z, Klionsky DJ (2007) Autophagosome formation: core machinery and adaptations. Nat Cell Biol 9:1102–1109

167. Xing S, Zhang Y, Li J, Zhang J, Li Y, Dang C, Li C, Fan Y, Yu J, Pei Z, Zeng J (2012) Beclin 1 knockdown inhibits autophagic activation and prevents the secondary neurodegenerative damage in the ipsilateral thalamus following focal cerebral infarction. Autophagy 8:63–76
168. Xiong Y, Mahmood A, Chopp M (2013) Animal models of traumatic brain injury. Nat Rev Neurosci 14:128–142
169. Yamamoto H, Kakuta S, Watanabe TM, Kitamura A, Sekito T, Kondo-Kakuta C, Ichikawa R, Kinjo M, Ohsumi Y (2012) Atg9 vesicles are an important membrane source during early steps of autophagosome formation. J Cell Biol 198:219–233
170. Yang DS, Stavrides P, Mohan PS, Kaushik S, Kumar A, Ohno M, Schmidt SD, Wesson D, Bandyopadhyay U, Jiang Y, Pawlik M, Peterhoff CM, Yang AJ, Wilson DA, St George-Hyslop P, Westaway D, Mathews PM, Levy E, Cuervo AM, Nixon RA (2011) Reversal of autophagy dysfunction in the TgCRND8 mouse model of Alzheimer's disease ameliorates amyloid pathologies and memory deficits. Brain 134:258–277
171. Yazdankhah M, Farioli-Vecchioli S, Tonchev AB, Stoykova A, Cecconi F (2014) The autophagy regulators Ambra1 and Beclin 1 are required for adult neurogenesis in the brain subventricular zone. Cell Death Dis 5:e1403. doi:10.1038/cddis.2014.358
172. Yu WH, Cuervo AM, Kumar A, Peterhoff CM, Schmidt SD, Lee JH, Mohan PS, Mercken M, Farmery MR, Tjernberg LO, Jiang Y, Duff K, Uchiyama Y, Näslund J, Mathews PM, Cataldo AM, Nixon RA (2005) Macroautophagy – a novel beta-amyloid peptide-generating pathway activated in Alzheimer's disease. J Cell Biol 171:87–98
173. Yu WH, Kumar A, Peterhoff C, Shapiro Kulnane L, Uchiyama Y, Lamb BT, Cuervo AM, Nixon RA (2004) Autophagic vacuoles are enriched in amyloid precursor protein-secretase activities: implications for beta-amyloid peptide over-production and localization in Alzheimer's disease. Int J Biochem Cell Biol 36:2531–2540
174. Yue Z, Jin S, Yang C, Levine AJ, Heintz N (2003) Beclin 1, an autophagy gene essential for early embryonic development, is a haploinsufficient tumor suppressor. Proc Natl Acad Sci U S A 100:15077–15082
175. Zhang H, Duan C, Yang H (2015) Defective autophagy in Parkinson's disease: lessons from genetics. Mol Neurobiol 51:89–104
176. Zhang M, Shan H, Chang P, Wang T, Dong W, Chen X, Tao L (2014) Hydrogen sulfide offers neuroprotection on traumatic brain injury in parallel with reduced apoptosis and autophagy in mice. PLoS One 9:e87241. doi:10.1371/journal.pone.0087241
177. Zhang X, Yan H, Yuan Y, Gao J, Shen Z, Cheng Y, Shen Y, Wang RR, Wang X, Hu WW, Wang G, Chen Z (2013) Cerebral ischemia-reperfusion-induced autophagy protects against neuronal injury by mitochondrial clearance. Autophagy 9:1321–1333
178. Zhang YB, Li SX, Chen XP, Yang L, Zhang YG, Liu R, Tao LY (2008) Autophagy is activated and might protect neurons from degeneration after traumatic brain injury. Neurosci Bull 24:143–149
179. Zhao T, Severijnen LA, van der Weiden M, Zheng PP, Oostra BA, Hukema RK, Willemsen R, Kros JM, Bonifati V (2013) FBXO7 immunoreactivity in alpha-synuclein-containing inclusions in Parkinson disease and multiple system atrophy. J Neuropathol Exp Neurol 72:482–488
180. Zhou X, Babu JR, da Silva S, Shu Q, Graef IA, Oliver T, Tomoda T, Tani T, Wooten MW, Wang F (2007) Unc-51-like kinase 1/2-mediated endocytic processes regulate filopodia extension and branching of sensory axons. Proc Natl Acad Sci U S A 104:5842–5847
181. Zhou Z, Hu J, Passafaro M, Xie W, Jia Z (2011) GluA2 (GluR2) regulates metabotropic glutamate receptor-dependent long-term depression through N-cadherin-dependent and cofilin-mediated actin reorganization. J Neurosci 31:819–833

Aspects of Autophagy in Inflammatory Bowel Disease

Ferenc Sipos and Györgyi Műzes

Abstract Crohn's disease and ulcerative colitis the main clinical phenotypes of inflammatory bowel disease are polygenic immune disorders with multifactorial etiology. Recent genome-wide association studies have highlighted on the importance of the autophagy pathway. Thus, autophagy is now widely considered as a key regulator mechanism with the capacity to integrate several aspects of Crohn's disease pathogenesis. Chronic, unchecked inflammation has widely been suggested to trigger carcinogenesis. In addition, accumulating evidence indicates that the aberrantly altered process of autophagy is definitely involved in carcinogenesis, as well. Toll-like receptors sensing cell-derived pattern/danger-associated molecules also have the capacity to promote tumor development and immune escape. However, both TLR- and autophagy-related signals may exert tumor suppressor mechanisms mainly in a cell-specific and context-dependent manner. Though the precise impact of autophagy in inflammatory bowel disease and on inflammation (colitis)-associated cancer has not yet been clarified, it may indicate a novel promising therapeutic aspect.

Abbreviations

AIEC	Adherent-invasive *Escherichia coli*
AKT	Protein kinase B (PKB)
AOM	Azoxymethane
APCs	Antigen-presenting cells
APP	Amyloid-β precursor protein
Atg	Autophagy-related
ATP	Adenosine triphosphate
Bcl-2	B-cell lymphoma 2
Bcl-xL	B-cell lymphoma-extra large
Bif-1	Bax-binding protein-1

F. Sipos (✉) • G. Műzes
2nd Department of Internal Medicine, Semmelweis University,
Szentkirályi Street 46, 1088 Budapest, Hungary
e-mail: dr.siposf@gmail.com; muzes.gyorgyi@med.semmelweis-univ.hu

M.C. Maiuri, D. De Stefano (eds.), *Autophagy Networks in Inflammation*,
Progress in Inflammation Research, DOI 10.1007/978-3-319-30079-5_12

CAC	Colitis-associated cancer
CALCOCO2/NDP52	Autophagy receptor calcium binding and coiled-coil domain 2
CAMP	Cyclic adenosine monophosphate
CARD	Caspase activation and recruitment domain
CD	Crohn's disease
COX	Cyclooxygenase
DAMP	Damage-associated molecular pattern
DCs	Dendritic cells
DSS	Dextran sulfate sodium
EGFR	Epidermal growth factor receptor
ER	Endoplasmatic reticulum
FAK	Focal Adhesion Kinase
FIP200	FAK family kinase-interacting protein of 200 kDa
GI	Gastrointestinal
GO	Graphene oxide
GTP	Guanosine-5′-triphosphate
HMGB-1	High-mobility-group B-1
HSP	Heat-schock protein
IBD	Inflammatory bowel disease
ICD	Immunogenic cell death
IECs	Intestinal epithelial cells
IFN	Interferon
IKK	Inhibitor of κB-kinase
IL	Interleukine
IP3	Inositol triphosphate
IRAKs	Interleukin-1 receptor-associated kinases
IRE1/XBP1	Inositol-requiring enzyme 1/X-box binding protein 1
IRF	Interferon regulatory factor
IRGM	Immunity-related GTPase family M protein
JNKs	c-Jun N-terminal kinases
LC3	Microtubule-associated protein 1A/1B-light chain 3
LP	Lamina propria
LPS	Lipopolysaccharide
LRR	Leucine-rich repeat
MAPK	Mitogen-activated protein kinases
MDP	N-acetyl-muramyl-peptide
MHC	Major histocompatibility complex
mTORC1	Mammalian TOR complex 1
MyD88	Myeloid differentiation primary response gene 88
NF-κB	Nuclear factor-κB
NLRP3	NOD-like receptor family, pyrin domain containing 3
NLRs	NOD-like receptors
NOD2	Nucleotide-binding oligomerization domain containing protein 2

ODN	Oligodeoxynucleotide
PAMPs	Pathogen-associated molecular patterns
PARP	Poly-ADP-ribose polymerase
pDCs	Plasmocytoid DCs
PG	Prostaglandin
PI3K	Phosphatidylinositol-4,5-bisphosphate 3-kinase
PRRs	Pattern recognition receptors
PTPN22	Protein tyrosine phosphatase non-receptor type 22
RIG	Retinoic acid-inducible gene 1
RIP2	Receptor interacting protein 2
RLRs	RIG-I-like receptors
ROS	Reactive oxygen species
SNP	Single nucleotide polymorphisms
STAT	Signal transducer and activator of transcription
TCD	Tolerogenic cell death
TIR	Toll/interleukin-1 receptor
TLR	Toll-like receptor
TNF	Tumor necrosis factor
TOR	Target of rapamycin
TRAF	TNF-receptor-associated factor
TRAM	Toll-like receptor 4 adaptor protein
Treg	Regulatory T cells
TRIF	TIR-domain-containing adapter-inducing interferon-β
UC	Ulcerative colitis
UPR	Unfolded protein response
UVRAG	UV radiation resistance-associated gene protein.

1 Introduction

1.1 Inflammatory Bowel Disease: Pathophysiological Background

Crohn's disease (CD) and ulcerative colitis (UC), the main clinical phenotypes of (idiopathic, relapsing-remitting) inflammatory bowel disease (IBD) are systemic disorders affecting the GI-tract with frequent extraintestinal manifestations and associated autoimmune conditions [136]. IBD is considered as a polygenic immune disorder with complex multifactor etiology. Generally, IBD is arising in susceptible individuals in whom upon environmental triggers a sustained disturbed, deleterious mucosal immune reaction is provoked towards commensal microbiota [79]. In chronic inflammatory conditions, when organs with large epithelial surfaces are affected, like in IBD the epithelial barrier function is critical for the disease onset. Since the epithelium is densely inhabited by a resident microbial flora the role of native immunity is particularly appreciated in recognising and distinguishing commensal enteric bacteria from the invading ones, and thus, in maintaining tolerance and homeostasis [79]. Subsequently, the chronic unrestrained inflammatory response that occurs in IBD is mainly driven by a desintegrated host immune regulatory network. In IBD development the host genetic susceptibility represents an important etiologic factor. In CD the genetic component is strongly indicated by familial aggregation, and further, by an approx. 26-fold greater population-based sibling risk, and an approx. 30–35 % of concordance rate in monozygotic twins [2, 175]. In CD pathogenesis genome-wide association studies highlighted on certain earlier not really suspected biological pathways, such as autophagy [79]. Many of the recently identified genetic risk loci in Crohn's disease are related to various cell types and pathways, suggesting the involvement of fairly different aspects of host immune responses in the IBD phenotype [79]. Missing heritability in CD cannot be simply explained by genetic alterations [79]. Moreover, the fact of the worldwide considerable increase in disease incidence and prevalence emphasizes the importance of additional, environmental and epigenetic contributions [100, 129, 130].

The interplay of genes regulating immune functions is strongly affected by the environment, especially gut resident microbiota. On the basis of genetic alterations in CD impaired sensing and handling of intracellular bacteria by the innate immunity, that is closely interrelated with the autophagic and unfolded protein pathways seem to be the most relevant pathophysiologic features [185].

Interestingly, association with epithelial barrier genes seems specific to UC, the converse of nucleotide-binding oligomerization domain containing protein 2 (NOD2), while the autophagy genes are Crohn's-specific. These observations correlate with UC being confined to the superficial layers of the colon, while the transmural inflammation of CD is caused by defects in cellular innate immunity and bacterial handling in the deeper layers of the colonic wall [131].

1.2 Relation of Autophagy Machinery to IBD

Autophagy is deeply implicated in the regulation of numerous physiologic functions including cell development and differentiation, survival and senescence, and it also affects fundamentally the inflammatory process, and the innate and adaptive arms of immune responses [96]. On a basal level intact autophagy serves constantly and constitutively as a critical adaptive and surveillance mechanism in maintaining cellular homeostasis [116].

Nevertheless, autophagy is inducible in response to different cellular metabolic stress conditions in order to preserve cell viability. Further, autophagy is upregulated in cases of protein aggregation and accumulation of misfolded proteins, i.e. when the structural remodeling is mandatory. In respect of innate immunity, however, autophagy plays an essential role during infections by degrading intracellular pathogens [96, 140]. By compromising cellular fitness defective autophagy has been ultimately related to several chronic inflammatory disease conditions, such as

IBD, like CD and cancer, neurodegeneration, and infectious disorders [21, 96, 116]. Generally autophagy deficiency is closely related to accelerated tumorigenesis. In autophagy-incompetent cells upon induced oxidative stress cell-autonomous mechanisms are exhibited in forms of accumulated DNA damage and chromatin instability [108]. However, inflammatory events as a non-cell-autonomous mechanism along with defective apoptosis could independently contribute to malignant transformation and cancer progression, partly by favouring cell necrosis [33]. Similar situation has been found in human IBD with high risk of malignancy, and in experimental cases of *Atg5*$^{-/-}$ or *Atg7*$^{-/-}$ mice displaying abnormalities resembling human IBD [20].

Autophagy and stress-responsive cellular degradation pathways of intrinsic and extrinsic apoptosis can fundamentally alter, activate or inhibit each other via an extensive molecular crosstalk, and in fact, cell destiny is determined by their actual functional status and interplay [56]. Their crosstalk is primarily regulated by the current status of the ATG6/Beclin-1 complex, a Bcl-2/Bcl-xL interacting element, since Bcl2 is a potent autophagy inhibitor. Dissociation of this complex can be achieved by Toll-like receptor (TLR) adaptors (MyD88, TRIF), or activation of mitogen activated phosphokinase (MAPK)-JNK cascade, as well as by translocation of the damage-associated molecular pattern (DAMP) protein high-mobility-group B (HMGB)-1 [56, 62]. There is also diverse interaction between autophagy and the NF-kB signaling pathways through positive and negative feedback regulatory loops [112]. Recently, a missense mutation in the autophagy receptor calcium binding and coiled-coil domain 2 (CALCOCO2/NDP52) gene has been identified, controlling also the NF-κB signaling downstream of TLRs [169]. The tumor suppressor *p53* gene exerts a typical dual role in autophagy regulation, depending primarily on its subcellular, nuclear or cytoplasmic distribution [62, 112].

1.3 NOD-Like Receptors and Crohn's Disease

NOD-like receptors (NLRs) are pattern recognition receptors (PRRs) and belong to the family of innate immune receptors sensing pathogen-associated molecular patterns (PAMPs). NOD2 is constitutively expressed intracellularly in macrophages and dendritic cells, and to lesser extent in intestinal epithelial cells (IECs) and T cells. The centrally located motifs of NLRs are referred to NOD domains, that are interacted with the caspase activation and recruitment domain (CARD) ones. NOD2 recognizes N-acetyl-muramyl-peptide (MDP), a bacterial peptidoglycan component, and upon activation the induced receptorial conformation changes result a multiprotein, the inflammasome (NLRP3). Ligation of NOD2 triggers recruitment of the adaptor protein RIP2 causing a tumor necrosis factor (TNF) receptor-associated factor (TRAF)-6-mediated ubiquitination of inhibitor of κB-kinase gamma (IKKγ; NEMO), and hence results in activation of downstream signaling pathways implicating nuclear factor-κB (NF-κB), MAPKs and proinflammatory caspases [76, 145]. The Crohn's disease-associated *NOD2* genetic variants are located in the leucine-rich repeat (LRR) region of NOD2, i.e. in the ligand-binding domain of this intracellular PRR [70]. The altered amino acid sequence is related either to insertion resulting in a frame-shift mutation, or to non-synonymous single nucleotide polymorphisms (SNPs) resulting in amino acid exchanges. The more commonly observed genetic variants (of missense or nonsense mutations) in CD are the SNP8 (R702W), SNP12 (G908R), and SNP13 (L1007fsC), respectively, however a number of rare *NOD2* variants have also been discovered, being localized again almost exclusively to the LRR region [70, 124]. Upon MDP ligation the CD-associated "loss-of-function" NOD2 variants abrogate receptor interacting protein 2 (RIP2) binding, and so fail to activate NF-κB [1, 16]. Further, NOD2 is involved in the modulation of TLR signaling, as well. Thus, in case of CD-related gene polymorphisms the TLR2-induced NF-κB activation is also decreased [64, 181]. On the other hand NOD2 has a pivotal role in direct antibacterial defence by the induced release of defensins. NOD2$^{-/-}$ mice and patients with the CD NOD2 variants display diminished expression of antimicrobial α-defensins in Paneth cells, that contributes to impaired antibacterial capacity and decreased epithelial barrier function [85, 182]. In contrast to hypomorphic functions the frame-shift gene mutation variant encodes a "gain-of-function" by actively suppressing interleukin-10 (IL-10) transcription [121].

1.4 Autophagy and Crohn's Disease

The autophagy machinery in IBD represents a recently developed pathway fundamentally contributing to the pathogenesis [96]. Functional polymorphisms of the autophagy genes *ATG16L1* (T300A) and immunity-related GTPase family M protein *(IRGM;* C313T) have been found as definite risk factors for CD [60, 110, 132, 144]. The ATG16L1 protein is widely expressed in IECs, and also in macrophages and lymphocytes. The ubiquitous *ATG16L1* seems to be fundamental in selective autophagy, i.e. in xenophagy, nonetheless its defect has only been described within the gut [81]. In CD patients homozygous for the risk *ATG16L1* allele the "loss-of-function" deficiency due to failures of autophagosome formation results in impared engulfment and degradation of cytoplasmic content (microbes), defective presentation of bacterial antigens to CD4+ T cells, and further, in alterations of Paneth cell granule formation causing a disrupted granule exocytosis [20, 97, 148, 164]. Additionally, ATG16L1 deficient Paneth cells in CD display a "gain-of-function" defect by increasing expression of inflammatory cytokines [20, 97]. A coding polymorphism (Thr300Ala) of ATG16L1 has recently been shown to decrease selective autophagy resulting in altered cytokine signaling and decreased antibacterial defense [90]. Moreover, upon stimulation with NOD2 ligands or with lipopolysaccharides (LPS) through TLR4, macrophages and myeloid cells with the ATG16L1 risk variant generate high levels of reactive oxygen species (ROS), and respond with inflammasome overactivation leading to enhanced IL-1β and IL-18 production via MyD88 and TRIF-dependent activation of caspase-1 [97, 148, 164].

Generally, aberrant activation of PRR signaling pathways may result critically severe inflammation. *IRGM* is the only human gene representative for innate immunity-related GTPases, necessary for γ-interferon (IFNγ)-mediated resistance to intracellular pathogens [104, 105, 168]. During initiation of autophagy IRGM expression is essentially required for the proper clearance of bacteria. The risk polymorphism of *IRGM* due to the impaired protein expression can lead to functional abnormalities in xenophagy [65, 132, 144]. Since IRGM is possibly regulated in a cell specific manner the CD risk allele may cause cell specific phenotypes.

1.5 NOD2 and Autophagy

Functionally NOD2 is closely associated with autophagy, and yet interacts mechanically (i.e. immunoprecipitated) with ATG16L1, therefore autophagy seems to be a key factor in CD [28, 164, 170]. Autophagy is mainly activated due to sensors of the innate immunity, i.e. by PRR signaling upon recognition of PAMPs (MDP, LPS, ss/ds RNA, methylated DNA/CpG), but it could also be induced by DAMPs (like ATP, ROS, and misfolded proteins), pathogen receptors (like CD46), inhibitor of NF-κB (IKK), JNK and HMGB proteins [56, 96, 97, 140]. Sensory PRR-molecules include TLRs, NLRs and RIG-I-like receptors (RLRs). Induction of NOD2 in dendritic and epithelial cells by bacterial ligands and leaving bacteria results in ATG16L1-dependent formation of autophagic vacuoles. However, the NOD2 variants of CD lack this activity, and further MDP-induced autophagy is also absent in cells with the ATG16L1 risk variant, suggesting that both NOD2 and ATG16L1 co-localized on plasma membrane are required for an optimal innate immune signaling [66, 170]. In addition, a NOD2-dependent failure in autophagy-induction and consequently a diminished bacterial killing was found for *Salmonella typhimurium*, *Shigella flexneri*, and enteroadherent invasive *E. coli* (AIEC) [28, 89, 170]. The normal NOD2, but not the CD-associated variants recruits ATG16L1 to the plasma membrane preferentially at the bacterial entry side, so physiologically NOD2 is critical for engulfing invading pathogens by autophagosomes [28, 66]. Furthermore, in dendritic cells NOD2-dependent autophagy is also essential for the appropriate antigen processing and presentation and a subsequent induction of CD4+ T-cells [28]. Dendritic cells from CD patients with either NOD2 or ATG16L1 variants display a failure to translocate bacteria to lysosomes and relocate class II major histocompatibility complex (MHC) to cell surface, as well [170]. In intestinal dendritic cell-epithelial cell interactions, autophagy deficiency leads to decreased antigen sampling, increased dendritic cell maturation and a more pro-inflammatory type of dendritic cells [165]. When the disease-associated *ATG16L1* and *NOD2* alleles are present in combination, a synergistic genetic epistasis, i.e. an increase in CD susceptibility was observed, underscoring the importance of a signaling crosstalk regarding the inflammasome and autophagy [146]. Monocytes of CD patients display enhanced phagocytosis associated with the presence of ATG16L1 and NOD2 variants [184]. Moreover, it has been demonstrated that loss of protein tyrosine phosphatase non-receptor type 22 (PTPN22) renders monocytes more reactive towards bacterial products [163], which also could be part of the pathophysiology of CD.

1.6 Endoplasmic Reticulum Stress and Autophagy

The unfolded protein response (UPR) induced by endoplasmatic reticulum (ER) stress represents another pathway in IBD pathophysiology [77]. Genetically ER stress is associated with both forms of IBD and occurs upon excessive accumulation of misfolded or unfolded proteins in the ER, leading to UPR especially in cells with high secretory capacity, like goblet cells and Paneth cells [77, 81]. UPR is regulated by different pathways (and related transcription factors) with the preference of the inositol-requiring enzyme 1/X-box binding protein 1 (IRE1/XBP1) axis [179]. Via this axis there is a conserved link between innate immunity (TLR and NOD signaling) and the UPR [179]. Genome-wide association studies revealed the role of *XBP1* SNPs in IBD-related ER-stress [11, 78]. Decreased or absent XBP1 function in the IEC compartment through IRE1 hyperactivation results in uncontrolled ER-stress, i.e. a proinflammatory overactivation, and further in dysfunction and premature apoptotic depletion of Paneth cells, with the consequent impaired handling of the microbiota [78]. Under ER stress autophagy is induced via JNK (downstream of IRE1), which is overactivated by the hypomorphic XBP1 [37, 123]. However, even defective autophagy *per se* is able to provoke ER stress, especially when the ATG7 protein involved in regulation of autophagosome formation is also depressed [186]. Regarding PI3K there is an antagonistic action, since in UPR it is responsible for the activation of XBP1, but in the contrary autophagy is suppressed by the canonical AKT-TOR pathway [106, 129, 130]. In IBD, IECs presumably are affected both by impaired UPR signaling and aberrant autophagy, but their exact interplay needs to be further clarified.

1.7 Autophagy-Dependent Effects of Gut Microbiota on Crohn's Disease

The intestinal microbiota, which normally colonize mucosal surfaces in symbiotic mutualism with the host is unique and quite stable over time [9]. The basic challenge for the intestinal immune recognition is the requirement of a simultaneous delicate balance between tolerance and responsiveness towards microbes. Several data suggest the existence of immune tolerance to antigens of the individual own bacterial flora, whereas its breakdown definitely contributes to IBD pathogenesis [53, 164].

The innate immunity ensures a primary host response to microbial invasion, which induces an inflammatory process to localize the infection and prevent systemic dissemination of pathogens. The key elements of this process are PRRs including TLRs, NLRs, RNA helicases, C-type lectin receptors, and cytosolic DNA sensors, which sense evolutionarily conserved PAMPs of microbiota. The detection of PAMPs by PRRs triggers sequential activation of intracellular signaling pathways resulting in induction of a wide range of cytokines and chemokines that unite the early host response to infection [45]. If pathogens cannot be eliminated, they may elicit chronic inflammation, which may partly mediated via TLRs. Additionally, chronic inflammation has long been suggested to trigger tissue tumorous transformation. Indeed, a higher incidence colitis-associated cancer (CAC) has been observed in IBD patients [45].

In CD there is a profound and complex host defect in sensing and responding intestinal (lumenal and mucosal) microbiome. Accordingly, reprogramming in the microbial composition, i.e. a significant decreased load of commensal, protective resident bacteria (like *Bifidobacteria, Lactobacilli* and *Firmicutes*) along with the impaired immunity against the putative pathogenic (harmful) ones (such as *Bacteriodetes*, and *Proteobacteria*, including *E. coli*) provoke a deleterious inflammatory condition, corresponding to CD [166]. The exact nature of the distinct mucosal flora (dysbiosis), however has not yet been fully clearified.

Specific strains of *E. coli* (termed AIEC) in CD affect especially the epithelial layer with the ability to adhere, invade and replicate in IECs, and further, a subpopulation even resides and survives within macrophages, and thereby induces increased production of tumor necrosis factor-α [54, 89]. ATG16L1 and IRGM-deficient autophagosomes promote the AIEC survival as well [89]. Moreover, in the presence of CD-associated NOD2 variants or hypomorphic XBP1 dendritic cells exhibit diminished intracellular bacterial killing [28]. Recently, it has been demonstrated that AIEC can up-regulate the levels of microRNAs (i.e.: MIR30C and MIR130A) in intestinal epithelial cells (IECs) to reduce autophagy [119]. Based on these data, it is hypothesized, that AIEC possesses the capacity to circumvent innate immune responses leading to activation of NF-κB [41].

Epigenetic factors, such as microRNAs like MIR106B and MIR93 have been found to reduce levels of ATG16L1 and autophagy, hence prevent autophagy-dependent eradication of intracellular bacteria [103]. Thus, regarding the host interactions with microbes genetic risk factors of CD functionally render pathways of the innate immunity to converge to a deeply impaired autophagic process.

2 Colitis-Associated Cancer: Role of Autophagy and TLRs

As early, as in ancient times Hippocrates and Galenus had already realized the similarity between inflammation and cancer, and hypothesized that cancer evolved from inflammatory lesions [171]. In 1863, Rudolf Virchow observed a close etiologic relation of chronic inflammation to carcinogenesis, realizing that tumors possess a typical "lymphoreticular infiltrate" [10, 82]. The first evidence of antitumoral effects of microbial products is dated to the beginning of the eighteenth century when Deider reported that infection in patients with cancer could be accompanied by the remission of the malignancies [52]. In the 1890s, William B. Coley, a surgeon from New York, observed that repeated injections of a mixture of bacterial toxins served as an efficient antitumoral therapeutic agent [183]. Later, in 1943, lipopolysaccharide (LPS) was discovered as the "hemorrhage-producing fraction" of the Coley's lysate, accounted for its antitumoral effects [52]. After the discovery of TLRs, their ligands and signaling pathways, it has been found that microbe-derived factors act by stimulating TLR signaling and activating both the innate and adaptive immune responses to enhance anti-tumor immunity [32].

In 2000, Hanahan and Weinberg proposed a model to define the six hallmarks of carcinogenesis [61]. Generally inflammation is required to fight microbial infections, heal wounds, and maintain tissue homeostasis, however, it could lead to cancer. Inflammation, the seventh hallmark of cancer may affect all phases of tumor development, including tumor initiation, promotion, invasion and metastatic dissemination, and also evade the immune system. Inflammation acts as a cellular stressor and may trigger DNA damage or genetic instability, and, further, chronic inflammation can provoke genetic mutations and epigenetic mechanisms that promote malignant cell transformation [12, 91]. Upon inflammation a peculiar tissue microenvironment is induced with the capacity to tolerate tumor cell growth and metastasis by altering immunoregulatory mechanism, and thus making the immune system incapable to destroy tumor cells [12].

Accumulating evidence indicates that the modulated process of autophagy is definitely involved in carcinogenesis representing one of the distinctive functional characteristics (hallmarks) of cancer cells [101]. In addition, autophagy contributes to tumor development by supporting survival and self-renewal of cancer stem cells as well [101]. Nonetheless, TLRs binding cell-derived PAMP/DAMP molecules also have the capacity to promote carcinogenesis and immune escape. TLRs are usually expressed in immunocompetent cells, though several types of cancer cells have also been reported to display these innate immune receptors [188]. In general, the danger signals via PRRs are critical players in inducing innate and adaptive responses of immunity. On the other hand, however, both TLR- and autophagy-related signals may exert tumor suppressor mechanisms in a cell-specific and context-dependent manner.

The role of autophagy has been radically expanded, and this machinery is considered not only a fundamental eukaryotic cellular homeostatic process but as an integral component of immune system, as well controlling infection, inflammation and immune responses [36]. Recent studies have documented that TLRs and autophagy are interrelated in response to PAMPs/DMAPs, furthermore there is a regulatory signaling cross-talk among them [34, 35].

In chronic inflammatory conditions, when organs with large epithelial surfaces are affected, like in IBD the epithelial barrier function is critical for the disease onset. Since the epithelium is densely inhabited by a resident microbial flora the role of native immunity is particularly appreciated in recognising and distinguishing commensal enteric bacteria from the invading ones, and thus, in maintaining tolerance and homeostasis. Subsequently, the chronic unrestrained inflammatory response that occurs in IBD is mainly driven by a desintegrated host immune regulatory network and, further is definitely responsible for the increased susceptibility to colorectal cancer.

TLRs are involved in the maintenance and functioning of the epithelial barrier integrity in the gut regulating the MyD88 adaptor protein. Thereby TLRs may display a protective function in the control of intestinal inflammation and inflammation-associated cancer [8]. CAC are considered as typical examples of inflammation-related cancers. However, tumors usually appear after several years of active disease, with a cumulative lifetime risk of 18–20 % in UC, and up to 8 % in CD [22, 40, 147]. Indeed, recent epidemiological data indicate that over 25 % of all cancers are related to chronic infection and other unresolved inflammation [176]. Current result indicate that TLRs have a potential role in microbiota-associated gastrointestinal cancer metastasis through the recognition of microbiota ligands, initiating inflammation, and promoting tumorigenesis [104, 105].

Interestingly, IBD patients also have an increased susceptibility to other malignancies, like lymphomas/leukemias, hepatocellular carcinoma suggesting that the local inflammation could have not only intestinal but also systemic tumor-promoting effects, or the genetic alterations that affect inflammatory and immune homeostasis in IBD also predispose the patients to cancer in other tissues [23, 43]. In IBD, the increased susceptibility to extraintestinal tumors could also be related to the immunosuppressive treatment. However, the types of tumors increasingly found in IBD patients are different from those observed in transplant patients under immunosuppression [6, 162].

Both intrinsic and extrinsic inflammatory pathways are linked to carcinogenesis. Intrinsic inflammation is mainly initiated by mutations leading to oncogene activation as well to inactivation of tumor suppressors. The extrinsic pathway by terms of infection or inflammation increases cancer risk. Although in IBD patients inflamed intestinal cells have already had the colorectal cancer-related genetic abnormalities before developing dysplasia, in CAC genetic alterations seem only to be a secondary step rather than a primary cause of carcinogenesis [174]. It is likely that abnormalities in PRR signaling lead to dysregulated expression of genes and enzymes involved in cell proliferation, apoptosis, and DNA repair prior to the gene alterations. Frequent alternative cycles of mucosal injury and repair in the presence of tumorigenic cytokines, chemokines, and prostaglandins may also predispose to genetic mutations, which increase cancer risk [88, 125].

Epithelial regeneration and myofibroblast activation, two major events of wound-healing, are strongly influenced by TLR signaling. A contribution of TLR signals to regeneration can be found in the intestine [46, 139]. It has also been reported that TLR-mediated MyD88 signaling in macrophages of the lamina propria (LP) regulates crypt stem cell differentiation and epithelial proliferation through cyclooxygenase (COX)- 2 and prostaglandin (PG)E2 expression [18, 139]. TLR4 activation has also been shown to induce IEC proliferation via induction of EGFR ligands [17, 68], moreover, in inflammatory circumstances the surface expression of TLR2 and

-4 may be enhanced leading to IECs responsiveness to their ligands [99, 142]. Based on these results, it seems that abnormal TLR signaling may induce enhanced epithelial proliferation and thus may contribute to colitis-associated carcinogenesis.

2.1 Autophagy and Malignancy

In tumorigenesis a typical dual-faced role of autophagy has been proposed. On one side it may be critical for cancer cell survival and progression, in particular under stressful situations, however, it may also elicit tumor death signaling pathways. Direction of autophagy toward cytoprotection or tumor cell suppression, thus the pro-survival or pro-death function is context-dependent, and influenced by several intra- and extracellular factors, such as involved tissues, surrounding microenvironment, genetic background, and stages of tumor development, nevertheless its precise relation to cancer networks has not yet been fully elucidated [31, 95, 109]. Regarding cell death the involvement of autophagy either in apoptosis (programmed, type I death) or in non-apoptotic or necrotic death, and their possible interactions are rather complicated [101]. In tumor cells autophagy usually displays a critical, programmed pro-survival function by inhibiting apoptosis or suppressing necrotic death, including programmed (or regulated) cell necrosis of caspase-independent necroptosis, and poly-ADP-ribose polymerase (PARP)-mediated necrosis [156]. Howbeit, in cases of autophagy deficiency no tumor suppression, but accelerated tumorigenesis can be manifestated. In case of induced oxidative stress in autophagy-incompetent cells, so-called cell-autonomous mechanisms are exhibited in forms of accumulated DNA damage and chromatin instability [108]. Nonetheless, like a non-cell-autonomous mechanism inflammatory processes along with defective apoptosis could also independently contribute to cancer progression, in part by favoring cell necrosis [33]. Similar situation has been found in human IBD with high risk of malignancy, and in experimental cases of atg5$^{-/-}$ or atg7$^{-/-}$ mice displaying inflammatory Paneth cell abnormalities resembling human IBD [20, 21].

In certain human cancers the *Beclin1 (ATG6)* gene, a Bcl-2/Bcl-xL interacting element has been found to be monoallelically lost, and confirmed that it functions as a haploinsufficient tumor suppressor [4]. However, the suppressive function of Beclin1 may be tissue-specific, since in colorectal and gastric carcinomas even its higher expression has been detected [3]. In addition to Beclin1, alterations of other autophagy-associated genes, e.g. atg4, atg5, UVRAG, or Bax-binding protein-1 (Bif-1) have also been detected in other cancer types, indicating that the suppression of tumors is attributed to different autophagy elements. Nonsense mutations of UVRAG, and downregulation of Bif-1 have been documented in colon and gastric adenocarcinomas [29, 73, 83]. Hypothetically, via excessively induced autophagy increased autophagic flux could promote non-apoptotic (programmed, type II) autophagic cell death, acting like a tumor suppressor [87]. Autophagy stimulates oncogene-induced senescense as well, providing another possible barrier against malignant transformation [187]. Nevertheless, there is no direct evidence regarding the realistic anti-tumor capacity of autophagy.

In human cancers constitutive activation of the Ras- and PI3K/Akt-mTOR pathway is a common phenomenon, and mTOR complex 1 (mTORC1) seems to be the main negative regulator of autophagy [59, 62]. The p53 gene plays a dual role (i.e. stimulatory or suppressive) in autophagy regulation, depending primarily on its subcellular, nuclear or cytoplasmic distribution [26].

Both stress-responsive cellular degradation pathways of intrinsic and extrinsic apoptosis and of autophagy can fundamentally affect (activate or inhibit) each other via an extensive molecular crosstalk, and in fact, cell death is determined by their actual functional status and interplay [56, 108]. Their crosstalk is regulated primarily by the current status of the Bcl-2/Beclin1 complex, dissociation of which can be achieved upon activation of MAPK-JNK or translocation of the DAMP protein HMGB-1 [56]. NF-κB plays also a critical role in malignant transformation, and in the majority of different tumor cells its constitutive, chronic activation has been observed. There is also a complex interaction between autophagy and the NF-κB signaling pathways via positive and negative feedback regulatory loops [172]. The important autophagy selective substrate p62 acts as an adaptor protein to regulate NF-κB, as well [117].

2.2 *Toll-Like Receptors and Intestinal Epithelial Cells*

In vitro data has demonstrated hyporesponsiveness of IECs to TLR ligands [114, 126]. Antigen-presenting cells (APCs) in the LP also seem to be unresponsive to TLR ligands [161]. Under physiologic conditions, TLR3, -7, -8, and -9 are expressed in endosomes, or basolateral membrane (TLR5), where these TLRs are not exposed to pathogens unless microbiota get into the cells or invade mucosa [45]. Apical epithelial TLR9 activation by bacterial DNA fragments has been reported to take part in colonic homeostasis [93]. These findings underline a unique feature of TLRs (and other PRRs) in IECs that establishes immune tolerance to the commensal flora of the colonic mucosal interface.

In addition, epithelial TLRs contribute to balancing the composition of luminal microorganisms by regulating the secretion of different antimicrobial peptides and mucosal IgA. TLR9$^{-/-}$ mice have impaired expression of cryptidin (α-defensin) compared to wild type mice [93]. Signaling through TLR2, -3, and -4 have all been implicated with the expression of β-defensins in IECs [173, 178]. Several TLR signaling in IECs induces B cell-activating factors leading to immunoglobulin class switch recombination in B cells of the LP without T cell activation, resulting in IgA secretion [155]. Moreover, activation of TLR3 and -4 has been found to induce epithelial expression of an epithelial immunoglobulin transporter (polymeric immunoglobulin receptor) that enhances luminal IgA secretion [19, 154].

TLR signaling can be classified into classical/canonical and alternative/noncanonical pathways [12]. All TLRs, but TLR3 utilize the MyD88-dependent signaling pathway to induce the expression of proinflammatory cytokine genes [111]. TLR3 exclusively uses the TRIF pathway [111]. The classical inflammatory signaling pathway is mainly activated through MyD88, which, in turn, recruits interleukin-1 receptor-associated kinases (IRAKs) and TRAF-6 [167]. TRAF6 activates transforming growth factor-activated kinase 1 that phosphorylates and activates the inhibitor of kappa light polypeptide gene enhancer in B-cells kinase (IKK) complex, finally resulting in the release and translocation of NF-κB into the nucleus, thereby inducing the production of TNF-α, IL-1, and IL-6, the key mediators of (intestinal) proinflammatory responses [5, 80, 122]. However, TLR3 and some of the TLR4 signals utilize the TRIF adaptor molecule signaling independently of MyD88. This alternative pathway culminates in the activation of TRAF3 and interferon regulatory factor (IRF)-3, resulting in the secretion of type I IFNs, even in the gut [128]. TLR4 is unique among the TLRs as it can activate two distinct signaling pathways: the classical pathway (through Toll-interleukin 1 receptor domain-containing adapter protein/TIRAP/and MyD88) and the alternative pathway (via TRIF and toll-like receptor 4 adaptor protein/TRAM/) [12].

2.3 TLRs and Malignancy

Previous studies have indicated that certain TLRs are present in different cancers and cancer cell lines. In colorectal cancers TLR3, -4, -5, -7, and -8 have been found to be expressed [12]. Other TLRs (including TLR7-9) are expressed in human colon carcinoma cells HCT15, SW620 or HT29 [49, 50, 69].

TLR expression in tumor cells appears to promote tumorigenesis by facilitating survival and migration within the tumorous microenvironment characterized by chronic inflammation and PAMPs [135]. However, studies have also established that boostering TLRs and downstream mediators, such as type I IFNs may shift the balance from immunotolerance to antitumoral effects [138]. Therefore in cancer cells a controversial role of TLR signaling pathways has been proposed.

TLRs may act as tumor promoting factors especially by transmitting proinflammatory, anti-apoptotic, proliferative or profibrogenic signals either in the tumor cells or the tumorous microenvironment. Enhancement of the signaling pathway of transcription factor NF-κB is one of the major mode of tumor-promoting actions of TLRs. TLR activation upregulates several tumorigenic inflammatory cytokines (e.g.: IL-1β, TNFα, IL-6) in an NF-κB dependent manner [58, 137, 138]. TLR signaling is also involved in the inhibition of apoptosis. NF-κB is considered as a relevant anti-apoptotic pathway controlling the expression of anti-apoptotic genes and restricting the activation of pro-apoptotic pathways [39, 180].

In colorectal cancer TLR-induced NF-κB activation has been found to facilitate tumor cell survival [46]. Furthermore, in the MC26 mouse colon cancer cell line TLR4 activation mediated resistance of tumor cells to cytotoxic T cell-mediated cell death, finally favoring tumor growth [69]. The TLR-mediated promotion of wound healing may also lead to cancer development. After dextran sulfate sodium (DSS)-mediated injury, TLR2 and TLR4 activation facilitates epithelial repair via the MyD88-dependent pathway [46], and TLR-MyD88 signaling also regulates the expression of epiregulin, which may contribute to colon cancer development [180]. In mice chronic inflammation arising from the bowel was found to induce thymic involution and regulatory T (Treg) cell suppression [44]. These events are suggested to enhance inflammation-mediated processes, and worsen IBD [42].

Based on the existing connection between TLR-signaling and Treg cells [133], the concept that in IBD uncontrolled inflammation weakens Treg-mediated inhibition and increases the risk for inflammation-associated carcinogenesis may represent a realistic idea. Controversial data exist regarding the role of TLR2 in CAC. In a TLR2-deficient azoxymethane (AOM)-DSS murine model increased tumor development and higher IL-6, IL-17A and phospho-signal transducer and activator of transcription (STAT)-3 levels were reported [102], while no differences in CAC between wild-type and TLR2-deficient AOM-DSS colitic animals were found [149].

The pro-tumorigenic role of TLR4 in CAC is well established. The intestinal microbiota, which normally colonize mucosal surfaces in symbiotic mutualism with the host is unique and quite stable over time [9]. In the colon, where there is a constant interaction between microbiota and IECs, TLR4 deletion significantly reduces inflammation and tumor size in a CAC-model of AOM-DSS mice [47]. Additionally, overexpression of the constitutively active TLR4 exhibits a higher sensitivity to CAC in a transgenic mouse model [48]. Other studies also support the results that both the deletion of the TLR4 adaptor MyD88 molecule and the depletion of TLR4 activating gut microbiome reduce colon cancer development [98, 141].

The antigen-presenting capacity of tumor cells is poor, therefore antitumoral immune responses usually depend on professional antigen-presenting cells (APCs) like dendritic cells (DCs) [127]. DCs have been in the focus of cancer research for their ability to initiate potent antitumoral immunity. The lack of DC activation often resulting by inhibitory signals from cancer cells may also induce immune tolerance via T cell deletion or Tregs [127], and thus favors tumor progression. TLR-activated DCs can mediate antitumoral effects through antigen presentation, T cell activation, and direct cytotoxicity on tumor cells [38, 51]. TLR5 activation on DCs as well as TLR9-stimulated plasmocytoid DCs promote antitumoral immunity [30, 120].

It is hypothesized that DC-mediated tumor cell killing triggers a more efficient antigen presentation to cytotoxic T cells, thus amplifying antitumoral responses. Activation of TLRs on DCs regulates T cell activation not only via MHC II and co-stimulatory molecules, but also through TLR-induced signals in DCs that block the suppressive effect of Tregs in an IL-6-dependent manner [133]. TLR8 activation, moreover, can directly inhibit Treg function, hence support antitumoral immunity [134].

Recently, the modulatory effect of TLR5-dependent signaling was assayed in a mouse xenograft model of human colon cancer [143]. The lack of MyD88 or TLR5 expression was found to promote tumor growth and inhibit necrosis [143]. On the contrary, however, TLR5 activation by peritumoral flagellin treatment substantially increased tumor necrosis, leading to significant tumor regression [143].

Within the TLR family TLR9 is specifically stimulated upon sequence- and methylation-dependent DNA signaling. Self-DNA and oligonucleotides containing unmethylated CpG motifs are also senzed by and activate TLR9. Modifications in the structure of nucleic acids influence their immunomodulatory, i.e. agonistic or suppressive, as well as pro- or anti-tumorigenic capacity [50, 160]. TLR9 activation by synthetic CpG-ODN agonists has also demonstrated antitumor activity in xenograft models of murine colon cancer [63]. Moreover, TLR9 agonists induce type I IFN secretion in DCs finally resulting in cytotoxic DCs, activated NK cells and cytotoxic T cells with a remarkable antitumor immune response [75, 86].

2.4 Autophagy and TLRs: A Bidirectional Communication

2.4.1 TLRs in Regulation of Autophagy

The crosstalk between TLRs and autophagy leads to the activation of innate immunity. A recently discovered ability of TLRs means that upon engagement almost all receptor prototypes are able to promote canonical form of autophagy, whereas some of them also stimulate LC3-associated phagocytosis [113] in innate immune cells like macrophages, dendritic cells, and neutrophils indicating the involvement of these pathways in cellular defense [35, 150, 159]. It has also been suggested that TLRs may have an intrinsic capability to induce autophagy [34]. Within innate immunity phagocytosis represents a basic protective mechanism, and TLR signaling in macrophages clearly links the autophagic pathway to phagocytosis by stimulation of cognate transduction signals [150]. Furthermore, autophagy now can be considered as an effector of TLR signaling [36].

Nevertheless, in plasmocytoid DCs (pDCs) upon ligation of TLR7 no autophagy induction has been detected [92]. DCs display high level of basal autophagy and permit little or no induction of autophagy upon other immunological stimulation. On the other hand competing signaling pathways could also be activated and thus, inhibit autophagy induction [34].

TLRs initiate the common NF-κB/MAPK (extracellular signal-regulated kinase/ERK/, p38 and JNK) and the IRF3/7 signaling pathways [94]. TLR-induced autophagy is mainly depend on the adaptor proteins MyD88 and TRIF [35, 157]. In addition, TLR signaling forces the interaction of MyD88 and TRIF with Beclin-1 and promotes the dissociation of Beclin-1 from the binding complex with Bcl-2 [94]. Furthermore, ubiquitination of Beclin-1 via TRAF6 enhanced TLR4-induced autophagy, while upon action of the deubiquitinating enzyme A20 the opposite process occurred [157, 158]. Activation of NF-κB downstream of TLR stimulation is more likely an inhibitory element of autophagy regulation.

2.4.2 Autophagy in Regulation of TLRs

In general autophagy exerts cytoprotective effect when cells are under any stressful conditions, therefore the involvement of autophagy in regulation of TLR-mediated proinflammatory responses is not surprising [72]. Autophagy mainly exerts suppressive effect on the induction of inflammatory responses [36]. As a cellular strategy it may influence inflammation directly by the breakdown of invading microorganisms, and further by the degradation of the adaptor proteins MyD88 and TRIF [71, 84]. Overexpression of aggregate-prone TLR adaptors may result in formation of large aggregates in the cytoplasm. Nonetheless, autophagy mainly has a suppressive effect on TLR signaling, but this action may become reversed in pDCs [72].

Several autophagy-related proteins negatively regulate the TLR-induced signals. In response to LPS stimulation of TLR4 Atg16L1-deficient macrophages produce large amounts of IL-1β and IL-18 via the excessive activation of caspase-1 [148]. Furthermore, in macrophages defective autophagy due to LC3B or Beclin1 deficiency resulted in accumulation of abnormal mitochondria correlating with increased reactive oxygen species (ROS) production.

Autophagy definitely facilitates the sequestration of endogenous viral or self-antigens into autophagosomes and their delivery to MHC class II antigens resulting in MHC II-restricted presentation of cytoplasmic antigens to T cells [152]. However, it has also been found that similar to antigen presentation the autophagic machinery can deliver PAMPs to endosomal TLRs indicating that autophagy is not only a TLR-effector mechanism, but may promote recognition of PAMPs by TLRs, thus initiate an innate immune response, upstream of TLR action [92].

2.5 TLRs and Autophagy in Colitis-Associated Cancer

Within experimental conditions CT26 colon cancer cells in mice upon treatment with graphene oxide (GO) displayed TLR activation and autophagy induction. GO phagocytosed by cancer cells led to simultaneous triggering of autophagy and TLR4 and TLR9 signaling pathways. Autophagy induced by GO was regulated via adaptors of MyD88- and TRAF6. Injection of GO to mice suppressed tumor progression, and further increased immunity, cell death, and autophagy in cancer cells [25].

Unmethylated CpG-ODNs belonging to DAMPs are recognized by TLR9 expressed mainly by immune cells, but also present on several cancer cells. Recently, upon proteomics analysis of different tumor cells several proteins (including those of the autophagic process) modulated by bacterial CpG motifs have been identified [15]. The CpG-TLR9 pathway is known to display several similarities with that of autophagy. CpG-ODN was found to trigger autophagy in tumor cell lines from colon cancers in a TLR9-dependent manner, thus extending the link between TLRs and autophagy in cancer [14]. Besides the potential of autophagy-induced cell death autophagy promotes the MHC II-related presentation of endogenous cytosolic proteins as well, therefore in tumors bacterial CpG motifs could force the presentation of tumor antigens, thus facilitate antitumor immunity [153].

3 Therapeutic Aspects of Autophagy

3.1 Autophagy in IBD Therapy

Unquestionably, autophagy can be considered as an apparently difficult regulatory network, being in close connection with several signal transduction pathways and cellular programs. Principle elements of immunological autophagy include the direct cell-autonomous pathogen elimination, the regulation of PRRs, and inflammasome activation, and the cytoplasmic antigen processing for MHC presentation to T cells. In CD functional consequences of the underlying autophagy-related gene defects *(ATG16L1, IRGM, NOD2, XBP1),* in particular the inappropriate stimulation of antimicrobial and inflammasome pathways eventually result in uncontrolled inflammation. Therefore, autophagy in CD is predicted as a key regulator mechanism with the capacity to integrate several aspects of disease pathogenesis.

Theoretically the complex autophagy signaling in CD offers a promising novel therapeutic target, since due to its induction potentially not only the load of cytoinvasive bacteria, and the perturbated immune responses, but the resulting inflammatory process, as well may simultaneously be reduced. Thus, autophagy boosting would represent an efficient biologic manipulation, and could provide an alternative therapeutic option. Several candidate pathways, e.g. inhibition of mTOR, decrease of ER-stress, lowering of inositol triphosphate (IP3), etc. could be considered.

Caspase dysfunction is also known to be associated with IBD and mucosal inflammation [13]. Under conditions of cellular stress in which caspase activation occurs, ATG16L1 T300A-mediated autophagy is particularly impaired [118]. Interestingly, a protective missense SNP in the amyloid-β precursor protein (APP) encoding gene has been shown to alter cleavage of full-length APP by aspartyl protease, suggesting that alterations of proteolytic cleavage could be a common feature of disease-associated SNPs [74]. Small molecule development to treat complex diseases like IBD may focus on disruption of SNP-dependent protease–substrate interactions, suggesting a promising strategy of therapeutic agent development in addition to compounds that can enhance autophagy.

Recently, NOD2 has been found to be a 1,25-dihydroxyvitamin-D target gene [177]. This observation links vitamin-D signaling to autophagy. Stimulation of NOD2 expression by 1,25-dihydroxyvitamin-D implies that it would boost autophagy at least in part by enhancing NOD2 function. Additionally, recent work has shown that 1,25-dihydroxyvitamin-D-stimulated CAMP production enhanced autophagy in mycobacteria-infected macrophages [189]. The effects of 1,25-dihydroxyvitamin-D-induced CAMP on autophagy may be at least partially independent of NOD2 function [177]. This raises the possibility that enhanced CAMP expression may be sufficient to induce clearance of intracellular pathogens despite mutations in the NOD2 pathway common in CD.

Recently, the suppression of autophagy by appendicitis and appendectomy in the distal colon has been shown in mice [24]. It has been hypothesized that the upregulation of autophagy-associated genes (i.e.: IRGM, FIP200, ATG04A) could be a reflection of complex compensatory changes that led to the pronounced autophagy suppression, inducing lesser antigen processing, thus leading to lesser cross-reactive immunity between microbes and self-antigens, finally ameliorating colitis.

On the other hand, however, much cautiousness is required regarding its pleiotropic physiological repertoire, since pharmacologic autophagy modulation can initiate additional biologic effects not expected in CD. Further detailed functional analyses of the CD-associated genetic polymorphisms are needed to explore and define more precisely the subcellular and molecular basis of the crosstalk between autophagy and the innate immune axis, hopefully allowing the introduction of selective new therapeutic approaches into daily practice.

3.2 Autophagy in Colitis-Associated Cancer Therapy

In view of immune surveillance selective, specific and effective eradication of cancer cells by a subsequent active host immune response serving as a widespread therapy option has still been remained unsolved.

Current therapies for cancer mainly are based on chemotherapeutic drugs that kill transformed, dividing cells or block cell division, but unfortunately these treatments may also attack normal proliferating cells, including immunocompetent ones. However, targeted immune responses (immunotherapy) to tumors may be specific, thus making the possibility to avoid normal cell injury. According to therapeutic vaccines killed tumor cells or tumor antigens can efficiently induce anticancer immunity.

So far less attention has been paid on the possible subcellular and molecular impact of chemotherapy-induced cell death regarding induction of host immune responses.

As our knowledge regarding the biological functions of autophagy and TLRs increases, the cross-talk of these pathways in cancer seems to be a critical aspect. There is no doubt that processes of autophagy and TLR-signaling are apparently difficult regulatory networks, being in close relation with several other signal transduction pathways and cellular programs. Autophagy deeply determines cell survival, thus interacts with types of cell death, such as apoptosis, necrosis, and necroptosis, and when it is extreme, *per se* contributes to cell destiny. Recently, significant advances have been achieved in understanding the importance of autophagy in immune responses [152]. Notably based upon the discovered bidirectional TLRs–autophagy and autophagy–TLRs communications, autophagy is now considered as a fully integrated element of immunity [35, 72]. Principle functions of immunological autophagy include direct elimination of pathogens, contribution to processing and MHC II-restricted presentation of cytoplasmic antigens, and critical involvement in regulation of T and B cell homeostasis, immune tolerance and inflammatory signaling [36]. An inflammatory response initiated by innate immunity is essential to stimulate protective immunity, in particular in the context of anticancer immunosurveillance. However, excessive induction of PRR-signaling and the subsequent inflammatory environment unequivocally predispose to carcinogenesis [55].

Data regarding the role of autophagy and TLRs in carcinogenesis are rather conflicting, since these pathways may be pro-tumorigenic, i.e. critical for cancer cell survival and progression, however, they may also be anti-tumorigenic, i.e. evoking tumor death.

Despite of their context-dependent, "dual-faced" actions both complex mechanisms can be considered as possible promising though challenging therapeutic targets either in cancer treatment or prevention. Nevertheless, the exact interplay of autophagy and PRR-signaling within cancer network and its relation to tumor immunity has not yet been clarified.

In autophagy-competent tumor cells upon response to different chemo- and radiotherapies increase in autophagy is often induced, representing an adaptive survival mechanism, and further, resistance to cancer treatment could also be evoked. Therefore it has been hypothesized that concurrent pharmacologic inhibition of autophagy (using e.g. lysosome-inhibitor drugs, like chloroquine, rapamycin or hydroxychloroquine) as an adjuvant may sensitize tumor cells to a spectrum of anticancer drugs [26, 27, 192]. In case of autophagy-deficient tumors, however, due to their extreme susceptibility predominantly metabolic stress and DNA-damage-inducing therapeutic protocols are suggested.

Hence, depending on the type of malignancy autophagy induction could provide an alternative therapeutic option as well [26, 27, 192]. For autophagy boosting several candidate pathways (e.g. inhibition of mTOR, decrease of endoplasmatic reticulum/ER/-stress, lowering of IP3, etc.) might be acknowledged. It has also been found that TLR-mediated signaling induced by the cell wall skeleton of Bacillus Calmette-Guerin has a radiosensitizing effect on colon cancer cells through the induction of autophagy, and thus may reflect another therapeutic strategy in this type of cancer [190]. In addition, autophagy, when excessive can potentially act as an active cell death machinery, presumably along with inherent defects of apoptosis, so induction of autophagy by antitumor drugs may be considered as an efficient cytotoxic manipulation [107].

Recently, a new aspect of anticancer drug actions has been proved, indicating that premortem autophagy in tumor cells induced by optimal release of DAMP molecules is required to immunogenic cell death (ICD) following chemotherapy. Consequently, autophagy deficiency significantly restricted the ability of cancer cells to induce an adaptive anti-tumor immune response [115]. Particular DAMPs, like calreticulin, HMGB1, heat-schock protein (HSP)70/HSP90, and adenosine triphosphate (ATP) released from tumor cells largely determine whether the cell death is immunogenic (ICD), and thus elicits protective immunity, or tolerogenic (TCD), and so contributes to tumoral progression [57]. In respect of PRR-TLRs the chaperon HSPs interact mainly with TLR4, and to lesser extent with TLR2, and thus promote engulfment of dying cells, facilitate tumor antigen processing and presentation by DCs and priming of T cells. The complex intersection of autophagy and DAMPs is fundamentally involved in regulation of cell death and also means a critical event for evoking a subsequent tumor-specific immune response in cancer [67].

Thus, upon translation of basic knowledge regarding the cross-talk of autophagy and PRRs into practice it is reasonable to speculate that their cell and context-dependent modulation should provide potential therapeutic targets in cancer. In the last few years nanomedicine has turned into rapidly growing research area, particularly for anticancer applications. Several nanomedicine techniques, including lipid-based drug carriers have received clinical approval, and along with polymeric ones are now undergoing clinical evaluation [7]. Autophagy of nanoparticles deeply influences their fate after endocytosis, therefore, as a drug-carrier it could modulate their therapeutic effects [191]. One of the expected extensions of these promising new drug-carrier manipulations is to improve their antitumoral actions in a more selective manner [151].

On the other hand, however, much cautiousness is required due to the controversial effects of these pathways in carcinogenesis. Furthermore, it is important to consider the pleiotropic physiological repertoires, since their pharmacologic manipulation can initiate additional, yet unexpected biological effects. In different types of cancer, more detailed and precise cellular explorations are warranted to understand the many faces of TLRs and autophagy. Clarifying the relationship amongst autophagy, TLR-signaling, inflammatory microenvironment, and immunogenicity in balancing of cell fate hopefully may allow the introduction of new therapeutic approaches into daily practice.

References

1. Abbott DW, Wilkins A, Asara JM, Cantley LC (2004) The Crohn's disease protein, NOD2, requires RIP2 in order to induce ubiquitinylation of a novel site on NEMO. Curr Biol 14:2217–2227
2. Abraham C, Cho JH (2009) Inflammatory bowel disease. N Engl J Med 361:2066–2078. doi:10.1056/NEJMra0804647
3. Ahn CH, Jeong EG, Lee JW et al (2007) Expression of beclin-1, an autophagy-related protein, in gastric and colorectal cancers. APMIS 115:1344–1349
4. Aita VM, Liang XH, Murty VV et al (1999) Cloning and genomic organization of beclin 1, a candidate tumor suppressor gene on chromosome 17q21. Genomics 59:59–65
5. Akira S, Uematsu S, Takeuchi O (2006) Pathogen recognition and innate immunity. Cell 124:783–801
6. Albugami M, Kiberd B (2013) Malignancies: pre and post transplantation strategies. Transplant Rev (Orlando). doi:10.1016/j.trre.2013.12.002
7. Allen TM, Cheng WW, Hare JI, Laginha KM (2006) Pharmacokinetics and pharmacodynamics of lipidic nano-particles in cancer. Anticancer Agents Med Chem 6:513–523
8. Aviello G, Corr SC, Johnston DG, O'Neill LA, Fallon PG (2014) MyD88 adaptor-like (Mal) regulates intestinal homeostasis and colitis-associated colorectal cancer in mice. Am J Physiol Gastrointest Liver Physiol 306(9):G769–G778. doi:10.1152/ajpgi.00399.2013
9. Bäckhed F, Ley RE, Sonnenburg JL, Peterson DA, Gordon JI (2005) Host-bacterial mutualism in the human intestine. Science 307:1915–1920
10. Balkwill F, Mantovani A (2001) Inflammation and cancer: back to Virchow? Lancet 357:539–545
11. Barrett JC, Hansoul S, Nicolae DL et al (2008) Genome-wide association defines more than 30 distinct susceptibility loci for Crohn's disease. Nat Genet 40:955–962. doi:10.1038/ng.175

12. Basith S, Manavalan B, Yoo TH, Kim SG, Choi S (2012) Roles of toll-like receptors in cancer: a double-edged sword for defense and offense. Arch Pharm Res 35:1297–1316. doi:10.1007/s12272-012-0802-7
13. Becker C, Watson AJ, Neurath MF (2013) Complex roles of caspases in the pathogenesis of inflammatory bowel disease. Gastroenterology 144:283–293. doi:10.1053/j.gastro.2012.11.035
14. Bertin S, Pierrefite-Carle V (2008) Autophagy and toll-like receptors: a new link in cancer cells. Autophagy 4:1086–1089
15. Bertin S, Samson M, Pons C et al (2008) Comparative proteomics study reveals that bacterial CpG motifs induce tumor cell autophagy in vitro and in vivo. Mol Cell Proteomics 7:2311–2322
16. Bonen DK, Ogura Y, Nicolae DL, Inohara N, Saab L, Tanabe T, Chen FF, Foster SJ, Duerr RH, Brant SR, Cho JH, Nuñez G (2003) Crohn's disease-associated NOD2 variants share a signaling defect in response to lipopolysaccharide and peptidoglycan. Gastroenterology 124:140–146
17. Brandl K, Sun L, Neppl C, Siggs OM, Le Gall SM, Tomisato W, Li X, Du X, Maennel DN, Blobel CP, Beutler B (2010) MyD88 signaling in nonhematopoietic cells protects mice against induced colitis by regulating specific EGF receptor ligands. Proc Natl Acad Sci U S A 107:19967–19972. doi:10.1073/pnas.1014669107
18. Brown SL, Riehl TE, Walker MR, Geske MJ, Doherty JM, Stenson WF, Stappenbeck TS (2007) Myd88-dependent positioning of Ptgs2-expressing stromal cells maintains colonic epithelial proliferation during injury. J Clin Invest 117:258–269
19. Bruno ME, Rogier EW, Frantz AL, Stefka AT, Thompson SN, Kaetzel CS (2010) Regulation of the polymeric immunoglobulin receptor in intestinal epithelial cells by Enterobacteriaceae: implications for mucosal homeostasis. Immunol Invest 39:356–382. doi:10.3109/08820131003622809
20. Cadwell K, Liu JY, Brown SL, Miyoshi H, Loh J, Lennerz JK, Kishi C, Kc W, Carrero JA, Hunt S, Stone CD, Brunt EM, Xavier RJ, Sleckman BP, Li E, Mizushima N, Stappenbeck TS, Virgin HW 4th (2008) A key role for autophagy and the autophagy gene Atg16l1 in mouse and human intestinal Paneth cells. Nature 456:259–263. doi:10.1038/nature07416
21. Cadwell K, Patel KK, Komatsu M, Virgin HW 4th, Stappenbeck TS (2009) A common role for Atg16L1, Atg5 and Atg7 in small intestinal Paneth cells and Crohn disease. Autophagy 5:250–252
22. Canavan C, Abrams KR, Mayberry J (2006) Meta-analysis: colorectal and small bowel cancer risk in patients with Crohn's disease. Aliment Pharmacol Ther 23:1097–1104
23. Castro FA, Liu X, Försti A, Ji J, Sundquist J, Sundquist K, Koshiol J, Hemminki K (2013) Increased risk of hepatobiliary cancers after hospitalization for autoimmune disease. Clin Gastroenterol Hepatol 12(6):1038–1045.e7. doi:10.1016/j.cgh.2013.11.007
24. Cheluvappa R, Luo AS, Grimm MC (2014) Autophagy suppression by appendicitis and appendectomy protects against colitis. Inflamm Bowel Dis 20:847–855. doi:10.1097/MIB.0000000000000034
25. Chen GY, Chen CL, Tuan HY et al (2014) Graphene oxide triggers toll-like receptors/autophagy responses in vitro and inhibits tumor growth in vivo. Adv Healthc Mater. doi:10.1002/adhm.201300591
26. Chen N, Debnath J (2010) Autophagy and tumorigenesis. FEBS Lett 584:1427–1435
27. Chen N, Karantza-Wadsworth V (2009) Role and regulation of autophagy in cancer. Biochim Biophys Acta 1793:1516–1523
28. Cooney R, Baker J, Brain O, Danis B, Pichulik T, Allan P, Ferguson DJ, Campbell BJ, Jewell D, Simmons A (2010) NOD2 stimulation induces autophagy in dendritic cells influencing bacterial handling and antigen presentation. Nat Med 16:90–97. doi:10.1038/nm.2069
29. Coppola D, Khalil F, Eschrich SA et al (2008) Down-regulation of Bax-interacting factor-1 in colorectal adenocarcinoma. Cancer 113:2665–2670
30. Cubillos-Ruiz JR, Engle X, Scarlett UK et al (2009) Polyethylenimine-based siRNA nanocomplexes reprogram tumor-associated dendritic cells via TLR5 to elicit therapeutic antitumor immunity. J Clin Invest 119:2231–2244

31. Dalby KN, Tekedereli I, Lopez-Berestein G, Ozpolat B (2010) Targeting the prodeath and prosurvival functions of autophagy as novel therapeutic strategies in cancer. Autophagy 6:322–329
32. de Kivit S, Tobin MC, Forsyth CB, Keshavarzian A, Landay AL (2014) Regulation of intestinal immune responses through TLR activation: implications for pro- and prebiotics. Front Immunol 5:60. doi:10.3389/fimmu.2014.00060. eCollection 2014
33. Degenhardt K, Mathew R, Beaudoin B, Bray K, Anderson D, Chen G, Mukherjee C, Shi Y, Gélinas C, Fan Y, Nelson DA, Jin S, White E (2006) Autophagy promotes tumor cell survival and restricts necrosis, inflammation, and tumorigenesis. Cancer Cell 10:51–64
34. Delgado MA, Deretic V (2009) Toll-like receptors in control of immunological autophagy. Cell Death Differ 16:976–983
35. Delgado MA, Elmaoued RA, Davis AS, Kyei G, Deretic V (2008) Toll-like receptors control autophagy. EMBO J 27:1110–1121
36. Deretic V, Saitoh T, Akira S (2013) Autophagy in infection, inflammation and immunity. Nat Rev Immunol 13:722–737
37. Ding WX, Ni HM, Gao W, Yoshimori T, Stolz DB, Ron D, Yin XM (2007) Linking of autophagy to ubiquitin-proteasome system is important for the regulation of endoplasmic reticulum stress and cell viability. Am J Pathol 171:513–524
38. Drobits B, Holcmann M, Amberg N et al (2012) Imiquimod clears tumors in mice independent of adaptive immunity by converting pDCs into tumor-killing effector cells. J Clin Invest 122:575–585
39. Dutta J, Fan Y, Gupta N, Fan G, Gélinas C (2006) Current insights into the regulation of programmed cell death by NF-kappaB. Oncogene 25:6800–6816
40. Eaden JA, Abrams KR, Mayberry JF (2001) The risk of colorectal cancer in ulcerative colitis: a meta-analysis. Gut 48:526–535
41. Eaves-Pyles T, Allen CA, Taormina J, Swidsinski A, Tutt CB, Jezek GE, Islas-Islas M, Torres AG (2008) Escherichia coli isolated from a Crohn's disease patient adheres, invades, and induces inflammatory responses in polarized intestinal epithelial cells. Int J Med Microbiol 298:397–409
42. Erdman SE, Poutahidis T (2010) Roles for inflammation and regulatory T cells in colon cancer. Toxicol Pathol 38:76–87
43. Farrell RJ, Ang Y, Kileen P, O'Briain DS, Kelleher D, Keeling PW, Weir DG (2000) Increased incidence of non-Hodgkin's lymphoma in inflammatory bowel disease patients on immunosuppressive therapy but overall risk is low. Gut 47:514–519
44. Faubion WA, De Jong YP, Molina AA et al (2004) Colitis is associated with thymic destruction attenuating CD4+25+ regulatory T cells in the periphery. Gastroenterology 126:1759–1770
45. Fukata M, Arditi M (2013) The role of pattern recognition receptors in intestinal inflammation. Mucosal Immunol 6:451–463. doi:10.1038/mi.2013.13
46. Fukata M, Chen A, Klepper A, Krishnareddy S, Vamadevan AS, Thomas LS, Xu R, Inoue H, Arditi M, Dannenberg AJ, Abreu MT (2006) Cox-2 is regulated by Toll-like receptor-4 (TLR4) signaling: Role in proliferation and apoptosis in the intestine. Gastroenterology 131:862–877
47. Fukata M, Hernandez Y, Conduah D et al (2009) Innate immune signaling by Toll-like receptor-4 (TLR4) shapes the inflammatory microenvironment in colitis-associated tumors. Inflamm Bowel Dis 15:997–1006
48. Fukata M, Shang L, Santaolalla R et al (2011) Constitutive activation of epithelial TLR4 augments inflammatory responses to mucosal injury and drives colitis-associated tumorigenesis. Inflamm Bowel Dis 17:1464–1473
49. Fűri I, Sipos F, Germann TM et al (2013) Epithelial toll-like receptor 9 signaling in colorectal inflammation and cancer: clinico-pathogenic aspects. World J Gastroenterol 19:4119–4126
50. Fűri I, Sipos F, Spisák S et al (2013) Association of self-DNA mediated TLR9-related gene, DNA methyltransferase, and cytokeratin protein expression alterations in HT29-cells to DNA fragment length and methylation status. ScientificWorldJournal 2013:293296. doi:10.1155/2013/293296. eCollection 2013

51. Garaude J, Kent A, van Rooijen N, Blander JM (2012) Simultaneous targeting of toll- and nod-like receptors induces effective tumor-specific immune responses. Sci Transl Med 4:120ra16. doi:10.1126/scitranslmed.3002868
52. Garay RP, Viens P, Bauer J, Normier G, Bardou M, Jeannin JF, Chiavaroli C (2007) Cancer relapse under chemotherapy: why TLR2/4 receptor agonists can help. Eur J Pharmacol 563:1–17
53. Garrett WS, Gordon JI, Glimcher LH (2010) Homeostasis and inflammation in the intestine. Cell 140:859–870. doi:10.1016/j.cell.2010.01.023
54. Glasser AL, Boudeau J, Barnich N, Perruchot MH, Colombel JF, Darfeuille-Michaud A (2001) Adherent invasive Escherichia coli strains from patients with Crohn's disease survive and replicate within macrophages without inducing host cell death. Infect Immun 69:5529–5537
55. Gonda TA, Tu S, Wang TC (2009) Chronic inflammation, the tumor microenvironment and carcinogenesis. Cell Cycle 8:2005–2013
56. Gordy C, He YW (2012) The crosstalk between autophagy and apoptosis: where does this lead? Protein Cell 3:17–27. doi:10.1007/s13238-011-1127-x
57. Green DR, Ferguson T, Zitvogel L, Kroemer G (2009) Immunogenic and tolerogenic cell death. Nat Rev Immunol 9:353–363
58. Grivennikov S, Karin E, Terzic J et al (2009) IL-6 and Stat3 are required for survival of intestinal epithelial cells and development of colitis-associated cancer. Cancer Cell 15:103–113
59. Guertin DA, Sabatini DM (2007) Defining the role of mTOR in cancer. Cancer Cell 12:9–22
60. Hampe J, Franke A, Rosenstiel P, Till A, Teuber M, Huse K, Albrecht M, Mayr G, De La Vega FM, Briggs J, Günther S, Prescott NJ, Onnie CM, Häsler R, Sipos B, Fölsch UR, Lengauer T, Platzer M, Mathew CG, Krawczak M, Schreiber S (2007) A genome-wide association scan of nonsynonymous SNPs identifies a susceptibility variant for Crohn disease in ATG16L1. Nat Genet 39:207–211
61. Hanahan D, Weinberg RA (2000) The hallmarks of cancer. Cell 100:57–70
62. He C, Klionsky DJ (2009) Regulation mechanisms and signaling pathways of autophagy. Annu Rev Genet 43:67–93. doi:10.1146/annurev-genet-102808-114910
63. Heckelsmiller K, Rall K, Beck S et al (2002) Peritumoral CpG DNA elicits a coordinated response of CD8 T cells and innate effectors to cure established tumors in a murine colon carcinoma model. J Immunol 169:3892–3899
64. Hedl M, Li J, Cho JH, Abraham C (2007) Chronic stimulation of Nod2 mediates tolerance to bacterial products. Proc Natl Acad Sci U S A 104:19440–19445
65. Henderson P, Stevens C (2012) The role of autophagy in Crohn's disease. Cells 1:492–519. doi:10.3390/cells1030492
66. Homer CR, Richmond AL, Rebert NA, Achkar JP, McDonald C (2010) ATG16L1 and NOD2 interact in an autophagy-dependent antibacterial pathway implicated in Crohn's disease pathogenesis. Gastroenterology 139:1630–1641, 1641.e1–1641.e2. doi:10.1053/j.gastro.2010.07.006
67. Hou W, Zhang Q, Yan Z et al (2013) Strange attractors: DAMPs and autophagy link tumor cell death and immunity. Cell Death Dis 4, e966. doi:10.1038/cddis.2013.493
68. Hsu D, Fukata M, Hernandez YG, Sotolongo JP, Goo T, Maki J, Hayes LA, Ungaro RC, Chen A, Breglio KJ, Xu R, Abreu MT (2010) Toll-like receptor 4 differentially regulates epidermal growth factor-related growth factors in response to intestinal mucosal injury. Lab Invest 90:1295–1305. doi:10.1038/labinvest.2010.100
69. Huang B, Zhao J, Li H et al (2005) Toll-like receptors on tumor cells facilitate evasion of immune surveillance. Cancer Res 65:5009–5014
70. Hugot JP, Chamaillard M, Zouali H, Lesage S, Cézard JP, Belaiche J, Almer S, Tysk C, O'Morain CA, Gassull M, Binder V, Finkel Y, Cortot A, Modigliani R, Laurent-Puig P, Gower-Rousseau C, Macry J, Colombel JF, Sahbatou M, Thomas G (2001) Association of NOD2 leucine-rich repeat variants with susceptibility to Crohn's disease. Nature 411:599–603

71. Inomata M, Niida S, Shibata K, Into T (2012) Regulation of Toll-like receptor signaling by NDP52-mediated selective autophagy is normally inactivated by A20. Cell Mol Life Sci 69:963–979
72. Into T, Inomata M, Takayama E, Takigawa T (2012) Autophagy in regulation of Toll-like receptor signaling. Cell Signal 24:1150–1162
73. Ionov Y, Nowak N, Perucho M, Markowitz S, Cowell JK (2004) Manipulation of nonsense mediated decay identifies gene mutations in colon cancer Cells with microsatellite instability. Oncogene 23:639–645
74. Jonsson T, Atwal JK, Steinberg S, Snaedal J, Jonsson PV, Bjornsson S, Stefansson H, Sulem P, Gudbjartsson D, Maloney J, Hoyte K, Gustafson A, Liu Y, Lu Y, Bhangale T, Graham RR, Huttenlocher J, Bjornsdottir G, Andreassen OA, Jönsson EG, Palotie A, Behrens TW, Magnusson OT, Kong A, Thorsteinsdottir U, Watts RJ, Stefansson K (2012) A mutation in APP protects against Alzheimer's disease and age-related cognitive decline. Nature 488:96–99. doi:10.1038/nature11283
75. Kadowaki N, Antonenko S, Liu YJ (2001) Distinct CpG DNA and polyinosinic-polycytidylic acid double-stranded RNA, respectively, stimulate CD11c- type 2 dendritic cell precursors and CD11c+dendritic cells to produce type I IFN. J Immunol 166:2291–2295
76. Kanneganti TD, Lamkanfi M, Núñez G (2007) Intracellular NOD-like receptors in host defense and disease. Immunity 27:549–559
77. Kaser A, Blumberg RS (2009) Endoplasmic reticulum stress in the intestinal epithelium and inflammatory bowel disease. Semin Immunol 21:156–163. doi:10.1016/j.smim.2009.01.001
78. Kaser A, Lee AH, Franke A, Glickman JN, Zeissig S, Tilg H, Nieuwenhuis EE, Higgins DE, Schreiber S, Glimcher LH, Blumberg RS (2008) XBP1 links ER stress to intestinal inflammation and confers genetic risk for human inflammatory bowel disease. Cell 134:743–756. doi:10.1016/j.cell.2008.07.021
79. Kaser A, Zeissig S, Blumberg RS (2010) Inflammatory bowel disease. Annu Rev Immunol 28:573–621. doi:10.1146/annurev-immunol-030409-101225
80. Kawai T, Akira S (2005) Pathogen recognition with Toll-like receptors. Curr Opin Immunol 17:338–344
81. Khor B, Gardet A, Xavier RJ (2011) Genetics and pathogenesis of inflammatory bowel disease. Nature 474:307–317. doi:10.1038/nature10209
82. Kim EH, Hong KS, Hong H, Hahm KB (2011) Detouring the undesired route of helicobacter pylori-induced gastric carcinogenesis. Cancers (Basel) 3:3018–3028. doi:10.3390/cancers3033018
83. Kim MS, Jeong EG, Ahn CH et al (2008) Frameshift mutation of UVRAG, an autophagy-related gene, in gastric carcinomas with microsatellite instability. Hum Pathol 39:1059–1063
84. Knaevelsrud H, Simonsen A (2010) Fighting disease by selective autophagy of aggregate-prone proteins. FEBS Lett 584:2635–2645
85. Kobayashi KS, Chamaillard M, Ogura Y, Henegariu O, Inohara N, Nuñez G, Flavell RA (2005) Nod2-dependent regulation of innate and adaptive immunity in the intestinal tract. Science 307:731–734
86. Krieg AM (2007) Development of TLR9 agonists for cancer therapy. J Clin Invest 117:1184–1194
87. Kroemer G, Levine B (2008) Autophagic cell death: the story of a misnomer. Nat Rev Mol Cell Biol 9:1004–1010
88. Kundu JK, Surh YJ (2008) Inflammation: gearing the journey to cancer. Mutat Res 659:15–30. doi:10.1016/j.mrrev.2008.03.002
89. Lapaquette P, Glasser AL, Huett A, Xavier RJ, Darfeuille-Michaud A (2010) Crohn's disease-associated adherent-invasive E. coli are selectively favoured by impaired autophagy to replicate intracellularly. Cell Microbiol 12:99–113. doi:10.1111/j.1462-5822.2009.01381.x
90. Lassen KG, Kuballa P, Conway KL, Patel KK, Becker CE, Peloquin JM, Villablanca EJ, Norman JM, Liu TC, Heath RJ, Becker ML, Fagbami L, Horn H, Mercer J, Yilmaz OH, Jaffe JD, Shamji AF, Bhan AK, Carr SA, Daly MJ, Virgin HW, Schreiber SL, Stappenbeck TS,

Xavier RJ (2014) Atg16L1 T300A variant decreases selective autophagy resulting in altered cytokine signaling and decreased antibacterial defense. Proc Natl Acad Sci U S A 111:7741–7746. doi:10.1073/pnas.1407001111
91. Lazebnik Y (2010) What are the hallmarks of cancer? Nat Rev Cancer 10:232–233
92. Lee HK, Lund JM, Ramanathan B, Mizushima N, Iwasaki A (2007) Autophagy-dependent viral recognition by plasmacytoid dendritic cells. Science 315:1398–1401
93. Lee J, Mo JH, Katakura K, Alkalay I, Rucker AN, Liu YT, Lee HK, Shen C, Cojocaru G, Shenouda S, Kagnoff M, Eckmann L, Ben-Neriah Y, Raz E (2006) Maintenance of colonic homeostasis by distinctive apical TLR9 signalling in intestinal epithelial cells. Nat Cell Biol 8:1327–1336. doi:10.1038/ncb1500
94. Lee MS, Kim YJ (2007) Signaling pathways downstream of pattern-recognition receptors and their cross talk. Annu Rev Biochem 76:447–480
95. Levine B (2007) Cell biology: autophagy and cancer. Nature 446:745–747
96. Levine B, Kroemer G (2008) Autophagy in the pathogenesis of disease. Cell 132:27–42. doi:10.1016/j.cell.2007.12.018
97. Levine B, Mizushima N, Virgin HW (2011) Autophagy in immunity and inflammation. Nature 469:323–335. doi:10.1038/nature09782
98. Li Y, Kundu P, Seow SW et al (2012) Gut microbiota accelerate tumor growth via c-jun and STAT3 phosphorylation in APCMin/+ mice. Carcinogenesis 33:1231–1238
99. Lin Y, Lee H, Berg AH, Lisanti MP, Shapiro L, Scherer PE (2000) The lipopolysaccharide-activated toll-like receptor (TLR)-4 induces synthesis of the closely related receptor TLR-2 in adipocytes. J Biol Chem 275:24255–24263
100. Loftus EV Jr (2004) Clinical epidemiology of inflammatory bowel disease: Incidence, prevalence, and environmental influences. Gastroenterology 126:1504–1517
101. Lorin S, Hamaï A, Mehrpour M, Codogno P (2013) Autophagy regulation and its role in cancer. Semin Cancer Biol 23:361–379
102. Lowe EL, Crother TR, Rabizadeh S et al (2010) Toll-like receptor 2 signaling protects mice from tumor development in a mouse model of colitis-induced cancer. PLoS One 5, e13027. doi:10.1371/journal.pone.0013027
103. Lu C, Chen J, Xu HG, Zhou X, He Q, Li YL, Jiang G, Shan Y, Xue B, Zhao RX, Wang Y, Werle KD, Cui R, Liang J, Xu ZX (2014) MIR106B and MIR93 prevent removal of bacteria from epithelial cells by disrupting ATG16L1-mediated autophagy. Gastroenterology 146:188–199. doi:10.1053/j.gastro.2013.09.006
104. Lu Q, Ding H, Li W (2013) Role of Toll-like receptors in microbiota-associated gastrointestinal cancer metastasis. J Cancer Res Ther 9(Suppl):S142–S149. doi:10.4103/0973-1482.122509
105. Lu XC, Tao Y, Wu C, Zhao PL, Li K, Zheng JY, Li LX (2013) Association between variants of the autophagy related gene--IRGM and susceptibility to Crohn's disease and ulcerative colitis: a meta-analysis. PLoS One 8(11), e80602. doi:10.1371/journal.pone.0080602
106. Lum JJ, Bauer DE, Kong M, Harris MH, Li C, Lindsten T, Thompson CB (2005) Growth factor regulation of autophagy and cell survival in the absence of apoptosis. Cell 120:237–248
107. Maiuri MC, Zalckvar E, Kimchi A, Kroemer G (2007) Self-eating and self-killing: crosstalk between autophagy and apoptosis. Nat Rev Mol Cell Biol 8:741–752
108. Mathew R, Karantza-Wadsworth V, White E (2007) Role of autophagy in cancer. Nat Rev Cancer 7:961–967
109. Mathew R, Kongara S, Beaudoin B, Karp CM, Bray K, Degenhardt K, Chen G, Jin S, White E (2007) Autophagy suppresses tumor progression by limiting chromosomal instability. Genes Dev 21:1367–1381
110. McCarroll SA, Huett A, Kuballa P, Chilewski SD, Landry A, Goyette P, Zody MC, Hall JL, Brant SR, Cho JH, Duerr RH, Silverberg MS, Taylor KD, Rioux JD, Altshuler D, Daly MJ, Xavier RJ (2008) Deletion polymorphism upstream of IRGM associated with altered IRGM expression and Crohn's disease. Nat Genet 40:1107–1112. doi:10.1038/ng.215

111. Medzhitov R, Preston-Hurlburt P, Kopp E et al (1998) MyD88 is an adaptor protein in the hToll/IL-1 receptor family signaling pathways. Mol Cell 2:253–258
112. Mehrpour M, Esclatine A, Beau I, Codogno P (2010) Autophagy in health and disease. 1. Regulation and significance of autophagy: an overview. Am J Physiol Cell Physiol 298:C776–C785. doi:10.1152/ajpcell.00507.2009
113. Mehta P, Henault J, Kolbeck R, Sanjuan MA (2014) Noncanonical autophagy: one small step for LC3, one giant leap for immunity. Curr Opin Immunol 26C:69–75
114. Melmed G, Thomas LS, Lee N, Tesfay SY, Lukasek K, Michelsen KS, Zhou Y, Hu B, Arditi M, Abreu MT (2003) Human intestinal epithelial cells are broadly unresponsive to Toll-like receptor 2-dependent bacterial ligands: implications for host-microbial interactions in the gut. J Immunol 170:1406–1415
115. Michaud M, Martins I, Sukkurwala AQ et al (2011) Autophagy-dependent anticancer immune responses induced by chemotherapeutic agents in mice. Science 334:1573–1577
116. Mizushima N, Levine B, Cuervo AM, Klionsky DJ (2008) Autophagy fights disease through cellular self-digestion. Nature 451:1069–1075. doi:10.1038/nature06639
117. Moscat J, Diaz-Meco MT, Wooten MW (2007) Signal integration and diversification through the p62 scaffold protein. Trends Biochem Sci 32:95–100
118. Murthy A, Li Y, Peng I, Reichelt M, Katakam AK, Noubade R, Roose-Girma M, DeVoss J, Diehl L, Graham RR, van Lookeren Campagne M (2014) A Crohn's disease variant in Atg16l1 enhances its degradation by caspase 3. Nature 506:456–462. doi:10.1038/nature13044
119. Nguyen HT, Dalmasso G, Müller S, Carrière J, Seibold F, Darfeuille-Michaud A (2014) Crohn's disease-associated adherent invasive Escherichia coli modulate levels of microRNAs in intestinal epithelial cells to reduce autophagy. Gastroenterology 146:508–519. doi:10.1053/j.gastro.2013.10.021
120. Nierkens S, den Brok MH, Garcia Z et al (2011) Immune adjuvant efficacy of CpG oligonucleotide in cancer treatment is founded specifically upon TLR9 function in plasmacytoid dendritic cells. Cancer Res 71:6428–6437
121. Noguchi E, Homma Y, Kang X, Netea MG, Ma X (2009) A Crohn's disease-associated NOD2 mutation suppresses transcription of human IL10 by inhibiting activity of the nuclear ribonucleoprotein hnRNP-A1. Nat Immunol 10:471–479. doi:10.1038/ni.1722
122. O'Neill LA (2006) How Toll-like receptors signal: what we know and what we don't know. Curr Opin Immunol 18:3–9
123. Ogata M, Hino S, Saito A, Morikawa K, Kondo S, Kanemoto S, Murakami T, Taniguchi M, Tanii I, Yoshinaga K, Shiosaka S, Hammarback JA, Urano F, Imaizumi K (2006) Autophagy is activated for cell survival after endoplasmic reticulum stress. Mol Cell Biol 26:9220–9231
124. Ogura Y, Bonen DK, Inohara N, Nicolae DL, Chen FF, Ramos R, Britton H, Moran T, Karaliuskas R, Duerr RH, Achkar JP, Brant SR, Bayless TM, Kirschner BS, Hanauer SB, Nuñez G, Cho JH (2001) A frameshift mutation in NOD2 associated with susceptibility to Crohn's disease. Nature 411:603–606
125. Ono M (2008) Molecular links between tumor angiogenesis and inflammation: inflammatory stimuli of macrophages and cancer cells as targets for therapeutic strategy. Cancer Sci 99:1501–1506. doi:10.1111/j.1349-7006.2008.00853.x
126. Otte JM, Cario E, Podolsky DK (2004) Mechanisms of cross hyporesponsiveness to Toll-like receptor bacterial ligands in intestinal epithelial cells. Gastroenterology 126:1054–1070
127. Palucka K, Banchereau J (2012) Cancer immunotherapy via dendritic cells. Nat Rev Cancer 12:265–277
128. Pandey S, Agrawal DK (2006) Immunobiology of Toll-like receptors: emerging trends. Immunol Cell Biol 84:333–341
129. Park JH, Wacholder S, Gail MH, Peters U, Jacobs KB, Chanock SJ, Chatterjee N (2010) Estimation of effect size distribution from genome-wide association studies and implications for future discoveries. Nat Genet 42:570–575. doi:10.1038/ng.610

130. Park SW, Zhou Y, Lee J, Lu A, Sun C, Chung J, Ueki K, Ozcan U (2010) The regulatory subunits of PI3K, p85alpha and p85beta, interact with XBP-1 and increase its nuclear translocation. Nat Med 16:429–437. doi:10.1038/nm.2099
131. Parkes M (2012) Evidence from genetics for a role of autophagy and innate immunity in IBD pathogenesis. Dig Dis 30:330–333. doi:10.1159/000338119
132. Parkes M, Barrett JC, Prescott NJ, Tremelling M, Anderson CA, Fisher SA, Roberts RG, Nimmo ER, Cummings FR, Soars D, Drummond H, Lees CW, Khawaja SA, Bagnall R, Burke DA, Todhunter CE, Ahmad T, Onnie CM, McArdle W, Strachan D, Bethel G, Bryan C, Lewis CM, Deloukas P, Forbes A, Sanderson J, Jewell DP, Satsangi J, Mansfield JC, Wellcome Trust Case Control Consortium, Cardon L, Mathew CG (2007) Sequence variants in the autophagy gene IRGM and multiple other replicating loci contribute to Crohn's disease susceptibility. Nat Genet 39:830–832
133. Pasare C, Medzhitov R (2003) Toll pathway-dependent blockade of CD4+CD25+ T cell-mediated suppression by dendritic cells. Science 299:1033–1036
134. Peng G, Guo Z, Kiniwa Y et al (2005) Toll-like receptor 8-mediated reversal of CD4+ regulatory T cell function. Science 309:1380–1384
135. Pikarsky E, Porat RM, Stein I et al (2004) NF-kappaB functions as a tumour promoter in inflammation-associated cancer. Nature 431:461–466
136. Podolsky DK (2002) Inflammatory bowel disease. N Engl J Med 347:417–429
137. Popivanova BK, Kitamura K, Wu Y et al (2008) Blocking TNF-alpha in mice reduces colorectal carcinogenesis associated with chronic colitis. J Clin Invest 118:560–570
138. Pradere JP, Dapito DH, Schwabe RF (2013) The Yin and Yang of Toll-like receptors in cancer. Oncogene 33:3485–3495. doi:10.1038/onc.2013.302
139. Pull SL, Doherty JM, Mills JC, Gordon JI, Stappenbeck TS (2005) Activated macrophages are an adaptive element of the colonic epithelial progenitor niche necessary for regenerative responses to injury. Proc Natl Acad Sci U S A 102:99–104
140. Rabinowitz JD, White E (2010) Autophagy and metabolism. Science 330:1344–1348. doi:10.1126/science.1193497
141. Rakoff-Nahoum S, Medzhitov R (2007) Regulation of spontaneous intestinal tumorigenesis through the adaptor protein MyD88. Science 317:124–127
142. Rehli M, Poltorak A, Schwarzfischer L, Krause SW, Andreesen R, Beutler B (2000) PU.1 and interferon consensus sequence-binding protein regulate the myeloid expression of the human Toll-like receptor 4 gene. J Biol Chem 275:9773–9781
143. Rhee SH, Im E, Pothoulakis C (2008) Toll-like receptor 5 engagement modulates tumor development and growth in a mouse xenograft model of human colon cancer. Gastroenterology 135:518–528
144. Rioux JD, Xavier RJ, Taylor KD, Silverberg MS, Goyette P, Huett A, Green T, Kuballa P, Barmada MM, Datta LW, Shugart YY, Griffiths AM, Targan SR, Ippoliti AF, Bernard EJ, Mei L, Nicolae DL, Regueiro M, Schumm LP, Steinhart AH, Rotter JI, Duerr RH, Cho JH, Daly MJ, Brant SR (2007) Genome-wide association study identifies new susceptibility loci for Crohn disease and implicates autophagy in disease pathogenesis. Nat Genet 39: 596–604
145. Rosenstiel P, Jacobs G, Till A, Schreiber S (2008) NOD-like receptors: ancient sentinels of the innate immune system. Cell Mol Life Sci 65:1361–1377. doi:10.1007/s00018-008-7502-y
146. Rosenstiel P, Sina C, Franke A, Schreiber S (2009) Towards a molecular risk map – recent advances on the etiology of inflammatory bowel disease. Semin Immunol 21:334–345. doi:10.1016/j.smim.2009.10.001
147. Rubin DC, Shaker A, Levin MS (2012) Chronic intestinal inflammation: inflammatory bowel disease and colitis-associated colon cancer. Front Immunol 3:107. doi:10.3389/fimmu.2012.00107
148. Saitoh T, Fujita N, Jang MH, Uematsu S, Yang BG, Satoh T, Omori H, Noda T, Yamamoto N, Komatsu M, Tanaka K, Kawai T, Tsujimura T, Takeuchi O, Yoshimori T, Akira S (2008)

Loss of the autophagy protein Atg16L1 enhances endotoxin-induced IL-1beta production. Nature 456:264–268. doi:10.1038/nature07383
149. Salcedo R, Worschech A, Cardone M et al (2010) MyD88-mediated signaling prevents development of adenocarcinomas of the colon: role of interleukin 18. J Exp Med 207:1625–1636
150. Sanjuan MA, Dillon CP, Tait SW et al (2007) Toll-like receptor signalling in macrophages links the autophagy pathway to phagocytosis. Nature 450:1253–1257
151. Sapra P, Allen TM (2002) Internalizing antibodies are necessary for improved therapeutic efficacy of antibody-targeted liposomal drugs. Cancer Res 62:7190–7194
152. Schmid D, Munz C (2007) Innate and adaptive immunity through autophagy. Immunity 27:11–21
153. Schmid D, Pypaert M, Münz C (2007) Antigen-loading compartments for major histocompatibility complex class II molecules continuously receive input from autophagosomes. Immunity 26:79–92
154. Schneeman TA, Bruno ME, Schjerven H, Johansen FE, Chady L, Kaetzel CS (2005) Regulation of the polymeric Ig receptor by signaling through TLRs 3 and 4: linking innate and adaptive immune responses. J Immunol 175:376–384
155. Shang L, Fukata M, Thirunarayanan N, Martin AP, Arnaboldi P, Maussang D, Berin C, Unkeless JC, Mayer L, Abreu MT, Lira SA (2008) Toll-like receptor signaling in small intestinal epithelium promotes B-cell recruitment and IgA production in lamina propria. Gastroenterology 135:529–538. doi:10.1053/j.gastro.2008.04.020
156. Shen HM, Codogno P (2012) Autophagy is a survival force via suppression of necrotic cell death. Exp Cell Res 318:1304–1308
157. Shi CS, Kehrl JH (2008) MyD88 and Trif target Beclin 1 to trigger autophagy in macrophages. J Biol Chem 283:33175–33182
158. Shi CS, Kehrl JH (2010) TRAF6 and A20 regulate lysine 63-linked ubiquitination of Beclin-1 to control TLR4-induced autophagy. Sci Signal 3:ra42. doi:10.1126/scisignal.2000751
159. Shin DM, Yuk JM, Lee HM et al (2010) Mycobacterial lipoprotein activates autophagy via TLR2/1/CD14 and a functional vitamin D receptor signalling. Cell Microbiol 12:1648–1665
160. Sipos F, Műzes G, Fűri I et al (2014) Intravenous administration of a single-dose free-circulating DNA of colitic origin improves severe murine DSS-colitis. Pathol Oncol Res 20:867–877. doi:10.1007/s12253-014-9766-x
161. Smythies LE, Sellers M, Clements RH, Mosteller-Barnum M, Meng G, Benjamin WH, Orenstein JM, Smith PD (2005) Human intestinal macrophages display profound inflammatory anergy despite avid phagocytic and bacteriocidal activity. J Clin Invest 115:66–75
162. Sodemann U, Bistrup C, Marckmann P (2011) Cancer rates after kidney transplantation. Dan Med Bull 58(12):A4342, PMID: 22142571
163. Spalinger MR, Lang S, Vavricka SR, Fried M, Rogler G, Scharl M (2013) Protein tyrosine phosphatase non-receptor type 22 modulates NOD2-induced cytokine release and autophagy. PLoS One 8, e72384. doi:10.1371/journal.pone.0072384
164. Stappenbeck TS, Rioux JD, Mizoguchi A, Saitoh T, Huett A, Darfeuille-Michaud A, Wileman T, Mizushima N, Carding S, Akira S, Parkes M, Xavier RJ (2011) Crohn disease: a current perspective on genetics, autophagy and immunity. Autophagy 7:355–374. doi:10.4161/auto.7.2.13074
165. Strisciuglio C, Duijvestein M, Verhaar AP, Vos AC, van den Brink GR, Hommes DW, Wildenberg ME (2013) Impaired autophagy leads to abnormal dendritic cell-epithelial cell interactions. J Crohns Colitis 7:534–541. doi:10.1016/j.crohns.2012.08.009
166. Swidsinski A, Ladhoff A, Pernthaler A, Swidsinski S, Loening-Baucke V, Ortner M, Weber J, Hoffmann U, Schreiber S, Dietel M, Lochs H (2002) Mucosal flora in inflammatory bowel disease. Gastroenterology 122:44–54
167. Takeda K, Akira S (2004) TLR signaling pathways. Semin Immunol 16:3–9

168. Taylor GA (2007) IRG proteins: key mediators of interferon-regulated host resistance to intracellular pathogens. Cell Microbiol 9:1099–1107
169. Till A, Lipinski S, Ellinghaus D, Mayr G, Subramani S, Rosenstiel P, Franke A (2013) Autophagy receptor CALCOCO2/NDP52 takes center stage in Crohn disease. Autophagy 9:1256–1257. doi:10.4161/auto.25483
170. Travassos LH, Carneiro LA, Ramjeet M, Hussey S, Kim YG, Magalhães JG, Yuan L, Soares F, Chea E, Le Bourhis L, Boneca IG, Allaoui A, Jones NL, Nuñez G, Girardin SE, Philpott DJ (2010) Nod1 and Nod2 direct autophagy by recruiting ATG16L1 to the plasma membrane at the site of bacterial entry. Nat Immunol 11:55–62. doi:10.1038/ni.1823
171. Trinchieri G (2012) Cancer and inflammation: an old intuition with rapidly evolving new concepts. Annu Rev Immunol 30:677–706. doi:10.1146/annurev-immunol-020711-075008
172. Trocoli A, Djavaheri-Mergny M (2011) The complex interplay between autophagy and NF-κB signaling pathways in cancer cells. Am J Cancer Res 1:629–649
173. Uehara A, Fujimoto Y, Fukase K, Takada H (2007) Various human epithelial cells express functional Toll-like receptors, NOD1 and NOD2 to produce anti-microbial peptides, but not proinflammatory cytokines. Mol Immunol 44:3100–3111
174. Ullman TA, Itzkowitz SH (2011) Intestinal inflammation and cancer. Gastroenterology 140:1807–1816. doi:10.1053/j.gastro.2011.01.057
175. Van Limbergen J, Wilson DC, Satsangi J (2009) The genetics of Crohn's disease. Annu Rev Genomics Hum Genet 10:89–116. doi:10.1146/annurev-genom-082908-150013
176. Vendramini-Costa DB, Carvalho JE (2012) Molecular link mechanisms between inflammation and cancer. Curr Pharm Des 18:3831–3852
177. Verway M, Behr MA, White JH (2010) Vitamin D, NOD2, autophagy and Crohn's disease. Expert Rev Clin Immunol 6:505–508. doi:10.1586/eci.10.31
178. Vora P, Youdim A, Thomas LS, Fukata M, Tesfay SY, Lukasek K, Michelsen KS, Wada A, Hirayama T, Arditi M, Abreu MT (2004) Beta-defensin-2 expression is regulated by TLR signaling in intestinal epithelial cells. J Immunol 173:5398–5405
179. Walter P, Ron D (2011) The unfolded protein response: from stress pathway to homeostatic regulation. Science 334:1081–1086. doi:10.1126/science.1209038
180. Wang CY, Mayo MW, Korneluk RG, Goeddel DV, Baldwin AS Jr (1998) NF-kappaB anti-apoptosis: induction of TRAF1 and TRAF2 and c-IAP1 and c-IAP2 to suppress caspase-8 activation. Science 281:1680–1683
181. Watanabe T, Kitani A, Murray PJ, Strober W (2004) NOD2 is a negative regulator of Toll-like receptor 2-mediated T helper type 1 responses. Nat Immunol 5:800–808
182. Wehkamp J, Salzman NH, Porter E, Nuding S, Weichenthal M, Petras RE, Shen B, Schaeffeler E, Schwab M, Linzmeier R, Feathers RW, Chu H, Lima H Jr, Fellermann K, Ganz T, Stange EF, Bevins CL (2005) Reduced Paneth cell alpha-defensins in ileal Crohn's disease. Proc Natl Acad Sci U S A 102:18129–18134
183. Wiemann B, Starnes CO (1994) Coley's toxins, tumor necrosis factor and cancer research: a historical perspective. Pharmacol Ther 64:529–564. doi:10.1016/0163-7258(94)90023-X
184. Wolfkamp SC, Verseyden C, Vogels EW, Meisner S, Boonstra K, Peters CP, Stokkers PC, te Velde AA (2014) ATG16L1 and NOD2 polymorphisms enhance phagocytosis in monocytes of Crohn's disease patients. World J Gastroenterol 20:2664–2672. doi:10.3748/wjg.v20.i10.2664
185. Xavier RJ, Podolsky DK (2007) Unravelling the pathogenesis of inflammatory bowel disease. Nature 448:427–434
186. Yang L, Li P, Fu S, Calay ES, Hotamisligil GS (2010) Defective hepatic autophagy in obesity promotes ER stress and causes insulin resistance. Cell Metab 11:467–478. doi:10.1016/j.cmet.2010.04.005
187. Young AR, Narita M, Ferreira M et al (2009) Autophagy mediates the mitotic senescence transition. Genes Dev 23:798–803

188. Yu L, Chen S (2008) Toll-like receptors expressed in tumor cells: targets for therapy. Cancer Immunol Immunother 57:1271–1278
189. Yuk JM, Shin DM, Lee HM, Yang CS, Jin HS, Kim KK, Lee ZW, Lee SH, Kim JM, Jo EK (2009) Vitamin D3 induces autophagy in human monocytes/macrophages via cathelicidin. Cell Host Microbe 6:231–243. doi:10.1016/j.chom.2009.08.004
190. Yuk JM, Shin DM, Song KS, Lim K, Kim KH, Lee SH, Kim JM, Lee JS, Paik TH, Kim JS, Jo EK (2010) Bacillus calmette-guerin cell wall cytoskeleton enhances colon cancer radiosensitivity through autophagy. Autophagy 6:46–60
191. Zhang X, Dong Y, Zeng X, Liang X, Li X, Tao W, Chen H, Jiang Y, Mei L, Feng SS (2014) The effect of autophagy inhibitors on drug delivery using biodegradable polymer nanoparticles in cancer treatment. Biomaterials 35:1932–1943
192. Zhou S, Zhao L, Kuang M et al (2012) Autophagy in tumorigenesis and cancer therapy: Dr. Jekyll or Mr. Hyde? Cancer Lett 323:115–127

Target Autophagy as a Novel Therapeutic Strategy in Autoimmune Diseases

Maud Wilhelm and Sylviane Muller

Abstract Autophagy is a normal physiological process that plays a pivotal role for cell survival, differentiation, development, and homeostasis. Selective or not, canonical or non-canonical, autophagy processes are considerably more complex than originally thought. Depending on favourable or unfavourable cell environment conditions, the autophagy machinery will promote both cell survival and cell death, thus maintaining a decisive balance between manufacture of cellular components and breakdown of damaged or superfluous organelles and other cellular constituents, for example. Autophagy displays complex, still-debated, interwoven links with several other degradative pathways, such as apoptosis and proteasome-mediated systems. Among its many cellular regulatory functions that have been experimentally proven or that are anticipated, autophagy decisively controls immunity and inflammation, and any impaired autophagy signaling can potentially lead to autoimmune-related diseases. Here we review recent progresses that have been made in deciphering existing links between autophagy and autoimmunity. We further discuss how targeting certain hot spots of autophagy processes with appropriate tools might influence the course of autoimmune diseases by controlling both innate and adaptive immune responses, which are improperly oriented in these settings.

M. Wilhelm
CNRS, Immunopathology and Therapeutic Chemistry/Laboratory of Excellence MEDALIS, Institut de Biologie Moléculaire et Cellulaire, University of Strasbourg, Strasbourg, France
e-mail: M.Wilhelm@ibmc-cnrs.unistra.fr

S. Muller (✉)
CNRS, Immunopathology and Therapeutic Chemistry/Laboratory of Excellence MEDALIS, Institut de Biologie Moléculaire et Cellulaire, University of Strasbourg, Strasbourg, France

UPR3572 CNRS, Immunopathologie et Chimie Thérapeutique, Institut de Biologie Moléculaire et Cellulaire, 15 rue René Descartes, Strasbourg 67000, France
e-mail: S.Muller@ibmc-cnrs.unistra.fr

M.C. Maiuri, D. De Stefano (eds.), *Autophagy Networks in Inflammation*, Progress in Inflammation Research, DOI 10.1007/978-3-319-30079-5_13

Abbreviations

3-MA	3-methyladenine
AEP	Asparagine endopeptidase
APC	Antigen-presenting cell
ATG	Autophagy-related protein
BECLIN1/beclin-1	BCL-2 interacting myosin/moesin-like coiled-coil protein 1
CD	Crohn's disease
CLIP	Class-II associated invariant chain peptide
CMA	Chaperone-mediated autophagy
CQ/HCQ	Chloroquine/hydroxychloroquine
DC	Dendritic cell
DN	Double negative
DRAM1	DNA damage-regulated autophagy modulator1
ds	Double-stranded
DSG	Deoxyspergualin
HSP	Heat shock protein
HSPA8/HSC70	Heat shock cognate protein of 70 KDa
IFN	Interferon
IL	Interleukin
IRGM	Immunity-related GTPase M
ITP	Immune thrombocytopenia
KO	Knockout
LAMP-2A	Lysosome-associated membrane protein-2A
LNC	Lymph node cells
LPS	Lipopolysaccharide
MAP1LC3/LC3	Microtubule-associated protein light chain 3
MHCI/II	Major histocompatibility complex class I or MHC class II
MIIC	Major histocompatibility complex class II compartment
MOA	Mode of action
MRL	Murphy Roths large
MS	Multiple sclerosis
mTOR	Mammalian target of rapamycin
NZB/W	(NZBxNZW) F1
PBMCs	Peripheral blood mononuclear cells
PC	Plasma cell
RA	Rheumatoid arthritis
ROS	Reactive oxygen species
SLE	Systemic lupus erythematosus
SQSTM1/p62	Sequestosome 1
TCR	T cell receptor
Th1/Th2	T helper type 1 and type 2
TLR	Toll-like receptor
TNF	Tumor necrosis factor
UPS	Ubiquitin-proteasome system.

1 Introduction

Autoimmune diseases are not considered as orphan diseases. In general they are even not regarded as rare since as a whole they affect millions people worldwide. As a result of genetic influence, which is mostly polygenic, or environmental and metabolic factors, there is some disequilibrium regarding their incidence or severity in some parts of the world or in a particular group of people. It remains that in the collective perception, they are viewed as a common group of diseases. It is true that it has been estimated that autoimmune diseases are among the top ten leading causes of death among women in all age groups up to 65 years. In fact, under the term autoimmune diseases, there are more than eighty illnesses caused by autoimmunity. Some of them are rare, either as an entity (e.g. Crohn's disease/CD; primary biliary cirrhosis, myasthenia gravis, immune thrombocytopenic purpura) or by the form they display in affected patients (neuropsychiatric systemic lupus erythematosus, ocular myasthenia gravis, psoriatic arthritis). Also some individuals may have more than one autoimmune disorder at the same time, which complicates the task of follow-up and treatment, and makes each case unique. There is no known prevention for most autoimmune disorders, and in general there is no specific treatment. Despite the complexity and uniqueness of cellular and molecular pathways that are altered in different autoimmune conditions, investigating these mechanisms is very rewarding for scientists. Such studies can effectively reveal new elements and interacting partners of the immune system as well as unexpected abnormalities linked to autoimmune features. These findings can then inspire researchers to design novel strategies of possible intervention developed to mislead and correct the defective immune system.

2 Autoimmune Diseases

Autoimmunity does not systematically leads to autoimmune diseases. In autoimmunity, the patient's immune system is activated against the body's own components and only in certain conditions involving genetic, environmental, and hormonal elements, will the individual develop illness, which is often chronic, debilitating, and life-threatening. A large number of autoimmune diseases are recognized. They are said "organ-specific" when they are restricted to certain organs such as thyroid (e.g. Graves' disease, autoimmune thyroiditis, Hashimoto's disease), pancreas (e.g. type 1 diabetes in which insulin-producing beta cells are destroyed) and muscles (myasthenia gravis) or involve a particular tissue in different places (e.g. Goodpasture's disease, which affects the basement membrane in the lung and kidney). In contrast, they are classified as "systemic" when they implicate a variety of organs and tissues in the whole body. The most emblematic representative of the large family of systemic autoimmune diseases is systemic lupus erythematosus (SLE) in which heart, joints, skin, lungs, blood vessels, liver, kidneys, and nervous system can be affected. In fact, between these two commonly described families, there is no sharp delineation. Thus scleroderma, also known as systemic sclerosis, which is a chronic systemic autoimmune disease characterized by hardening of the skin, also affects blood vessels, muscles, and internal organs in severe forms. A continuing debate and a matter of controversy remain about when a disease should be considered autoimmune. Within the usually reported list of somewhat 80 autoimmune diseases that are currently described [36], very few in fact do respond to the strict Witebsky's postulates formulated in 1957 and modified 35 years later [126]. The passive transfer of T lymphocytes, which should lead to disease development in the recipient, is generally hardly observed.

According to the American Autoimmune Related Diseases Association, autoimmune diseases affect up to 50 million Americans. The overall cumulative prevalence of all autoimmune diseases is around 5 %, with about 3 % for males and 7 % for females [36]. There is a sexual dimorphism among autoimmune diseases with a well-established disequilibrium toward the female population. This female bias occurs in 59 % of autoimmune diseases, probably in relation with hormonal influence and X-chromosome encoded genes. In general the onset for autoimmune diseases occurs in young people (20–29 year age-group).

Deciphering the molecular and cellular mechanisms leading to immune tolerance breaking and evolution toward autoimmune disease remains a vast area of investigations in the scientific and clinical community. Nowadays, no universal signature could be identified, and clues are largely lacking regarding the reasons of their tropism as well as on the elements triggering their initiation and maintenance. Relatively few is also known regarding the events governing the successive periods of flares and remission occurring in certain autoimmune diseases such as SLE. The multifactorial and polymorphic nature of most autoimmune diseases dramatically complicates their diagnosis and the treatment that can be applied to mitigate the symptoms. Except in very rare cases, the treatments are largely palliative and do not target the cause of illness. Although immense progresses have been made over the last decades leading to patients' survival rates that have considerably augmented, innovative therapeutic solutions are still awaiting that would combine efficacy, selectivity -and thus less secondary effects- and reliability. Without adapted treatment, the quality-of-life can be relatively poor in autoimmune patients and decreases as the disease evolves (fatigue, pain, fever associated to specific symptoms). Unfortunately, the medications required to minimize symptoms and slow-down inflammatory syndrome (i.e. corticosteroids, immunosuppressive drugs and tumor necrosis factor (TNF)-α blockers used for long-term periods) induce an alteration of the whole immune system leading to intestinal bleeding, kidney failure, increased blood pressure, insomnia, depression, psychosis, osteoporosis, muscle loss, and diabetes, not to mention overwhelming repetitive infection episodes and cancer development. In certain autoimmune diseases such as those affecting the central nervous system, or in anti-phospholipid syndrome that can be associated to SLE, the therapeutic solutions are limited, not specific, and unfortunately sometimes inefficient [9, 35, 40, 44]. Intense research is currently ongoing to develop novel immunomodulatory strategies based on molecular targets that are engaged in deregulated autoimmune processes and can be specifically re-orientated. In this context, a better knowledge of cellular and molecular mechanisms that underline autoimmune responses and most particularly the homeostasis and regulation of autoimmune cells is central. Although the picture is immensely complex, studying the autophagic process, which is involved in the establishment and maintenance of immune tolerance and the proper effectiveness of the immune system, has particular importance in autoimmunity and might reveal decisive hot spots for therapeutic intervention.

3 Autophagy and Its Implication in Autoimmune Diseases

Autophagy is a lysosome-based physiological process, which in basal conditions occurs at low levels to continuously degrade unwanted cytoplasmic constituents and generate substrates for energy production. During oxidative stress, hypoxia or nutritional starvation, its level raises to allow cell survival. Autophagy represents therefore a major hub involved in cellular homeostasis [4, 37, 89, 111, 124]. It also plays a pivotal role in differentiation of many lineages, including adipocytes, erythrocytes and lymphocytes, and tissue remodelling [10, 62, 90, 91, 106, 117]. Under specific environmental conditions, however, autophagy can also mediate cell death and it is mechanistically important to distinguish autophagic cell death, which refers to cell death "by" autophagy from cell death "with" autophagy [58, 83, 128, 134]. Thus, recent studies suggest that autophagy and apoptosis processes are closely nested and share cross-talk between signal transduction elements. It has been shown in particular that certain autophagy-related (ATG) proteins play dual roles in autophagy and apoptosis regulation. This is the case of ATG5 and its binding partner ATG12, BCL-2 interacting myosin/moesin-like coiled-coil protein 1 (BECLIN1/beclin-1), the mammalian ortholog of yeast *Atg6*/vacuolar protein sorting (Vps)-30 that acts during the formation of autophagosomes by interacting with the class III PI3K pathway, and microtubule-associated-protein light chain 3 (MAP1LC3/LC3) a mammalian ortholog of yeast *Atg8*, for example [48, 56, 71, 84]. Other forms of cell death are also interconnected with autophagy, such as necrosis, necroptosis (regulated Fas-dependent, caspase-independent non-apoptotic cell death), and pyroptosis (caspase-1-dependent cell death) [128].

Three main types of autophagy have been identified and can be distinguished by both their physiological functions and the mechanisms they use to deliver cytoplasmic cargo to lysosomes (Fig. 1a). They are macroautophagy, microautophagy and chaperone-mediated autophagy or CMA [17, 27, 50, 111]. In fact, many more forms of autophagy have been described. Mention can be made, for example, of aggrephagy (for aggregated proteins), mitophagy (for mitochondria), ribophagy (for ribosomes), pexophagy (for peroxisomes), reticulophagy (for the endoplasmic reticulum, ER), and xenophagy (for pathogens). Thus, we now realize that while originally viewed as a nonselective (random) cytoplasmic degradation system, autophagy actually participates in a highly selective and tightly regulated process of substrate delivery.

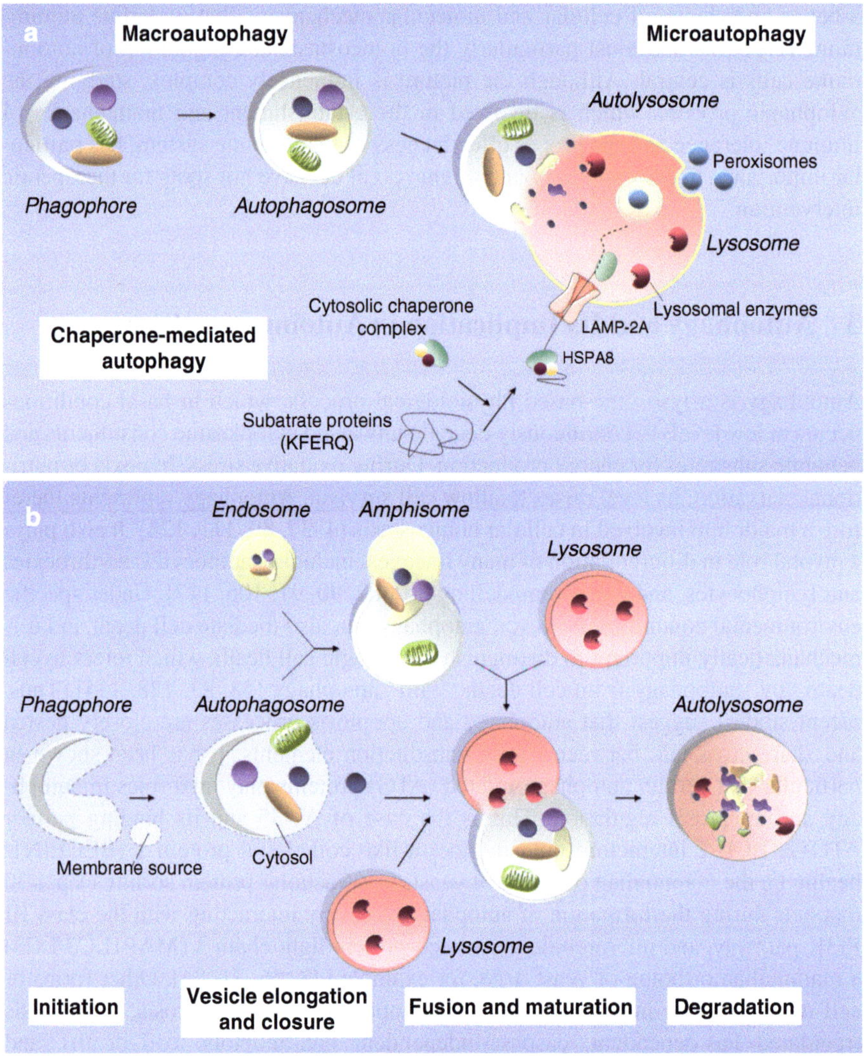

Fig. 1 Schematic depiction of autophagic pathways. (**a**) The three main autophagy axes, macroautophagy, microautophagy and CMA. The process of macroautophagy is initiated with the formation of the so-called isolation membrane. This structure is elongated to engulf cytosolic materials, forming a characteristic double-membrane structure termed autophagosome. The latter next fuses with a lysosome to become an autolysosome, after which the engulfed material is degraded. The molecular pathways regulating autophagy are highly conserved from yeast to higher eukaryotic cells. In CMA, proteins carrying the pentapeptide KFERQ-like signal sequence are recognized by the HSPA8 chaperone, which then associates to LAMP-2A, triggering its oligomerization. This event permits to the targeted protein to be translocated into the lysosome lumen through a process that requires HSPA8. Microautophagy involves the direct sequestration of cellular components by the lysosome through invagination of the lysosomal membranes; (**b**) Main steps of the macroautophagic process

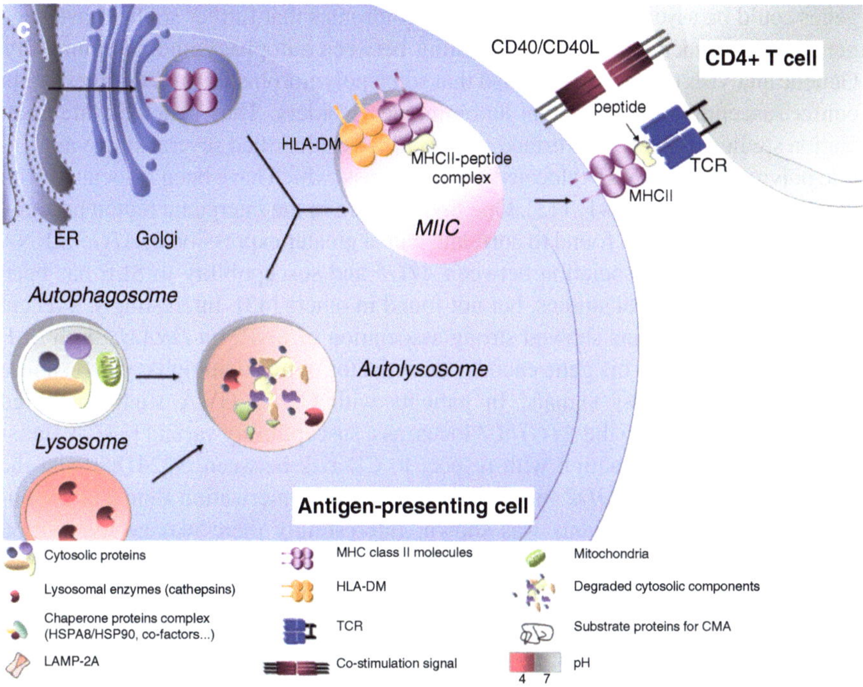

Fig. 1 (continued) (**c**) Autophagy as the major sources of peptides for presentation by MHCII molecules to T cells. *Abbreviations*: *CMA* chaperone-mediated autophagy, *ER* endoplasmic reticulum, *HLA* human leukocytes antigen, *HSPA8/HSC70* heat shock cognate protein of 70 KDa, *LAMP-2A* lysosome-associated membrane protein-2A, *MIIC* major histocompatibility complex class II compartment, *MHCII* major histocompatibility complex class II, *TCR* T cell receptor

Macroautophagy (commonly referred as "autophagy", which can in some cases create confusion in the literature) remains the major autophagic process through its ability to massively entrap macromolecules and entire organelles. The latter are captured into double-membrane autophagosomes where they are degraded. It therefore represents an alternative mechanism of proteasomal degradation, which rather treats short-lived intracellular proteins, although a cross-talk that is being increasingly understood, has been described to occur between the ubiquitin-proteasome system (UPS) and macroautophagy [19, 52, 57, 69, 124]. The fusion of autophagosomes with lysosomes leads to the formation of autolysosomes in which engulfed cellular constituents -including lipid droplets and protein aggregates- are degraded by lysosomal glycosidases, proteases, lipases and sulfatases (Fig. 1b). Concerning the CMA process, proteins containing a specific peptide motif biochemically related to KFERQ are recognized by the HSPA8/HSC70 chaperone protein prior being internalized and degraded in lysosomes (Fig. 1a). By contrast, in microautophagy, cytosolic components are directly taken up by invaginations of the lysosomal membrane (Fig. 1a).

Autophagic pathways are genetically regulated by proteins belonging to the *ATG* gene family and are well characterized in yeast and mammals [14, 55, 61, 94, 110, 135]. ATG proteins are evolutionary conserved and each of them has a specific function during autophagy. It is mainly through the discovery that certain *ATG* genes could be associated to autoimmune syndromes that further studies have been generated to understand the links existing between autophagy and autoimmunity. Genetic analyses effectively reported that some polymorphisms in *ATG* genes might confer susceptibility to different autoimmune disorders. Thus genome-wide association studies (GWAS) performed in SLE patients identified several single nucleotide polymorphisms (SNPs) located on *ATG* genes, which have been associated with the disease occurrence [41, 113]. One SNP located in the intergenic region between *ATG5* and *PRDM1* was found to correlate with a greater expression of *ATG5* mRNA [159]. The genetic association between *ATG5* and susceptibility to SLE has been confirmed in individual studies, but not found in others [43]. Interestingly, a recent meta-analysis in Asians showed strong association of SNPs on *DRAM1* with SLE susceptibility [156]. This gene encodes an activator of macroautophagy in response to p53-mediated stress signals. In patients with CD, a GWA study identified rs2241880, mapping to the *ATG16L1* locus, as a susceptibility variant [34]. A statistically significant interaction with respect to CD risk between rs2241880 and the established *CARD15/NOD2* (nucleotide-binding oligomerization domain containing 2) susceptibility variants was shown. Interestingly there was no association between rs2241880 and ulcerative colitis, another closely related inflammatory bowel disease. Recent data showed that *Atg16L1* mutant mice are resistant to intestinal disease induced by the model bacterial pathogen *Citrobacter rodentium* [82]. The hyperimmune phenotype and protective effects developed in these mice were lost in *Atg16L1/Nod2* double-mutant mice, indicating that the susceptibility from *Nod2*-deficiency is dominant over the benefit of *Atg16L1* deficiency. ATG16L1 is central in the autophagosome formation, being part of the ATG12-ATG5 complex, which is required for the recruitment of MAP1LC3 [94]. Removal of ATG16L1 abrogates the ability of cells to form autophagosomes [130]. More recently it was described that the variant protein that contains a Thr → Ala substitution at position 300 is highly sensitive to cleavage by caspase 3, which is activated during cell stress [105]. Destruction of ATG16L1^{T300A} impaired autophagy and increased release of pro-inflammatory cytokines TNF-α and IL-1β. Several SNPs have been described in association with CD, notably in the so called immunity-related GTPase family M (*IRGM*) gene [30, 77]. The results indicated that autophagy gene-*IRGM* polymorphisms confer susceptibility to CD but not ulcerative colitis, especially in Europeans. IRGM is a member of the interferon-inductible GTPase family conferring autophagic defence against intracellular pathogens like *M. Tuberculosis*. IRGM controls the latter by enhancing mycobacterial phagosome maturation [137].

Altogether these data argue for a strong impact of autophagy elements in several aspects of immunity, including protection to infectious agents and control of inflammatory and autoimmune responses, as well as in tumorigenesis and cancer. Paradoxically, it is only recently that experimental studies based on cellular and molecular investigation shed some light on the involvement of autophagy in immunity. A number of comprehensive review articles have been recently published on this topic with a particular emphasis on the role of autophagy in infection and inflammation [10, 22, 23, 32, 66, 112, 123, 124]. The present review mainly focuses on autophagy in autoimmunity, in relation with possible manipulation of immune system by small molecules and peptides in order to divert deleterious immune responses and at least partly restore impaired tolerance to self.

Innate immune responses importantly influence the adaptive immunity in the induction and regulation of autoimmune diseases. In innate immunity, autophagy works at different levels, notably by controlling activation and release of certain cytokines and chemokines [22, 23, 32, 47, 129]. Autophagy would activate the secretion of TNFα, interleukin (IL)-6, IL-8 and type I interferon (IFN) while it controls the production of IL-1α and β (the latter by regulating inflammasome activation and by targeting pro-IL-1β for degradation), IL-18 and type I IFN. In turn, some secreted cytokines influence autophagy. Thus, T helper type 1 (Th1) and pro-inflammatory cytokines such as IFN-γ (via IRGM), TNFα, IL-1α and β, IL-23, reactive oxygen species (ROS) and engagement of some TLRs (mechanisms that are still poorly understood) induce autophagy. TWEAK (the TNF-like weak inducer of apoptosis, in C2C12 myotubes), IL-2 in $CD4^+$ T cells, IL-6 in peripheral blood mononuclear cells (PBMCs) and TGF-β in hepatocarcinoma cell lines also promote autophagy. Conversely, Th2 and regulatory cytokines such as IL-4, IL-13 and IL-10, via an effect on STAT-3 or −6 pathways and the serine/threonine-protein kinase (AKT) pathway were found to activate mammalian target of rapamycin (mTOR), which inhibits the serine/threonine protein kinase ULK1 and therefore autophagosome formation [33, 47]. Via its effect on cytokine secretion, particularly in antigen-presenting cells (APCs), autophagy represents a pivotal regulator of immune responses [10, 23, 32, 66, 106, 124, 129].

Although not yet recognized to such a level of crucial importance in current text books, autophagy in fact exerts profound effects on different aspects of adaptive immunity. It is a major player in thymic selection of T cells, affecting also T cell homeostasis, repertoire and polarization, survival of B cells, immune tolerance, and antigen presentation.

The discovery that autophagy is a key regulatory element for delivering self-antigens to major histocompatibility complex II (MHCII) molecules has been a critical turning point [21, 116, 158]. At the time of this finding, it was established classically that MHCI molecules presented peptides from intracellular source proteins to T cells while MHCII molecules presented antigenic peptides from exogenous and membrane proteins. The overall picture of T cell activation by MHCII peptide was thus considerably reconsidered and new nexus between immune response and cellular stress, cell metabolism, cell nutrient and cell environment were suggested and analysed further. Incidentally, it is interesting to note that following experiments in which potent macroautophagy inhibitors acting on PI 3-kinase activity, i.e. wortmannin, LY294002 and 3-methyladenine (3-MA) were incubated with macrophage cell line BMC-2 transfected with Eα52-68-eGFP (a peptide fragment issued from transmembrane protein I-Eα) and shown to have no effect, it was concluded that macroautophagy was not a mechanism for cytoplasmic expressed proteins to gain access to the luminal peptide binding site of MHCII molecules [20]. At that time conflicting data were published, which could result from the inherent properties of the antigen that was studied, its half-life and intracellular (vesicular or not) trafficking, and the type of APCs [24, 64, 116]. More recent data have shown that in APCs that are less proteolytically active than other cells such as macrophages, cleavage by lysosomal cysteine proteases – generally known as cathepsins – of particles and proteins that finally reach autolysosomes give rise to protein fragments, which will constitute the major source of peptides for MHCII molecules (Fig. 1c). Lysosomes and autolysosomes have a pH of 4–4.5, which is optimum for cathepsins. Thus, and of importance in the context of autoimmunity, MHCII molecules can bind peptides generated from endogenous antigens that are generated by lysosomal proteolysis. Such endogeneous antigens can be from membranous, cytoplasmic (including vesicle components) or nuclear origin and can have trafficked into the endo-lysosomal network via several forms of autophagy for subsequent processing and presentation by MHCII molecules to promote $CD4^+$ T cells priming [7, 104]. Interestingly, in their pioneer work, Stevanovic, Rammensee and coll. already demonstrated that the induction of autophagy by starvation altered the balance of active proteases in lysosomes [21], which as a matter of consequence, can change the quality of peptides that are loaded onto MHCII molecules. Over the last decade, the role and regulation of specific proteases on the liberation and processing of self-antigens has been studied extensively [148, 150] and it was shown in particular that a distinct set of cathepsins is at work in different APCs, e.g. dendritic cells (DCs) and B cells [8, 81]. There are also multiple mechanisms (including gene up-regulation or down-regulation governed by the environment), that are involved for controlling proteases activity, even in individual endosomes, and strongly affect antigen presentation [21, 148]. Endo-lysosomal proteases are thus key players to generate antigens that *in fine* will be presented to T cells. Via a stepwise process involving asparagine endopeptidase (AEP) also known as legumain, cystatin C, specific cathepsins and other still unspecified proteases, endo-lysosomal proteases act for processing the invariant (Ii) chain linked to MHCII molecule into class-II associated invariant chain peptide (CLIP), thus generating peptide-receptive MHCII molecules in which the CLIP peptide is exchanged for a high affinity peptide by the enzyme HLA-DM (Fig. 1c) prior its transport to the cell surface of APCs for display to $CD4^+$ T cells [107]. Endo-lysosomal proteases, including AEP, also act to generate epitopes that will be presented by functional MHCII molecules [15, 86, 148]. In the many examples of antigens that have been examined so far, stability was found to be a determining factor that influences antigen presentation. Furthermore because the cleavage via cathepsins can liberate epitopes but also destroy some others, cathepsins regulation is even more strategic for defining the final panel of antigenic peptides that are delivered.

Finally, another important role of endo-lysosomal proteases in antigen-presentation lies to their influence on TLR-receptor signaling. Initially claimed while observing the effect of chloroquine (CQ) on TLR9 signaling [38, 85], it has been demonstrated later that endo-lysosomal proteases also activate endosomal TLRs 3, 7, and 8 [80] and that the mode of action was not the one proposed in the first studies. In fact, whether for TLR9 or for endosomal TLRs, endo-lysosomal proteases would act by converting the receptor from a non-signaling full-length form to a shorter form deleted from an N-terminal region [26, 118]. Although the precise mechanisms that are behind this effect -notably considering the specific proteases that are involved- are still a continuing matter of debates, it remains that such an effect can be strategic as TLR-signaling is central for DC maturation that dictates protease activity and consequently influences the quality of peptides that are presented onto MHCII molecules. These data highlight the importance of TLRs in autophagy processes in conjunction with both innate (see above; [153]) and adaptive immunity.

The importance of autophagy in immunity also came from experiments performed with mice or cells that have been manipulated to under-express *Atg* genes. Using this strategy, associated to our growing knowledge of genes that appear defective in some individuals, it has been possible to better approach the potential role of some ATG proteins and establish some links with human diseases [12, 45, 79]. Thus, using mice with a B-cell-specific deletion of *Atg5*, a gene implicated in the elongation of autophagosome membrane, it has been shown that in autophagy-deficient B-cell progenitors the transition from the pro-B to the pre-B cell stage in the bone marrow was defective [87]. Studies of mice in which *Atg5* was conditionally deleted in B lymphocytes revealed further that this gene is essential for plasma cells (PC) homeostasis [16]. Class-switch did occur in these mice but antibody responses were strongly decreased after specific immunisation, parasitic infection and mucosal inflammation. These data and others [119] highlight the importance of ATG5 not only in early B cell development but also in late B cell activation and PC differentiation. Conditional deletion of essential autophagy genes *Atg5* [139], *Atg7* [46, 122], *Atg3* [46] also showed that macroautophagy is critical to the survival of peripheral T cells. Some *Atg* genes are important in infection setting. Thus, using mouse embryonic fibroblasts (MEFs) lacking human *ATG16L1* or murine *Atg7*, *Atg9a*, or *Atg14* [109] showed the importance of ATG16L1, ATG7 and ATG16L1, but not of ATG9A and ATG14, in the IFN-γ-induced recruitment of the immunity-related GTPases to the intracellular pathogen T. *gondii*. A number of examples in different forms of autophagy processes, including macroautophagy, CMA, and mitophagy have been described in which autophagy genes have been deleted or over-expressed, in some cases in specific tissues. Examples are *Pink1/parkin* knockout (KO) mice, the *Atg16L1* mutant and *Atg16L1/Nod2* double-mutant mice described above, *Sqstm1/p62/A170* (encoding SQSTM1 multifunctional protein, also known as signaling adaptor/scaffold protein) mutant mice, conditional deletion models invalidating *Beclin-1* or *Vps34*, to quote just a few. Some mutations affecting binding partners of key elements of autophagy pathways were also introduced. Thus, deletion of the gene encoding lysosome-associated membrane protein-2 (LAMP-2A) in T cells was shown recently to cause deficient in vivo responses to immunization or infection with L. *monocytogenes* [147]. In these mice, CMA in T cells was found to be altered with age. It should be mentioned here that mice invalidated for HSPA8 are not viable, as are *Beclin-1* KO mice that die in utero or *Atg5* KO mice that die within 24 h after birth due at least in part to deficient amino acid production.

At this stage of our thoughts, it seems important to insist on the fact that if investigations with such mutated mice provide decisive information, it remains that in general much more additional observations are needed to establish direct links between autophagy and certain pathologies, since mutations and polymorphisms of the *ATG* (or Atg in mice) genes can have many indirect effects as described above. Consistent with these considerations, important caveats have also been warranted

Table 1 List of autoimmune diseases with autophagy failures

Autoimmune diseases	Associated genes	Cellular dysfonctions	References
CD	ATG16L1		[34]
	IRGM		[30, 77]
SLE	ATG5		[41, 159]
	DRAM1		[156]
	PRDM1		[159]
		MaA increased in T cells from MRL/lpr and NZB/W mice and from patients: autophagic vacuoles over-represented (WB, EM)[a]	[31]
		MaA deregulated in naïve CD4⁺T cells from patients : autophagosome-associated marker MAP1LC3 increased (WB)	[1]
		MaA hyper-activated in B cells from NZB/W mice and naïve B cells of patients; autophagosomes number increased (FACS, FM)	[13]
		MaA activated in macrophages from lupus-prone mice and patients : ATG5, ATG12 and BECN1 expression increased	[67]
		Increased HSPA8 expression in B and T cells of MRL/lpr mice (WB, FACS, PCR)	[114]
		Increased LAMP-2A and CTSD expression in B cells of MRL/lpr mice; lysosomes are defective in MRL/lpr mice (WB, FACS, Q-PCR, in vitro assay for CMA)	[78]
RA	ATG5		[113]
	ATG7		[70]
	BECN1		[70]
		MaA activated in osteoclasts from patients : BECN1 and ATG7 expression increased (WB)	[70]
		Autophagic process increased in synovial fibroblast : p62 and MAP1LC3 expression increased (WB, FM)	[49]
PM		MaA activated in muscle fiber : MAP1LC3, CTSD and CTSB expression increased (WB)	[108]
MS	ATG5		[2]
		MaA deregulated in T cells : ATG5 expression increased (WB, PCR)	[2]
Type 1 diabetes		MaA diminished in diabetic mouse heart : MAP1LC3 and ATG5/12 expression reduced (WB, FM)	[151, 154]

Abbreviations : *ATG* autophagy related-gene, *BECN1* beclin-1, *CD* Crohn's disease *CMA* chaperone-mediated autophagy, *CTSB* cathepsins B, *CTSD* cathepsins D, *DRAM1* damage-regulated autophagy modulator, *EM* electron microscopy, *FM* fluorescence microscopy, *HSPA8* heat shock protein A8, *IRGM* Immunity-related GTPase family M protein, *LAMP-2A* lysosomal-associated membrane protein 2A, *MaA* macroautophagy, *MAP1LC3* microtubule-associated protein light chain 3, *MS* multiple sclerosis, *PCR* polymerase chain reaction, *PM* polymyositis, *PRDM1* positive regulatory domain I-binding factor 1, *RA* rheumatoid arthritis, *SLE* systemic lupus erythematosus, *WB* Western blot
[a]The method used to evaluate these changes is given in parentheses

regarding the interpretation of data that can be generated using RNA interference-based KO of *Atg* mRNAs in mammalian cell lines [138].

The close relationships between autophagy and immunity reported above easily explain that any deregulation of autophagy machinery can affect various aspects of immune responses and lead to autoimmunity development [32, 72, 121]. Enhanced autophagy, allowing survival of self-reactive lymphocytes, can promote autoimmunity. Moreover, autophagy, which produces autoantigens through intracellular protein digestion can participate in the initiation or maintenance of autoimmunity. In addition to SNPs and susceptibility genes, a number of studies have highlighted that expression of some genes related to autophagic process is modified during autoimmunity. In rheumatoid arthritis (RA), it has been shown that both *ATG7* and *BECLIN-1* gene expression is increased in osteoclasts from patients [70]. Atg7 expression was found to be increased by pro-inflammatory cytokine TNF-α, a critical element for the pathogenesis through the regulation of synovial inflammation. Other studies have also demonstrated that in autoimmune demyelination syndrome and in multiple sclerosis (MS), *ATG5* gene expression is also significantly elevated compared to healthy controls [2].

Based on genetic evidences, potential links between autophagy and autoimmunity have been suggested for a decade. In general, however, experimental arguments at the cellular and molecular level showing a role of autophagy in the initiation and/or progression of autoimmune diseases are still scarce (Table 1). In SLE patients and two genetically unrelated mouse models of lupus, namely MRL/*lpr* and (NZBxNZW)F1 (NZB/W) mice, we showed in a seminal report that autophagy is deregulated in T lymphocytes [31]. Autophagic vacuoles were found to be over-represented in T cells indicating that autophagy is hyperactivated. This deregulation was even more obvious when T cells were stimulated by chemical activators of T cell receptor (TCR)-related signalling pathways. The elevated autophagic compartment was not found in all T cells but was restricted to a subset of them. As autophagy is known to be involved in cell survival, these results suggest that autophagy could promote the survival of autoreactive T cells during the disease. A few months after our results came out, independent studies were published describing some deregulated autophagy features in lupus T and B cells. Alessandri et al. [1] showed an increase of the autophagosome-associated MAP1LC3-II isoform in T cells, which mainly occurred in naïve CD4 T cells isolated from SLE patients. These results, which confirm our own data, suggest that there is an intrinsic deregulation of autophagic activity in SLE T cells. The authors proposed another interpretation in concluding that SLE T cells are resistant to macroautophagy induction and could thus become more prone to apoptosis. They came to this conclusion by re-stimulating T cells with rapamycin or with autologous (pro-autophagic) serum. It is possible, however, that SLE T cells are already at the maximum level of autophagosome loading and that re-exposure to their own serum had no further effect on autophagic activity. In any case, these data confirm the pro-autophagic role of SLE serum on normal T cells. Pierdominici and her colleagues also observed that the increase of autophagy was correlated with disease activity scores, important information that could be exploited in future therapeutic strategies [1, 120, 121].

More recent studies have reinforced and extended the pioneered works described above. Thus, for the first time, Clarke et al. [13] showed in NZB/W mice that macroautophagy activation also occurs in B cells, and more particularly in early developmental and transitional stages of B cell development (before disease onset). In patients with lupus, autophagy was also activated compared to healthy individuals, and again this activation occurred mainly in naïve B cells. When autophagy inhibitors such as 3-MA, bafilomycin A1 or CQ were used, plasmablast differentiation and survival hardly occurred. These findings must be related to the overproduction of autoantibodies in the serum of lupus prone mice and patients with lupus. In their study, the authors confirmed that in addition to B cells, autophagy was increased in T cells from lupus patients, and that in both cases, this activation could be correlated to disease activity. Li et al. [67] also described convincing results demonstrating that compared to controls, autophagy was significantly activated in the macrophages collected from an induced mouse model of lupus (BALB/c mice that develop a lupus-like disease after administration in Freund's adjuvant of homologous activated lymphocyte-derived DNA) and in the PBMCs of patients with lupus. Adoptive transfer of *Beclin*-1 KO macrophages significantly ameliorates the clinical conditions of recipient mice (decrease of proteinuria levels, reduction of typical renal complex deposition, amelioration of glomerulonephritis) as well as the biological features (decrease of serum anti-dsDNA antibody levels and circulating pro-inflammatory cytokines IL-6 and TNF-α, as measured by ELISA).

A few studies have highlighted the role of autophagy in other autoimmune diseases, notably in human RA [49, 70, 151] and in experimental autoimmune encephalomyelitis, a model of MS [6]. Autophagy appears to be activated in osteoclasts from patients with RA and regulates osteoclasts differentiation [70]. This increased autophagic process, also found in RA synovial fibroblast compared to osteoarthritis synovial fibroblast by Kato et al. [49] correlates with a reduced apoptosis level in RA synovial tissues [152]. It was concluded from these observations that the activation of autophagy induced by overproduced TFN-α leads to the reduction of apoptosis in joints and more importantly causes the survival of synovial fibroblasts, which are responsible for the pathology. This again highlights the dual effect of autophagy, which is cytoprotective when it eliminates misfolded or too abundant cellular components, but in excess, can become deleterious and generate negative effects.

4 Targeting Autophagy for Intervention in Autoimmune Diseases

A number of recent findings underlined the pivotal role of macroautophagy in the control of muscle mass, and misregulation of autophagy has been described in myopathies and muscular dystrophies [131]. Information in relation to possible autophagy process dysfunction is scarce, however, regarding patients with fibromyalgia, for example, or with polymyositis [73, 143], a rare disease with an autoimmune component which is characterized by inflammation and degeneration of the muscles. On the other hand, autophagy defects have been observed (or suspected) in several autoimmune settings, including CD, SLE, possibly RA and MS (Table 1), as well as in inflammatory syndromes, notably in pulmonary diseases [88]. It is strongly anticipated that in all these situations, modulation of autophagy, in order to re-establish a proper flux regulation in particular, might rescue alterations and improve the clinical status of treated patients.

As underlined recently [32], some molecules used for years to treat inflammatory and autoimmune diseases have been found much later to target one or another type of autophagy processes. Nowadays, in fact, there are very few specific compounds targeting precise steps of autophagy pathways, and even a single pathway in particular [3], and quite surprisingly, the targets of some autophagy regulators that are widely prescribed to patients are not really known. This is the case, in particular, of CQ and hydroxychloroquine (HCQ) or of dexamethasone, which mode of action (MOA) is still being debated (see below).

A number of comprehensive review articles have recently exhaustively covered various aspects, structural and functional, of families of compounds, activators and inhibitors, which have been generated to modulate autophagy directly or indirectly [5, 11, 29, 32, 45, 125, 127, 148, 149]. Evaluated in rigorously calibrated assays performed both in vitro and in vivo [53, 93], some of these small molecules might prove to be relevant to modulate autoimmune diseases in appropriate settings. In the examples shown in the next section we will limit ourselves to a few pharmacological regulators of autophagy with established or promising clinical efficacy in autoimmune diseases.

Before providing a short description of these selected pharmacological autophagy modulators, several conceptual and practical comments should be made. Firstly, this field of possible intervention is new (or newly rediscovered) and autophagy processes, which are complex and somehow confusing, are not well perceived by decision makers of technology companies and Big Pharmas, of course even less by the general public and informed users. Communication including education towards professionals and patients is certainly much easier when, for example, one describes the activity of a therapeutic antibody specific for a soluble molecule or a surface receptor that is raised in inflammatory and autoimmune conditions. Important efforts of clarification and simplification have thus to be made as it was the cases some decades ago for apoptosis.

Secondly, it is well appreciated for a long time in the field that pharmacological small molecules rarely exert their action on only one single target. This is the case of HCQ and dexamethasone, for example, and many others (see below). These multi-target effects can explain their strong efficacy, but they also complicate the description of the said-molecule and of its safety file.

Thirdly, it is often argued that small molecules (<900–1000 daltons) and short peptides (<20–40 amino acid residues) will be eliminated rapidly from the body and therefore will have a too short period of possible action. This statement regarding pharmacokinetics and pharmacodynamics of molecules may be correct but if it is the case, there are numerous carrier systems or novel devices that increase molecule bioavailability and trafficking leading to improving their efficacy. It should be noted here that conversely, their low molecular mass can be an advantage when the desired objective is to develop a strategy supposed to target the central nervous system, for example [44].

Fourthly, and most importantly, solubility of small molecules and peptides remains a limiting factor, as it is also the case of antibodies and fusion proteins that are designed and produced for therapeutic purposes. This aspect has to be taken into consideration at the very early stage of molecule selection as in general, it cannot be solved easily in the downstream steps of development.

On the other hand, pharmacological small molecules and peptides display a number of advantageous properties that makes them excellent therapeutics, notably for autoimmune diseases. In addition to their synthesis and production that can be highly optimized, and in some cases remarkably simple in comparison to some biologics, and automatable, small molecules and peptides selected as active components of pharmaceutical compositions are characterized by their stability and robustness, easy handling, the relatively low doses that have to be administrated to patients and their cost, which remains reasonable with regard to most biologics. Small molecules and short peptides are not immunogenic per se, another considerable advantage for treating patients with chronic autoimmune diseases [132].

Finally, it must be stressed that, as it is the case for all new therapies that emerge, standardized and universalized animal models of the related human disease have to be developed -if they do not already exist-, a consensus position regarding the most promising modality to be tested has to be established, and formation of a cooperative international network of committed clinical investigators has to be gathered to evaluate these new therapies in a pre-designed rigorous fashion.

5 Existing Pharmacological Regulators of Autophagy

Herein, we briefly describe the characteristics of some chemical molecules that are established pharmacological regulators of autophagy (Fig. 2) and are given to patients with autoimmune diseases. Further details on these and other compounds can be found in recent reviews and articles [5, 11, 29, 32, 125, 127, 140, 146].

CQ and HCQ These two small molecules are lipophilic weak bases that easily pass through the lipid cell membrane and preferentially concentrate in acidic cytoplasmic vesicles. As lysosomotropic agent, they raise intralysosomal pH, leading to defective autophagic protein degradation. CQ/HCQ may also affect peptide degradation within lysosomes due to the pH effect on lysosomal cathepsins and therefore the entire process of antigen presentation by MHC molecules in the MIIC compartment leading to activation of autoreactive T cells. HCQ is used for years in the treatment of inflammatory autoimmune diseases, SLE, RA and Sjögren's syndrome. CQ has been shown to reduce the severity of experimental autoimmune encephalomyelitis, a model for MS, and the mechanism of action that was previously known to involve in part regulatory T cells has been recently established in much more details [144]. CQ and HCQ also operate by interacting directly with TLR ligands [59]. Other characteristics of CQ and derivatives, such as radiosensitising and chemosensitising

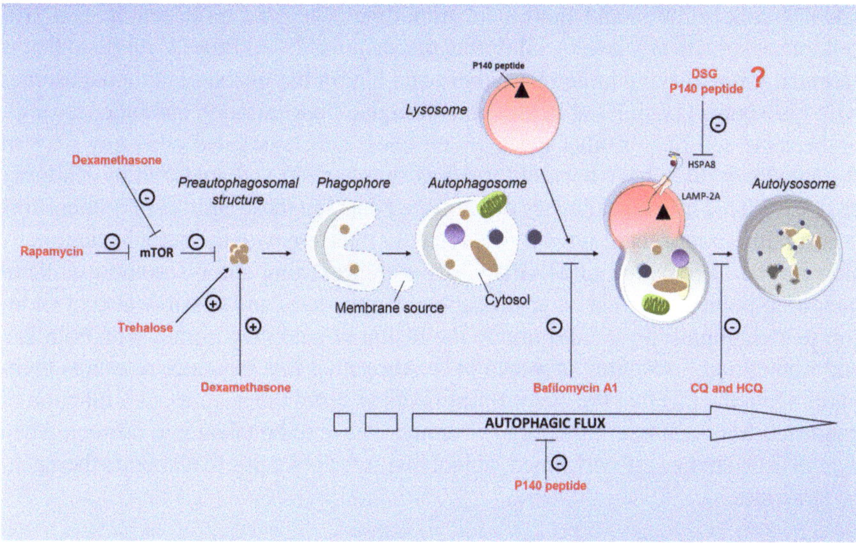

Fig. 2 Pharmacological regulators of autophagy**.** A diagram illustrating possible sites of intervention of pharmacological autophagy regulators. From the left to the right: rapamycin and dexamethasone inhibit the kinase activity of mTOR, leading to the upregulation of macroautophagy. Dexamethasone is also known as acting on pre-autophagosomal structure. Trehalose, the target of which still remains debated, is an activator of autophagy through an mTOR-independent pathway. Bafilomycin A1 prevents maturation of autophagic vacuoles by inhibiting fusion between autophagosomes and lysosomes. It acts by inhibiting vacuolar H+ ATPase. P140 peptide (▲), the uptake into B lymphocytes by clathrin-mediated endocytosis and homing into lysosomes has been demonstrated after administration to mice, and DSG, both interact with HSPA8 in vitro and alter intralysosomal pH. P140 provokes the accumulation of autophagy markers p62/sequestosome 1 and MAP1LC3-II in MRL/lpr B cells, consistent with a down-regulation of autophagic flux. This peptide affects both CMA and macroautophagy. CQ and HCQ are lysosomotropic agents that prevent endosomal acidification. They accumulate inside endosomes and lysosomes, leading to inhibition of lysosomal enzymes, which requires an acidic pH, defective fusion of endosomes and lysosomes and maturation of autolysosomes. *Abbreviations*: *CMA* chaperone-mediated autophagy, *CQ* chloroquine, *DSG* 15-deoxyspergualin, *HCQ* hydroxychloroquine, *HSPA8* heat shock protein 8, *LAMP-2A* lysosome-associated membrane protein-2A, *MAP1LC3* microtubule-associated protein light chain 3, *mTOR* mammalian target of rapamycin

properties also receive attention in anti-cancer indications [45]. It should be reminded, however, that CQ/HCQ toxicity, in particular in the eye (cornea and macula) and in the occurrence of cardiomyopathies [142], remains a major break. Observed ocular toxicity is related to the total cumulative dose rather than the daily dose; therefore it becomes a serious potential problem in the cases of long-term use. A number of HCQ analogs and mimics have been tentatively designed that keep the molecule activity without secondary effects. Ongoing research should provide such safe molecules in the future.

Bafilomycin A This compound isolated from *Streptomyces* sp. is a member of the plecomacrolide sub-class of macrolide antibiotics. Early studies showed that at a

100 nM-concentration and short incubation time (1 h), in a rat hepatoma H-4-II-E cell line, it specifically acts by inhibiting the vacuolar H^+ ATPase (V-ATPase) that is essential for acidifying lumen lysosomes and blocks the fusion of autophagosomes with lysosomes [155]. Used at the same or higher concentration and other settings, in the same cell line or other types of cell lines, effects targeted other key steps of the autophagy axes have been observed as summarized and analyzed by Klionsky et al. [54]. This in-depth analysis of published data led these authors to propose that at early time-points, bafilomycin could mainly interfere with the autophagic flux by slowing the degradation of MAP1LC3-II within existing autolysosomes, while at later time-points, its effect on acidification of lysosomes and possibly also of endosomes and amphisomes could impair the fusion of autophagosomes with both late endosomes and lysosomes as shown [42]. Altogether this sequence of events highlights again the fact that the pleiotropic effects of certain molecules, as a function of concentration, treatment time, or environment, have to be taken into account when mechanistic studies are performed, notably with the objective to elaborate therapeutic strategies.

P140 peptide/Lupuzor This 21-mer linear peptide encompassing the sequence 131–151 of the spliceosomal U1-70 K protein and containing a phosphoserine residue at position 140, was found to be safe and significantly ameliorated lupus patients' clinical status when administrated subcutaneously in the presence of mannitol as excipient [102, 103, 160]. All appropriate preclinical studies were done in the widely used MRL/lpr model, a mouse that develops a strong and rapid lupus disease. The capacity of P140 to ameliorate biological and clinical parameters in these mice, and to enhance their survival, was demonstrated in a robust manner [99, 133]. After P140 treatment, an accumulation of autophagy markers SQSTM1 and MAP1LC3 was observed in MRL/lpr B cells, consistent with a down-regulation of autophagic flux [114]. CMA was also recently found to be a target of P140 and it was demonstrated that P140 peptide inhibitory effect on CMA is likely tied to its ability to interact with HSPA8 [115] and to alter the composition of HSPA8 heterocomplexes [78]. Expression of both HSPA8 and the limiting CMA component LAMP-2A, which is increased in MRL/lpr B cells, is down-regulated after treating mice with P140 peptide. It was shown further that P140, but not the non-phosphorylated peptide that is not protective against disease development in mice [99], uses the clathrin-dependent endo-lysosomal pathway to enter into MRL/lpr B lymphocytes and accumulates in the lysosomal lumen where it may directly hamper lysosomal HSPA8 chaperoning functions, and also destabilize LAMP-2A in lysosomes as a result of its effect on HSP90. This dual effect may interfere with the endogenous (auto)antigen processing and loading to MHCII molecules and as a consequence, lead to the lower activation of autoreactive T cells that was previously shown experimentally [100, 101].

Interestingly, earlier work also indicated that ex vivo, P140 does not induce proliferation of human peripheral T cells (in contrast to the non-phosphorylated form that does) but generates secretion of high levels of regulatory cytokine IL-10 in cell cultures [98]. This observation and others generated in our own studies might indicate that beside its effect on autophagy processes, P140 might also act as a so-called 'peptide altered ligand' of the TCR. Our first studies showed that the nominal peptide 131–151 contains an epitope that is effectively recognized by $CD4^+$ T cells from MRL/lpr and NZBxW mice [96, 97]. The phosphate moiety introduced at position 140 in the P140 peptide might have no effect on MHC presentation (as experimentally demonstrated) but induce qualitatively different activation of T cells with changes in cytokine production and T cell responsiveness [98, 99]. Altogether, these considerations point out again the multi-target functions of efficient immunomodulator molecules. Thus, in the case of P140 peptide, both specific $CD4^+$ T cell clones recognizing the sequence 131–151 of U1-70 K protein and T cell clones with a broader specificity for various self-components generated in autolysosomes and lysosomes and loaded onto MHC class II molecules in the MIIC compartment (Fig. 1c), could be simultaneously involved in the mechanism of peptide action.

15-Deoxyspergualin (DSG) This compound (1-amino-19-guanidino-11--hydroxy-4, 9, 12-triazanona-decane-10, 1–3-dione) is a synthetic analogue of spergualin, a natural product of the bacterium *Bacillus laterosporus*. A long list of more stable analogs have been designed, synthesized and evaluated over years. 15-DSG is a potent immunosuppressant, which showed immunosuppressive activity both in vitro and in vivo, affecting B lymphocytes, T lymphocytes and macrophage/monocyte functions. It was shown to bind to the EEVD domain of HSPA8, a site that is apparently different from the one(s) recognized by P140 peptide [140], with an affinity of approximately 4 μM, and increase its ATPase activity of 20–40 %. It also binds to HSP90. 15-DSG blocks the NF-κB pathway and antigen presentation, causing alteration in the activation of immune cells, notably monocytes, DCs and T cells. It also inhibits AKT activation and phosphatidylcholine synthesis [51]. DSG was also shown to suppress the progression of polyclonal B cell activation and lupus nephropathy in lupus-prone MRL/lpr mice. In patients, in an first short clinical trial, two of three patients treated with DSG showed infectious episodes and the trial was interrupted [74]. Later, another phase-I/II study including a total of 21 patients was engaged [75]. After the first DSG injection, one patient was excluded from the study due to renal failure. Five patients dropped out due to adverse events or serious adverse events including fever, leukopenia, oral candidiasis, herpes zoster or pneumonia. Eleven of 20 patients achieved partial (4) or complete responses (7), 8 were judged as treatment failures and 1 patient was not assessable. In the 12 patients who completed all nine cycles, proteinuria was statistically decreased and the Selena-SLEDAI SLE responder score was decreased from 17.6 to 11.7. These data led the authors to conclude that although the number of patients still remained small, the improvement of their clinical status, particularly their proteinuria, was encouraging, supporting further investigations with large cohorts. At this stage, however, and although some promising data were also obtained in patients with anti-neutrophil cytoplasmic autoantibodies-associated vasculitis and cancer conditions, careful studies designed to better characterize toxicity and side-effects generated by DSG will be determining [63].

Dexamethasone This potent immunosuppressive drug is widely used to treat many different inflammatory and autoimmune conditions such as inflammatory bowel diseases (ulcerative colitis and CD), RA, SLE, chronic skin conditions (e.g. dermatitis *herpetiformis*, pemphigus, severe psoriasis and seborrheic dermatitis). It is also given in severe allergic conditions and certain types of cancer. Its MOA is multiple, complex and still a matter of some controversies. It was found in particular that dexamethasone induces the expression of a gene encoding the stress response protein Dig2/RTP801/REDD1 [95], and the elevation of Dig2/RTP801/REDD1, a negative regulator of mTOR signaling pathway, contributes to the induction of macroautophagy. It should be mentioned herein that depending on the dose and the type of cells, the effect of dexamethasone on Dig2/RTP801/REDD1 is not equivalent (less dependence at high dexamethasone dose, for example). Other dexamethasone effects were described. Thus, dexamethasone was shown to increase expression of several autophagy genes, including *ATG5*, *MAP1LC3*, *BECLIN1n* and *SQSTM1*, and to trigger 5' AMP-activated protein kinase-dependent mitochondrial fragmentation associated with increased levels of dynamin-1-like protein, a GTPase that regulates mitochondrial fission [145]. Thus, certain steps of the mitophagy axis would be targeted by dexamethasone as well. The anti-inflammatory actions of dexamethasone are also thought to involve phospholipase A_2 inhibitory proteins, lipocortins, which control the biosynthesis of potent mediators of inflammation such as prostaglandins and leukotrienes.

Rapamycine/sirolimus This macrolide antibiotic is a safe and well-tolerated drug clinically used for rejection prophylaxis in renal transplantation. It is also used as immunosuppressant and anti-fungal agent. It forms a complex with the immunophilin FKBP-12 and inhibits the kinase activity of mTOR complex 1 (mTORC1), leading thus to autophagy induction (mTORC2 is largely resistant to rapamycin). It regulates mitochondrila transmembrane potential and calcium influx. Its potent effect on the development of nephritis in NZB/W mice was shown [76]. Twelve-week-old female NZB/W mice were treated by oral gavage for 20 weeks with rapamycin (3 mg/kg body weight). Rapamycin treatment markedly reduced proteinuria, improved renal function, decreased serum anti-dsDNA antibody levels and diminished splenomegaly. Rapamycin-treated mice had near normal renal histology, with marked reduction in glomerular immune deposition and the infiltration by T cells, B cells and macrophages. These data were reinforced by recent mechanistic findings published independently [141]. In humans, rapamycin treatment showed some benefit in the treatment of nine SLE patients with refractory disease [28]. In a recent prospective open-label study based on 59 patients and 54 matched healthy subjects (for a total of 274 visits), rapamycin was shown to mainly block IL-4 production and necrosis of double negative (DN) T cells in patients with SLE. In addition, rapamycin was found to enhance FoxP3 expression in $CD25^+/CD4^+$ T cells and expand $CD25^+CD19^+$ B cells, suggesting that mTOR activation can trigger IL-4 production and necrosis of DN T cells in active SLE [60]. Further investigation in large cohorts of patients with lupus and also in patients with other immune-mediated disorders, including type 1 diabetes and RA are awaited for consolidating these data. If we only take into account the role exerted by rapamycin on the autophagy flux (Fig. 2), and considering that basal autophagy seems to be activated in different subsets of lymphocytes in murine and human lupus (Table 1), rapamycin administration should not be beneficial in lupus. It might even make the illness more severe. This leads us to conclude that rapamycin probably modulates another pathway and not autophagy as main target.

Recently a randomized trial was conducted to investigate the efficacy and safety of rapamycin treatment in adults with chronic immune thrombocytopenia (ITP), an acquired autoimmune disease characterized by an autoantibody-mediated destruction and impaired platelet production [68]. Two groups of 40 patients were examined, the control one that received cyclosporine A plus prednisone and the experimental one that received rapamycin plus prednisone. The overall response was similar in both groups. However, sustained response was more pronounced in the experimental group than in the control group. Both groups showed similar incidence of adverse events (7 % vs. 11 %). The experimental group experienced a significant rise in $CD4^{+}CD25^{+}CD127^{low}$ regulatory T cells level, and there was a strong correlation between the levels of regulatory T cells and TGF-β after the treatment. From these data it was concluded that rapamycin plus low dose prednisone could provide a new promising option for therapy of ITP.

6 Future Prospects and Concluding Remarks

The list of components described briefly above is far to be exhaustive. Excellent recent review articles gave much more structural and functional details on many other molecules (small molecules and peptides), some of them that are already administrated as therapeutics and some others that are under evaluation in autoimmune patients or included in preclinical studies in pertinent animal models [25, 39, 136]. The information we summarized herein underlines that most, if not all of the molecules, exhibit complex pleiotropic properties, and can notably influence different autophagy pathways (e.g. mTOR-dependent and -independent) as well as other quality-control mechanisms affecting the cell live/death balance. Several widely used molecules can exert dual (sometimes opposite) effects on upstream and downstream molecular events of the autophagy axes. It should be kept in mind also that the large majority of these molecules have been initially evaluated in cell culture conditions (some are issued from cellular screens) and it has been seen that their MOA largely depends on the selected cell type (immortalized cell lines, primary cells; cancer cells or non-cancer cells), concentration, and time of exposure. As underlined recently [32], these considerations are fundamental to analyze the conclusions that can be raised with most caution.

Nowadays, a number of pharmaceuticals approved in the European Union and USA, and in regular clinical use for alternative indications, inhibit autophagy and may therefore be novel treatments for autoimmune diseases. Chemical drugs acting on autophagy and/or other pivotal cellular pathways are also often evaluated in association to reinforce their efficacy while lowering dosage to minimize deleterious side effects. Based on our increasing understanding of the physiological autophagy mechanisms and of their dysfunctions in pathological settings [18, 65, 92, 120, 147, 157], we dare believe that molecules that very specifically target key elements of the autophagy process will emerge and, with a minimum of side effects, will efficiently modulate debilitating autoimmune diseases that today affect more than 3 % of the general population worldwide.

Acknowledgments The authors would like to thank Jean-Paul Briand, Frédéric Gros and Jean-Louis Pasquali for useful discussion and a critical reading of the manuscript. Research in the SM's laboratory is financially supported by the French Centre National de la Recherche Scientifique and the Laboratory of Excellence Medalis (ANR-10-LABX-0034), Initiative of Excellence (IdEx), Strasbourg University. MW was supported by a Region Alsace and ImmuPharma-France doctoral fellowship

Conflicts of Interests SM is a consultant of ImmuPharma. MW declares no conflicts of interests

References

1. Alessandri C, Barbati C, Vacirca D et al (2012) T lymphocytes from patients with systemic lupus erythematosus are resistant to induction of autophagy. FASEB J 26:4722–4732. doi:10.1096/fj.12-206060
2. Alirezaei M, Fox HS, Flynn CT et al (2009) Elevated ATG5 expression in autoimmune demyelination and multiple sclerosis. Autophagy 5:152–158. doi:10.4161/auto.5.2.7348
3. Anguiano J, Garner TP, Mahalingam M et al (2013) Chemical modulation of chaperone-mediated autophagy by retinoic acid derivatives. Nat Chem Biol 9:374–382. doi:10.1038/nchembio.1230
4. Awan MUF, Deng Y (2014) Role of autophagy and its significance in cellular homeostasis. Appl Microbiol Biotechnol 98:5319–5328. doi:10.1007/s00253-014-5721-8
5. Baek K-H, Park J, Shin I (2012) Autophagy-regulating small molecules and their therapeutic applications. Chem Soc Rev 41:3245–3263. doi:10.1039/c2cs15328a
6. Bhattacharya A, Parillon X, Zeng S et al (2014) Deficiency of autophagy in dendritic cells protects against experimental autoimmune encephalomyelitis. J Biol Chem 289:26525–26532. doi:10.1074/jbc.M114.575860
7. Blum JS, Wearsch PA, Cresswell P (2013) Pathways of antigen processing. Annu Rev Immunol 31:443–473. doi:10.1146/annurev-immunol-032712-095910
8. Burster T, Beck A, Tolosa E et al (2004) Cathepsin G, and not the asparagine-specific endoprotease, controls the processing of myelin basic protein in lysosomes from human B lymphocytes. J Immunol 172:5495–5503. doi:10.4049/jimmunol.172.9.5495
9. Carrithers MD (2014) Update on disease-modifying treatments for multiple sclerosis. Clin Ther. doi:10.1016/j.clinthera.2014.08.006
10. Cenci S (2014) Autophagy, a new determinant of plasma cell differentiation and antibody responses. Mol Immunol 62:289–295. doi:10.1016/j.molimm.2014.02.008
11. Cheong H, Lu C, Lindsten T, Thompson CB (2012) Therapeutic targets in cancer cell metabolism and autophagy. Nat Biotechnol 30:671–678. doi:10.1038/nbt.2285
12. Choi AMK, Ryter SW, Levine B (2013) Autophagy in human health and disease. N Engl J Med 368:1845–1846. doi:10.1056/NEJMc1303158

13. Clarke AJ, Ellinghaus U, Cortini A et al (2014) Autophagy is activated in systemic lupus erythematosus and required for plasmablast development. Ann Rheum Dis. doi:10.1136/annrheumdis-2013-204343
14. Codogno P, Mehrpour M, Proikas-Cezanne T (2012) Canonical and non-canonical autophagy: variations on a common theme of self-eating? Nat Rev Mol Cell Biol 13:7–12. doi:10.1038/nrm3249
15. Colbert JD, Matthews SP, Miller G, Watts C (2009) Diverse regulatory roles for lysosomal proteases in the immune response. Eur J Immunol 39:2955–2965. doi:10.1002/eji.200939650
16. Conway KL, Kuballa P, Khor B et al (2013) ATG5 regulates plasma cell differentiation. Autophagy 9:528–537. doi:10.4161/auto.23484
17. Cuervo AM (2004) Autophagy: many paths to the same end. Mol Cell Biochem 263:55–72. doi:10.1023/B:MCBI.0000041848.57020.57
18. Cuervo AM, Macian F (2014) Autophagy and the immune function in aging. Curr Opin Immunol 29:97–104. doi:10.1016/j.coi.2014.05.006
19. Cuervo AM, Wong E (2014) Chaperone-mediated autophagy: roles in disease and aging. Cell Res 24:92–104. doi:10.1038/cr.2013.153
20. Dani A, Chaudhry A, Mukherjee P et al (2004) The pathway for MHCII-mediated presentation of endogenous proteins involves peptide transport to the endo-lysosomal compartment. J Cell Sci 117:4219–4230. doi:10.1242/jcs.01288
21. Dengjel J, Schoor O, Fischer R et al (2005) Autophagy promotes MHC class II presentation of peptides from intracellular source proteins. Proc Natl Acad Sci U S A 102:7922–7927. doi:10.1073/pnas.0501190102
22. Deretic V (2012) Autophagy: an emerging immunological paradigm. J Immunol 189:15–20. doi:10.4049/jimmunol.1102108
23. Deretic V, Saitoh T, Akira S (2013) Autophagy in infection, inflammation and immunity. Nat Rev Immunol 13:722–737. doi:10.1038/nri3532
24. Dörfel D, Appel S, Grünebach F et al (2005) Processing and presentation of HLA class I and II epitopes by dendritic cells after transfection with in vitro-transcribed MUC1 RNA. Blood 105:3199–3205. doi:10.1182/blood-2004-09-3556
25. Dowdle WE, Nyfeler B, Nagel J et al (2014) Selective VPS34 inhibitor blocks autophagy and uncovers a role for NCOA4 in ferritin degradation and iron homeostasis in vivo. Nat Cell Biol 16:1069–1079. doi:10.1038/ncb3053
26. Ewald SE, Lee BL, Lau L et al (2008) The ectodomain of Toll-like receptor 9 is cleaved to generate a functional receptor. Nature 456:658–662. doi:10.1038/nature07405
27. Feng Y, He D, Yao Z, Klionsky DJ (2014) The machinery of macroautophagy. Cell Res 24:24–41. doi:10.1038/cr.2013.168
28. Fernandez D, Bonilla E, Mirza N et al (2006) Rapamycin reduces disease activity and normalizes T cell activation-induced calcium fluxing in patients with systemic lupus erythematosus. Arthritis Rheum 54:2983–2988. doi:10.1002/art.22085
29. Fleming A, Noda T, Yoshimori T, Rubinsztein DC (2011) Chemical modulators of autophagy as biological probes and potential therapeutics. Nat Chem Biol 7:9–17. doi:10.1038/nchembio.500
30. Glas J, Seiderer J, Bues S et al (2013) IRGM variants and susceptibility to inflammatory bowel disease in the German population. PLoS One 8:e54338. doi:10.1371/journal.pone.0054338
31. Gros F, Arnold J, Page N et al (2012) Macroautophagy is deregulated in murine and human lupus T lymphocytes. Autophagy 8:1113–1123. doi:10.4161/auto.20275
32. Gros F, Muller S (2014) Pharmacological regulators of autophagy and their link with modulators of lupus disease. Br J Pharmacol 171:4337–4359. doi:10.1111/bph.12792
33. Gutierrez MG, Master SS, Singh SB et al (2004) Autophagy is a defense mechanism inhibiting BCG and Mycobacterium tuberculosis survival in infected macrophages. Cell 119:753–766. doi:10.1016/j.cell.2004.11.038

34. Hampe J, Franke A, Rosenstiel P et al (2007) A genome-wide association scan of nonsynonymous SNPs identifies a susceptibility variant for Crohn disease in ATG16L1. Nat Genet 39:207–211. doi:10.1038/ng1954
35. Hanly JG (2014) Diagnosis and management of neuropsychiatric SLE. Nat Rev Rheumatol 10:338–347. doi:10.1038/nrrheum.2014.15
36. Hayter SM, Cook MC (2012) Updated assessment of the prevalence, spectrum and case definition of autoimmune disease. Autoimmun Rev 11:754–765. doi:10.1016/j.autrev.2012.02.001
37. He C, Klionsky DJ (2009) Regulation mechanisms and signaling pathways of autophagy. Annu Rev Genet 43:67–93. doi:10.1146/annurev-genet-102808-114910
38. Hong Z, Jiang Z, Liangxi W et al (2004) Chloroquine protects mice from challenge with CpG ODN and LPS by decreasing proinflammatory cytokine release. Int Immunopharmacol 4:223–234. doi:10.1016/j.intimp.2003.12.006
39. Huss M, Wieczorek H (2009) Inhibitors of V-ATPases: old and new players. J Exp Biol 212:341–346. doi:10.1242/jeb.024067
40. Inglese M, Petracca M (2014) Therapeutic strategies in multiple sclerosis: a focus on neuroprotection and repair and relevance to schizophrenia. Schizophr Res. doi:10.1016/j.schres.2014.04.040
41. International Consortium for Systemic Lupus Erythematosus Genetics (SLEGEN, Harley JB, Alarcón-Riquelme ME et al (2008) Genome-wide association scan in women with systemic lupus erythematosus identifies susceptibility variants in ITGAM, PXK, KIAA1542 and other loci. Nat Genet 40:204–210. doi:10.1038/ng.81
42. Jahreiss L, Menzies FM, Rubinsztein DC (2008) The itinerary of autophagosomes: from peripheral formation to kiss-and-run fusion with lysosomes. Traffic Cph Den 9:574–587. doi:10.1111/j.1600-0854.2008.00701.x
43. Järvinen TM, Hellquist A, Zucchelli M et al (2012) Replication of GWAS-identified systemic lupus erythematosus susceptibility genes affirms B-cell receptor pathway signalling and strengthens the role of IRF5 in disease susceptibility in a Northern European population. Rheumatol Oxf Engl 51:87–92. doi:10.1093/rheumatology/ker263
44. Jeltsch-David H, Muller S (2014) Neuropsychiatric systemic lupus erythematosus: pathogenesis and biomarkers. Nat Rev Neurol 10:579–596. doi:10.1038/nrneurol.2014.148
45. Jiang P, Mizushima N (2014) Autophagy and human diseases. Cell Res 24:69–79. doi:10.1038/cr.2013.161
46. Jia W, He Y-W (2011) Temporal regulation of intracellular organelle homeostasis in T lymphocytes by autophagy. J Immunol 186:5313–5322. doi:10.4049/jimmunol.1002404
47. Jones SA, Mills KHG, Harris J (2013) Autophagy and inflammatory diseases. Immunol Cell Biol 91:250–258. doi:10.1038/icb.2012.82
48. Kang R, Zeh HJ, Lotze MT, Tang D (2011) The Beclin 1 network regulates autophagy and apoptosis. Cell Death Differ 18:571–580. doi:10.1038/cdd.2010.191
49. Kato M, Ospelt C, Gay RE et al (2014) Dual role of autophagy in stress-induced cell death in rheumatoid arthritis synovial fibroblasts. Arthritis Rheumatol 66:40–48. doi:10.1002/art.38190
50. Kaushik S, Cuervo AM (2012) Chaperone-mediated autophagy: a unique way to enter the lysosome world. Trends Cell Biol 22:407–417. doi:10.1016/j.tcb.2012.05.006
51. Kawada M, Masuda T, Ishizuka M, Takeuchi T (2002) 15-Deoxyspergualin inhibits Akt kinase activation and phosphatidylcholine synthesis. J Biol Chem 277:27765–27771. doi:10.1074/jbc.M200318200
52. Kirkin V, McEwan DG, Novak I, Dikic I (2009) A role for ubiquitin in selective autophagy. Mol Cell 34:259–269. doi:10.1016/j.molcel.2009.04.026
53. Klionsky DJ, Abdalla FC, Abeliovich H et al (2012) Guidelines for the use and interpretation of assays for monitoring autophagy. Autophagy 8:445–544. doi:10.4161/auto.19496
54. Klionsky DJ, Elazar Z, Seglen PO, Rubinsztein DC (2008) Does bafilomycin A1 block the fusion of autophagosomes with lysosomes? Autophagy 4:849–850. doi:10.4161/auto.6845
55. Klionsky DJ, Emr SD (2000) Autophagy as a regulated pathway of cellular degradation. Science 290:1717–1721. doi:10.1126/science.290.5497.1717

56. Konishi A, Arakawa S, Yue Z, Shimizu S (2012) Involvement of Beclin 1 in engulfment of apoptotic cells. J Biol Chem 287:13919–13929. doi:10.1074/jbc.M112.348375
57. Korolchuk VI, Menzies FM, Rubinsztein DC (2010) Mechanisms of cross-talk between the ubiquitin-proteasome and autophagy-lysosome systems. FEBS Lett 584:1393–1398. doi:10.1016/j.febslet.2009.12.047
58. Kroemer G, Levine B (2008) Autophagic cell death: the story of a misnomer. Nat Rev Mol Cell Biol 9:1004–1010. doi:10.1038/nrm2529
59. Kuznik A, Bencina M, Svajger U et al (2011) Mechanism of endosomal TLR inhibition by antimalarial drugs and imidazoquinolines. J Immunol 186:4794–4804. doi:10.4049/jimmunol.1000702
60. Lai Z-W, Borsuk R, Shadakshari A et al (2013) Mechanistic target of rapamycin activation triggers IL-4 production and necrotic death of double-negative T cells in patients with systemic lupus erythematosus. J Immunol 191:2236–2246. doi:10.4049/jimmunol.1301005
61. Lamb CA, Yoshimori T, Tooze SA (2013) The autophagosome: origins unknown, biogenesis complex. Nat Rev Mol Cell Biol 14:759–774. doi:10.1038/nrm3696
62. Lee E, Koo Y, Ng A et al (2014) Autophagy is essential for cardiac morphogenesis during vertebrate development. Autophagy 10:572–587. doi:10.4161/auto.27649
63. Lee YH, Lee HS, Choi SJ et al (2011) Efficacy and safety of tacrolimus therapy for lupus nephritis: a systematic review of clinical trials. Lupus 20:636–640. doi:10.1177/0961203310389486
64. Leung CS, Haigh TA, Mackay LK et al (2010) Nuclear location of an endogenously expressed antigen, EBNA1, restricts access to macroautophagy and the range of CD4 epitope display. Proc Natl Acad Sci U S A 107:2165–2170. doi:10.1073/pnas.0909448107
65. Levine B, Kroemer G (2008) Autophagy in the pathogenesis of disease. Cell 132:27–42. doi:10.1016/j.cell.2007.12.018
66. Levine B, Mizushima N, Virgin HW (2011) Autophagy in immunity and inflammation. Nature 469:323–335. doi:10.1038/nature09782
67. Li B, Yue Y, Dong C et al (2014) Blockade of macrophage autophagy ameliorates activated lymphocytes-derived DNA induced murine lupus possibly via inhibition of proinflammatory cytokine production. Clin Exp Rheumatol 32:705–714
68. Li J, Wang Z, Dai L et al (2013) Effects of rapamycin combined with low dose prednisone in patients with chronic immune thrombocytopenia. Clin Dev Immunol 2013:548085. doi:10.1155/2013/548085
69. Lilienbaum A (2013) Relationship between the proteasomal system and autophagy. Int J Biochem Mol Biol 4:1–26
70. Lin N-Y, Beyer C, Giessl A et al (2013) Autophagy regulates TNFα-mediated joint destruction in experimental arthritis. Ann Rheum Dis 72:761–768. doi:10.1136/annrheumdis-2012-201671
71. Li W, Zou W, Yang Y et al (2012) Autophagy genes function sequentially to promote apoptotic cell corpse degradation in the engulfing cell. J Cell Biol 197:27–35. doi:10.1083/jcb.201111053
72. Lleo A, Invernizzi P, Selmi C et al (2007) Autophagy: highlighting a novel player in the autoimmunity scenario. J Autoimmun 29:61–68. doi:10.1016/j.jaut.2007.06.003
73. Lloyd TE (2010) Novel therapeutic approaches for inclusion body myositis. Curr Opin Rheumatol 22:658–664. doi:10.1097/BOR.0b013e32833f0f4a
74. Lorenz H-M, Grunke M, Wendler J et al (2005) Safety of 15-deoxyspergualin in the treatment of glomerulonephritis associated with active systemic lupus erythematosus. Ann Rheum Dis 64:1517–1519. doi:10.1136/ard.2005.035329
75. Lorenz H-M, Schmitt WH, Tesar V et al (2011) Treatment of active lupus nephritis with the novel immunosuppressant 15-deoxyspergualin: an open-label dose escalation study. Arthritis Res Ther 13:R36. doi:10.1186/ar3268
76. Lui SL, Yung S, Tsang R et al (2008) Rapamycin prevents the development of nephritis in lupus-prone NZB/W F1 mice. Lupus 17:305–313. doi:10.1177/0961203307088289

77. Lu XC, Tao Y, Wu C et al (2013) Association between variants of the autophagy related gene – IRGM and susceptibility to Crohn's disease and ulcerative colitis: a meta-analysis. PLoS One 8:e80602. doi:10.1371/journal.pone.0080602
78. Macri C, Wang F, Tasset I et al (2015) Modulation of deregulated chaperone-mediated autophagy by a phopsphopeptide. Autophagy 11:472–486. http://dx.doi.org/10.1080/15548627.2015.1017179
79. Majai G, Kiss E, Tarr T et al (2014) Decreased apopto-phagocytic gene expression in the macrophages of systemic lupus erythematosus patients. Lupus 23:133–145. doi:10.1177/0961203313511557
80. Manoury B (2013) Proteases: essential actors in processing antigens and intracellular toll-like receptors. Front Immunol. doi:10.3389/fimmu.2013.00299
81. Manoury B, Mazzeo D, Fugger L et al (2002) Destructive processing by asparagine endopeptidase limits presentation of a dominant T cell epitope in MBP. Nat Immunol 3:169–174. doi:10.1038/ni754
82. Marchiando AM, Ramanan D, Ding Y et al (2013) A deficiency in the autophagy gene Atg16L1 enhances resistance to enteric bacterial infection. Cell Host Microbe. doi:10.1016/j.chom.2013.07.013
83. Marino G, Niso-Santano M, Baehrecke EH, Kroemer G (2014) Self-consumption: the interplay of autophagy and apoptosis. Nat Rev Mol Cell Biol 15:81–94. doi:10.1038/nrm3735
84. Marquez RT, Xu L (2012) Bcl-2:Beclin 1 complex: multiple, mechanisms regulating autophagy/apoptosis toggle switch. Am J Cancer Res 2:214–221
85. Matsumoto F, Saitoh S-I, Fukui R et al (2008) Cathepsins are required for Toll-like receptor 9 responses. Biochem Biophys Res Commun 367:693–699. doi:10.1016/j.bbrc.2007.12.130
86. Matthews SP, Werber I, Deussing J et al (2010) Distinct protease requirements for antigen presentation in vitro and in vivo. J Immunol 184:2423–2431. doi:10.4049/jimmunol.0901486
87. Miller BC, Zhao Z, Stephenson LM et al (2008) The autophagy gene ATG5 plays an essential role in B lymphocyte development. Autophagy 4:309–314
88. Mizumura K, Cloonan SM, Haspel JA, Choi AMK (2012) The emerging importance of autophagy in pulmonary diseases. Chest 142:1289–1299. doi:10.1378/chest.12-0809
89. Mizushima N (2007) Autophagy: process and function. Genes Dev 21:2861–2873. doi:10.1101/gad.1599207
90. Mizushima N, Komatsu M (2011) Autophagy: renovation of cells and tissues. Cell 147:728–741. doi:10.1016/j.cell.2011.10.026
91. Mizushima N, Levine B (2010) Autophagy in mammalian development and differentiation. Nat Cell Biol 12:823–830. doi:10.1038/ncb0910-823
92. Mizushima N, Levine B, Cuervo AM, Klionsky DJ (2008) Autophagy fights disease through cellular self-digestion. Nature 451:1069–1075. doi:10.1038/nature06639
93. Mizushima N, Yoshimori T, Levine B (2010) Methods in mammalian autophagy research. Cell 140:313–326. doi:10.1016/j.cell.2010.01.028
94. Mizushima N, Yoshimori T, Ohsumi Y (2011) The role of Atg proteins in autophagosome formation. Annu Rev Cell Dev Biol 27:107–132. doi:10.1146/annurev-cellbio-092910-154005
95. Molitoris JK, McColl KS, Swerdlow S et al (2011) Glucocorticoid elevation of dexamethasone-induced gene 2 (Dig2/RTP801/REDD1) protein mediates autophagy in lymphocytes. J Biol Chem 286:30181–30189. doi:10.1074/jbc.M111.245423
96. Monneaux F, Briand JP, Muller S (2000) B and T cell immune response to small nuclear ribonucleoprotein particles in lupus mice: autoreactive CD4(+) T cells recognize a T cell epitope located within the RNP80 motif of the 70 K protein. Eur J Immunol 30:2191–2200. doi:10.1002/1521-4141(2000)30:8<2191::AID-IMMU2191>3.0.CO;2-R
97. Monneaux F, Dumortier H, Steiner G et al (2001) Murine models of systemic lupus erythematosus: B and T cell responses to spliceosomal ribonucleoproteins in MRL/Fas(lpr) and (NZB x NZW)F(1) lupus mice. Int Immunol 13:1155–1163. doi:10.1093/intimm/13.9.1155
98. Monneaux F, Hoebeke J, Sordet C et al (2005) Selective modulation of CD4+ T cells from lupus patients by a promiscuous, protective peptide analog. J Immunol 175:5839–5847. doi:10.4049/jimmunol.175.9.5839

99. Monneaux F, Lozano JM, Patarroyo ME et al (2003) T cell recognition and therapeutic effect of a phosphorylated synthetic peptide of the 70 K snRNP protein administered in MRL/lpr mice. Eur J Immunol 33:287–296. doi:10.1002/immu.200310002
100. Monneaux F, Parietti V, Briand J-P, Muller S (2004) Intramolecular T cell spreading in unprimed MRL/lpr mice: importance of the U1-70 k protein sequence 131–151. Arthritis Rheum 50:3232–3238. doi:10.1002/art.20510
101. Monneaux F, Parietti V, Briand J-P, Muller S (2007) Importance of spliceosomal RNP1 motif for intermolecular T-B cell spreading and tolerance restoration in lupus. Arthritis Res Ther 9:R111. doi:10.1186/ar2317
102. Muller S, Monneaux F, Schall N et al (2008) Spliceosomal peptide P140 for immunotherapy of systemic lupus erythematosus: results of an early phase II clinical trial. Arthritis Rheum 58:3873–3883. doi:10.1002/art.24027
103. Muller S, Wallace D (2014) The importance of implementing proper selection of excipients in lupus clinical trials. Lupus. doi:10.1177/0961203314525249
104. Münz C (2012) Antigen processing for MHC class II presentation via autophagy. Front Antigen Present Cell Biol 3:9. doi:10.3389/fimmu.2012.00009
105. Murthy A, Li Y, Peng I et al (2014) A Crohn's disease variant in Atg16l1 enhances its degradation by caspase 3. Nature 506:456–462. doi:10.1038/nature13044
106. Nedjic J, Aichinger M, Emmerich J et al (2008) Autophagy in thymic epithelium shapes the T-cell repertoire and is essential for tolerance. Nature 455:396–400. doi:10.1038/nature07208
107. Neefjes J, Jongsma MLM, Paul P, Bakke O (2011) Towards a systems understanding of MHC class I and MHC class II antigen presentation. Nat Rev Immunol 11:823–836. doi:10.1038/nri3084
108. Nogalska A, D'Agostino C, Terracciano C et al (2010) Impaired autophagy in sporadic inclusion-body myositis and in endoplasmic reticulum stress-provoked cultured human muscle fibers. Am J Pathol 177:1377–1387. doi:10.2353/ajpath.2010.100050
109. Ohshima J, Lee Y, Sasai M et al (2014) Role of mouse and human autophagy proteins in IFN-γ-induced cell-autonomous responses against Toxoplasma gondii. J Immunol 192:3328–3335. doi:10.4049/jimmunol.1302822
110. Ohsumi Y (2014) Historical landmarks of autophagy research. Cell Res 24:9–23. doi:10.1038/cr.2013.169
111. Okamoto K (2014) Organellophagy: eliminating cellular building blocks via selective autophagy. J Cell Biol 205:435–445. doi:10.1083/jcb.201402054
112. Oliva L, Cenci S (2014) Autophagy in plasma cell pathophysiology. Front Immunol 5:103. doi:10.3389/fimmu.2014.00103
113. Orozco G, Eyre S, Hinks A et al (2011) Study of the common genetic background for rheumatoid arthritis and systemic lupus erythematosus. Ann Rheum Dis 70:463–468. doi:10.1136/ard.2010.137174
114. Page N, Gros F, Schall N et al (2011) HSC70 blockade by the therapeutic peptide P140 affects autophagic processes and endogenous MHCII presentation in murine lupus. Ann Rheum Dis 70:837–843. doi:10.1136/ard.2010.139832
115. Page N, Schall N, Strub J-M et al (2009) The spliceosomal phosphopeptide P140 controls the lupus disease by interacting with the HSC70 protein and via a mechanism mediated by γδ T cells. PLoS One 4:e5273. doi:10.1371/journal.pone.0005273
116. Paludan C, Schmid D, Landthaler M et al (2005) Endogenous MHC class II processing of a viral nuclear antigen after autophagy. Science 307:593–596. doi:10.1126/science.1104904
117. Pampliega O, Orhon I, Patel B et al (2013) Functional interaction between autophagy and ciliogenesis. Nature 502:194–200, doi:10.1038/nature12639
118. Park B, Brinkmann MM, Spooner E et al (2008) Proteolytic cleavage in an endolysosomal compartment is required for activation of Toll-like receptor 9. Nat Immunol 9:1407–1414. doi:10.1038/ni.1669
119. Pengo N, Scolari M, Oliva L et al (2013) Plasma cells require autophagy for sustainable immunoglobulin production. Nat Immunol 14:298–305. doi:10.1038/ni.2524

120. Pierdominici M, Barbati C, Vomero M et al (2014) Autophagy as a pathogenic mechanism and drug target in lymphoproliferative disorders. FASEB J 28:524–535. doi:10.1096/fj.13-235655
121. Pierdominici M, Vomero M, Barbati C et al (2012) Role of autophagy in immunity and autoimmunity, with a special focus on systemic lupus erythematosus. FASEB J 26:1400–1412. doi:10.1096/fj.11-194175
122. Pua HH, Guo J, Komatsu M, He Y-W (2009) Autophagy is essential for mitochondrial clearance in mature T lymphocytes. J Immunol 182:4046–4055. doi:10.4049/jimmunol.0801143
123. Puleston DJ, Simon AK (2014) Autophagy in the immune system. Immunology 141:1–8. doi:10.1111/imm.12165
124. Ravikumar B, Sarkar S, Davies JE et al (2010) Regulation of mammalian autophagy in physiology and pathophysiology. Physiol Rev 90:1383–1435. doi:10.1152/physrev.00030.2009
125. Renna M, Jimenez-Sanchez M, Sarkar S, Rubinsztein DC (2010) Chemical inducers of autophagy that enhance the clearance of mutant proteins in neurodegenerative diseases. J Biol Chem 285:11061–11067. doi:10.1074/jbc.R109.072181
126. Rose NR, Bona C (1993) Defining criteria for autoimmune diseases (Witebsky's postulates revisited). Immunol Today 14:426–430. doi:10.1016/0167-5699(93)90244-F
127. Rubinsztein DC, Codogno P, Levine B (2012) Autophagy modulation as a potential therapeutic target for diverse diseases. Nat Rev Drug Discov 11:709–730. doi:10.1038/nrd3802
128. Ryter SW, Mizumura K, Choi AMK (2014) The impact of autophagy on cell death modalities. Int J Cell Biol 2014:502676. doi:10.1155/2014/502676
129. Saitoh T, Akira S (2010) Regulation of innate immune responses by autophagy-related proteins. J Cell Biol 189:925–935. doi:10.1083/jcb.201002021
130. Saitoh T, Fujita N, Jang MH et al (2008) Loss of the autophagy protein Atg16L1 enhances endotoxin-induced IL-1beta production. Nature 456:264–268. doi:10.1038/nature07383
131. Sandri M, Coletto L, Grumati P, Bonaldo P (2013) Misregulation of autophagy and protein degradation systems in myopathies and muscular dystrophies. J Cell Sci 126:5325–5333. doi:10.1242/jcs.114041
132. Schall N, Muller S (2015) Resetting the autoreactive immune system with a therapeutic peptide in lupus. Lupus 24:412–418. doi: 10.1177/0961203314556138
133. Schall N, Page N, Macri C et al (2012) Peptide-based approaches to treat lupus and other autoimmune diseases. J Autoimmun 39:143–153. doi:10.1016/j.jaut.2012.05.016
134. Shen S, Kepp O, Kroemer G (2012) The end of autophagic cell death? Autophagy 8:1–3. doi:10.4161/auto.8.1.16618
135. Shibutani ST, Yoshimori T (2014) A current perspective of autophagosome biogenesis. Cell Res 24:58–68. doi:10.1038/cr.2013.159
136. Shoji-Kawata S, Sumpter R, Leveno M et al (2013) Identification of a candidate therapeutic autophagy-inducing peptide. Nature 494:201–206. doi:10.1038/nature11866
137. Singh SB, Davis AS, Taylor GA, Deretic V (2006) Human IRGM induces autophagy to eliminate intracellular mycobacteria. Science 313:1438–1441. doi:10.1126/science.1129577
138. Staskiewicz L, Thorburn J, Morgan MJ, Thorburn A (2013) Inhibiting autophagy by shRNA knockdown: cautions and recommendations. Autophagy 9:1449–1450. doi:10.4161/auto.24895
139. Stephenson LM, Miller BC, Ng A et al (2009) Identification of Atg5-dependent transcriptional changes and increases in mitochondrial mass in Atg5-deficient T lymphocytes. Autophagy 5:625–635. doi:10.4161/auto.5.5.8133
140. Stricher F, Macri C, Ruff M, Muller S (2013) HSPA8/HSC70 chaperone protein: structure, function, and chemical targeting. Autophagy 9:1937–1954. doi:10.4161/auto.26448
141. Stylianou K, Petrakis I, Mavroeidi V et al (2011) The PI3K/Akt/mTOR pathway is activated in murine lupus nephritis and downregulated by rapamycin. Nephrol Dial Transplant 26:498–508. doi:10.1093/ndt/gfq496
142. Sumpter MD, Tatro LS, Stoecker WV, Rader RK (2012) Evidence for risk of cardiomyopathy with hydroxychloroquine. Lupus 21:1594–1596. doi:10.1177/0961203312462757
143. Temiz P, Weihl CC, Pestronk A (2009) Inflammatory myopathies with mitochondrial pathology and protein aggregates. J Neurol Sci 278:25–29. doi:10.1016/j.jns.2008.11.010

144. Thomé R, Issayama LK, DiGangi R et al (2014) Dendritic cells treated with chloroquine modulate experimental autoimmune encephalomyelitis. Immunol Cell Biol 92:124–132. doi:10.1038/icb.2013.73
145. Troncoso R, Paredes F, Parra V et al (2014) Dexamethasone-induced autophagy mediates muscle atrophy through mitochondrial clearance. Cell Cycle 13:2281–2295. doi:10.4161/cc.29272
146. Tsvetkov AS, Miller J, Arrasate M et al (2010) A small-molecule scaffold induces autophagy in primary neurons and protects against toxicity in a Huntington disease model. Proc Natl Acad Sci U S A 107:16982–16987. doi:10.1073/pnas.1004498107
147. Valdor R, Mocholi E, Botbol Y et al (2014) Chaperone-mediated autophagy regulates T cell responses through targeted degradation of negative regulators of T cell activation. Nat Immunol 15:1046–1054. doi:10.1038/ni.3003
148. Van Kasteren SI, Overkleeft HS (2014) Endo-lysosomal proteases in antigen presentation. Curr Opin Chem Biol 23C:8–15. doi:10.1016/j.cbpa.2014.08.011
149. Vidal RL, Matus S, Bargsted L, Hetz C (2014) Targeting autophagy in neurodegenerative diseases. Trends Pharmacol Sci. doi:10.1016/j.tips.2014.09.002
150. Villadangos JA, Bryant RA, Deussing J et al (1999) Proteases involved in MHC class II antigen presentation. Immunol Rev 172:109–120. doi:10.1111/j.1600-065X.1999.tb01360.x
151. Xu K, Xu P, Yao J-F et al (2013) Reduced apoptosis correlates with enhanced autophagy in synovial tissues of rheumatoid arthritis. Inflamm Res 62:229–237. doi:10.1007/s00011-012-0572-1
152. Xu X, Kobayashi S, Chen K et al (2013) Diminished autophagy limits cardiac injury in mouse models of type 1 diabetes. J Biol Chem 288:18077–18092. doi:10.1074/jbc.M113.474650
153. Xu Y, Jagannath C, Liu X-D et al (2007) Toll-like receptor 4 is a sensor for autophagy associated with innate immunity. Immunity 27:135–144. doi:10.1016/j.immuni.2007.05.022
154. Yamahara K, Yasuda M, Kume S et al (2013) The role of autophagy in the pathogenesis of diabetic nephropathy. J Diabetes Res 2013:193757. doi:10.1155/2013/193757
155. Yamamoto A, Tagawa Y, Yoshimori T et al (1998) Bafilomycin A1 prevents maturation of autophagic vacuoles by inhibiting fusion between autophagosomes and lysosomes in rat hepatoma cell line, H-4-II-E cells. Cell Struct Funct 23:33–42
156. Yang W, Tang H, Zhang Y et al (2013) Meta-analysis followed by replication identifies loci in or near CDKN1B, TET3, CD80, DRAM1, and ARID5B as associated with systemic lupus erythematosus in Asians. Am J Hum Genet 92:41–51. doi:10.1016/j.ajhg.2012.11.018
157. Yang Z, Klionsky DJ (2010) Eaten alive: a history of macroautophagy. Nat Cell Biol 12:814–822. doi:10.1038/ncb0910-814
158. Zhou D, Li P, Lin Y et al (2005) Lamp-2a facilitates MHC class II presentation of cytoplasmic antigens. Immunity 22:571–581. doi:10.1016/j.immuni.2005.03.009
159. Zhou X, Lu X, Lv J et al (2011) Genetic association of PRDM1-ATG5 intergenic region and autophagy with systemic lupus erythematosus in a Chinese population. Ann Rheum Dis 70:1330–1337. doi:10.1136/ard.2010.140111
160. Zimmer R, Scherbarth HR, Rillo OL et al (2013) Lupuzor/P140 peptide in patients with systemic lupus erythematosus: a randomised, double-blind, placebo-controlled phase IIb clinical trial. Ann Rheum Dis. doi:10.1136/annrheumdis-2012-202460

Autophagy Networks in Cardiovascular Diseases

Lorena García, Mario Bustamante, Marcela Pizarro, Clara Quiroga, Andrea E. Rodriguez, Francisco Westermeier, Izela Bernal-Sore, Pablo F. Castro, Sergio Lavandero, and Rodrigo Troncoso

Abstract Cardiovascular system is responsible of delivering all nutrients and oxygen that may request the organism, and because of its importance exist an almost complete knowledge about how it works. However, the major incidence of diseases in the world are related to the cardiovascular system, for this reason is vital to detect the effect and causality of this type of diseases. Due to autophagy is necessary to maintain the structure and function of cardiac cells, this process has been deeply study in cardiomyocytes, cardiac fibroblast, endothelial cells and vascular smooth muscle cells. Optimal autophagy activity is critical to the maintenance of cardiovascular homeostasis; deregulated autophagy levels contribute to development of heart disease. In this chapter, we discuss the relationship between autophagy networks and cardiovascular diseases.

L. García (✉) • M. Bustamante • M. Pizarro • A.E. Rodriguez • F. Westermeier
I. Bernal-Sore
Advanced Center for Chronic Diseases (ACCDiS), Facultad Ciencias Químicas y Farmacéuticas and Facultad Medicina, Universidad de Chile, Santiago 838049, Chile
e-mail: logarcia@uchile.cl

C. Quiroga • P.F. Castro
Advanced Center for Chronic Diseases, Facultad Medicina, Pontificia Universidad Catolica de Chile, Santiago, Chile

S. Lavandero
Advanced Center for Chronic Diseases (ACCDiS), Facultad Ciencias Químicas y Farmacéuticas and Facultad Medicina, Universidad de Chile, Santiago 838049, Chile

Cardiology Division, Department of Internal Medicine, University of Texas Southwestern Medical Center, Dallas, TX, USA

R. Troncoso (✉)
Instituto de Nutrición y Tecnología de los Alimentos (INTA), Universidad de Chile, Santiago, Chile

Advanced Center for Chronic Diseases (ACCDiS), Facultad Ciencias Químicas y Farmacéuticas and Facultad Medicina, Universidad de Chile, Santiago 838049, Chile
e-mail: rtroncoso@inta.uchile.cl

M.C. Maiuri, D. De Stefano (eds.), *Autophagy Networks in Inflammation*, Progress in Inflammation Research, DOI 10.1007/978-3-319-30079-5_14

Abbreviations

7-KC	7-Ketocholesterol
AGEs	Advanced glycation end-products
AKT (PKB)	Protein kinase B
AMPK	AMP-activated protein kinase
ANF	Atrial natriuretic factor
Ang II	Angiotensin II
AT1 receptor	Angiotensin II receptor, type 1
AT2 receptor	Angiotensin II receptor, type 2
ATG	Autophagy-related gene
ATP	Adenosine triphosphate
Aβ	β-amyloid
Bad	Bcl-2-associated death promoter
BAECs	Bovine aortic endothelial cells
Bcl-2	B-cell lymphoma 2
CFs	Cardiac fibroblasts
CMFs	Cardiac myofibroblasts
Col-I	Type I collagen
CryAB	αB-crystallin (CryAB)
DOX	Doxorubicin
DRCM	Desmin-related cardiomyopathy
eIF2α	Eukaryotic Initiation Factor 2α
ER	Endoplasmic reticulum
UNEAC	Unificar nomenclatura en ambos casos
HDACs	Histone deacetylases
HF	Heart failure
HFD	High Fat Diet
HNE	4-hydroxynonenal
HUVECs	Human umbilical veins endothelial cells
I/R	Ischemic/Reperfusion
IFN-γR	Interferon-γ receptor
IL-1β	Interleukin-1β
JNK	c-Jun N-terminal kinase
LAMP2	Lysosome-associated membrane protein 2
LC3	Microtubule-associated protein 1 light chain 3
LDL	Low-density lipoprotein
MAPK	Mitogen-activated protein kinases
MDs	Mitochondrial diseases
MIF	Macrophage migration inhibitory factor
miRNAs	MicroRNAs
MMPs	Matrix metalloproteinases
mTOR	Mammalian target of rapamycin
mTORC1	mTOR complex 1

NADPH	Nicotinamide adenine dinucleotide phosphate
NO	Nitric oxide
OPN	Osteopontin
Ox-LDL	Oxidized LDL particles
p62/SQSTM1	Sequestosome 1
PDGF	Platelet-derived growth factor
PERK	Protein kinase RNA-like endoplasmic reticulum kinase
PI3K	Phosphatidylinositol 3-kinase
POAF	Postoperative atrial fibrillation
POVPC	1-palmitoyl-2-(5-oxovaleroyl)-sn-glycero-3-phosphocholine
PPARγ	Peroxisome proliferator-activated receptor-γ
ROS	Reactive oxygen species
Scar	Fibrous tissue formation
SMCs	Smooth muscle cells
TAC	Transverse aortic constriction
TFA	Trans fatty-acids
TFEB	Transcription factor EB
TGFβ1	Transforming growth factor β1
TNFα	Tumor necrois factor α
TRPV1	Transient receptor potential vanilloid subfamily 1
ULK1	UNC-51-like kinase 1
UPR	Unfolded protein response
VEGF	Vascular endothelial growth factor
VSMCs	Vascular smooth muscle cells

1 Introduction

The cardiovascular system comprises the heart and blood vessels, including arteries, veins, and capillaries, both systemic and pulmonary. The cardiovascular system is a closed tubular system in which the blood is propelled by the heart. The heart is a mechanical pump, its main function is to propel blood throughout the tissues of the body, carrying nutrients and oxygen and removing carbon dioxide and other metabolic waste, and through the lungs, delivering carbon dioxide and accepting oxygen. The heart is an incredibly resilient organ, marked by approximately 2.5 billion contractions over a 70 year lifetime. As a consequence, the heart is a robust consumer of energy, requiring a constant supply of oxygen and metabolic fuels in order to sustain contractile function. Energy reserves in the heart are limited, sufficient only to support contraction for a very few seconds; as a result, energy must be produced continually by catabolism of a variety of energy substrates [78]. The heart is a "metabolic omnivore", capable of metabolizing free fatty acids, glucose, lactate, pyruvate, ketone bodies, and amino acids. Under normal resting conditions, metabolism is mainly oxidative, with free fatty acids and glucose being the major sources of energy. The preferred substrate depends on arterial substrate concentrations (dietary conditions), hormonal factors (mainly insulin), and workload. However, glycolytic ATP production through conversion of glucose to lactate is independent of oxygen, thus glucose is the preferred substrate under hypoxic conditions such as ischemia and increased workload [78]. Much of cardiovascular disease centers on blood vessels, which can be afflicted by atherosclerotic change, calcification, inflammation, vasomotor dysfunction, hypertrophic thickening of the vessel walls, and more. Arteries, which carry blood from the left heart to the body and from the right heart to the lungs, are thick-walled, with a muscular contraction that helps propel blood downstream. Smaller caliber arteries are termed arterioles, which feed ultimately into capillaries where gas and small molecule exchange takes place. Conversely, veins carry blood back to the heart. Macroautophagy (herein referred as "autophagy") is a catabolic process involved in protein degradation, replacement of organelles, and no selective degradation of cytoplasmic components during stress or nutrient starvation. Autophagy begins with the formation of autophagosomes, a double membrane intracellular structure that surrounds origin reticular cytoplasmic contents and eventually fuses with lysosomes for load degradation. Materials degraded within these autolysosomes are recruited to anabolic reactions and maintain energy levels and provide macromolecules for the synthesis of higher order structures (nucleic acids, proteins or organelles), thus maintaining cell metabolism, homeostasis, and survival [69]. Despite its key role in the survival, autophagy also contributes to cell death when activated or inefficient excess, as occurs during the development of organs and tissues or in certain disease states. Indeed, diverse studies have shown that autophagic flux contributes to the pathogenesis of cardiovascular diseases, diabetes, inflammatory disorders, infection, and cancer [72]. However, despite considerable evidence linking autophagic activity to heart failure (HF) progression, uncertainty remains regarding whether increased autophagy is an epiphenomenon or a causative factor.

2 Heart

2.1 Cardiomyocytes

The myocardium comprises long-lived, largely post-mitotic cardiomyocytes. Therefore, despite ongoing controversy regarding the regenerative capacity of adult heart, elucidation of cellular mechanisms underlying cardiomyocyte function, viability, and cellular homeostasis has a pivotal role in the design of new therapeutics in cardiovascular medicine. Autophagy is important to maintain cardiomyocyte function and viability. Also, autophagy provides a critical means for intracellular self-renewal, energy repletion, and substrate recycling through degradation of dysfunctional or misfolded proteins and aged and/or damaged organelles.

Cardiomyocyte function and survival rely critically on the presence of basal levels of autophagy [71]. In a model of controlled cardiomyocyte-specific Atg5 deficiency, abrogation of basal autophagy provoked precipitous declines in cardiac structure and performance [106]. In this context where autophagic flux is silenced, pressure overload triggers rapid-onset cardiac hypertrophy, left ventricular dilation, and diminished cardiac output [106]. Thus, constitutive autophagy controls cardiomyocyte size and function, and is a protective mechanism in hemodynamic stress [106]. Further, mutation of LAMP2 protein, characteristic of Danon disease, triggers a severe and progressive cardiomyopathy, stemming from defective fusion of autophagosomes with lysosomes [82]. On the other hand, the long-term consequences of Atg5-deficiency in the heart includes cardiac hypertrophy and diminished cardiac output with age, resulting from accumulation of defective proteins and organelles [142]. Together, these facts highlight the vital housekeeping role for cardiomyocyte autophagy as a mechanism of protein and organelle surveillance and quality control.

2.2 *Cardiac Fibroblasts*

The most abundant cells in heart are cardiac fibroblasts (CFs) comprising about 70 % of all cardiac cells. They main role is to maintain the structural integrity of the heart through controlled proliferation and extracellular matrix turnover [117, 153]. Most of cardiac diseases possess an inflammatory component, so, in response to injury or stress, CFs or infiltrating immune cells respond by secreting growth factors and proinflammatory cytokines as TGFβ1, TNFα and IL-1β and can adopt a specialized phenotype, the cardiac myofibroblasts (CMFs) [17, 135], in response to TGFβ1 activity. CMFs promote increased synthesis of collagen I and III, fibronectin, MMPs and fibrous tissue formation (scar) in the area of damage [116, 145]. After healing, CMFs died by apoptosis and disappear from affected area [115]. However, in chronic cardiac disease with inflammatory components, both CFs and CMFs overreact increasing the rate of fibrotic tissue formation, increasing CMFs survival and perpetuating the inflammatory-fibrogenic cascade.

In fibroblasts, autophagy is necessary to both normal growth and fibrotic disorders [22]. There are some evidence indicating that autophagy might play a protective role for CFs. We have previously shown that the increase of autophagic flux in CFs exposed to β_2-adrenergic stimulation reduces the deleterious effect of high adrenergic stimulation and correlates with an enhanced degradation of collagen [4]. In line with these results, the generation of mice deficient in autophagic protein Beclin 1 showed an increased collagen-I deposition; interestingly TGFβ1 which is known to induces profibrotic phenotype also induces autophagy and thus suppress aberrant accumulation of Col-I [59]. On the other hand, Ghavami et al [36] showed that trans fatty-acids (TFA) from diverse origin diminished CMFs survival by apoptosis. Furthermore, TFA-induced apoptosis is dependent on the activation of autophagy by TFA [36].

2.3 Vascular Smooth Muscle Cells

Vascular smooth muscle cells (VSMCs) are the main constitutive stromal cells of the vascular wall, assuming a variety of different structural and physiological functions. Its principal function is contraction which permits regulation of vessel tone and diameter and thus control of blood pressure and blood flow distribution. VSMCs are a highly plastic cell type that, under different kind of stimuli, switch their phenotypic state from a differentiated-contractile one to a less differentiated-synthetic state characterized by an increased ability to proliferate, migrate and synthesize proteins of the extracellular matrix as collagen [80]; to a macrophage-like phenotype [2]; or even to an osteoblastic lineage [136]. This chameleon-like feature of VSMCs has been extensively studied and it is common to all cardiovascular diseases where they are involved as hypertension and atherosclerosis and vascular calcification [37, 65, 88].

Between all cardiovascular diseases, atherosclerosis is the leading cause of heart disease [14, 144]. Several immune cells as monocytes, macrophages, T cells, mast cells and dendritic cells among others [43] are involved in the genesis and development of this chronic inflammatory disease. Activated macrophages in the vascular wall secrete pro-inflammatory cytokines as IL-1β and TNFα. It is this pro-inflammatory milieu that induces proliferation and migration of VSMCs from the media to the intima layer, where produce extracellular matrix molecules, including interstitial collagen and elastin, and form a fibrous cap that covers the plaque [76]. The increased VSMC content of atherosclerotic lesions is associated with increased plaque stability [152].

Autophagy has been linked to several cardiovascular diseases [69, 90] and atherosclerosis is not an exemption [87]. Autophagy can be triggered by reactive oxygen species [127], oxidized lipoproteins [102, 103, 110], ER stress [164] and hypoxia [30, 131], all of which are found in atheroma plaque. In regard with VSMCs, autophagy has been observed since long ago. A work published in 1961 by Geer et al. titled "*The fine structure of human atherosclerotic lesions*", the authors made some interesting observations about some double membrane structures containing cytoplasmic dense material in VSMCs present at atherosclerotic lesions which now could be easily recognized as autophagosomal structures [35]. In spite of this and other similar observation (see references in [125]), autophagy in VSMCs in atherosclerosis has not received enough attention for both, technical issues and widely used animal models that does not resembles human atherosclerosis as they show fewer VSMCs than lesion and atheroma from humans and other animal models [1, 87].

According to stimuli triggering autophagy in atherosclerosis, VSMCs autophagy can be initiated indirectly or directly by ROS. It has been shown that an increase in mitochondrial superoxide production as result of stimulation of cultured bovine aortic VSMCs with inorganic phosphate induce autophagy; this increase in autophagy play a protective role inhibiting the calcification of VSMCs [19]. Indirectly, oxidized lipids as 4-hydroxynonenal (HNE) and 1-palmitoyl-2-(5-oxovaleroyl)-sn-glycero-3-phosphocholine (POVPC) are both accumulated at atheroma plaques *in vivo* [126] and both has been shown to induce autophagy in VSMCs (increased LC3-II formation) [50], in a process possibly mediated by induction of ER-stress, phosphorylation of PERK and eIF2α of the unfolded protein response (UPR) and activation of the stress kinases p38 and JNK [41]. HNE is present at oxidized LDL particles (ox-LDL) together with the oxysterol 7-ketocholesterol (7-KC), that has been registered as a potent stimulator of autophagy and apoptosis in VSMCs [84]. The mechanisms by which 7-KC induces autophagy include the up-regulation of Nox4 expression, increased intracellular hydrogen peroxide levels, and inhibited autophagy-related gene 4B (ATG4) activity, indicating that also is a ROS-mediated autophagy induction [47]. This data is according to the notion that both 7-KC and HNE promote a protective form of autophagy triggered by ER-stress.

Autophagy can also be induced in VSMCs by cytokines as TNFα, osteopontin (OPN) and growth factors as PDGF-BB, but with despair results. Cytokines seems to induce mal-adaptive autophagy that ends-up with cell death by apoptosis playing a relevant role in atherosclerosis (TNFα) and abdominal aortic aneurysm (OPN) [55, 170]. Even both cytokines have a deleterious effect over cell survival, the mechanisms by they induce autophagy is different. While TNFα induces JNK and AKT kinases and this increase Beclin-1 expression [55], OPN stimulates autophagy directly through integrin/CD44 and p38 MAPK-mediated pathways in VSMCs [170]. The case of PDGF-BB is quite different. PDGF is known to stimulate VSMCs proliferation and migration during atherosclerosis course, increasing the stability of atheroma plaque. The phenotype switch induced by PDGF-BB is regulated by autophagy as cultured cells stimulated with the growth factor showed increased LC3-II abundance and LC3 puncta formation [124]. Likewise, inhibition of autophagy with 3-methyladenine, spautin-1 or bafilomycin inhibited the synthetic phenotype [124]. In this case, autophagy induction is protective as it stabilizes the fibrous cap that covers the plaque. On the other hand, pharmacological activation of autophagy can inhibit foam cell formation derived from VMSCs. This is how, activation of transient receptor potential vanilloid subfamily 1 (TRPV1) or telmisartan acting through AT2 receptor, induce AMPK-dependent autophagy [73, 74]. Also, telmisartan increase PPARγ expression and decreased lipid droplet accumulation [74].

In spite of the large knowledge regarding the development of atheroma plaques, and the efforts made with diverse therapies [14, 77], the pathophysiological mechanisms of atherosclerosis still are under a deep investigation and some new theories and therapeutic strategies have arisen [76], and the modulation of autophagy in VSMCs is proposed as a new therapeutic target [86, 107], as it can modulate the phenotype switching seen in VSMCS not only in atherosclerosis but hypertension and vascular calcification, and also improves survival of VMSCs, the latter being important for the stability of the plaque [148].

2.4 Endothelial Cells

In 1973, was reported by Jaffe et al. the first successful *in vitro* culture of human umbilical veins endothelial cells (HUVECs) [54], initiating explosive growth in research vascular biology and leading insights into angiogenesis, vasculogenesis, and tumor biology [104]. Even when in 1966 Nobel Laureate Lord Adrian Florey considered the endothelial cell as a "sheet of nucleated cellophane" to protect the vascular wall [29, 56], the current knowledge regarding endothelial function has been associated with several processes that underlie its pivotal role in the vascular homeostasis [104, 105] such as constitutive, anticoagulant, and anti-inflammatory functions as well as nonconstitutive, activated, thromboregulatory activities. Furthermore, these cells secrete angiocrines, which are able to modulate embryonic organogenesis, hematopoiesis, metastasis, and lung and liver regeneration. On the other hand, the endothelium has also been related with platelet activity, leukocyte adhesion and the regulation/maintenance of vascular tone in response to humoral, neural, and mechanical stimuli by synthesizing and releasing vasoactive substances such as endothelin-1 (vasoconstrictor) or a potent vasodilator substance identified as nitric oxide (NO) [33, 53].

The term "endothelial dysfunction" has been used to refer to several pathological conditions, including altered anticoagulant and anti-inflammatory properties of the endothelium, impaired modulation of vascular growth, and dysregulation of vascular remodeling [12]. However, in much of the literature this term has also been used to refer to an impairment of endothelium-dependent vasorelaxation caused by a loss of NO bioactivity in the vessel wall [31]. Several studies have shown endothelial dysfunction associated with atherosclerosis, hypertension [114, 129] and diabetes type II [5, 91, 138]. Moreover, it has also been related with alterations in specific organs such as heart [42, 48, 68, 140], kidney [10, 11, 58] or liver [147]. In this context, one of the key cellular processes involved in the pathogenesis of some cardiovascular diseases listed above is both oxidant stress and production of reactive oxygen species (ROS) [12]. ROS are a family of molecules including molecular oxygen and its derivatives produced in all aerobic cells. Excessive production of ROS, outstripping endogenous antioxidant defense mechanisms, has been implicated in processes in which they oxidize biological macromolecules, such as DNA, protein, carbohydrates, and lipids [12]. The main sources of ROS described in vascular cells include the arachidonic acid pathways enzymes lipoxygenase and cyclooxygenase, cytochrome p450s, NADH/NADPH oxidases, NO synthase, peroxidases, and the mitochondrial respiration [12]. Mitochondrial diseases (MDs) are a clinically heterogeneous group of disorders characterized by impairment of the respiratory chain function associated both altered oxidative phosphorylation and oxidative stress, which are one of the key factors responsible for endothelial dysfunction [98, 165]. The interface between stress adaptation and cell death is important for understanding redox biology and disease pathogenesis, being the autophagy (or self-eating) a pivotal sensor of redox signaling at this switch in cellular responses [70].

The functional significance of autophagy in human cardiovascular disease pathogenesis remains unclear [121]. However, the rapid advancement regarding the mechanisms and regulation of autophagy has placed this process at the center of current research in major human disorders [85]. Thus, preclinical studies have identified autophagy as a process that can be activated during vascular disorders, including ischemia–reperfusion injury of the heart and other organs, cardiomyopathy, myocardial injury, and atherosclerosis [121]. In this context, autophagy plays dual roles in cardiovascular diseases through adaptive or maladaptive regulation. Physiological autophagy serves as a protective mechanism to maintain normal cardiovascular function. However, impaired autophagy contributes to disease development [90]. In endothelial cells has been reported that initial progression of Alzheimer disease and endothelial autophagy induced by vascular β-amyloid (Aβ) could be related with impairment of neurovascular regeneration [45]. In addition, activation of AMP-activated protein kinase (AMPK) by mitochondria-derived reactive oxygen species (ROS) is required for autophagy in cultured of bovine aortic endothelial cells (BAECs) [150]. Moreover, HUVECs were induced an autophagic as well as an apoptotic response exposed to Kringle 5 (K5), a fragment of plasminogen and described as potent angiogenesis inhibitor [108] in a similar way to human vascular endothelial cell line (EAhy926) exposed to endostatin [15]. Despite that the evidence described is associated with deleterious effects of autophagy, there are reports related with protective roles of this process in the injury induced by advanced glycation end-products (AGEs) [154, 155] or associated with increased VEGF-induced angiogenesis in HUVEC [25].

In conclusion, a better understanding of the function of autophagy and potentially involves both adaptive and maladaptive outcomes in the vascular system could provide new therapeutic avenues for disease prevention or control.

3 Autophagy in Cardiovascular Diseases

3.1 Cardiac Hypertrophy

The heart is a highly plastic organ capable of growth or shrinkage in response to changes in physiological or pathological demand. The heart undergoes hypertrophy in response to mechanical overload, which can be induced by high blood pressure or a loss of myocardial tissue after myocardial infarction. Initially, the ventricular hypertrophy is a compensatory mechanism to decrease wall stress and to increase cardiac output. However, when the ventricular hypertrophy progresses, the heart become dilated, contractile function decline, and occur a heart failure. Indeed, this progressive course of disease occurs commonly in patients with hypertension or ischemic heart disease [69].

Cardiac hypertrophy is characterized by thickening myocardium and decreasing in heart chamber volume. The hypertrophic growth of cardiomyocytes is initiated by endocrine, paracrine, and autocrine factors that stimulate a wide array of membrane receptors [151]. Their activation results in the triggering of multiple cytoplasmic signal transduction cascades, which ultimately affects nuclear factors and the regulation of gene expression [101]. This produces an enhanced protein synthesis increasing the expression of fetal genes and forming of new sarcomeres. However, it is necessary a remodeling of existing cellular elements, therefore catabolic processes like autophagy are also activated [151].

3.2 *Basal Autophagy*

The presence of basal levels of autophagy is necessary to maintain the structure and function of the cardiomyocyte. The abrogation of basal autophagy in a model of cardiac-specific Atg5 deficiency in mice provokes cardiac hypertrophy and left ventricular dilatation. Furthermore, the hearts of those mice shows disorganized sarcomere structure and mitochondrial misalignment, which interfere with its contractile function [106]. Cardiomyocyte hypertrophy is also observed when the Atg7 levels are reduced. The knockdown of ATG7 in rat neonatal cardiomyocytes induces the morphological and biochemical features of *in vitro* hypertrophy [106]. Thus, inhibition of basal autophagy provokes cardiomyocyte hypertrophy *in vivo* and *in vitro* models.

3.3 *Adaptive or Maladaptive Autophagy*

Several investigators have reported that a decrease in autophagy facilitates the cardiac hypertrophy in response to hypertrophic stimulus. Models where autophagy is silenced, pressure overload triggers rapid-onset cardiac hypertrophy, left ventricular dilation and diminished cardiac output. Indeed, autophagy plays a beneficial role in the heart in response to pressure overload or isoproterenol [106].

The mTOR-signaling inhibitor rapamycin activates autophagy and prevents cardiac hypertrophy in several models. In mice and neonatal cardiomyocytes, rapamycin inhibits completely the cardiac hypertrophy induces by thyroid hormones [66]. Additionally, rapamycin significantly decreases the level of the cardiac hypertrophy induced by pressure overload [40]. Rapamycin reverts changes in alpha-myosin heavy chain and sarcoplasmic reticulum Ca^{2+} ATPase that occurs in cardiomyocytes hypertrophied, improving the cardiac parameters in mice with decompensated cardiac hypertrophy [89]. The inhibition of cardiac hypertrophy is also achieved with the activation of AMPK, which stimulate autophagy decreasing the mTORC1 signaling [73, 75].

In contrast, other group of publications evidence that autophagy is required for hypertrophic growth of the myocardium. It seems paradoxical that a mechanism of protein degradation activates during cell growth, but the hypertrophic remodeling requires the processing of existing cellular elements. Cardiac-specific Beclin1 overexpression in an *in vivo* model of pressure-overload promotes rapid transition to cardiac failure. Conversely, diminishing the autophagic response in a Beclin1 haploinsufficient mouse attenuates pathological remodeling induced by afterload stress [13, 171]. It has proposed that the maladaptative autophagy is controlled by Histone deacetylases (HDACs) during the pathological cardiac remodeling. Trichostatin A, a HDAC inhibitor, abolish the hypertrophic growth and attenuate the activation of the autophagy associated. Moreover, in animals with preexisting hypertrophy, the inhibition of HDAC reverted ventricular mass and normalized the ventricular function [13].

Therefore, there are discrepancies regarding to the benefic role of autophagy in cardiac hypertrophy. Such discrepancies may be attributed to differences in experimental settings. For instance, autophagy is adaptive under mild cardiac hypertrophy, whereas it becomes maladaptive under severe pressure overload.

3.4 *Regression of Cardiac Hypertrophy*

Hypertrophy can be reversed when the cardiac wall stress is reduced, a process termed regression. The reduction of hemodynamic stress, by example with antihypertensive treatment, induces regression of cardiac hypertrophy and improve the cardiac parameters. Regression of cardiac hypertrophy is accompanied by activation of sets of genes, including those involved in protein degradation [32].

Autophagy is activated in a model of hypertrophy regression where the aortic constriction is follow by deconstriction. The activation of autophagy in this model is mediated by FOXO1, which increase the expression of autophagy genes and the autophagosome formation [44]. Autophagy is also induced during the regression of cardiac hypertrophy generated by a continuous infusion of Ang II. Regression of cardiac hypertrophy induced by Ang II and pressure overload were attenuated in Atg5-deficient mice [111]. Those findings suggest that autophagy is necessary for regression of hypertrophy, probably to reduce the cellular components unnecessary.

3.5 *Regulation of Autophagy by MicroRNAs*

MicroRNAs (miRNAs) are endogenous small RNA molecules, which regulate post-transcriptionally the expression of their target gene. Generally, miRNAs inhibit protein synthesis by either repressing the translation or triggering the degradation of their mRNA targets [28].

Recent evidence has shown evidence that miRNAs control the autophagy activation during the cardiac hypertrophic growth. Hypertrophic conditions upregulate expression of miR-212 and miR-132 in cardiomyocytes. The cardio-specific overexpression of the miR-212 and miR-132 leads to pathological hypertrophy, while null mice of these miRNAs are protected of pressure overload-induced heart failure. Both miR-212 and miR-132 target FoxO3, an anti-hypertrophic and pro-autophagic transcription factor [146].

An opposite effect is observed with miR-34a, which is reduced in a rat model of Ang II-induced hypertrophy. miR-34a antagonizes Ang II-stimulated hypertrophy, whereas inhibition of miR-34a expression aggravated Ang II-stimulated hypertrophy. This miRNA decreases autophagy, binding the UTR of ATG9A inhibiting its protein expression and its activity in autophagy [52]. A similar effect is observed with the downregulation of miR-30a, which aggravates pressure overload-induced cardiomyocyte hypertrophy by activating autophagy through inhibition of the beclin-1 expression

[163]. Downregulation of miR-30 is also observed in Ang II-stimulated hypertrophy [112]. Therefore, the effect of miRNA in hypertrophic growth and autophagy is divergent; some miRNAs have a positive effect and others a negative effect.

3.6 Control of Autophagy by Cytokines and Its Role in Cardiac Hypertrophy

Cardiac hypertrophy is a multifactorial process where several extracellular signals are involved in the growth of myocardial tissue. However, a few numbers of investigations have established the contribution of inflammatory cytokines in the cardiac hypertrophy and autophagy activation. Angiotensin (Ang) II, the major effector peptide of renin–angiotensin system, causes hypertrophy of cardiac myocytes and mitogenesis of cardiac fibroblasts via AT1 receptor. Ang II induces the "fetal program" (induction of skeletal α-actin and ANF) and induces expression of the angiotensinogen and TGF-βI genes [122]. Indeed, TGF-β1 is necessary for Ang II–mediated cardiac hypertrophy [128]. On the other hand, Ang II induces vascular injury in part by activating innate and adaptive immunity. The treatment with the immunosuppressive agent cyclosporine A, protects against the Ang II–induced myocardial damage [95]. Moreover, Ang II-treated IFN-γR knockout mice exhibited reduced cardiac hypertrophy, reduced cardiac macrophage and T-cell infiltration, less fibrosis, and less electric remodeling independent of blood pressure changes [81]. Therefore, inflammatory signals produced by lymphocytes, like IFN-γ, play a role in cardiac damage induced by Ang II. Unfortunately, it is unknown whether IFN-γ has an effect in the activation of autophagy in cardiac cells.

Macrophage migration inhibitory factor (MIF) is a proinflammatory cytokine expressed in several cell types, including monocytes/macrophages, vascular smooth muscle, and cardiomyocytes. MIF is involved in the pathogenesis of several inflammatory diseases, and it has shown to be cardioprotective under various pathological conditions, including ischaemia-reperfusion injury [96]. Recently, it has shown that MIF antagonizes myocardial hypertrophy and fibrosis in a model of myocardial hypertrophy by TAC, maintaining a redox homeostasis and attenuating the activation of hypertrophic signaling pathways [63]. Furthermore, MIF deficiency increase the cardiac hypertrophy Induced by pressure overload and inhibit the autophagy. Rapamycin administration mitigated the exacerbated hypertrophic responses in MIF knockout mice, in agreement with a protective role of autophagy in cardiac hypertrophy pathology [159]. MIF controls the activation of autophagy, through to the inhibition of mTOR to protect against cardiac hypertrophic responses [159].

3.7 Ischemic/Reperfusion Injury

With the pass of the years autophagy, has been become in a key factor to develop of ischemic heart disease, heart failure and I/R injury [69]. Since autophagy is key to supply of essential nutrients during ischemia, to remove damaged mitochondria and to protect from apoptosis and other major injuries, but its overactivation can destroy essential cellular components and lead to cell death. These opposite effects of autophagy are considered by an important number of researchers like a double-edged sword [69].

In 2005, Yan et al. reported the activation of autophagy in cardiomyocytes in a porcine model of chronic I/R. They demonstrated I/R induces an increase in cathepsin D expression, in conjunction with an increase in autophagic proteins beclin-1 and LC3-II [160]. Also, it has been reported that samples from patients with dilated cardiomyopathy, a classical ischemic disease, there is an increase in myofibrillar disorganization and an evident vacuolization, in addition to an increase in lysosomal activity and cardiomyocytes loss [132].

However, there is few evidence that related the development of I/R injury, autophagy and inflammatory processes in heart. One of the most studied inflammatory phenomenon related to I/R events in heart is the atherosclerosis. More than 80 % of acute myocardial infarcts are the result of coronary atherosclerosis. This pathology is a chronic inflammatory disease of arteries and involves the development of plaques in the vessel walls. The rupture of these atherosclerotic plaques could produce thrombosis, myocardial infarction and death and its stability is mainly dependent from function and activity of two cellular lineages: macrophages and smooth muscle cells (SMCs) [21]. The different factors involved between autophagy and atherosclerosis was reviewed in [21], but its direct relationship with I/R injury development is still in the land of speculation.

Recently, Emanuel et al. determined that oxidized low-density lipoproteins (LDLs) and cholesterol crystals, commonly founded in lipid core of atherosclerotic plaques, produce lysosomal dysfunction in macrophages. Cells isolated from atherosclerotic plaques and culture maintained, showed alterations in lysosomal pH, proteolytic capacity and morphology. In contrast, when they increased biogenesis and lysosomal function, through the overexpression of the lysosomal master key transcription factor EB (TFEB), they observed an increase in pro-degradative response and protection against atherogenic lipid [27]. These results suggest that lysosomal function is impaired in atherosclerosis and that induction of a lysosomal function has antiatherogenic effects in macrophages and are complementary within those described by Razani et al. who showed dysfunctional autophagy involved in plaque formation. In mice, the macrophage-specific haploinsufficiency of ATG5 showed an increase in plaques formation and therefore a crucial role in atheroprotection [120].

Using transmission electron microscopy, it was observed that damage SMCs in experimental or human plaques presents an increase in cellular vacuolization suggesting therefore an increase in autophagy [62]. Despite this observation, is not fully clear yet whether this autophagy protects SMCs. In these cells, has been reported that the autophagy inducers free cholesterol or 4-hydroxynonenal protects against cell death [50, 156] and cell death induced by statins is reduced by 7-ketocholesterol, a recognized inducer of autophagy [84]. Protection of SMCs through autophagy activation might stabilize atherogenic plaque and prevent coronary artery syndromes, I/R injuries and sudden death.

Nowadays, there still is a pending task to determinate a direct relationship between autophagy, inflammation and I/R injuries.

3.8 Diabetic Cardiomyopathy

Two-thirds of the diabetic patients die from heart disease or stroke, which highlights the importance of understanding and treating cardiovascular complications of diabetes including diabetic cardiomyopathy [61]. Insulin signaling activates the PI3K-AKT-mTORC1 pathway not only to stimulate protein synthesis, but also to concurrently inhibit autophagy [92, 100]. It was thus hypothesized that insulin deficiency (type 1 diabetes) or insulin resistance (type 2 diabetes) would increase autophagic activity [93]. Recent studies have demonstrated that autophagy is inhibited in the heart of type 1 diabetic mice [154, 155, 168] and several forms of metabolic syndrome and type 2 diabetic animal models [46, 158]. Specifically, diabetes reduced the protein levels of LC3-II.

Cardiomyocytes isolated from type 2 diabetic *db/db* mice and (High Food Diet) HFD-induced obese mice exhibits a blunted autophagic response [83, 130] suggesting that the inhibition of autophagy may contribute to diabetic cardiomyopathy. However, some recent reports conflict with these findings. Mellor et al. reported that increased myocardial autophagic flux in fructose diet-induced type 2 diabetic mice resulted in pathological remodeling of the heart [94]. These results suggest that in type 2 diabetes, increased autophagy may serve as a compensatory response to insulin resistance, because the degradation of unnecessary cellular components are essential for maintaining normal cellular architecture and function.

3.9 Heart Failure

Heart failure is a multifactorial syndrome, which derives from a wide range of diseases. Is a progressive disease characterized by adverse ventricular remodeling, which involves changes in the balance between protein synthesis and protein degradation. The role of autophagy in heart failure is less clear, but multiple lines reveal that autophagy is potently induced in this pathology [99]. Evidence for autophagy in human heart disease emerged first from tissue sample of dilated cardiomyopathy [132]. In a model of pressure overload induced surgically by transverse aortic constriction (TAC), they have reported that autophagic activity increases rapidly after TAC, with a peak at 72 h, and is maintained at elevated levels for at least 3–4 weeks [171]. The degree of autophagic activity correlates with the magnitude of hypertrophic growth and with the rate of transition to heart failure [171], and steady-state levels of autophagic flux correlate with heart mass [13]. Transgenic mice with cardiomyocyte-restricted over-expression of Beclin 1, a rate-limiting protein in the autophagic cascade, manifest increased autophagic activity in the setting of elevated afterload and a correspondingly amplified pathological remodeling response, including ventricular dilation, systolic dysfunction, and early mortality [171]. Ultra structural analyses revealed numerous autophagic vacuoles containing cytoplasmatic material and organelles that were localized within degenerated cardiomyocytes. In dilated cardiomyopathy, autophagy appeared to be associated not only with degradation of damage intracellular organelles but also with progressive destruction of cardiomyocytes [132]. Miyata and collaborators studied the hamster model of human dilated cardiomyopathy. In this model, heart failure develops progressively. Ultra structural analysis of these heart revealed that many of the cardiomyocytes contained autophagy vacuoles and degraded mitochondria [99].

So, autophagy may facilitate hypertrophic growth and allow the sustenance of greater degrees of hypertrophy; this, in turn, promotes the emergence of systolic dysfunction and heart failure. But, complete abrogation of the catabolic response is similarly maladptative. Inactivation of gene coding for ATG5 triggers rapid on-set heart failure [106].

On the other hands, the "energy crisis" of heart failure stimulates robust activation of autophagy [70]. In advanced heart failure myocardial, ATP levels drop to 30–40 % of control [6, 137], which activates AMPK signaling. Recent studies show that AMPK can directly up-regulated autophagy by phosphorylating ULK1, an upstream kinase involved in autophagy initiation [26]. Moreover, fatty acid oxidation and oxidative phosphorylation inevitably generate ROS. When detoxifying system is over helmed, excessive ROS cause oxidative damage to proteins, lipids and organelles, which can directly activate autophagy. Activation of autophagy in this pathology may exacerbate the metabolic derangements characteristic of the syndrome. Excessive autophagy may trigger nonspecific degradation of essential metabolic enzymes and mitochondria, contributing to the crisis of energy. Autophagic cell death, or programmed cell death type II, may be a significant contributor to the pathogenesis of heart failure [64, 123].

3.10 Atrial Fibrillation

Postoperative atrial fibrillation (POAF) is a common surgical complication. In patients post-coronary bypass surgery several histological abnormalities, such as interstitial fibrosis and vacuolization, have been described in atrial samples from patients developing POAF [34]. This ultrastructural remodeling has been associated with the establishment of a pro-arrhythmic substrate. Electron micrographs of atrial tissue from patients with POAF showed significant accumulation of autophagic vesicles and lipofuscin deposits. Total protein ubiquitination was similar in the patients with and without POAF, but LC3B processing was markedly reduced in those with POAF, suggesting a selective impairment in autophagic flux in patients developing POAF. This study provides novel evidence that ultrastructural atrial remodeling characterized by impaired cardiac autophagy is present in patients developing POAF after coronary artery bypass surgery [34].

3.11 Anti-cancer Drugs Induced Cardiomyopathy

Cancer chemotherapy, particularly anthracyclines, as long been associated with significant cardiotoxicity [166]. Doxorubicin (DOX) is one of the most widely used and successful antitumor drugs, however, it is well known that is cardiotoxic [97].

Although intensive investigation, the underlying mechanisms responsible for DOX-induced cardiotoxicity have not been completely elucidate, but the ROS production is considered a major culprit of cardiomyocyte damage triggered by doxorubicin [141]. DOX-induced cardiotoxicity may present as either acute or chronic cardiomyopathy. The chronic cardiotoxicity is dose-dependent. In this case, the patient may develop dilated cardiomyopathy many years after receiving the last doxorubicin treatment and may lead to cardiac dysfunction and eventually to severe heart failure and death [149, 162].

In recent years a number of studies have emerged have focused on the role of autophagy in doxorubicin-induced cardiotoxicity [57, 79]. Several studies have shown that DOX treatment affects autophagy *in vitro* and *in vivo* [24]. Lu and colleagues were the first to report the effects of DOX on cardiac autophagy demonstrating that DOX stimulated autophagy *in vitro* and *in vivo* [79] and four subsequent studies supported these findings [16, 23, 60, 157]. However, two studies reported that DOX reduces autophagy [57, 133]. The discrepancies between studies may be due to a cell type specific effect. But the one consistency between all the studies is that reversal of the DOX-induced effect is beneficial in protecting against DOX-induced cytotoxicity.

The stimulation of autophagy by DOX has been shown to involve multiple mechanisms [24]. DOX induce autophagy via depletion of the transcription factor GATA4 and Bcl-2 [60], activation of S6K1 (p70 S6 kinase 1), down-regulation of insulin signaling [157], increased expression of Atg5, Atg12 and Bad [16]. It has also been reported that DOX-induced stimulation of autophagy involved up-regulation of several *atg* genes including *atg12*, *atg7*, *atg4* and Beclin1 and down-regulation of Bcl-2 [134].

Most *in vitro* and *in vivo* studies during the past several decades have suggested that DOX-induced cardiac toxicity are associated with cardiomyocyte apoptosis but, it has not been determined how specifically autophagy may contribute to inducing apoptosis in DOX-induced cardiotoxicity.

3.12 Aging

Aging is characterized by a progressive deterioration of cells and organs. It is to a large extent related to macromolecular damage by mitochondria ROS mostly affecting neurons and cardiomyocytes. These cells are not replaced during life. Continuous removal of exhausted components and replacement with newly synthesized ones, ensure cellular homeostasis and delay the aging process. However, the rate of autophagosome formation, the maturation and the efficiency fusion with lysosome as well as the proteolytic activity of lysosomes decline with age [18]. The inability of autophagy and other cellular degradation mechanisms to completely remove damaged structures results in the progressive accumulation of "garbage", including cytosolic aggregates and defective mitochondria. The progress of these changes seems to result in enhanced oxidative stress, decreased ATP production and collapse of the cellular catabolic machinery, which eventually is incompatible with survival [20].

Studies point toward defective formation of autophagosomes being related to signaling-mediated deregulation of macroautophagy rather than to a primary defect in any of the molecular components that participate in this process. In particular, the effects of (age-related) oxidative stress on the insulin receptor-signaling pathway seem to play a critical role in decreased macroautophagy in old organisms [49]. The slow accumulation of lipofuscin within lysosomes depresses autophagy in the aging heart. The integrity of the autophagosomal-lysosomal network appears to be a critical in the progression of aging [119].

Aging is a complicated pathophysiological processes accompanied with a wide array of biological adaptation, including progressive myocardial remodeling and deteriorated cardiac reserve [8, 9, 67, 161]. Aging is often accompanied by geometric and functional changes in the heart. Aging induces cardiac hypertrophy and fibrosis and decreased cardiac contractility [51]. Loss of autophagy governed by mTOR was demonstrated to accelerate aging [18, 38, 142, 167]. Levels of Beclin 1, Atg5 and LC3-II/LC3-I ratios are decreased in aged hearts and levels of p62 are increased [51]. Rapamycin reduces aging-induced cardiomyocyte contractile and intracellular Ca^{2+} dysfunction [51].

3.13 Genetic Diseases

3.13.1 Desmin-Related Cardiomyopathy (DRCM)

DRCM is a cardiomyopathy cause by a missense mutation in the αB-crystallin (CryAB) gene and is characterized by accumulation of misfolded proteins. When Tannous et al. infected neonatal rat ventricular myocytes with virus expressing mutant human CryAB-R120G they observed two fold increase in autophagic activity [143]. The transgenic over-expression of the mutant desmin CryAB-R120G in mice as well as *in vitro* up-regulates p62 mRNA and protein levels which protects cardiomyocytes from misfolded protein induced cell injury and death by maintaining responsive autophagosome formation and autophagy [169]. But when CryAB-R120G mice were crossed with Beclin $1^{+/-}$ mice, autophagy was blunted and heart failure progression increased. This was associated with an acceleration of ventricular dysfunction and early mortality. As ATG7 induces basal autophagy [113], sustained ATG7 expression rescues impaired autophagy in the CryAB-R120G heart with decreased cardiac hypertrophy and prolonged survival suggesting autophagy activation would be a viable therapeutic strategy for ameliorating desmin-related cardiomyopathy [7]. These findings suggest that the activation of autophagy in this setting is beneficial in attenuating progression of protein misfolding cardiomyopathies [39].

3.13.2 Glycogen Storage Disease-Related Cardiomyopathy

Glycogen storage disease can present as hypertrophic cardiomyopathy [3, 109]. This is particularly the case for Danon disease, a condition characterized by defective autophagosome-lysosome fusion owing to a mutation in the lysosomal membrane receptor Lamp2 (lysosome-associated membrane protein 2). Consequent accretion of unprocessed autophagosomes provokes cardiomyopathy. In a mouse model of Pompe disease, a disorder marked by defective metabolism of glycogen due to insufficiency of lysosomal acid alpha-glucosidase, suppression of the initiation steps of autophagy by inactivating Atg7 facilitates successful enzyme replacement therapy [118]. A novel Lamp2-positive dilated cardiomyopathy has also been reported [139]. This late-onset cardiomyopathy is characterized by increased autophagic vacuoles along with clinical features suggestive of Danon disease, yet LAMP-2 gene mutations are lacking [139].

4 Conclusions

Autophagic activity has been reported in each of the diverse tissues and cell types that constitute the circulatory system. Abundant evidence indicates that this response participates in a wide range of cellular responses to both physiologic and disease-related events (Fig. 1).

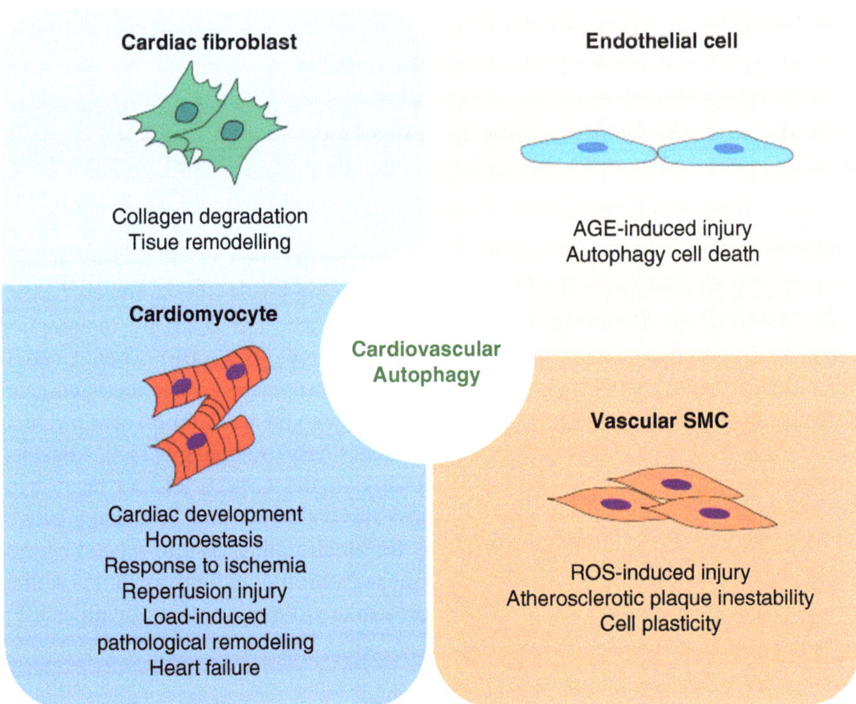

Fig. 1 Autophagy in the cardiovascular system. Autophagic activity occurs in all cell types within the cardiovascular system. This activity contributes to a wide range of cellular events in normal physiology, growth, and development and in disease-related pathophysiology

A great deal of data from preclinical models demonstrates that excessive autophagy elicited by pathological stimuli, such as pressure overload or I/R, is maladaptive and promotes cell death. Conversely, basal levels of constitutive autophagy are essential to maintain proteostasis, and elimination of this means of protein quality control triggers rapid cell death. In other words, understanding of the context-dependent role of autophagic flux in disease promotion and disease antagonism is emerging. These insights follow precedents in oncology, where a similar requirement of finely tuned autophagic activation exists. Our vision for the future includes elucidation of the autophagic circuitry in the cardiovascular systems such that precise tuning of its actions can be accomplished for therapeutic gain. A comprehensive view of myocardial autophagy will be obligatory, as strategies for suppressing excessive activation of pathological pathways must always be precisely regulated to avoid disrupting homeostatic mechanisms. Major challenges remain, but patients with heart disease are likely to benefit from these efforts.

Acknowledgements This work was supported by FONDECYT (grant 1161156 to S.L., 11130285 to R.T., 3140532 to F.W., 3160287 to M.B. 11140470 to C.Q. and 1140713 to L.G.), CONICYT FONDAP 15130011 to S.L., P.C., R.T. and L.G.).

References

1. Allahverdian S et al (2012) Contribution of monocyte-derived macrophages and smooth muscle cells to arterial foam cell formation. Cardiovasc Res 95:165–172
2. Andreeva ER et al (1997) Subendothelial smooth muscle cells of human aorta express macrophage antigen in situ and in vitro. Atherosclerosis 135:19–27
3. Arad M et al (2005) Glycogen storage diseases presenting as hypertrophic cardiomyopathy. N Engl J Med 352:362–372
4. Aranguiz-Urroz P et al (2011) Beta(2)-adrenergic receptor regulates cardiac fibroblast autophagy and collagen degradation. Biochim Biophys Acta 1812:23–31
5. Balletshofer BM et al (2000) Endothelial dysfunction is detectable in young normotensive first-degree relatives of subjects with type 2 diabetes in association with insulin resistance. Circulation 101:1780–1784
6. Beer M et al (2002) Absolute concentrations of high-energy phosphate metabolites in normal, hypertrophied, and failing human myocardium measured noninvasively with (31) P-SLOOP magnetic resonance spectroscopy. J Am Coll Cardiol 40:1267–1274
7. Bhuiyan MS et al (2013) Enhanced autophagy ameliorates cardiac proteinopathy. J Clin Invest 123:5284–5297
8. Boengler K et al (2008) Cardioprotection by ischemic postconditioning is lost in aged and STAT3-deficient mice. Circ Res 102:131–135
9. Boengler K et al (2009) Loss of cardioprotection with ageing. Cardiovasc Res 83:247–261
10. Bolton CH et al (2001) Endothelial dysfunction in chronic renal failure: roles of lipoprotein oxidation and pro-inflammatory cytokines. Nephrol Dial Transplant 16:1189–1197
11. Brodsky SV et al (2002) Endothelial dysfunction in ischemic acute renal failure: rescue by transplanted endothelial cells. Am J Physiol Renal Physiol 282:F1140–F1149
12. Cai H, Harrison DG (2000) Endothelial dysfunction in cardiovascular diseases: the role of oxidant stress. Circ Res 87:840–844
13. Cao DJ et al (2011) Histone deacetylase (HDAC) inhibitors attenuate cardiac hypertrophy by suppressing autophagy. Proc Natl Acad Sci U S A 108:4123–4128

14. Charo IF, Taub R (2011) Anti-inflammatory therapeutics for the treatment of atherosclerosis. Nat Rev Drug Discov 10:365–376
15. Chau YP et al (2003) Endostatin induces autophagic cell death in EAhy926 human endothelial cells. Histol Histopathol 18:715–726
16. Chen K et al (2011) Caloric restriction mimetic 2-deoxyglucose antagonizes doxorubicin-induced cardiomyocyte death by multiple mechanisms. J Biol Chem 286:21993–22006
17. Chen W, Frangogiannis NG (2013) Fibroblasts in post-infarction inflammation and cardiac repair. Biochim Biophys Acta 1833:945–953
18. Cuervo AM et al (2005) Autophagy and aging: the importance of maintaining "clean" cells. Autophagy 1:131–140
19. Dai XY et al (2013) Phosphate-induced autophagy counteracts vascular calcification by reducing matrix vesicle release. Kidney Int 83:1042–1051
20. De Keulenaer GW, Brutsaert DL (2009) The heart failure spectrum: time for a phenotype-oriented approach. Circulation 119:3044–3046
21. De Meyer GR, Martinet W (2009) Autophagy in the cardiovascular system. Biochim Biophys Acta 1793:1485–1495
22. Del Principe D et al (2013) Fibroblast autophagy in fibrotic disorders. J Pathol 229:208–220
23. Dimitrakis P et al (2012) Effects of doxorubicin cancer therapy on autophagy and the ubiquitin-proteasome system in long-term cultured adult rat cardiomyocytes. Cell Tissue Res 350:361–372
24. Dirks-Naylor AJ (2013) The role of autophagy in doxorubicin-induced cardiotoxicity. Life Sci 93:913–916
25. Du J et al (2012) Role of autophagy in angiogenesis in aortic endothelial cells. Am J Physiol Cell Physiol 302:C383–C391
26. Egan DF et al (2011) Phosphorylation of ULK1 (hATG1) by AMP-activated protein kinase connects energy sensing to mitophagy. Science 331:456–461
27. Emanuel R et al (2014) Induction of lysosomal biogenesis in atherosclerotic macrophages can rescue lipid-induced lysosomal dysfunction and downstream sequelae. Arterioscler Thromb Vasc Biol 34:1942–1952
28. Fabian MR et al (2010) Regulation of mRNA translation and stability by microRNAs. Annu Rev Biochem 79:351–379
29. Florey L (1966) The endothelial cell. Br Med J 2:487–490
30. Folco EJ et al (2014) Moderate hypoxia potentiates interleukin-1beta production in activated human macrophages. Circ Res 115:875–883
31. Forstermann U (2010) Nitric oxide and oxidative stress in vascular disease. Pflugers Arch 459:923–939
32. Friddle CJ et al (2000) Expression profiling reveals distinct sets of genes altered during induction and regression of cardiac hypertrophy. Proc Natl Acad Sci U S A 97:6745–6750
33. Furchgott RF (1996) The 1996 Albert Lasker Medical Research Awards. The discovery of endothelium-derived relaxing factor and its importance in the identification of nitric oxide. JAMA 276:1186–1188
34. Garcia L et al (2012) Impaired cardiac autophagy in patients developing postoperative atrial fibrillation. J Thorac Cardiovasc Surg 143:451–459
35. Geer JC et al (1961) The fine structure of human atherosclerotic lesions. Am J Pathol 38:263–287
36. Ghavami S et al (2012) Autophagy regulates trans fatty acid-mediated apoptosis in primary cardiac myofibroblasts. Biochim Biophys Acta 1823:2274–2286
37. Gomez D, Owens GK (2012) Smooth muscle cell phenotypic switching in atherosclerosis. Cardiovasc Res 95:156–164
38. Gottlieb RA et al (2009) Cardioprotection requires taking out the trash. Basic Res Cardiol 104:169–180
39. Gottlieb RA, Mentzer RM Jr (2013) Autophagy: an affair of the heart. Heart Fail Rev 18:575–584

40. Ha T et al (2005) Attenuation of cardiac hypertrophy by inhibiting both mTOR and NFkappaB activation in vivo. Free Radic Biol Med 39:1570–1580
41. Haberzettl P, Hill BG (2013) Oxidized lipids activate autophagy in a JNK-dependent manner by stimulating the endoplasmic reticulum stress response. Redox Biol 1:56–64
42. Halcox JP et al (2002) Prognostic value of coronary vascular endothelial dysfunction. Circulation 106:653–658
43. Hansson GK, Hermansson A (2011) The immune system in atherosclerosis. Nat Immunol 12:204–212
44. Hariharan N et al (2013) Autophagy plays an essential role in mediating regression of hypertrophy during unloading of the heart. PLoS One 8:e51632. doi:10.1371/journal.pone.0051632
45. Hayashi S et al (2009) Alzheimer disease-associated peptide, amyloid beta40, inhibits vascular regeneration with induction of endothelial autophagy. Arterioscler Thromb Vasc Biol 29:1909–1915
46. He C et al (2012) Exercise-induced BCL2-regulated autophagy is required for muscle glucose homeostasis. Nature 481:511–515
47. He C et al (2013) 7-Ketocholesterol induces autophagy in vascular smooth muscle cells through Nox4 and Atg4B. Am J Pathol 183:626–637
48. Heitzer T et al (2001) Endothelial dysfunction, oxidative stress, and risk of cardiovascular events in patients with coronary artery disease. Circulation 104:2673–2678
49. Hierro JL et al (2006) Disturbance facilitates invasion: the effects are stronger abroad than at home. Am Nat 168:144–156
50. Hill BG et al (2008) Unsaturated lipid peroxidation-derived aldehydes activate autophagy in vascular smooth-muscle cells. Biochem J 410:525–534
51. Hua Y et al (2011) Chronic Akt activation accentuates aging-induced cardiac hypertrophy and myocardial contractile dysfunction: role of autophagy. Basic Res Cardiol 106:1173–1191
52. Huang J et al (2014) miR-34a modulates angiotensin II-induced myocardial hypertrophy by direct inhibition of ATG9A expression and autophagic activity. PLoS One 9:e94382. doi:10.1371/journal.pone.0094382
53. Ignarro LJ et al (1987) Endothelium-derived relaxing factor produced and released from artery and vein is nitric oxide. Proc Natl Acad Sci U S A 84:9265–9269
54. Jaffe EA et al (1973) Culture of human endothelial cells derived from umbilical veins. Identification by morphologic and immunologic criteria. J Clin Invest 52:2745–2756
55. Jia G et al (2006) Insulin-like growth factor-1 and TNF-alpha regulate autophagy through c-jun N-terminal kinase and Akt pathways in human atherosclerotic vascular smooth cells. Immunol Cell Biol 84:448–454
56. Kaiser L, Sparks HV Jr (1987) Endothelial cells. Not just a cellophane wrapper. Arch Intern Med 147:569–573
57. Kawaguchi T et al (2012) Prior starvation mitigates acute doxorubicin cardiotoxicity through restoration of autophagy in affected cardiomyocytes. Cardiovasc Res 96:456–465
58. Khosla UM et al (2005) Hyperuricemia induces endothelial dysfunction. Kidney Int 67:1739–1742
59. Kim SI et al (2012) Autophagy promotes intracellular degradation of type I collagen induced by transforming growth factor (TGF)-beta1. J Biol Chem 287:11677–11688
60. Kobayashi S et al (2010) Transcription factor GATA4 inhibits doxorubicin-induced autophagy and cardiomyocyte death. J Biol Chem 285:793–804
61. Kobayashi S, Liang Q (2015) Autophagy and mitophagy in diabetic cardiomyopathy. Biochim Biophys Acta 1852:252–261
62. Kockx MM et al (1998) Cell composition, replication, and apoptosis in atherosclerotic plaques after 6 months of cholesterol withdrawal. Circ Res 83:378–387
63. Koga K et al (2013) Macrophage migration inhibitory factor antagonizes pressure overload-induced cardiac hypertrophy. Am J Physiol Heart Circ Physiol 304:H282–H293
64. Kostin S et al (2003) Myocytes die by multiple mechanisms in failing human hearts. Circ Res 92:715–724

65. Kudryavtseva O et al (2013) Vascular smooth muscle cell phenotype is defined by Ca2+-dependent transcription factors. FEBS J 280:5488–5499
66. Kuzman JA et al (2007) Rapamycin prevents thyroid hormone-induced cardiac hypertrophy. Endocrinology 148:3477–3484
67. Lakatta EG (1999) Cardiovascular aging research: the next horizons. J Am Geriatr Soc 47:613–625
68. Landmesser U et al (2002) Vascular oxidative stress and endothelial dysfunction in patients with chronic heart failure: role of xanthine-oxidase and extracellular superoxide dismutase. Circulation 106:3073–3078
69. Lavandero S et al (2013) Cardiovascular autophagy: concepts, controversies, and perspectives. Autophagy 9:1455–1466
70. Lee J et al (2012) Autophagy, mitochondria and oxidative stress: cross-talk and redox signalling. Biochem J 441:523–540
71. Lee Y et al (2012) Regulation of autophagy by metabolic and stress signaling pathways in the heart. J Cardiovasc Pharmacol 60:118–124
72. Levine B, Kroemer G (2008) Autophagy in the pathogenesis of disease. Cell 132:27–42
73. Li BH et al (2014) TRPV1 activation impedes foam cell formation by inducing autophagy in oxLDL-treated vascular smooth muscle cells. Cell Death Dis 5:e1182. doi:10.1038/cddis.2014.146
74. Li BH et al (2015) Telmisartan-induced PPARgamma activity attenuates lipid accumulation in VSMCs via induction of autophagy. Mol Biol Rep 42:179–186
75. Li Y et al (2014) AMPK inhibits cardiac hypertrophy by promoting autophagy via mTORC1. Arch Biochem Biophys 558:79–86
76. Libby P et al (2011) Progress and challenges in translating the biology of atherosclerosis. Nature 473:317–325
77. Lobatto ME et al (2011) Perspectives and opportunities for nanomedicine in the management of atherosclerosis. Nat Rev Drug Discov 10:835–852
78. Lopaschuk GD, Jaswal JS (2010) Energy metabolic phenotype of the cardiomyocyte during development, differentiation, and postnatal maturation. J Cardiovasc Pharmacol 56:130–140
79. Lu L et al (2009) Adriamycin-induced autophagic cardiomyocyte death plays a pathogenic role in a rat model of heart failure. Int J Cardiol 134:82–90
80. Mack CP (2011) Signaling mechanisms that regulate smooth muscle cell differentiation. Arterioscler Thromb Vasc Biol 31:1495–1505
81. Marko L et al (2012) Interferon-gamma signaling inhibition ameliorates angiotensin II-induced cardiac damage. Hypertension 60:1430–1436
82. Maron BJ et al (2009) Clinical outcome and phenotypic expression in LAMP2 cardiomyopathy. JAMA 301:1253–1259
83. Marsh BJ et al (2007) Regulated autophagy controls hormone content in secretory-deficient pancreatic endocrine beta-cells. Mol Endocrinol 21:2255–2269
84. Martinet W et al (2004) 7-ketocholesterol induces protein ubiquitination, myelin figure formation, and light chain 3 processing in vascular smooth muscle cells. Arterioscler Thromb Vasc Biol 24:2296–2301
85. Martinet W et al (2007) Autophagy in cardiovascular disease. Trends Mol Med 13:482–491
86. Martinet W et al (2014) mTOR inhibition: a promising strategy for stabilization of atherosclerotic plaques. Atherosclerosis 233:601–607
87. Martinet W, De Meyer GR (2009) Autophagy in atherosclerosis: a cell survival and death phenomenon with therapeutic potential. Circ Res 104:304–317
88. McCarty MF, DiNicolantonio JJ (2014) The molecular biology and pathophysiology of vascular calcification. Postgrad Med 126:54–64
89. McMullen JR et al (2004) Inhibition of mTOR signaling with rapamycin regresses established cardiac hypertrophy induced by pressure overload. Circulation 109:3050–3055
90. Mei Y et al (2014) Autophagy and oxidative stress in cardiovascular diseases. Biochim Biophys Acta 1852:209–218

91. Meigs JB et al (2004) Biomarkers of endothelial dysfunction and risk of type 2 diabetes mellitus. JAMA 291:1978–1986
92. Meijer AJ, Codogno P (2006) Signalling and autophagy regulation in health, aging and disease. Mol Aspects Med 27:411–425
93. Meijer AJ, Codogno P (2007) Macroautophagy: protector in the diabetes drama? Autophagy 3:523–526
94. Mellor KM et al (2011) Myocardial autophagy activation and suppressed survival signaling is associated with insulin resistance in fructose-fed mice. J Mol Cell Cardiol 50:1035–1043
95. Mervaala E et al (2000) Cyclosporin A protects against angiotensin II-induced end-organ damage in double transgenic rats harboring human renin and angiotensinogen genes. Hypertension 35:360–366
96. Miller EJ et al (2008) Macrophage migration inhibitory factor stimulates AMP-activated protein kinase in the ischaemic heart. Nature 451:578–582
97. Minotti G et al (2004) Anthracyclines: molecular advances and pharmacologic developments in antitumor activity and cardiotoxicity. Pharmacol Rev 56:185–229
98. Minuz P et al (2012) Endothelial dysfunction and increased oxidative stress in mitochondrial diseases. Clin Sci (Lond) 122:289–297
99. Miyata S et al (2006) Autophagic cardiomyocyte death in cardiomyopathic hamsters and its prevention by granulocyte colony-stimulating factor. Am J Pathol 168:386–397
100. Mizushima N (2005) The pleiotropic role of autophagy: from protein metabolism to bactericide. Cell Death Differ 12(Suppl 2):1535–1541
101. Molkentin JD, Dorn GW (2001) Cytoplasmic signaling pathways that regulate cardiac hypertrophy. Annu Rev Physiol 63:391–426
102. Muller C et al (2011) Oxidized LDLs trigger endoplasmic reticulum stress and autophagy: prevention by HDLs. Autophagy 7:541–543
103. Muller C et al (2011) HDLs inhibit endoplasmic reticulum stress and autophagic response induced by oxidized LDLs. Cell Death Differ 18:817–828
104. Nachman RL (2012) Endothelium: from cellophane to orchestral maestro. J Clin Invest 122:796–797
105. Nachman RL, Jaffe EA (2004) Endothelial cell culture: beginnings of modern vascular biology. J Clin Invest 114:1037–1040
106. Nakai A et al (2007) The role of autophagy in cardiomyocytes in the basal state and in response to hemodynamic stress. Nat Med 13:619–624
107. Nemchenko A et al (2011) Autophagy as a therapeutic target in cardiovascular disease. J Mol Cell Cardiol 51:584–593
108. Nguyen TM et al (2007) Kringle 5 of human plasminogen, an angiogenesis inhibitor, induces both autophagy and apoptotic death in endothelial cells. Blood 109:4793–4802
109. Nishino I et al (2000) Primary LAMP-2 deficiency causes X-linked vacuolar cardiomyopathy and myopathy (Danon disease). Nature 406:906–910
110. Nowicki M et al (2007) No upregulation of lectin-like oxidized low-density lipoprotein receptor-1 in serum-deprived EA.hy926 endothelial cells under oxLDL exposure, but increase in autophagy. Eur J Cell Biol 86:605–616
111. Oyabu J et al (2013) Autophagy-mediated degradation is necessary for regression of cardiac hypertrophy during ventricular unloading. Biochem Biophys Res Commun 441:787–792
112. Pan W et al (2013) MiR-30-regulated autophagy mediates angiotensin II-induced myocardial hypertrophy. PLoS One 8:e53950. doi:10.1371/journal.pone.0053950
113. Pattison JS et al (2011) Atg7 induces basal autophagy and rescues autophagic deficiency in CryABR120G cardiomyocytes. Circ Res 109:151–160
114. Perticone F et al (2001) Prognostic significance of endothelial dysfunction in hypertensive patients. Circulation 104:191–196
115. Petrov VV et al (2008) TGF-beta1-induced cardiac myofibroblasts are nonproliferating functional cells carrying DNA damages. Exp Cell Res 314:1480–1494
116. Phan SH (2008) Biology of fibroblasts and myofibroblasts. Proc Am Thorac Soc 5:334–337

117. Porter KE, Turner NA (2009) Cardiac fibroblasts: at the heart of myocardial remodeling. Pharmacol Ther 123:255–278
118. Raben N et al (2010) Suppression of autophagy permits successful enzyme replacement therapy in a lysosomal storage disorder – murine Pompe disease. Autophagy 6:1078–1089
119. Rajawat YS et al (2009) Aging: central role for autophagy and the lysosomal degradative system. Ageing Res Rev 8:199–213
120. Razani B et al (2012) Autophagy links inflammasomes to atherosclerotic progression. Cell Metab 15:534–544
121. Ryter SW et al (2010) Autophagy in vascular disease. Proc Am Thorac Soc 7:40–47
122. Sadoshima J, Izumo S (1993) Molecular characterization of angiotensin II – induced hypertrophy of cardiac myocytes and hyperplasia of cardiac fibroblasts. Critical role of the AT1 receptor subtype. Circ Res 73:413–423
123. Saijo M et al (2004) Cardiomyopathy with prominent autophagic degeneration, accompanied by an elevated plasma brain natriuretic peptide level despite the lack of overt heart failure. Intern Med 43:700–703
124. Salabei JK et al (2013) PDGF-mediated autophagy regulates vascular smooth muscle cell phenotype and resistance to oxidative stress. Biochem J 451:375–388
125. Salabei JK, Hill BG (2013) Implications of autophagy for vascular smooth muscle cell function and plasticity. Free Radic Biol Med 65:693–703
126. Salomon RG et al (2000) HNE-derived 2-pentylpyrroles are generated during oxidation of LDL, are more prevalent in blood plasma from patients with renal disease or atherosclerosis, and are present in atherosclerotic plaques. Chem Res Toxicol 13:557–564
127. Scherz-Shouval R, Elazar Z (2011) Regulation of autophagy by ROS: physiology and pathology. Trends Biochem Sci 36:30–38
128. Schultz Jel J et al (2002) TGF-beta1 mediates the hypertrophic cardiomyocyte growth induced by angiotensin II. J Clin Invest 109:787–796
129. Schulz E et al (2011) Oxidative stress and endothelial dysfunction in hypertension. Hypertens Res 34:665–673
130. Sciarretta S et al (2012) Rheb is a critical regulator of autophagy during myocardial ischemia: pathophysiological implications in obesity and metabolic syndrome. Circulation 125: 1134–1146
131. Sergin I et al (2014) Hypoxia in plaque macrophages: a new danger signal for interleukin-1beta activation? Circ Res 115:817–820
132. Shimomura H et al (2001) Autophagic degeneration as a possible mechanism of myocardial cell death in dilated cardiomyopathy. Jpn Circ J 65:965–968
133. Sishi BJ et al (2013) Autophagy upregulation promotes survival and attenuates doxorubicin-induced cardiotoxicity. Biochem Pharmacol 85:124–134
134. Smuder AJ et al (2013) Doxorubicin-induced markers of myocardial autophagic signaling in sedentary and exercise trained animals. J Appl Physiol 115:176–185
135. Souders CA et al (2009) Cardiac fibroblast: the renaissance cell. Circ Res 105:1164–1176
136. Speer MY et al (2009) Smooth muscle cells give rise to osteochondrogenic precursors and chondrocytes in calcifying arteries. Circ Res 104:733–741
137. Starling RC et al (1998) Human myocardial ATP content and in vivo contractile function. Mol Cell Biochem 180:171–177
138. Stehouwer CD et al (2002) Increased urinary albumin excretion, endothelial dysfunction, and chronic low-grade inflammation in type 2 diabetes: progressive, interrelated, and independently associated with risk of death. Diabetes 51:1157–1165
139. Sugimoto S et al (2007) LAMP-2 positive vacuolar myopathy with dilated cardiomyopathy. Intern Med 46:757–760
140. Suwaidi JA et al (2000) Long-term follow-up of patients with mild coronary artery disease and endothelial dysfunction. Circulation 101:948–954
141. Takemura G, Fujiwara H (2007) Doxorubicin-induced cardiomyopathy from the cardiotoxic mechanisms to management. Prog Cardiovasc Dis 49:330–352

142. Taneike M et al (2010) Inhibition of autophagy in the heart induces age-related cardiomyopathy. Autophagy 6:600–606
143. Tannous P et al (2008) Autophagy is an adaptive response in desmin-related cardiomyopathy. Proc Natl Acad Sci U S A 105:9745–9750
144. Tedgui A, Mallat Z (2006) Cytokines in atherosclerosis: pathogenic and regulatory pathways. Physiol Rev 86:515–581
145. Turner NA, Porter KE (2013) Function and fate of myofibroblasts after myocardial infarction. Fibrogenesis Tissue Repair 6:5. doi:10.1186/1755-1536-6-5
146. Ucar A et al (2012) The miRNA-212/132 family regulates both cardiac hypertrophy and cardiomyocyte autophagy. Nat Commun 3:1078. doi:10.1038/ncomms2090
147. Villanova N et al (2005) Endothelial dysfunction and cardiovascular risk profile in nonalcoholic fatty liver disease. Hepatology 42:473–480
148. Virmani R et al (2002) Vulnerable plaque: the pathology of unstable coronary lesions. J Interv Cardiol 15:439–446
149. Wallace KB (2007) Adriamycin-induced interference with cardiac mitochondrial calcium homeostasis. Cardiovasc Toxicol 7:101–107
150. Wang Q et al (2011) 2-Deoxy-D-glucose treatment of endothelial cells induces autophagy by reactive oxygen species-mediated activation of the AMP-activated protein kinase. PLoS One 6:e17234. doi:10.1371/journal.pone.0017234
151. Wang ZV et al (2010) Autophagy in hypertensive heart disease. J Biol Chem 285: 8509–8514
152. Watanabe M et al (2007) Distribution of inflammatory cells in adventitia changed with advancing atherosclerosis of human coronary artery. J Atheroscler Thromb 14:325–331
153. Weber KT (2004) Fibrosis in hypertensive heart disease: focus on cardiac fibroblasts. J Hypertens 22:47–50
154. Xie Y et al (2011) Protective role of autophagy in AGE-induced early injury of human vascular endothelial cells. Mol Med Rep 4:459–464
155. Xie Z et al (2011) Improvement of cardiac functions by chronic metformin treatment is associated with enhanced cardiac autophagy in diabetic OVE26 mice. Diabetes 60:1770–1778
156. Xu K et al (2010) Autophagy plays a protective role in free cholesterol overload-induced death of smooth muscle cells. J Lipid Res 51:2581–2590
157. Xu X et al (2012) Resveratrol attenuates doxorubicin-induced cardiomyocyte death via inhibition of p70 S6 kinase 1-mediated autophagy. J Pharmacol Exp Ther 341:183–195
158. Xu X et al (2013) Akt2 knockout preserves cardiac function in high-fat diet-induced obesity by rescuing cardiac autophagosome maturation. J Mol Cell Biol 5:61–63
159. Xu X et al (2014) Macrophage migration inhibitory factor deletion exacerbates pressure overload-induced cardiac hypertrophy through mitigating autophagy. Hypertension 63: 490–499
160. Yan L et al (2005) Autophagy in chronically ischemic myocardium. Proc Natl Acad Sci U S A 102:13807–13812
161. Yang X et al (2005) Views from within and beyond: narratives of cardiac contractile dysfunction under senescence. Endocrine 26:127–137
162. Yeh ET et al (2004) Cardiovascular complications of cancer therapy: diagnosis, pathogenesis, and management. Circulation 109:3122–3131
163. Yin X et al (2013) miR-30a downregulation aggravates pressure overload-induced cardiomyocyte hypertrophy. Mol Cell Biochem 379:1–6
164. Yorimitsu T et al (2006) Endoplasmic reticulum stress triggers autophagy. J Biol Chem 281:30299–30304
165. Yu E et al (2012) Mitochondria in vascular disease. Cardiovasc Res 95:173–182
166. Zbinden G et al (1978) Model systems for cardiotoxic effects of anthracyclines. Antibiot Chemother 23:255–270
167. Zhang C, Cuervo AM (2008) Restoration of chaperone-mediated autophagy in aging liver improves cellular maintenance and hepatic function. Nat Med 14:959–965

168. Zhao Y et al (2013) Heme oxygenase-1 prevents cardiac dysfunction in streptozotocin-diabetic mice by reducing inflammation, oxidative stress, apoptosis and enhancing autophagy. PLoS One 8:e75927. doi:10.1371/journal.pone.0075927
169. Zheng Q et al (2011) Autophagy and p62 in cardiac proteinopathy. Circ Res 109:296–308
170. Zheng YH et al (2012) Osteopontin stimulates autophagy via integrin/CD44 and p38 MAPK signaling pathways in vascular smooth muscle cells. J Cell Physiol 227:127–135
171. Zhu H et al (2007) Cardiac autophagy is a maladaptive response to hemodynamic stress. J Clin Invest 117:1782–1793

Aging

Eugenia Morselli and Alfredo Criollo

Abstract Autophagic turnover is a cellular catabolic process in which old cytoplasmic proteins and organelles, that would otherwise alter cell homeostasis and compromise cell viability, are delivered to lysosomes and degraded. Aged cells are characterized by decreased autophagic activity, which has been implicated in the pathogenesis of age-related diseases. In this work, we discuss how impaired autophagy in aged tissues leads to increased inflammatory response and promotes age-related diseases. Current research is focusing on genetic and pharmacological induction of autophagy as a possible treatment for age-related illnesses. In this chapter, we will review the molecular pathways altered with aging that lead to inhibited autophagy and, therefore, to the development of the aforementioned diseases. Knowledge of these signaling pathways will lead to the identification of cellular targets that can be used for the development of new pharmacological compounds.

Abbreviations

AMPK	AMP-activated protein kinase
ASC	Adaptor protein apoptosis-associated speck-like protein containing a CARD
ATG	Autophagy-related
BCL-2	B cell lymphoma 2
BECN1	Beclin 1

E. Morselli, PhD (✉)
Department of Physiology, Faculty of Biological Sciences, Pontificia Universidad Católica de Chile, Avda. Libertador Bernardo O'Higgins 340, 8331150 Santiago, Chile
e-mail: emorselli@bio.puc.cl

A. Criollo
Laboratory of Cellular and Molecular Biology, Research Institute of Dental Science and Advanced Center for Chronic Diseases (ACCDiS), Faculty of Dentistry, University of Chile, Santiago 8380492, Chile

M.C. Maiuri, D. De Stefano (eds.), *Autophagy Networks in Inflammation*, Progress in Inflammation Research, DOI 10.1007/978-3-319-30079-5_15

CARD	Caspase recruitment domain
CKD	Chronic kidney disease
CR	Caloric restriction
CVD	Cardiovascular diseases
EGCG	Epigallocatechin gallate
ER	Endoplasmic reticulum
FAK	Focal adhesion kinase
FIP200	Family-interacting protein of 200 kD
FOXO	Forkhead box O
ICAM-1	Intercellular adhesion molecule 1
IFN	Interferon
IGF-1	Insulin-like growth factor 1
IGF1R	Insulin-like growth factor receptor
IKK	IκB kinase
IL	Interleukin
iNOS	Inducible nitric oxide synthase
IP3	1,4,5-inositol trisphosphate
IP_3R	IP_3 receptor
IPS1	IFN-β promoter stimulator-1
JNK-1	c-Jun N-terminal kinase-1
MAP-LC3	Microtubule-associated protein 1 light chain 3
MCP-1	Monocyte chemoattractant protein-1
mTOR	Mechanistic target of Rapamycin
mTORC1	mTOR complex1
mTORC2	mTOR complex 2
NF-κB	Nuclear factor kappa beta
NLR	NOD-like receptor
PI(5)P	Phosphatidylinositol 5-phosphate
PI3K	Phosphoinositide 3-kinase
PINK1	PTEN-induced putative kinase
PKB	PtdIns3K-protein kinase B
PtdIns3K	Phosphatidylinositol 3 kinase
PTEN	Phosphatase and tensin homology
RIR	RIG-like receptors
ROS	Reactive oxygen species
SIRT1	Sirtuin 1
TNF-α	Tumor necrosis factor alpha
TSC	Tuberous sclerosis protein
ULK	unc-51 like autophagy activating kinase
UVRAG	UV radiation resistance-associated gene
VAMP8	Vesicle-associated membrane protein 8
VCAM-1	Vascular cell adhesion molecule 1
VPS	Vacuolar protein sorting
Vti1B	Vesicle transport through interaction with t-SNAREs homolog 1B

1 Introduction

Autophagy (from the Greek words *auto* 'self' and *phagein* 'to eat') comprises different processes – macroautophagy, microautophagy, and chaperone-mediated autophagy – that allow cells to digest their organelles through lysosomes. Macroautophagy, hereafter referred to as "autophagy", is the most studied autophagic process and will be the focus of this article. During autophagy, characteristic double-membrane vesicles called autophagosomes enclose portions of the cytoplasm and organelles. Each autophagosome fuses with a lysosome to form an autophagolysosome. Lysosomal hydrolases within the autophagolysosome then degrade the cytoplasmic material and this lysosomal digestion generates highly energetic compounds that favor cell survival in conditions of nutrient deprivation and represents the only mechanism through which certain cytosolic components can be recycled [64]. Basal levels of autophagy preserve cellular homeostasis and maintain quality control of essential cellular components through the elimination of damaged and old organelles, as well as the turnover of long-lived proteins [40, 107].

Extensive evidence shows that formation of autophagosomes, as well as their elimination, declines with aging, and leads to the accumulation of altered organelles and membranes, further enhancing the pro-aging process [92]. This chapter will describe the effect of autophagy reduction during aging and how the inhibition of this cellular pathway promotes inflammation and the development of age-associated diseases.

2 The Autophagic Machinery

From a molecular point of view, the autophagic process requires different autophagy-related (*atg*) genes for autophagosome formation. Autophagosome assembly starts at phagophore assembly sites, which most likely occur at the (ER) endoplasmic reticulum–mitochondria contact site in mammalian cells [46]. The phagophore represents the autophagosome precursor and requires class III phosphoinositide 3-kinase (PI3K) vacuolar protein sorting 34 (VPS34), which interacts with ATG6 (also known as Beclin 1/BECN1), ATG14, and VPS15 (p150) to form a large macromolecular complex [94]. In the early stage of autophagosome formation other ATG proteins are also needed, including ATG5, ATG12, ATG16, and focal adhesion kinase (FAK) family-interacting protein of 200 kD (FIP200). FIP200 interacts with ATG1 (also called unc-51 like autophagy activating kinase 1, ULK1), and the mammalian ortholog of ATG13 [121] to control autophagy induction [3, 135]. Interestingly, a recent published study identifies the phosphatidylinositol 5-phosphate (PI(5)P) as a regulator of autophagosomes biogenesis. PI(5)P acting through a non-canonical VPS34-independent mechanism promotes autophagosome formation regulating ATG5-ATG12 conjugation [129]. Elongation and expansion of the phagophore membrane requires two ubiquitination-like reactions. The process begins with ATG12 conjugation to ATG5 by the combined action of ATG7, which is similar to an E1 ubiquitin-activating enzyme, and that of ATG10, which is similar to an E2 ubiquitin-conjugating enzyme. The ATG12–ATG5 conjugate then interacts non-covalently with ATG16L, oligomerizing to form a large multimeric complex. In the second ubiquitin-like reaction, microtubule-associated protein 1 light chain 3 (MAP-LC3/ATG8/LC3), following cleavage at its C-terminus by proteases member of the ATG4 family, is conjugated to the lipid phosphatidylethanolamine by ATG7 (E1-like) and ATG3 (E2-like) to form LC3-II, which is inserted in the membrane of the autophagosome. This lipid conjugation leads to the conversion of the soluble form of LC3 (named LC3-I) to the autophagic vesicle-associated form, LC3-II [140]. Once the autophagosome is formed, ATG12–ATG5–ATG16L complexes are released in the cytoplasm to be reused for the biogenesis of new vesicles [64]. Several SNARE-like proteins, such as vesicle-associated membrane protein 8 (VAMP8) and vesicle transport through interaction with t-SNAREs homolog 1B (Vti1B), mediate the fusion of autophagosomes with lysosomes to create autolysosomes [49]. In the autolysosomes, lysosomal enzymes degrade the inner membrane and luminal cargo, generating new biomolecules that are finally recycled back to the cytoplasm [39].

3 Regulation of Mammalian Autophagy

The best-characterized regulators of autophagy are those that modulate the pathway in response to nutritional changes. The mechanistic target of Rapamycin (formerly known as mammalian target of Rapamycin, mTOR) is a serine/threonine protein kinase able to sense nutritional and energetic status [94]. mTOR functions as part of 2 distinct signaling complexes mTORC1 and mTORC2 (mTOR complex 1 and 2). Autophagy initiation is controlled by mTORC1; starvation conditions inhibit mTORC1 and promote the process of autophagy via activation of mTORC1 target proteins such as ATG13, ULK1, and ULK2 [94]. This mechanism is mimicked by Rapamycin, which inhibits mTORC1 and induces the autophagic pathway [88]. Nutrient deprivation also activates the energy sensor AMP-activated protein kinase (AMPK), inhibiting mTORC1 activity. Activation of AMPK reduces ULK1 phosphorylation, releases ULK1 from mTORC1 and promotes its mobilization to the phagophore assembly site, thus initiating the autophagic pathway [35, 61]. AMPK can also induce autophagy through TSC1/TSC2 (tuberous sclerosis complex 1/2)-mediated mTORC1 inhibition [33, 50]. mTORC1 inhibits autophagy when growth factors are elevated, integrating different upstream signals through the class I Phosphatidylinositol 3 kinase (PtdIns3K)-protein kinase B (PKB, also known as AKT) pathway. Tyrosine kinase receptors, once activated by growth factors, undergo autophosphorylation and stimulate the small GTPase Ras and class I PtdIns3K. Class I PtdIns3K then catalyzes the production of PtdIns(3)P at the plasma membrane. This signal allows the recruitment and subsequent activation of AKT at the plasma membrane. Importantly, AKT can also phosphorylate the subunit tuberous sclerosis protein (TSC) 2 of the TSC1/TSC2 complex, leading to the disruption of the heterodimer and promoting mTORC1 activation [94, 140]. Among the growth factors able to positively modulate mTORC1 activity, there is the insulin-like growth factor 1 (IGF-1), which binds to the insulin-like growth factor receptor 1 (IGF1R). The activation of this tyrosine kinase receptor leads to AKT phosphorylation and subsequent inhibition of the aforementioned TSC1/TSC2 complex, which activates mTORC1 and inhibits autophagy [56]. Modulation of the mTORC1 pathway can also occur through p53, a protein involved in cellular senescence and cancer development. p53 is activated following cellular stress responses; this signal activates AMPK and phosphatase and tensin homology (PTEN), a negative regulator of AKT [84]. Both these signals increase cellular autophagy. Interestingly, genetic and chemical inhibition of p53 also activates autophagy [114]. Another mechanism that regulates autophagy is the interaction between BECN1 and the apoptosis-related proteins B cell lymphoma 2 (BCL-2)-family members. In conditions of nutrient withdrawal c-Jun N-terminal kinase-1 (JNK-1)-mediated BCL-2 phosphorylation inhibits the interaction between BECN1 and BCL-2, a signal that promotes autophagy [132]. BECN1 can interact with others BCL-2 family proteins such as Bcl-X_L, and Mcl-1. The interaction suppresses BECN1 function, which is required for autophagosomes formation, leading to autophagy inhibition [32, 42, 110]. The autophagic pathway can also be induced by mechanisms independent of mTORC1 [101]. Reductions in the intracellular level of 1,4,5-inositol trisphosphate (IP_3), as well as downregulation or chemical inhibition of the IP_3 receptor (IP_3R), are a strong stimulus for autophagy induction [23]. Decreases in intracellular IP_3 lower ER Ca^{2+} uptake into the mitochondria, affecting mitochondrial respiration and upregulating autophagy [16]. This may not be the only process for IP_3 receptors-mediated autophagy; in fact, the binding of IP_3R to BECN1 also inhibits autophagy [128]. Another study has shown that the stress-activated kinase, IκB kinase (IKK) modulates autophagy in a nuclear factor kappa beta (NF-κB)-independent manner by activating AMPK and JNK1 [24]. Current research is focusing on highlighting the differences between mTORC1-dependent and mTORC1-independent autophagy, and the functional interactions between these two regulatory mechanisms.

4 Autophagy and Aging

As aging occurs, reduced expression of *Atg* genes and proteins involved in the autophagic pathway decrease, diminishing basal autophagy [92, 108]. This reduction is characterized by decreased formation and decreased elimination of autophagosomes. These processes lead to defective autophagic clearance and finally to the accumulation of altered organelles, membranes, proteins and other intracellular "waste" material in most tissues of aged organisms. This is linked to age-associated pathologies, including triglyceride accumulation, mitochondrial dysfunction, cancer, muscle degeneration, neurodegeneration and cardiac malfunction (Fig. 1) [9, 26, 60, 67, 71, 104]. The inefficient autophagosomal turnover is not only due to

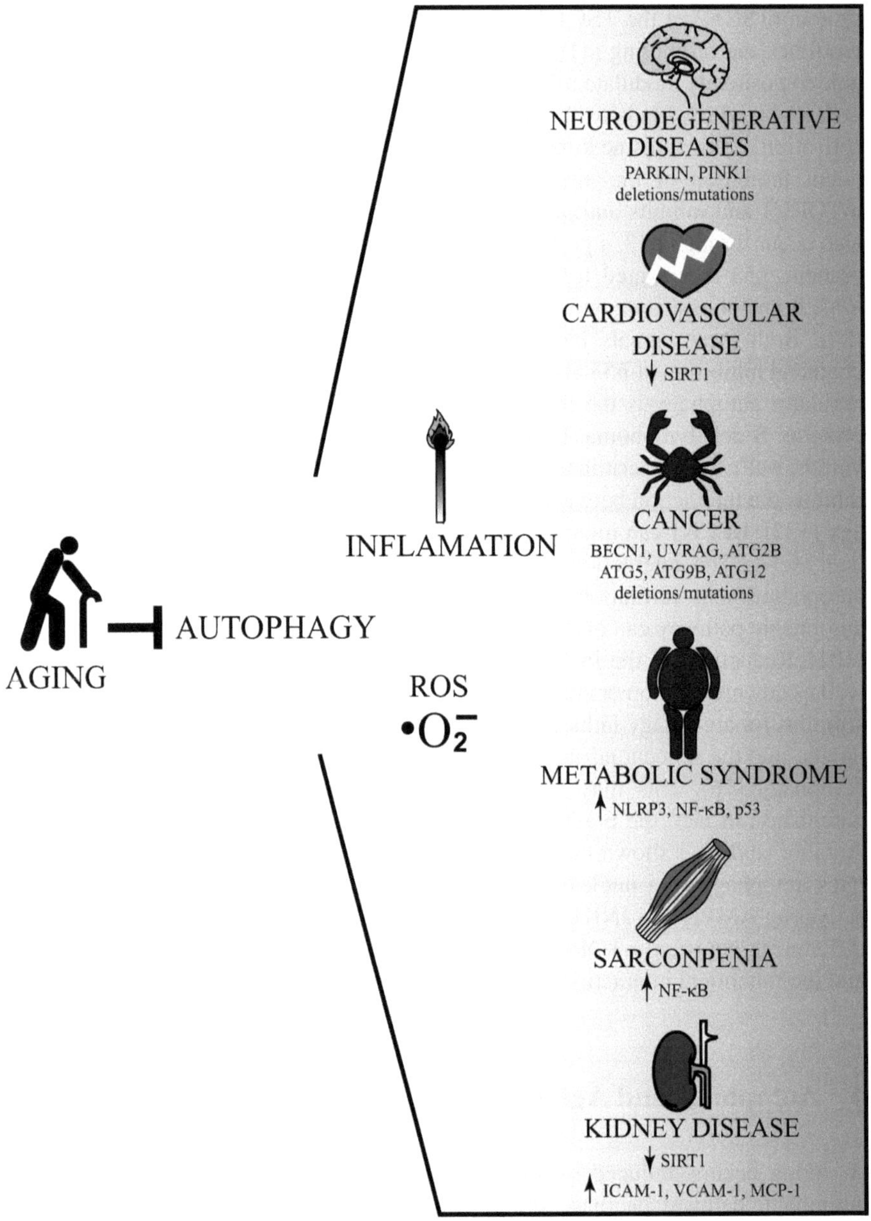

Fig. 1 As aging occurs basal autophagy decreases. This correlates with mutations/deletions/decrease in *Atg* genes and in proteins involved in the regulation of the autophagic pathway. This reduction leads to chronic inflammation, increase in intracellular ROS, defective autophagic clearance and finally to the accumulation of altered organelles, membranes, proteins, lipids and other intracellular "waste" material in tissues of aged organisms. This is linked to age-associated pathologies, including neurodegenerative diseases, cardiovascular diseases, cancer, metabolic syndrome, sarcopenia and kidney disease. See the main text for further details. *ATG* autophagy-related, *BECN1* beclin 1, *ICAM-1* intercellular adhesion molecule 1, *MCP-1* monocyte chemoattractant protein-1, *NF-κB* nuclear factor kappa beta, *NLRP3* NOD-like receptor protein 3, *PINK1* PTEN-induced putative kinase 1, *ROS* reactive oxygen species, *SIRT1* sirtuin 1, *UVRAG* UV radiation resistance-associated gene, *VCAM-1* vascular cell adhesion molecule 1

decreased proteolytic activity of lysosomes, which undergo striking changes that affect hydrolase activity as cells age, such as increased volume and impaired regulation of lysosomal pH [118], but is also associated with impaired autophagosome-lysosome fusion [28].

These age-related changes in basal autophagy could be secondary to age-related changes in metabolism. Indeed, regulation of autophagy by glucagon and insulin, which induce and inhibit autophagy respectively, is affected by age [34]. Conditions that stimulate autophagy, such as caloric restriction and exercise, delay the aging phenotype [139], suggesting that autophagy counteracts the aging process. Enhancement of the autophagic pathway has shown to increase healthy life span in different organisms from yeasts and worms to mice and primates. For example, in mice, overexpression of ATG5 is sufficient to enhance autophagy and extend lifespan [85]. Conversely, deletion or loss-of-function mutations in *Atg* genes creates short-lived mutants in *Saccharomyces cerevisiae* and decreases the life span of *Caenorhabditis elegans* [122] and of the fruit fly *Drosophila melanogaster* [67, 106]. In mice, total body inactivation of *Atg* genes is lethal; however, tissue-specific deletion of *Atg* genes promotes tissue degeneration in different organs, especially the brain [2]. All these studies argue for a causal link between autophagy and longevity; nevertheless, few studies have evaluated if correction of the aging-associated defect in autophagy is sufficient to counteract the age-related phenotype. Further research is required in this field.

5 Autophagy, Reactive Oxygen Species Production and Inflammation

Inflammation is a vital response of the organism to tissue stress, damage, and infection. It occurs in response to the loss of cellular and tissue homeostasis and contributes to host defense, tissue repair, and regulation of metabolism. During the inflammatory response, pro-inflammatory mediators, such as cytokines, chemokines, and eicosanoids are produced; these factors defend the host from pathogens and restore cellular and tissue homeostasis. Nevertheless, when the inflammatory response becomes chronic, its long-term effect is negative and is associated with aging and tissue degeneration [83]. Indeed, activation of the adaptive immune system in aging leads to a pro-inflammatory tissue phenotype, a process that has been named "inflammaging" [7, 38]. The "inflammasome" is a multi-protein complex composed by an inflammasome sensor molecule, the NOD-like receptor (NLR), the adaptor protein apoptosis-associated speck-like protein containing a CARD (ASC), and caspase-1. The inflammasome is a key component of pro-inflammatory pathways, and, through the activation of caspase-1, it promotes the maturation of the pro-inflammatory cytokines interleukin-1β (IL-1β) and interleukin-18 (IL-18) [66, 83]. Induction of autophagy counteracts inflammasome activation and IL-1β release. Indeed, mice deficient for ATG16L1, an essential component of the autophagic machinery, show increased production of IL-1β following pro-inflammatory stimulation [96]. However, the possible mechanisms by which autophagy controls inflammasome activation are unclear. One hypothesis is that autophagy is involved in the removal of ubiquitinated inflammasomes [102] and pro-IL-1β molecules [47]. Other studies suggest that the autophagic process inhibits inflammasome formation through the removal of damaged mitochondria, a process known as mitophagy, thereby limiting the release of pro-inflammatory reactive oxygen species (ROS) [143]. Together with a low-grade inflammatory response, oxidative stress and ROS production are the hallmark of the aging process [117]. During aging, with the decrease in basal autophagy and mitophagy (the process of autophagy of the mitochondria), the clearance of mitochondria declines and dysfunctional mitochondria accumulate, leading to chronic oxidative stress and cellular redox state imbalance. Oxidative stress provokes inflammatory responses and is associated with the pathogenesis of several age-related diseases, including neurodegenerative diseases and metabolic syndrome [21, 90]. Further, removal of ROS-producing and/or permeabilized mitochondria by the autophagic pathway prevents activation of the inflammasome component NLRP3, confirming the key role of autophagy in the regulation of inflammation [124]. Autophagy can mitigate inflammatory reactions through mechanisms other than the anti-inflammatory effect of mitochondria recycling. In apoptotic cells, autophagy contributes to the cleaning of apoptotic bodies, thereby inhibiting pro-inflammatory reactions [43, 68]. In addition, ATG12-ATG5 conjugates associate with the caspase recruitment domain (CARD) of RIG-like receptors (RLR) and IPS-1 (interferon (IFN)-β promoter stimulator-1). RLRs sense the cytoplasmic double-stranded RNA of RNA viruses and mediate signals to IPS-1, leading to type I INF production. Ectopic expression of ATG12 and ATG5, and therefore activation of the autophagic pathway, prevents RLR–mediated activation of the type I IFN promoter. Autophagy deficiency affects the RLR-IPS-1 signaling pathway, resulting in accumulation of IPS-1 protein, which leads to the activation of IPS-1–dependent immune responses [95]. In conclusion, since a condition of chronic inflammation is characteristic of pathological aging, the anti-inflammatory effects of autophagy could lead to health benefits.

6 Autophagy in Age-Related Diseases

6.1 Neurodegenerative Diseases

Alterations of the autophagic pathway are associated with a variety of neurodegenerative, age-related diseases. Parkinson's disease is characterized by marked microgliosis (activation of the microglia, cells that produce pro-inflammatory cytokines), as well as increased inducible nitric oxide synthase (iNOS), which promotes ROS production and disrupts autophagy [25]. This disruption in autophagy/mitophagy leads to accumulation of damaged mitochondria and production of ROS, enhancing the development of the disease [55]. Indeed, brain tissue of Parkinson's patients is characterized by increased numbers of damaged mitochondria [4]. Further, hereditary Parkinson's disease patients carry loss of function mutations in genes, such as PTEN-induced putative kinase 1 (PINK1) and Parkin, which are required for mitophagy [41]. Similarly, a condition of deficient mitophagy, due to a decrease in the protein Parkin, characterizes the cortex of individuals afflicted with Alzheimer's disease [91].

Other studies in Alzheimer's patients have reported the accumulation of neuronal autophagosomes and proteins in brain tissue as a result of the failure in autophagosome-lysosome fusion [17, 78, 100].

Another mechanism involved in Alzheimer's disease development is NLRP3-mediated IL-1β production in brain microglial cells, which appears to be promoted by fibrillary amyloid-β peptides [45, 112]. Although future research is required to confirm this hypothesis, increased IL-1β levels do correlate with decreased autophagy, which would promote an inflammatory response and accelerate the development of Alzheimer's disease.

In this context, defective mitophagy plays a pivotal role in the development of Parkinson and Alzheimer's disease, and therapeutic strategies leading to the induction of autophagy/mitophagy in neurodegenerative diseases are currently being tested.

6.2 Cardiovascular Diseases

The incidence of cardiovascular diseases (CVD) dramatically increases with age; CVD is a leading cause of morbidity and mortality in older persons. Age-associated structural and functional changes in the cardiovascular system include excessive accumulation of ROS and chronic inflammation, factors that depend on a decreased autophagic pathway [31, 54, 76, 113]. Further, disruption of lysosomal function contributes to the slow-down of autophagy in aged hearts [27, 30, 119]. Some of these effects appear to be correlated with age-associated decreases in sirtuin 1 (SIRT1), a protein that regulates oxidative stress, inflammation, and autophagy in the heart by regulating numerous other proteins, such as forkhead box Os (FOXOs), NF-κB, and mTOR [98, 99, 138]. SIRT1 exerts protective effects against cardiovascular aging and age-related CVDs by mediating multiple signaling pathways; thus, up-regulation or activation of SIRT1 has emerged as a promising avenue to retard aging and treat age-related CVD through upregulation of autophagy. Aging also causes arterial endothelial dysfunction that increases the risk of CVD. Older humans and rodents show arterial endothelial dysfunction, measured through arterial endothelium-dependent dilatation, increased oxidative stress and inflammation, which is associated with decreased expression of autophagy markers in arterial endothelial cells. Interestingly, in mouse models treatment with a pro-autophagic agent restores expression of autophagy markers and pro-inflammatory cytokines [65]. Once again, autophagy-enhancing strategies appear to have therapeutic efficacy for ameliorating age-associated arterial dysfunction and thus for the prevention of CVD.

6.3 Cancer

Several oncosuppressors stimulate autophagic flow, while oncoproteins inhibit it. This suggests that autophagy may prevent oncogenesis (reviewed in Morselli et al. [74]), and that the age-associated decline in autophagy contributes to malignancy. Nonetheless, it is still not completely clear how age-related autophagy defects contribute to tumor initiation. As previously mentioned, one function of the autophagic process is the maintenance of cellular homeostasis through the continuous recycling of old/damaged proteins and organelles. This process declines with aging and leads to the accumulation of harmful metabolic end products, oxidized proteins, and damaged organelles, such as mitochondria. Age-associated autophagy disruption induces damage in chromosomal DNA, leading to genomic instability and tumor development [37]. Further, autophagy inhibition enhances pro-inflammatory responses, which promote tumor development [126]. As previously mentioned, oncogenes tend to inhibit, while tumor suppressors activate autophagy, suggesting that autophagy needs to be down regulated for cancer proliferation [74]. For example, the inhibitory BCL-2/BECN1 interaction is an important mechanism in autophagy regulation, and the expression balance between the two proteins regulates autophagy induction. Monoallelic deletion of BECN1, and therefore autophagy inhibition, is associated with higher tumor frequency in multiple tissues, and mice with heterozygous deletion of BECN1 show increased susceptibility to tumors. In cancer patients, increased BECN1 expression is considered a good outcome, whereas enhanced BCL-2 expression is linked with poor prognosis [77, 137]. Although there are discrepancies based on the tissue analyzed, expression level of BECN1 seems to decline with age in brain and liver [22, 70, 103, 136]. Moreover, with aging, BECN1 protein can accumulate in non-functional, insoluble protein aggregates and thus be unable to trigger autophagy [103]. In addition to *Becn1*, genetic deletions or mutations in other autophagy-associated genes including UV radiation resistance-associated gene (*Uvrag*), *Atg2b*, *Atg5*, *Atg9b*, and *Atg12* are observed in different types of cancers, confirming that autophagy inhibition exacerbates cancer development [57, 58]. Further, lung specific inactivation of *Atg5* promotes the early phases of lung cancer, which correlates with enhanced infiltration by FOXP3(+) regulatory T cells. Depletion of these regulatory T cells inhibits this process, suggesting that autophagy deficiency modifies the tumor microenvironment and promotes oncogenesis [86, 87]. Interestingly, autophagy deficiency reduces the progression from adenoma to adenocarcinoma, finally improving the survival of tumor-bearing mice, thus suggesting a dual role of autophagy in cancer [86]. Work from the group of Eileen White support the idea of autophagy as a "double-edged sword" in cancer [134]. Through activation of the autophagic response, tumor cells can survive in stress conditions, limiting damage and maintaining cellular viability [133]. Indeed, Karsli-Uzunbas and colleagues show that acute autophagy inhibition in mice bearing RAS-driven non-small lung cell cancer, blocks tumor growth and promotes tumor cell death through increased p53 activity [59]. Modulation of autophagy is now considered a therapeutical strategy against cancer. Current work relates to determining time points, stage and type of cancers in which induction or inhibition of autophagy can be beneficial.

6.4 Metabolic Syndrome

As we age and autophagy decreases, abnormal accumulation of dysfunctional mitochondria and lipids leads to oxidative stress and inflammation, and increases the risk of developing the metabolic syndrome, a condition characterized by central (visceral) obesity, insulin resistance, impaired glucose tolerance or overt diabetes, hypertension, dyslipidemia and cardiovascular complications [19, 29, 141]. The decline in autophagic activity, and in particular of lipophagy (autophagic degradation of intracellular lipid), contributes to intracellular accumulation and expansion of lipid droplets and increased inflammation [108]. This mechanism represents a possible feedback loop for the perpetuation of the age related-metabolic syndrome [108]. Other studies have shown that metabolic stress can trigger inflammation through NLRP3 inflammasomes in adipose tissues, leading to age-associated metabolic syndrome and decreased autophagy [127]. Aging also promotes oxidative stress-related NF-κB activation in metabolic tissues, vascular systems, neurons, and glial cells, thus promoting metabolic syndrome and metabolic syndrome-related neural diseases [15]. Minamino et al. proposed that p53 represents one link between aging and abnormal metabolism [73]. p53 expression increases with aging and is upregulated in adipose tissue and endothelial cells when mice are fed a high-calorie diet [1, 73, 142]. These alterations trigger inflammatory responses and stimulate cytokines production leading to insulin resistance. Consistently, Minamino et al. found that p53 deficiency, which is known to promote autophagy [114], lowered inflammation and improved insulin sensitivity in genetic obese mice and in mice fed a high-calorie diet [73]. Overall, these studies show that activation of the autophagic pathway, using different approaches, can prevent or delay the development of age-induced metabolic syndrome.

6.5 Sarcopenia

Sarcopenia is defined as the age-related loss of muscle mass and function. This impairment is driven by age-related chronic, low-grade inflammation that is promoted by interleukin 6, C-reactive protein and tumor necrosis factor alpha (TNF-α), factors which affect vascular and mitochondrial function [115]. Indeed, muscle aging slows mitochondrial turnover and decreases mitochondrial activity, myocyte homeostasis and autophagy [72]. Old skeletal myocytes are characterized by intralysosomal accumulation of lipofuscin, a non-degradable pigment that promotes lysosomal and mitochondrial impairment leading to ROS production [116]. Increased ROS production accelerates lipofuscinogenesis, which diminishes lysosomal degradative capacity by preventing lysosomal enzymes from targeting and recycling functional autophagosomes, compromising the autophagic/mitophagic response [14, 120]. With increase in age, marked changes in hormone production, metabolism, and action occur. The development of age-related sarcopenia represents one of the effects of the changes in endocrine factor levels [97]. Recent research has shown that treatment with testosterone, which declines with aging because of a decreased basal hypothalamic-pituitary-gonadal response, combats sarcopenia in humans by repression of NF-κB, the expression of which is increased in studies of age-related systemic inflammation [125]. Interestingly, NF-κB activation inhibits autophagy in Ewing's sarcoma cancer cells [123], suggesting that a mechanism of NF-κB-mediated autophagy inhibition might occur in age-related sarcopenia as well. Further, reduced testosterone production in aged rat Leydig cells is associated with decreased autophagic activity [69]. Thyroid hormone levels also decrease with aging [82], thus promoting sarcopenia. This effect could be due to decreased autophagic processes since this hormone has been shown to stimulate autophagy; however, further research is needed to confirm this hypothesis.

6.6 Age-Associated Kidney Disease

As they age, most healthy individuals show a progressive decline in renal function [12, 20], as well as increased amounts of glomerular, vascular and interstitial scarring in the renal tissue [93]. Age represents a risk factor for the development of chronic kidney disease (CKD), which might lead to the development of diabetic nephropathy [51]. Individuals with CKD have enlarged kidneys with disintegrated cristae; as with other age-related diseases, these features are associated with increased mitochondrial oxidative stress [63]. Studies in 24 month-old mice, designed to determine the molecular mechanisms leading to the onset and/or progression of CKD, showed that decreased expression of SIRT1 in the kidneys is associated with morphological and functional alteration of the mitochondria. Caloric restriction (CR) stimulates the autophagic process and improves age-associated alterations seen in the mitochondria [63]. Kume et al. showed that this effect is dependent on SIRT1, since CR in Sirt1$^{+/-}$ mice failed to attenuate the age-associated alterations in the kidney. Importantly, these effects are dependent on the induction of the autophagic process, identifying CR- and SIRT1-enhanced autophagy in aged tissue of mammals as a therapeutic target for age-associated kidney damage. SIRT1 is also involved in the control of the inflammatory process, which, as already mentioned, is one of the pivotal mechanisms for the initiation and progression of age-related diseases. In a model of diabetic neuropathy, Kitada et al. [62] found that decreased expression of SIRT1 in the kidneys of obese, diabetic Wistar rats correlated with enhanced acetylation of NF-κB and increased expression of inflammation-related genes (intercellular adhesion molecule 1 (*Icam-1*), vascular cell adhesion molecule 1 (*Vcam-1*) and monocyte chemoattractant protein-1

(*Mcp-1*)). Notably, CR reversed this effect through the restoration of SIRT1 protein levels and autophagy [62]. These studies show that SIRT1-mediated autophagy regulation, which is altered in kidneys by age, is essential in the CR-mediated protection of aged kidneys [63]. In podocytes, levels of constitutive autophagy are higher than other body tissues. Hartleben et al. demonstrated that mice carrying a podocyte-specific deletion of *Atg5*, which abrogates the autophagic process, have increased ER and oxidative stress, which ultimately leads to irreversible podocyte injury and loss, and development of glomerulopathy and glomerulosclerosis [48]. This study showed that constitutive autophagy is required to maintain podocyte integrity. Since autophagy decreases with aging, stimulation of the autophagic process to maintain the high constitutive levels characteristic of this cell type, appears to be an important protective and therapeutical mechanism against podocyte aging and glomerular injury.

7 Anti-aging Dietary Habits

CR is the only dietary habit that has been showed to retard aging in different animal models such as yeast, worms, flies rodents and non-human primates [13]. CR is also a potent inducer of autophagy in virtually all species, including mammals [11, 63]. Consistent with the notion of slowed aging, CR significantly attenuates age-related increases of pro-inflammatory markers, such as pro-inflammatory cytokines and NF-κB, and, consequently, the onset of age-associated pathologies; including obesity-associated diseases, cancer and brain atrophy [8]. Multiple studies confirm that activation of the protein AMPK mediates these effects and that responsiveness to AMPK decreases with aging [89]. CR activates AMPK, which regulates pathways such as (i) autophagy, through ULK1 and SIRT1, (ii) inflammation, through inhibition of the NF-κB signaling, and (iii) oxidative stress, through stimulation of FOXO3 [36, 62, 111]. Importantly, inhibition of NF-κB, as well as activation of FOXO3, is also a pro-autophagic signal [18, 131]. These findings indicate that the beneficial effects of CR, through the stimulation of different signaling pathways, converge into the final effect on autophagy induction. Consistent with the idea that an increased autophagic response is required for the beneficial health effects of CR, our own study shows that inhibition of the autophagic pathway using siRNA targeting *Becn1* suppresses CR-related beneficial outcomes [75]. Some of the effects of CR are mimicked by the polyphenol resveratrol (3,5,4′-trihydroxystilbene), the most studied anti-aging phytochemical [81]. Resveratrol is a small molecule present in many plants such as grapes, as well as red wine [44], peanuts [130], cocoa [52] and various berries [80]. Resveratrol was discovered nearly 20 years ago, when it was proposed as responsible for the positive effects of red wine on health. Since that time, resveratrol has been linked to a myriad of physiological benefits, such as protection against age-related illnesses, cardiovascular disease, cancer and diabetes [6]. Resveratrol and the CR mechanism of autophagy partially overlap. For instance, resveratrol lowers pro-inflammatory markers via inhibition of the NF-κB signaling pathway and increases plasma antioxidant capacity [10]. Further, resveratrol indirectly activates SIRT1, mediating the anti-aging effect [79]. Through SIRT1 activation, resveratrol also induces autophagy, leading to cytoprotective and anti-aging effects. These physiological benefits are lost when essential autophagy modulators are genetically or pharmacologically inactivated, indicating once again that autophagy is required for resveratrol-mediated pro-health effects [75]. Other anti-aging phytochemicals of interest include: curcumin, epigallocatechin gallate (EGCG, contained in green tea), caffeine, quercetin (a flavonoid widely distributed in nature), epicatechin (a flavonoid found in cocoa, tea and grapes). All these compounds have been shown to have anti-aging effects, lower inflammatory markers and induce autophagy [5, 10, 53, 105, 109]. Although research still needs to confirm this hypothesis, it is tempting to speculate that autophagy is required for the anti-aging/physiological benefits of these compounds.

8 Conclusions

Autophagy declines with aging in a great variety of tissues; however, it is still unknown if this decrease occurs through similar signaling pathways in all the different tissues of our body. Studies have shown that inhibition of the autophagic pathway promotes the accumulation of altered protein and organelles in the cell, alters cellular homeostasis, promotes accumulation of ROS and stimulates an inflammatory response, finally leading to the development of a broad range of age-associated pathologies, including neurodegeneration, cancer, sarcopenia, renal failure and cardiovascular diseases. The tissue-specific mechanisms that underlie these mechanisms are currently being investigated to identify specific targets and treatments. Further, even though the causal link between autophagy and longevity has been confirmed in different studies, it is unknown if upregulation of the autophagic pathway is sufficient to counteract the age-related phenotype and protect against age-related diseases. Although current research indicates that upregulation of the autophagic pathway might be a therapeutic approach for the treatment of certain age-related diseases, disease and organ-specific pathways for the modulation of autophagy must be more precisely identified. These studies will allow the development of novel therapeutic interventions against age-related disorders.

Acknowledgments The authors thank Aaron P. Frank for critically reading the manuscript. E.M. is supported by the VRA Insertion and New Professors Program of Pontificia Universidad Católica de Chile. A.C. is supported by the FONDECYT grant No 1140908, CONICYT-FONDAP Advanced Center for Chronic Diseases (ACCDiS) No 15130011 and the PEW Latin American Fellow in Biomedical Science No 00002991.

References

1. Ahima RS (2009) Connecting obesity, aging and diabetes. Nat Med 15:996–997
2. Ahmed I, Liang Y, Schools S, Dawson VL, Dawson TM, Savitt JM (2012) Development and characterization of a new Parkinson's disease model resulting from impaired autophagy. J Neurosci 32:16503–16509
3. Alers S, Loffler AS, Paasch F, Dieterle AM, Keppeler H, Lauber K, Campbell DG, Fehrenbacher B, Schaller M, Wesselborg S, Stork B (2011) Atg13 and FIP200 act independently of Ulk1 and Ulk2 in autophagy induction. Autophagy 7:1423–1433
4. Alirezaei M, Kemball CC, Whitton JL (2011) Autophagy, inflammation and neurodegenerative disease. Eur J Neurosci 33:197–204
5. Babu PV, Si H, Liu D (2012) Epigallocatechin gallate reduces vascular inflammation in db/db mice possibly through an NF-kappaB-mediated mechanism. Mol Nutr Food Res 56:1424–1432
6. Baur JA, Sinclair DA (2006) Therapeutic potential of resveratrol: the in vivo evidence. Nat Rev Drug Discov 5:493–506
7. Baylis D, Bartlett DB, Patel HP, Roberts HC (2013) Understanding how we age: insights into inflammaging. Longevity Healthspan 2:8. doi:10.1186/2046-2395-2-8
8. Bendlin BB, Canu E, Willette A, Kastman EK, McLaren DG, Kosmatka KJ, Xu G, Field AS, Colman RJ, Coe CL, Weindruch RH, Alexander AL, Johnson SC (2011) Effects of aging and calorie restriction on white matter in rhesus macaques. Neurobiol Aging 32:2319.e1–2319.e11. doi:10.1016/j.neurobiolaging.2010.04.008
9. Bergamini E, Cavallini G, Donati A, Gori Z (2004) The role of macroautophagy in the ageing process, anti-ageing intervention and age-associated diseases. Int J Biochem Cell Biol 36:2392–2404
10. Bisht K, Wagner KH, Bulmer AC (2010) Curcumin, resveratrol and flavonoids as anti-inflammatory, cyto- and DNA-protective dietary compounds. Toxicology 278:88–100
11. Blagosklonny MV (2010) Linking calorie restriction to longevity through sirtuins and autophagy: any role for TOR. Cell Death Dis 1, e12. doi:10.1038/cddis.2009.17
12. Bolignano D, Mattace-Raso F, Sijbrands EJ, Zoccali C (2014) The aging kidney revisited: a systematic review. Ageing Res Rev 14:65–80
13. Bordone L, Guarente L (2005) Calorie restriction, SIRT1 and metabolism: understanding longevity. Nature reviews. Mol Cell Biol 6:298–305
14. Brunk UT, Terman A (2002) The mitochondrial-lysosomal axis theory of aging: accumulation of damaged mitochondria as a result of imperfect autophagocytosis. Eur J Biochem 269:1996–2002
15. Cai D, Liu T (2012) Inflammatory cause of metabolic syndrome via brain stress and NF-kappaB. Aging 4:98–115
16. Cardenas C, Miller RA, Smith I, Bui T, Molgo J, Muller M, Vais H, Cheung KH, Yang J, Parker I, Thompson CB, Birnbaum MJ, Hallows KR, Foskett JK (2010) Essential regulation of cell bioenergetics by constitutive InsP3 receptor Ca2+ transfer to mitochondria. Cell 142:270–283
17. Cataldo AM, Hamilton DJ, Barnett JL, Paskevich PA, Nixon RA (1996) Properties of the endosomal-lysosomal system in the human central nervous system: disturbances mark most neurons in populations at risk to degenerate in Alzheimer's disease. J Neurosci 16:186–199
18. Chang CP, Su YC, Lee PH, Lei HY (2013) Targeting NFKB by autophagy to polarize hepatoma-associated macrophage differentiation. Autophagy 9:619–621
19. Choi AM, Ryter SW, Levine B (2013) Autophagy in human health and disease. N Engl J Med 368:1845–1846
20. Coresh J, Astor BC, Greene T, Eknoyan G, Levey AS (2003) Prevalence of chronic kidney disease and decreased kidney function in the adult US population: Third National Health and Nutrition Examination Survey. Am J Kid Dis 41:1–12

21. Coskun P, Wyrembak J, Schriner SE, Chen HW, Marciniack C, Laferla F, Wallace DC (2012) A mitochondrial etiology of Alzheimer and Parkinson disease. Biochim Biophys Acta 1820:553–564
22. Cribbs DH, Berchtold NC, Perreau V, Coleman PD, Rogers J, Tenner AJ, Cotman CW (2012) Extensive innate immune gene activation accompanies brain aging, increasing vulnerability to cognitive decline and neurodegeneration: a microarray study. J Neuroinflammation 9:179. doi:10.1186/1742-2094-9-179
23. Criollo A, Maiuri MC, Tasdemir E, Vitale I, Fiebig AA, Andrews D, Molgo J, Diaz J, Lavandero S, Harper F, Pierron G, di Stefano D, Rizzuto R, Szabadkai G, Kroemer G (2007) Regulation of autophagy by the inositol trisphosphate receptor. Cell Death Diff 14:1029–1039
24. Criollo A, Senovilla L, Authier H, Maiuri MC, Morselli E, Vitale I, Kepp O, Tasdemir E, Galluzzi L, Shen S, Tailler M, Delahaye N, Tesniere A, De Stefano D, BenYounes A, Harper F, Pierron G, Lavandero S, Zitvogel L, Israel A, Baud V, Kroemer G (2010) The IKK complex contributes to the induction of autophagy. EMBO J 29:619–631
25. Croisier E, Graeber MB (2006) Glial degeneration and reactive gliosis in alpha-synucleinopathies: the emerging concept of primary gliodegeneration. Acta Neuropathol 112:517–530
26. Cuervo AM (2004) Autophagy: in sickness and in health. Trends Cell Biol 14:70–77
27. Cuervo AM (2004) Autophagy: many paths to the same end. Mol Cell Biochem 263:55–72
28. Cuervo AM (2008) Autophagy and aging: keeping that old broom working. Trends Genet 24:604–612
29. Cuervo AM, Bergamini E, Brunk UT, Droge W, Ffrench M, Terman A (2005) Autophagy and aging: the importance of maintaining “clean” cells. Autophagy 1:131–140
30. Cuervo AM, Dice JF (2000) When lysosomes get old. Exp Gerontol 35:119–131
31. Dai DF, Rabinovitch PS, Ungvari Z (2012) Mitochondria and cardiovascular aging. Circ Res 110:1109–1124
32. Decuypere JP, Parys JB, Bultynck G (2012) Regulation of the autophagic bcl-2/beclin 1 interaction. Cells 1:284–312
33. Dodson M, Darley-Usmar V, Zhang J (2013) Cellular metabolic and autophagic pathways: traffic control by redox signaling. Free Radic Biol Med 63:207–221
34. Donati A, Cavallini G, Paradiso C, Vittorini S, Pollera M, Gori Z, Bergamini E (2001) Age-related changes in the regulation of autophagic proteolysis in rat isolated hepatocytes. J Gerontol A Biol Sci Med Sci 56:B288–B293
35. Egan D, Kim J, Shaw RJ, Guan KL (2011) The autophagy initiating kinase ULK1 is regulated via opposing phosphorylation by AMPK and mTOR. Autophagy 7:643–644
36. Egan DF, Shackelford DB, Mihaylova MM, Gelino S, Kohnz RA, Mair W, Vasquez DS, Joshi A, Gwinn DM, Taylor R, Asara JM, Fitzpatrick J, Dillin A, Viollet B, Kundu M, Hansen M, Shaw RJ (2011) Phosphorylation of ULK1 (hATG1) by AMP-activated protein kinase connects energy sensing to mitophagy. Science 331:456–461
37. Fang EF, Scheibye-Knudsen M, Brace LE, Kassahun H, SenGupta T, Nilsen H, Mitchell JR, Croteau DL, Bohr VA (2014) Defective mitophagy in XPA via PARP-1 hyperactivation and NAD(+)/SIRT1 reduction. Cell 157:882–896
38. Franceschi C, Bonafe M, Valensin S, Olivieri F, De Luca M, Ottaviani E, De Benedictis G (2000) Inflamm-aging. An evolutionary perspective on immunosenescence. Ann NY Acad Sci 908:244–254
39. Furuta N, Amano A (2010) Cellular machinery to fuse antimicrobial autophagosome with lysosome. Commun Integr Biol 3:385–387
40. Galluzzi L, Pietrocola F, Levine B, Kroemer G (2014) Metabolic control of autophagy. Cell 159:1263–1276
41. Geisler S, Holmstrom KM, Treis A, Skujat D, Weber SS, Fiesel FC, Kahle PJ, Springer W (2010) The PINK1/Parkin-mediated mitophagy is compromised by PD-associated mutations. Autophagy 6:871–878

42. Germain M, Nguyen AP, Le Grand JN, Arbour N, Vanderluit JL, Park DS, Opferman JT, Slack RS (2011) MCL-1 is a stress sensor that regulates autophagy in a developmentally regulated manner. EMBO J 30:395–407
43. Green DR, Levine B (2014) To be or not to be? How selective autophagy and cell death govern cell fate. Cell 157:65–75
44. Gu X, Creasy L, Kester A, Zeece M (1999) Capillary electrophoretic determination of resveratrol in wines. J Agric Food Chem 47:3223–3227
45. Halle A, Hornung V, Petzold GC, Stewart CR, Monks BG, Reinheckel T, Fitzgerald KA, Latz E, Moore KJ, Golenbock DT (2008) The NALP3 inflammasome is involved in the innate immune response to amyloid-beta. Nat Immunol 9:857–865
46. Hamasaki M, Furuta N, Matsuda A, Nezu A, Yamamoto A, Fujita N, Oomori H, Noda T, Haraguchi T, Hiraoka Y, Amano A, Yoshimori T (2013) Autophagosomes form at ER-mitochondria contact sites. Nature 495:389–393
47. Harris J, Hartman M, Roche C, Zeng SG, O'Shea A, Sharp FA, Lambe EM, Creagh EM, Golenbock DT, Tschopp J, Kornfeld H, Fitzgerald KA, Lavelle EC (2011) Autophagy controls IL-1beta secretion by targeting pro-IL-1beta for degradation. J Biol Chem 286:9587–9597
48. Hartleben B, Godel M, Meyer-Schwesinger C, Liu S, Ulrich T, Kobler S, Wiech T, Grahammer F, Arnold SJ, Lindenmeyer MT, Cohen CD, Pavenstadt H, Kerjaschki D, Mizushima N, Shaw AS, Walz G, Huber TB (2010) Autophagy influences glomerular disease susceptibility and maintains podocyte homeostasis in aging mice. J Clin Invest 120:1084–1096
49. Hegedus K, Takats S, Kovacs AL, Juhasz G (2013) Evolutionarily conserved role and physiological relevance of a STX17/Syx17 (syntaxin 17)-containing SNARE complex in autophagosome fusion with endosomes and lysosomes. Autophagy 9:1642–1646
50. Hoyer-Hansen M, Jaattela M (2007) AMP-activated protein kinase: a universal regulator of autophagy? Autophagy 3:381–383
51. Hsu CY, Iribarren C, McCulloch CE, Darbinian J, Go AS (2009) Risk factors for end-stage renal disease: 25-year follow-up. Arch Intern Med 169:342–350
52. Hurst WJ, Glinski JA, Miller KB, Apgar J, Davey MH, Stuart DA (2008) Survey of the trans-resveratrol and trans-piceid content of cocoa-containing and chocolate products. J Agric Food Chem 56:8374–8378
53. Indra MR, Karyono S, Ratnawati R, Malik SG (2013) Quercetin suppresses inflammation by reducing ERK1/2 phosphorylation and NF kappa B activation in Leptin-induced Human Umbilical Vein Endothelial Cells (HUVECs). BMC Res Notes 6:275. doi:10.1186/1756-0500-6-275
54. Inuzuka Y, Okuda J, Kawashima T, Kato T, Niizuma S, Tamaki Y, Iwanaga Y, Yoshida Y, Kosugi R, Watanabe-Maeda K, Machida Y, Tsuji S, Aburatani H, Izumi T, Kita T, Shioi T (2009) Suppression of phosphoinositide 3-kinase prevents cardiac aging in mice. Circulation 120:1695–1703
55. Janda E, Isidoro C, Carresi C, Mollace V (2012) Defective autophagy in Parkinson's disease: role of oxidative stress. Mol Neurobiol 46:639–661
56. Jung CH, Ro SH, Cao J, Otto NM, Kim DH (2010) mTOR regulation of autophagy. FEBS Lett 584:1287–1295
57. Kang MR, Kim MS, Oh JE, Kim YR, Song SY, Kim SS, Ahn CH, Yoo NJ, Lee SH (2009) Frameshift mutations of autophagy-related genes ATG2B, ATG5, ATG9B and ATG12 in gastric and colorectal cancers with microsatellite instability. J Pathol 217:702–706
58. Karantza-Wadsworth V, Patel S, Kravchuk O, Chen G, Mathew R, Jin S, White E (2007) Autophagy mitigates metabolic stress and genome damage in mammary tumorigenesis. Genes Dev 21:1621–1635
59. Karsli-Uzunbas G, Guo JY, Price S, Teng X, Laddha SV, Khor S, Kalaany NY, Jacks T, Chan CS, Rabinowitz JD, White E (2014) Autophagy is required for glucose homeostasis and lung tumor maintenance. Cancer Discov 4:914–927

60. Keller JN, Dimayuga E, Chen Q, Thorpe J, Gee J, Ding Q (2004) Autophagy, proteasomes, lipofuscin, and oxidative stress in the aging brain. Int J Biochem Cell Biol 36:2376–2391
61. Kim J, Kundu M, Viollet B, Guan KL (2011) AMPK and mTOR regulate autophagy through direct phosphorylation of Ulk1. Nat Cell Biol 13:132–141
62. Kitada M, Takeda A, Nagai T, Ito H, Kanasaki K, Koya D (2011) Dietary restriction ameliorates diabetic nephropathy through anti-inflammatory effects and regulation of the autophagy via restoration of Sirt1 in diabetic Wistar fatty (fa/fa) rats: a model of type 2 diabetes. Exp Diab Res 2011:908185. doi:10.1155/2011/908185
63. Kume S, Uzu T, Horiike K, Chin-Kanasaki M, Isshiki K, Araki S, Sugimoto T, Haneda M, Kashiwagi A, Koya D (2010) Calorie restriction enhances cell adaptation to hypoxia through Sirt1-dependent mitochondrial autophagy in mouse aged kidney. J Clin Invest 120:1043–1055
64. Lamb CA, Yoshimori T, Tooze SA (2013) The autophagosome: origins unknown, biogenesis complex. Nat Rev Mol Cell Biol 14:759–774
65. LaRocca TJ, Henson GD, Thorburn A, Sindler AL, Pierce GL, Seals DR (2012) Translational evidence that impaired autophagy contributes to arterial ageing. J Physiol 590:3305–3316
66. Latz E, Xiao TS, Stutz A (2013) Activation and regulation of the inflammasomes. Nat Rev Immunol 13:397–411
67. Lee JH, Budanov AV, Park EJ, Birse R, Kim TE, Perkins GA, Ocorr K, Ellisman MH, Bodmer R, Bier E, Karin M (2010) Sestrin as a feedback inhibitor of TOR that prevents age-related pathologies. Science 327:1223–1228
68. Levine B, Kroemer G (2008) Autophagy in the pathogenesis of disease. Cell 132:27–42
69. Li WR, Chen L, Chang ZJ, Xin H, Liu T, Zhang YQ, Li GY, Zhou F, Gong YQ, Gao ZZ, Xin ZC (2011) Autophagic deficiency is related to steroidogenic decline in aged rat Leydig cells. Asian J Androl 13:881–888
70. Lipinski MM, Zheng B, Lu T, Yan Z, Py BF, Ng A, Xavier RJ, Li C, Yankner BA, Scherzer CR, Yuan J (2010) Genome-wide analysis reveals mechanisms modulating autophagy in normal brain aging and in Alzheimer's disease. Proc Natl Acad Sci U S A 107:14164–14169
71. Lopez-Otin C, Blasco MA, Partridge L, Serrano M, Kroemer G (2013) The hallmarks of aging. Cell 153:1194–1217
72. Marzetti E, Calvani R, Cesari M, Buford TW, Lorenzi M, Behnke BJ, Leeuwenburgh C (2013) Mitochondrial dysfunction and sarcopenia of aging: from signaling pathways to clinical trials. Int J Biochem Cell Biol 45:2288–2301
73. Minamino T, Orimo M, Shimizu I, Kunieda T, Yokoyama M, Ito T, Nojima A, Nabetani A, Oike Y, Matsubara H, Ishikawa F, Komuro I (2009) A crucial role for adipose tissue p53 in the regulation of insulin resistance. Nat Med 15:1082–1087
74. Morselli E, Galluzzi L, Kepp O, Marino G, Michaud M, Vitale I, Maiuri MC, Kroemer G (2011) Oncosuppressive functions of autophagy. Antioxid Redox Signal 14:2251–2269
75. Morselli E, Maiuri MC, Markaki M, Megalou E, Pasparaki A, Palikaras K, Criollo A, Galluzzi L, Malik SA, Vitale I, Michaud M, Madeo F, Tavernarakis N, Kroemer G (2010) Caloric restriction and resveratrol promote longevity through the Sirtuin-1-dependent induction of autophagy. Cell Death Dis 1, e10. doi:10.1038/cddis.2009.8
76. Nair S, Ren J (2012) Autophagy and cardiovascular aging: lesson learned from rapamycin. Cell Cycle 11:2092–2099
77. Nicotra G, Mercalli F, Peracchio C, Castino R, Follo C, Valente G, Isidoro C (2010) Autophagy-active beclin-1 correlates with favourable clinical outcome in non-Hodgkin lymphomas. Mod Pathol 23:937–950
78. Nixon RA, Wegiel J, Kumar A, Yu WH, Peterhoff C, Cataldo A, Cuervo AM (2005) Extensive involvement of autophagy in Alzheimer disease: an immuno-electron microscopy study. J Neuropathol Exp Neurol 64:113–122
79. Park SJ, Ahmad F, Philp A, Baar K, Williams T, Luo H, Ke H, Rehmann H, Taussig R, Brown AL, Kim MK, Beaven MA, Burgin AB, Manganiello V, Chung JH (2012) Resveratrol ameliorates aging-related metabolic phenotypes by inhibiting cAMP phosphodiesterases. Cell 148:421–433

80. Pastrana-Bonilla E, Akoh CC, Sellappan S, Krewer G (2003) Phenolic content and antioxidant capacity of muscadine grapes. J Agric Food Chem 51:5497–5503
81. Pearson KJ, Baur JA, Lewis KN et al (2008) Resveratrol delays age-related deterioration and mimics transcriptional aspects of dietary restriction without extending life span. Cell Metab 8:157–168
82. Peeters RP (2008) Thyroid hormones and aging. Hormones 7:28–35
83. Perry VH, Cunningham C, Holmes C (2007) Systemic infections and inflammation affect chronic neurodegeneration. Nat Rev Immunol 7:161–167
84. Pietrocola F, Izzo V, Niso-Santano M, Vacchelli E, Galluzzi L, Maiuri MC, Kroemer G (2013) Regulation of autophagy by stress-responsive transcription factors. Semin Cancer Biol 23:310–322
85. Pyo JO, Yoo SM, Ahn HH, Nah J, Hong SH, Kam TI, Jung S, Jung YK (2013) Overexpression of Atg5 in mice activates autophagy and extends lifespan. Nat Commun 4:2300. doi:10.1038/ncomms3300
86. Rao S, Tortola L, Perlot T et al (2014) A dual role for autophagy in a murine model of lung cancer. Nat Commun 5:3056
87. Rao S, Yang H, Penninger JM, Kroemer G (2014) Autophagy in non-small cell lung carcinogenesis: a positive regulator of antitumor immunosurveillance. Autophagy 10:529–531
88. Ravikumar B, Sarkar S, Davies JE et al (2010) Regulation of mammalian autophagy in physiology and pathophysiology. Physiol Rev 90:1383–1435
89. Reznick RM, Zong H, Li J, Morino K, Moore IK, Yu HJ, Liu ZX, Dong J, Mustard KJ, Hawley SA, Befroy D, Pypaert M, Hardie DG, Young LH, Shulman GI (2007) Aging-associated reductions in AMP-activated protein kinase activity and mitochondrial biogenesis. Cell Metab 5:151–156
90. Roberts CK, Sindhu KK (2009) Oxidative stress and metabolic syndrome. Life Sci 84:705–712
91. Rosen KM, Moussa CE, Lee HK, Kumar P, Kitada T, Qin G, Fu Q, Querfurth HW (2010) Parkin reverses intracellular beta-amyloid accumulation and its negative effects on proteasome function. J Neurosci Res 88:167–178
92. Rubinsztein DC, Marino G, Kroemer G (2011) Autophagy and aging. Cell 146:682–695
93. Rule AD, Amer H, Cornell LD, Taler SJ, Cosio FG, Kremers WK, Textor SC, Stegall MD (2010) The association between age and nephrosclerosis on renal biopsy among healthy adults. Ann Intern Med 152:561–567
94. Russell RC, Yuan HX, Guan KL (2014) Autophagy regulation by nutrient signaling. Cell Res 24:42–57
95. Saitoh T, Akira S (2010) Regulation of innate immune responses by autophagy-related proteins. J Cell Biol 189:925–935
96. Saitoh T, Fujita N, Jang MH, Uematsu S, Yang BG, Satoh T, Omori H, Noda T, Yamamoto N, Komatsu M, Tanaka K, Kawai T, Tsujimura T, Takeuchi O, Yoshimori T, Akira S (2008) Loss of the autophagy protein Atg16L1 enhances endotoxin-induced IL-1beta production. Nature 456:264–268
97. Sakuma K, Yamaguchi A (2012) Sarcopenia and age-related endocrine function. Int J Endocrinol 2012:127362
98. Salminen A, Kaarniranta K (2009) SIRT1: regulation of longevity via autophagy. Cell Signal 21:1356–1360
99. Salminen A, Kaarniranta K, Kauppinen A (2013) Crosstalk between oxidative stress and SIRT1: impact on the aging process. Int J Mol Sci 14:3834–3859
100. Salminen A, Kaarniranta K, Kauppinen A, Ojala J, Haapasalo A, Soininen H, Hiltunen M (2013) Impaired autophagy and APP processing in Alzheimer's disease: the potential role of Beclin 1 interactome. Prog Neurobiol 106–107:33–54
101. Sarkar S, Ravikumar B, Floto RA, Rubinsztein DC (2009) Rapamycin and mTOR-independent autophagy inducers ameliorate toxicity of polyglutamine-expanded huntingtin and related proteinopathies. Cell Death Diff 16:46–56

102. Shi CS, Shenderov K, Huang NN, Kabat J, Abu-Asab M, Fitzgerald KA, Sher A, Kehrl JH (2012) Activation of autophagy by inflammatory signals limits IL-1beta production by targeting ubiquitinated inflammasomes for destruction. Nat Immunol 13:255–263
103. Shibata M, Lu T, Furuya T, Degterev A, Mizushima N, Yoshimori T, MacDonald M, Yankner B, Yuan J (2006) Regulation of intracellular accumulation of mutant Huntingtin by Beclin 1. J Biol Chem 281:14474–14485
104. Shintani T, Klionsky DJ (2004) Autophagy in health and disease: a double-edged sword. Science 306:990–995
105. Si H, Fu Z, Babu PV, Zhen W, Leroith T, Meaney MP, Voelker KA, Jia Z, Grange RW, Liu D (2011) Dietary epicatechin promotes survival of obese diabetic mice and Drosophila melanogaster. J Nutr 141:1095–1100
106. Simonsen A, Cumming RC, Brech A, Isakson P, Schubert DR, Finley KD (2008) Promoting basal levels of autophagy in the nervous system enhances longevity and oxidant resistance in adult Drosophila. Autophagy 4:176–184
107. Singh R, Cuervo AM (2011) Autophagy in the cellular energetic balance. Cell Metab 13:495–504
108. Singh R, Cuervo AM (2012) Lipophagy: connecting autophagy and lipid metabolism. Int J Cell Biol 2012:282041
109. Sinha RA, Farah BL, Singh BK, Siddique MM, Li Y, Wu Y, Ilkayeva OR, Gooding J, Ching J, Zhou J, Martinez L, Xie S, Bay BH, Summers SA, Newgard CB, Yen PM (2014) Caffeine stimulates hepatic lipid metabolism by the autophagy-lysosomal pathway in mice. Hepatology 59:1366–1380
110. Tai WT, Shiau CW, Chen HL, Liu CY, Lin CS, Cheng AL, Chen PJ, Chen KF (2013) Mcl-1-dependent activation of Beclin 1 mediates autophagic cell death induced by sorafenib and SC-59 in hepatocellular carcinoma cells. Cell Death Dis 4, e485. doi:10.1038/cddis.2013.18
111. Takeda-Watanabe A, Kitada M, Kanasaki K, Koya D (2012) SIRT1 inactivation induces inflammation through the dysregulation of autophagy in human THP-1 cells. Biochem Biophys Res Commun 427:191–196
112. Tan MS, Yu JT, Jiang T, Zhu XC, Tan L (2013) The NLRP3 inflammasome in Alzheimer's disease. Mol Neurobiol 48:875–882
113. Taneike M, Yamaguchi O, Nakai A, Hikoso S, Takeda T, Mizote I, Oka T, Tamai T, Oyabu J, Murakawa T, Nishida K, Shimizu T, Hori M, Komuro I, Takuji Shirasawa TS, Mizushima N, Otsu K (2010) Inhibition of autophagy in the heart induces age-related cardiomyopathy. Autophagy 6:600–606
114. Tasdemir E, Maiuri MC, Galluzzi L, Vitale I, Djavaheri-Mergny M, D'Amelio M, Criollo A, Morselli E, Zhu C, Harper F, Nannmark U, Samara C, Pinton P, Vicencio JM, Carnuccio R, Moll UM, Madeo F, Paterlini-Brechot P, Rizzuto R, Szabadkai G, Pierron G, Blomgren K, Tavernarakis N, Codogno P, Cecconi F, Kroemer G (2008) Regulation of autophagy by cytoplasmic p53. Nat Cell Biol 10:676–687
115. Terjung RL, Zarzeczny R, Yang HT (2002) Muscle blood flow and mitochondrial function: influence of aging. Int J Sport Nutr Exerc Metab 12:368–378
116. Terman A, Brunk UT (2004) Myocyte aging and mitochondrial turnover. Exp Gerontol 39:701–705
117. Terman A, Brunk UT (2006) Oxidative stress, accumulation of biological 'garbage', and aging. Antioxid Redox Signal 8:197–204
118. Terman A, Dalen H, Eaton JW, Neuzil J, Brunk UT (2004) Aging of cardiac myocytes in culture: oxidative stress, lipofuscin accumulation, and mitochondrial turnover. Ann NY Acad Sci 1019:70–77
119. Terman A, Kurz T, Gustafsson B, Brunk UT (2008) The involvement of lysosomes in myocardial aging and disease. Curr Cardiol Rev 4:107–115
120. Terman A, Kurz T, Navratil M, Arriaga EA, Brunk UT (2010) Mitochondrial turnover and aging of long-lived postmitotic cells: the mitochondrial-lysosomal axis theory of aging. Antioxid Redox Signal 12:503–535
121. Tooze SA, Jefferies HB, Kalie E, Longatti A, McAlpine FE, McKnight NC, Orsi A, Polson HE, Razi M, Robinson DJ, Webber JL (2010) Trafficking and signaling in mammalian autophagy. IUBMB Life 62:503–508

122. Toth ML, Sigmond T, Borsos E, Barna J, Erdelyi P, Takacs-Vellai K, Orosz L, Kovacs AL, Csikos G, Sass M, Vellai T (2008) Longevity pathways converge on autophagy genes to regulate life span in Caenorhabditis elegans. Autophagy 4:330–338
123. Trocoli A, Djavaheri-Mergny M (2011) The complex interplay between autophagy and NF-kappaB signaling pathways in cancer cells. Am J Cancer Res 1:629–649
124. Tschopp J (2011) Mitochondria: sovereign of inflammation? Eur J Immunol 41:1196–1202
125. Urban RJ, Dillon EL, Choudhary S, Zhao Y, Horstman AM, Tilton RG, Sheffield-Moore M (2014) Translational studies in older men using testosterone to treat sarcopenia. Trans Am Clin Climatol Assoc 125:27–44
126. Vakkila J, Lotze MT (2004) Inflammation and necrosis promote tumour growth. Nat Rev Immunol 4:641–648
127. Vandanmagsar B, Youm YH, Ravussin A, Galgani JE, Stadler K, Mynatt RL, Ravussin E, Stephens JM, Dixit VD (2011) The NLRP3 inflammasome instigates obesity-induced inflammation and insulin resistance. Nat Med 17:179–188
128. Vicencio JM, Ortiz C, Criollo A, Jones AW, Kepp O, Galluzzi L, Joza N, Vitale I, Morselli E, Tailler M, Castedo M, Maiuri MC, Molgo J, Szabadkai G, Lavandero S, Kroemer G (2009) The inositol 1,4,5-trisphosphate receptor regulates autophagy through its interaction with Beclin 1. Cell Death Differ 16:1006–1017
129. Vicinanza M, Korolchuk VI, Ashkenazi A, Puri C, Menzies FM, Clarke JH, Rubinsztein DC (2015) PI(5)P regulates autophagosome biogenesis. Mol Cell 57:219–234
130. Wang KH, Lai YH, Chang JC, Ko TF, Shyu SL, Chiou RY (2005) Germination of peanut kernels to enhance resveratrol biosynthesis and prepare sprouts as a functional vegetable. J Agric Food Chem 53:242–246
131. Warr MR, Binnewies M, Flach J, Reynaud D, Garg T, Malhotra R, Debnath J, Passegue E (2013) FOXO3A directs a protective autophagy program in haematopoietic stem cells. Nature 494:323–327
132. Wei Y, Sinha S, Levine B (2008) Dual role of JNK1-mediated phosphorylation of Bcl-2 in autophagy and apoptosis regulation. Autophagy 4:949–951
133. White E (2015) The role for autophagy in cancer. J Clin Invest 125:42–46
134. White E, DiPaola RS (2009) The double-edged sword of autophagy modulation in cancer. Clin Cancer Res 15:5308–5316
135. Wirth M, Joachim J, Tooze SA (2013) Autophagosome formation – the role of ULK1 and Beclin1-PI3KC3 complexes in setting the stage. Semin Cancer Biol 23:301–309
136. Wohlgemuth SE, Julian D, Akin DE, Fried J, Toscano K, Leeuwenburgh C, Dunn WA Jr (2007) Autophagy in the heart and liver during normal aging and calorie restriction. Rejuvenation Res 10:281–292
137. Won KY, Kim GY, Kim YW, Song JY, Lim SJ (2010) Clinicopathologic correlation of beclin-1 and bcl-2 expression in human breast cancer. Hum Pathol 41:107–112
138. Xie J, Zhang X, Zhang L (2013) Negative regulation of inflammation by SIRT1. Pharmacol Res 67:60–67
139. Yang F, Chu X, Yin M, Liu X, Yuan H, Niu Y, Fu L (2014) mTOR and autophagy in normal brain aging and caloric restriction ameliorating age-related cognition deficits. Behav Brain Res 264:82–90
140. Yang Z, Klionsky DJ (2010) Mammalian autophagy: core molecular machinery and signaling regulation. Curr Opin Cell Biol 22:124–131
141. Yen WL, Klionsky DJ (2008) How to live long and prosper: autophagy, mitochondria, and aging. Physiology 23:248–262
142. Yokoyama M, Okada S, Nakagomi A, Moriya J, Shimizu I, Nojima A, Yoshida Y, Ichimiya H, Kamimura N, Kobayashi Y, Ohta S, Fruttiger M, Lozano G, Minamino T (2014) Inhibition of endothelial p53 improves metabolic abnormalities related to dietary obesity. Cell Rep 7:1691–1703
143. Zhou R, Yazdi AS, Menu P, Tschopp J (2011) A role for mitochondria in NLRP3 inflammasome activation. Nature 469:221–225

Pathophysiologic Role of Autophagy in Human Airways

Valentina Sica and Valentina Izzo

Abstract Lung diseases are among the most common and widespread disorders worldwide. They refer to many different pathological conditions affecting the pulmonary system in acute or chronic forms, such as asthma, chronic obstructive pulmonary disease, infections, cystic fibrosis, lung cancer and many other breath complications. Environmental, epigenetic and genetic co-factors are responsible for these pathologies that can lead to respiratory failure, and, even, ultimately death. Increasing evidences have highlighted the implication of the autophagic pathways in the pathogenesis of lung diseases and, in some cases, the deregulated molecular mechanisms underlying autophagy may be considered as potential new therapeutic targets. This chapter summarizes recent advances in understanding the pathophysiological functions of autophagy and its possible roles in the causation and/or prevention of human lung diseases.

Abbreviations

AAT	Alpha-1-antitrypsin
AATD	Alpha-1-antitrypsin deficiency
ALI	Acute lung injury

V. Sica
Ecole doctorale de Cancerologie, University of Paris SUD, Paris, France

"Apoptosis, Cancer & Immunity", Team 11, INSERM U1138 Cordeliers Research Center, 15 rue Ecole de Médecine, 75006 Paris, France

Cell Biology and Metabolomics Platform, Gustave Roussy Comprehensive Cancer Center, Pavillon de Recherche 1, 114 rue Édouard-Vaillant, 94805 Villejuif, France

V. Izzo (✉)
"Apoptosis, Cancer & Immunity", Team 11, INSERM U1138 Cordeliers Research Center, 15 rue Ecole de Médecine, 75006 Paris, France

Cell Biology and Metabolomics Platform, Gustave Roussy Comprehensive Cancer Center, Pavillon de Recherche 1, 114 rue Édouard-Vaillant, 94805 Villejuif, France
e-mail: valentina.izzo@upmc.fr

M.C. Maiuri, D. De Stefano (eds.), *Autophagy Networks in Inflammation*, Progress in Inflammation Research, DOI 10.1007/978-3-319-30079-5_16

ALT-E	Alternaria-associated asthma
ARDS	Acute respiratory distress syndrome
Atg	Autophagy-related
ATP	Adenosine triphosphate
Bcl-2	B-cell lymphoma 2
BMP	Bone morphogenetic protein
BMPR2	BMP receptor type-II
BRAF	B-Raf proto-oncogene
CAV-1	Caveolin-1
CD274	Cluster of differentiation 274 (known as Programmed death-ligand 1, PD-L1 or B7 homolog 1, B7-H1)
CF	Cystic Fibrosis
CFTR	Cystic Fibrosis Transmembrane Conductance Regulator
COPD	Chronic obstructive pulmonary disease
CRC	Murine colorectal carcinoma
CS	Cigarette smoke
ECM	Extracellular matrix
EGFR	Epidermal growth factor receptor
Egr-1	Early growth response protein 1
EMT	Epithelial-to-mesenchymal transition
ER	Endoplasmic Reticulum
F508del-CFTR	Deletion of phenylalanine in position 508 of the CFTR
FEV1	Forced expiratory volume in 1 second
FF	Fibroblastic foci
FMD	Myofibroblast differentiation
FoxO3	Forkhead box O3
FOXP3	Forkhead box P3
H_2O_2	Hydrogen peroxide
HDAC6	Histone deacetylase 6
HH	Hedgehog
HO-1	Heme oxygenase-1
IFN	Interferon
IFT20	Intraflagellar transport protein 20 homolog
IL	Interleukin
ILD	Interstitial lung disease
IPF	Idiopathic pulmonary fibrosis
KRAS	Kirsten rat sarcoma viral oncogene homolog
LC3 (MAP1LC3)	Microtubule-associated protein 1 light chain 3*
LPS	Lipopolysaccharide
MCC	Mucociliary clearance
MMP	Matrix metalloproteinases
mTOR	Mammalian target of rapamycin
MUC5AC	Mucin 5AC
MyD88	Myeloid differentiation primary response gene 88
NK	Natural killer

NO	Nitric oxide
NSCLC	Human non-small cell lung carcinoma
OFD1	Oral facial digital syndrome
p62/SQSTM1	Sequestosome 1
PAH	Pulmonary arterial hypertension
PARK2	Parkin RBR E3 ubiquitin protein ligase
PASMCs	Pulmonary artery smooth muscle cells
PH	Pulmonary hypertension
PI3K	Class III-phosphoinositide 3-kinase
PINK	PTEN-induced putative kinase
PTEN	Phosphatase and tensin homolog
ROS	Reactive oxygen species
Rtp801	Known as Redd1 (regulated in development and DNA damage responses 1)
SIRT6	Sirtuin 6
SNPs	Single Nucleotide Polymorphisms
STK11 (LKB1)	Serine/threonine kinase 11
TFEB	Transcription factor EB
TG2	Transglutaminase type 2
TGF-β1	Transforming growth factor-β1
Th	T helper
TLR4	Toll-like receptor 4
TSC	Tuberous sclerosis complex
WHO	World Health Organization
α-SMA	Smooth muscle-α actin

1 Introduction

Lung diseases are some of the most common medical conditions in the world. The lung has the principal aim to mediate gas exchange [60]. For this reason, the lung can be subjected to several insults, belonging to the environment (inspiration of foreign matter, particles, smoke), reactive oxygen species (ROS) production, biological origins (e.g., viruses, bacteria), changes in O_2 tension, and mechanical stresses (e.g., mechanical ventilation). It is possible to discriminate between diseases affecting: (I) the airways (asthma, chronic obstructive pulmonary disease, chronic bronchitis, emphysema, acute bronchitis and cystic fibrosis); (II) the interstitium (sarcoidosis, idiopathic pulmonary fibrosis, autoimmune diseases, pneumonias and pulmonary edemas); (III) the blood vessels (pulmonary embolism and hypertension); the pleura (pleural effusion, pneumothorax and mesothelioma); (IV) the chest wall (obesity hypoventilation syndrome and neuromuscular disorders). The development of lung diseases can be associated to both acute and chronic exposure to such insults. However, in most conditions, a favouring genetic is necessary [60]. Yet, the lung has various inducible defence mechanisms to protect itself. First, constitutive and inducible stress protein and antioxidant defences; second, innate immune responses; third, pro- and anti-apoptotic mechanisms [84, 85, 103]. Several studies have recently pinpointed the emerging role of macroautophagy (more often and hereby referred to as autophagy) in lung homeostasis and diseases. Autophagy is a catabolic process that involves the sequential sequestration of cytoplasmic material within double-membraned vesicles (autophagosomes), the fusion of autophagosomes with lysosomes, and the degradation of autophagosomal cargoes (as well as of structural autophagosomal components) by lysosomal hydrolases [26]. Autophagy is mediated by a genetically encoded, evolutionary conserved machinery that is connected to most, if not all, major biochemical processes of the cell, including core metabolic circuitries as well as signal transduction pathways initiated by plasma membrane receptors [18]. Basically, autophagy responds to three major organismal needs: (1) it preserves cellular homeostasis in physiological conditions; (2) it plays a key role in cellular adaptation to stressful stimuli; and (3) it participates in the communication of states of the danger to the whole organism [21]. Indeed, autophagy continuously operates to mediate the disposal of potentially dangerous structures that may otherwise accumulate in the cytoplasm as a consequence of normal cellular activities, like old (and damaged) organelles or protein aggregates [64]. Moreover, the autophagic flux is highly responsive to situations in which intracellular or extracellular homeostasis is perturbed, which generally involves either an increased offer of autophagic substrates (as it occurs in the course of viral infection) or an increased need for autophagic functions or products (as it occurs in response to nutrient deprivation) [90]. In both these settings, proficient autophagic responses are required for the optimal adaptation of cells to stress, as demonstrated in experiments involving pharmacological inhibitors of autophagy or the depletion of essential components of the autophagic machinery [46]. Finally, autophagy is required for cells experiencing so-called "oncogenic stress" (i.e., the boost of cellular functions driven by activating mutations in one oncogene or loss-of-function mutation in one tumor suppressor gene) to become senescent (a cell-intrinsic oncosuppressive mechanism) while secreting immunostimulatory cytokines and expressing on their surface ligands for activatory natural killer (NK)-cell receptors (hence triggering a cell-extrinsic mechanism of tumor suppression) [55]. Along similar lines, cancer cells succumbing to a peculiar form of apoptosis known as "immunogenic cell death" are able to recruit antigen-presenting cells and hence trigger an adaptive immune response only if they secrete ATP as they die, a process that requires proficient autophagic responses [42, 45]. It should be noted that autophagy has also been causally implicated in some instances of cell death, especially in lower organisms like *Drosophila melanogaster* [13, 17]. However, in mammals autophagy mainly mediates robust cytoprotective functions, and – when cellular homeostasis is irremediably compromised – contributes to the maintenance of organismal homeostasis by playing a role in danger signalling. In line with this notion, defects in the autophagic machinery have been associated with a wide panel of human pathologies, including (but not limited to) malignant diseases, neurodegenerative disorders, as well as cardiovascular, renal and pulmonary conditions [86]. An accurate description of the autophagy pathway and its role in immunity and inflammation has been provided in several previous chapters of this book; therefore, here we will focus on the impact of autophagic in the etiology and treatment of human pulmonary diseases.

2 Acute Lung Injury

Acute lung injury (ALI) and the acute respiratory distress syndrome (ARDS) describe clinical syndromes of acute respiratory failure with substantial morbidity and mortality. ALI is characterised by acute inflammation that causes disruption of the lung endothelial and epithelial barriers. The ALI cellular features include loss of alveolar–capillary membrane integrity, excessive transepithelial neutrophil migration, and release of pro-inflammatory, cytotoxic mediators. The treatment of ALI is predominantly based on ventilatory strategies [35]. However, prolonged exposure to high oxygen therapy (hyperoxia) can result in lung injury [7]. Few studies are present in the literature concerning the role of autophagy in ALI, even so these works support the hypothesis that activation of autophagy has a protective role in this disease. It has been demonstrated that prolonged hyperoxia, which causes characteristic lung injury in mice, induced the increase of LC3II expression. Moreover, in pulmonary epithelial cells, the genetic depletion of LC3 sentitizes the cells to hyperoxia-induced cell death suggesting that LC3 activation confers cytoprotection in oxygen-dependent cytotoxicity [93]. Besides, the involvement of mitophagy has also been identified. The ability to resist hyperoxia is proportional to PTEN-induced putative kinase 1 (PINK1) expression. In fact, the *Pink1*$^{-/-}$ mice were more susceptible to hyperoxia when compared to wild-type mice. Furthermore, genetic deletion of PINK1 or PINK1 silencing in the lung endothelium cells increased susceptibility to hyperoxia *via* alterations in autophagy/mitophagy, proteasome activation, apoptosis and oxidant generation [108].

3 Chronic Obstructive Pulmonary Disease (COPD)

Chronic obstructive pulmonary disease (COPD) is a chronic inflammatory lung disease that causes breathing difficulty, cough, sputum production and dyspnoea. Emphysema and chronic bronchitis can contribute to COPD development. Emphysema is a condition resulting from a severe damage of air sacs (the alveoli). Chronic bronchitis is due to inflammation of the lining of the bronchial tubes. The lung damage that leads to COPD is caused by long-term exposure to irritating gases or particulate matter, most often from cigarette smoke (CS), air pollution or workplace exposure to dust, smoke or fumes. However, a genetic susceptibility to the disease should be considered as an important cofactor. Patients with COPD present increased risk of developing other pathologies, such as heart disease or lung cancer [53]. Multiple molecular mechanisms, not fully understood, participate to the

COPD evolution and, among others, the involvement of the autophagic pathway has been pointed out [3, 86]. In lung tissue from COPD patients, an increase of autophagic vacuoles as well as several autophagy markers (LC3, ATG4, ATG5/12, ATG7) expression has been detected [8]. These evidences are perhaps a result of defective autophagic flux. To corroborate this hypothesis, an increased accumulation of p62 and ubiquitinated proteins and a decreased expression levels of sirtuin 6 (SIRT6) have been evaluated in lung homogenates from COPD patients [92]. Kuwano and colleagues hypothesize that the insufficient autophagic clearance is involved in the accelerated cell senescence observed in COPD [16, 92]. The CS induces mitochondrial damage, accompanied by increased ROS production *in vitro*. The CS-induced mitophagy was inhibited by PINK1 and PARK2 knockdown, resulting in enhanced mitochondrial ROS production. Moreover, a decreased expression of PARK2 in COPD lungs compared with non-COPD lungs has been detected, suggesting that insufficient mitophagy is a part of the pathogenic sequence and cellular senescence of COPD [32]. In addition, a defective xenophagy has been observed in alveolar macrophages of smokers, suggesting that the deregulation of this selective process may contribute to recurrent infections [65]. In contrast, other findings indicate that autophagy has an opposite role in COPD favouring the pathological environment. It has been shown that Rtp801 (also known as Reddl) expression is increased in human emphysematous lungs and in lungs of mice exposed to CS, whereas Rtp801 knockout mice were protected against acute CS-induced lung injury. Rtp801 inhibits mammalian target of rapamycin (mTOR), by stabilizing the TSC1-TSC2 inhibitory complex. The inhibition of mTOR is linked to autophagy induction, but Rtp801 expression enhances oxidative stress-dependent cell death, amplifying the development of CS-induced lung injury [105]. Furthermore, the higher expression of autophagy proteins has been linked to lung epithelial cell death, airway dysfunction and emphysema in response to CS. Genetic depletion of LC3B *in vivo* (*Map1lc3B*$^{-/-}$ mice) suppressed cell death and emphysematous airspace enlargement during chronic CS exposure compared to the wild type mice [9]. More recently, the same group demonstrated that mitophagy regulates necroptosis, which contributes to the COPD pathogenesis. Mice deficient for *Pink1* were protected against mitochondrial dysfunction, airspace enlargement and mucociliary clearance (MCC) disruption during CS exposure [63]. Interestingly, they identified the contribution of a novel selective autophagy-dependent pathway that regulates cilia length, "ciliophagy", in the COPD pathophysiological evolution. Exposure to CS reduced cilia length and autophagy-impaired (*Beclin 1*$^{+/-}$ or *Map1lc3B*$^{-/-}$) mice resisted to the CS-induced cilia shortening *via* a mechanism involving histone deacetylase 6 (HDAC6) [48]. Accordingly, it has been shown that autophagy negatively regulate ciliogenesis by the degradation of the essential ciliary protein IFT20 [70]. Conversely, Hedgehog (HH) signalling from primary cilia promotes autophagy [70] and autophagy promotes ciliogenesis by degrading OFD1 (oral facial digital syndrome) at centriolar satellites [95]. Further studies are necessary to clarify the dual relationship between these processes [101]. In conclusion, these studies illustrate that the contribution of autophagy in COPD pathophysiology is complex and show a context-specific role depending on the cell type and tissue as well as on the different stimuli involved.

4 Interstitial Lung Disease (ILD)

Interstitial lung disease (ILD) is a general category that includes all lung diseases affecting the interstitium, the tissue and space that extends throughout both lungs. Among them the most common are Sarcoidosis and Idiopathic pulmonary fibrosis (IPF). Sarcoidosis is a systemic inflammatory disease caused by persistent reaction toward a stimulus (virus or antigens) that continues even when it is physiologically cleared from the body. Lung interstitium fibrosis is the first symptom in patients with Sarcoidosis. Conversely, IPF is characterized by specific fibrosis at interstitial level due to the increased extracellular matrix (ECM) protein deposition and hyper activation of myofibroblasts [10].

Recently, reduced LC3II expression and p62 accumulation has been found in lung tissue from IPF patients [72]. The reduced expression of the transcription factor FoxO3a in IPF fibroblasts could be the cause for the reduction in the levels of LC3 protein as the expression of this latter is positively stimulated by FoxO3a [30].

Furthermore, in fibroblast of IPF patients, decreased expression in Beclin-1 protein and increased expression of the anti-apoptotic protein Bcl-2 have been found, confirming a defect in the autophagy pathway at different level [81]. Moreover, fibroblastic foci (FF), that are the starting point for fibrogenesis, are enriched in ubiquitinated proteins and p62, confirming the insufficient autophagy at the basis of IPF pathogenesis [3].

Autophagy inhibition is able to induce acceleration of epithelial cell senescence and fibroblast to myofibroblast differentiation (FMD), which have a critical role in IPF development [3]. Transforming growth factor-β1 (TGF-β1) is one of the essential mediators of fibrosis since it stimulates fibroblasts to produce fibronectin and the smooth muscle-α actin (α-SMA), which is a myofibroblast marker. Autophagy has been associated to fibrosis through TGF-β1. In fact, genetic deletion of LC3 or Beclin 1 increases TGF-β1 activity as well as *in vivo* treatment with Rapamycin can protect from fibrosis [72]. TGF-β1 expression seems to be dependent on IL-17A, a proinflammatory cytokine involved in chronic inflammation and autoimmune disease. Blocking IL-17A might reduce the progression of fibrosis promoting the autophagic degradation of collagen [61].

Recently, lacking of matrix metalloproteinases-19 (MMP-19) has been associated with exacerbated fibrosis in the hyperplastic alveolar epithelium of IPF lungs [106]. Additionally, MMP-19-deficient mice exhibit diminished Atg4c protein expression, demonstrating a direct correlation between these two pathways [33]. Similar evidences from an independent group corroborate the role of autophagy in promoting FMD. In fact, Atg4b-deficient mice exhibited reduction in autophagic activity in lungs, collagen accumulation and increased protein levels of the myofibroblast biomarker α-SMA [6].

Pharmacological treatment with the alkaloid Barberine has been proposed for IPF monitoring because of its capacity to inhibit the activation of mTOR and to increase the expression of LC3 and Beclin 1 in an bleomycin *in vivo* model of airway-fibrosis [11]. Furthermore, the multiple tyrosine kinase inhibitor Nintedanib has recently been approved for the treatment of IPS for its anti-fibrotic effect. It has been shown that Nintedanib is able to reduce the expression of ECM proteins, fibronectin and collagen as well as to induce a Beclin 1 dependent, ATG7 independent autophagy [76].

5 Asthma

Asthma is a chronic respiratory disease affecting 300 million people worldwide. Asthma manifests through several symptoms including wheezing, breathlessness, and chest tightness. Asthmatic airways are characterized by chronic inflammation, eosinophil infiltration, epithelial fibrosis, mucus hyperproduction, and goblet cell hyperplasia [20].

It is considered as chronic allergic inflammatory disease, mostly mediated by a Th2 response, but an initial Th1-type immune response seems to be the trigger for the subsequent Th2-type response [82]. Thus, Th2 hyperactivation leads to persistent airway inflammation and the occurring of asthma phenotype [38].

Emerging evidences suggest that activation of autophagy is associated with reduced lung function in asthmatic patients. In particular electron microscopy analysis of fibroblast and epithelial cells from asthmatic patients showed increased autophagic hallmarks "such as double membrane autophagosomes" compared to healthy patients [75]. Unfortunately, at present, the role of autophagy in asthma is still unclear.

A recent study demonstrated that two Single Nucleotide Polymorphisms (SNPs), namely rs12201458 and rs510432 were associated with childhood asthma. In particular rs510432 localises at the promoter of ATG5 gene and could increase its expression in nasal epithelium of acute asthmatics compared to stable asthmatics and non-asthmatic patients [58]. Another intronic SNP variant (rs12212740) in ATG5 gene was also shown to be associated with pre-bronchodilator forced expiratory volume in 1 second (FEV1) in asthmatic patients [75].

ATG5 is an essential player in the initiation of autophagy, but its role in asthma pathogenesis is controversial. On one hand ATG5 could help viral elimination through the activation of Xenophagy, and on the other hand it negatively regulates the antiviral properties of type I interferon (IFN) inhibiting innate anti-virus immune responses [36, 90]. Together with these findings, lungs from conditional *Atg7* knockout mice manifest hyper-responsiveness to cholinergic stimuli, which is a common sign of asthma and chronic inflammatory diseases [31]. Asthma severity has been directly correlated with the level of autophagic response in the sputum granulocytes, peripheral blood cells and peripheral blood eosinophils of severe and non-severe asthmatic patients [5].

Autophagy is also involved in the maintenance of intracellular ROS homeostasis, and it has been well established that oxidative stress is associated with asthma so that exhaled levels of hydrogen peroxide (H_2O_2) and nitric oxide (NO) are currently used as predictors of asthma severity [68].

Chronic asthma is characterized by excessive ECM deposition and proliferation of myofibroblasts, leading to fibrosis in the airway wall [79]. The accumulation of fibrotic tissue is mostly due to the production of collagen A1 and fibronectin by the primary human airway smooth muscle through a mechanism autophagy-dependent that involves the TGFβ1. This response is reverted by the silencing of the major key autophagy-inducing gene Atg5 and Atg7 [104].

As already mentioned, asthma is a pathology mostly driven by Th2-type cytokines. Among them, IL-13 is extensively produced in activated $CD4^+$ Th2 lymphocytes and is overexpressed in the airway epithelium of asthmatic patients [47]. Here, IL-13 is thought to be responsible for epithelial hypertrophy, mucus hypersecretion, adventitial fibrosis and goblet cell hyperplasia [111]. It directly induces hypersecretion of mucin 5AC, oligomeric mucus/gel forming (MUC5AC) in airway epithelial cell and oxidant stress through a mechanism that is autophagy-dependent, as demonstrated *in vitro* by depletion of ATG5 or ATG14 in primary human tracheal-bronchial epithelial cells [15].

Autophagy might be involved in the pathophysiology of Alternaria (ALT-E)-associated asthma. ALT-E is an outdoor allergen able to activate autophagy, which in turn stimulates epithelial cells to release IL-18 [67]. This latter when produced is able to stimulate Th2 differentiation from naïve $CD4^+$ T-cells and IFN–γ production by Th1 cells. IL-18 level in serum of asthmatic patients might reflect the degree of disease exacerbation [94].

6 Cystic Fibrosis (CF)

Cystic Fibrosis (CF) is one of the most common lethal genetic diseases in Caucasian population. It is an autosomal recessive disease caused by mutation in the Cystic Fibrosis Transmembrane Conductance Regulator (CFTR) gene. Approximately 1 out of 20 Caucasians are carriers for mutation in this gene. Up to date over 2000 types of different mutations have been discovered and classified according to the degree of functional CFTR protein (http://www.genet.sickkids.on.ca/StatisticsPage.html; [27]). Among these, the most common one is the F508del-CFTR. Approximately 90 % of CF patients have at least one F508del-CFTR allele, and about 70 % are homozygous for it.

The CFTR channel is located at the apical surface of epithelial cells and it is deputized to move out Cl^- from the cell. Na^+ passes through the membranes passively, increasing the movement of water by osmosis. Loss of functional CFTR expression is thought to alter this homeostatic balance through the epithelial layer, leading to net volume depletion of mucus, increased viscosity, and ineffective bacterial clearance [43, 78]. Recurrent pulmonary infections in turn induce an increased inflammatory response and signalling, thus starting a vicious cycle of mucus retention, infection, and inflammation. Since the CFTR is localized in many organs, CF symptoms could go from malabsorption at pancreatic level and gastrointestinal obstruction to male infertility and liver disease. Nevertheless, the main cause of death remains persistent and untreatable pulmonary *Pseudomonas aeruginosa* infection.

Several recent studies have demonstrated an impairment of autophagy in CF. In fact, in epithelial cells, mutated/unfunctional CFTR causes increased ROS production with consequent increase in tissue transglutaminase type 2 (TG2) levels. TG2, in turn, leads to crosslinking of several targets including Beclin 1 [54, 57]. Beclin 1 interactome displaces from the Endoplasmic Reticulum (ER) leading to the sequestration of class III-phosphoinositide 3-kinase (PI3K) complex, accumulation of p62 with consequent inhibition of autophagosomes formation. The resulting accumulation of aggresomes leads to proteasome overload and may promote the accumulation of mutated CFTR in intracellular aggregates [54]. Restoration of Beclin 1 activity, depletion of p62 by genetic manipulation or treatment with autophagy-stimulatory proteostasis regulators, such as cystamine, functionally rescue the CFTR mutated protein at the apical surface of epithelial cells both *in vitro* and *in vivo* [54].

Heme oxygenases are enzymes involved in the catabolism of the heme ring to generate carbon monoxide, biliverdin-IXα, and ferrous iron. The inducible isoform Heme oxygenase-1 (HO-1) is activated in response to stress such as oxidative stress, hypoxia, heavy metals exposure and cytokines. HO-1, together with its enzymatic products, is able to inhibit apoptosis and related cell death pathways, conferring tissue protection in case of lung or vascular injury [66]. HO-1 could represent the link between CF and impaired autophagy since its expression is increased in human bronchial CF cells. This increase has been associated either to the reduction of apoptosis/injury during *P. aeruginosa* challenge either to the expression of inflammatory mediators [109]. Other evidences suggesting the cytoprotective role of HO-1 in CF showed that Lipopolysaccharide (LPS)-challenged CF macrophages fail to compartmentalize HO-1 to the cell surface and this mechanism seems to be dependent on the reduction in Caveolin-1 (CAV-1) expression [107]. In fact, when HO-1 localises at the plasma membrane, is able to form a complex with CAV-1, which in turn binds and detaches MyD88 from its complex with TLR4 thus terminating the cell death signal [99].

Autophagic clearance of bacteria (so-called Xenophagy) could also be impaired in case of disease, inducing increased bacterial infection that is one of the most frequent injuries in CF patients [90]. In fact it has been demonstrated that *Burkholderia cenocepacia* has the capacity to survive in F508del-CFTR macrophages since immediately after the engulfment, the bacteria resides on LC3-positive vacuoles that appear as arrested autophagosomes [98]. This capacity is directly correlated to the levels of p62, so that its depletion leads not only to a decreased bacterial survival in macrophages but also to the release of Beclin 1 from aggresomes allowing its recruitment to the *B. cenocepacia* vacuole and bacterial clearance via autophagy [2]. *B. cenocepacia* represents a serious threat for CF patients since the infection results in persistent lung inflammation and the bacteria are resistant to most of all available antibiotics [1].

Similar findings showed that pharmacological or molecular inhibition of autophagy reduces the clearance of intracellular *Pseudomonas aeruginosa* in vitro [37].

Treatment of CF mice with the mTOR inhibitor Rapamycin decreases bacterial burden in the lungs and drastically reduces signs of lung inflammation [1].

In a normal situation, autophagy can help not only removing polyubiquitinated protein but also controlling bacteria clearance; for these reasons novel strategies aimed at restoring autophagy are emerging as promising therapeutic approaches for CF patients [56].

7 Alpha-1-Antitrypsin Deficiency (AATD)

AATD is a hereditary disorder characterized by a low serum level of alpha-1-antitrypsin (AAT), a 52 kDa serine protease inhibitor, member of the serpin family [29]. AAT is essentially synthetized in the liver and secreted into the bloodstream, where it controls tissue degradation by the enzyme neutrophil elastase. The deficiency in AAT is associated with liver and lung disease due to the loss of anti-inflammatory and antiproteolytic functions. The majority of patients with AAT deficiency are homozygotes for a missense mutation ("PiZ mutation": lysine replaces glutamic acid at position 342) that alters protein folding. Mutant AAT molecules polymerize and aggregate in the ER of hepatocytes, forming large intrahepatocytic globules, the characteristic features of this disease. The proteasome is responsible for degrading the soluble form of ATT by means of ER-associated degradation while autophagy is involved in disposal of insoluble ATT polymers and aggregates [74]. In fact, a significant accumulation of autophagic vacuoles was found *in vitro* and *in vivo* in liver cells from AATD patients as well as in PiZ mouse model [96, 97]. Whereas in absence of autophagy the degradation of AAT was retarded [39]. Moreover, it has been demonstrated that the stimulation of autophagy by carbamazepine or rapamycin treatment or by liver-directed gene transfer of transcription factor EB (TFEB), a gene regulating lysosomal function and autophagy [89], reduce the hepatic amount of AAT as well as the hepatic fibrosis in mice expressing mutant AAT [28, 41, 71]. Although these results should be corroborated, altogether indicate that autophagy exerts a protective role in AATD and open a real possibility to treat AATD with pro-autophagic molecules.

8 Pulmonary Hypertension (PH)

Pulmonary hypertension (PH) was first identified in 1891 by Ernst von Romberg. PH is a severe and progressive disease that consists in increased blood pressure of lung vasculature and, often, can be a complication of chronic lung disease [88].

Since 2008 the pathology has been classified, by the World Health Organization (WHO), in five groups on the basis of mechanisms underlying the pathogenesis of the multiple types of PH.

The role of autophagy in pulmonary hypertension has mainly been described in correlation with pulmonary arterial hypertension (PAH), WHO Group I.

Little is known about the aetiology of PH, one of the most frequent genetic mutations causing idiopathic inherited form of PH is found in the gene encoding bone morphogenetic protein (BMP) receptor type-II (BMPR2).

In PAH, the pulmonary artery smooth muscle cells (PASMCs) proliferate excessively and are resistant to apoptosis. Chloroquine, a known inhibitor of autophagy flux, has been described as a drug preventing experimental PAH progression. The induction of PAH, by monocrotaline, in rat is associated with increased autophagy and decreased BMPR2 protein expression. The inhibition of autophagy by chloroquine ameliorates the level of BMPR2, inhibits the proliferation and stimulates apoptosis of rat PASMCs [52]. A recent publication [50] confirms that the inhibition of autophagy, by overexpressing mTOR, is a promising therapeutic strategy against PAH.

However, the role of autophagy in PH is still unclear and controversial, in fact, its protective role has been described in the initial phase of the pathogenesis of PH. Histochemical analysis of samples obtained from human PH lungs and mouse exposed to chronic hypoxia, showed an increase in the lipidated form of LC3 and in Egr-1, which regulates LC3 expression. Moreover, *LC3*$^{-/-}$ or *Egr-1*$^{-/-}$, but not *Beclin 1*$^{+/-}$ mice are more susceptible to PH and *in vitro* LC3 knockdown cells showed an increase of hypoxic cell proliferation, suggesting a role for LC3 in the adaptation during vascular remodelling under hypoxia [49].

9 Autophagy in the Etiology of Lung Cancer

In most organs, including the lung, autophagy robustly counteracts malignant transformation, *i.e.*, the conversion of a healthy cell into a (pre-)neoplastic cell, and several mechanisms related to the ability of autophagy to preserve cellular or organismal homeostasis account for such a pronounced oncosuppressive activity [19]. Indeed, besides being required for oncogene-induced senescence and anticancer immunosurveillance (see above) [112], autophagy promotes the maintenance of genomic integrity by multiple mechanisms [25]. First, it mediates the degradation of damaged mitochondria, which are prone to overproduce genotoxic ROS and other redox active entities of endogenous and exogenous origin [22]. Second, proficient autophagic responses appear to be required for optimal DNA damage responses [59]. Third, autophagy is involved in the disposal of potentially oncogenic retrotransposons and micronuclei [80]. Moreover, autophagy generally mediates anti-inflammatory effects, and chronic inflammation is known to accelerate oncogenesis (at least in some tissues, including the lung) [14]. Finally, it has been proposed that autophagy is required for the preservation of normal tissue architecture, in particular at the level of the stem-cell compartment [23]. Although little is known on the deregulation of stem cells in pulmonary carcinogenesis, it cannot be excluded that autophagic defects may promote malignant transformation in the lung also via this mechanism [69]. Conversely, the ability of autophagy to preserve genomic and redox homeostasis seems very relevant in the context of lung tumorigenesis, which in a significant proportion of cases is associated with tobacco smoking or exposure to environmental nanoparticles like asbestos crystals [65]. Indeed, the oncogenic effects of both smoking and asbestos have been linked to their ability to cause ROS overgeneration along with genetic/genomic defects and chronic inflammatory responses [12]. All these effects are limited, at least to some extent, by proficient autophagic responses.

Irrespective of the precise mechanisms whereby autophagy counteracts malignant transformation in the lung, various genetic interventions aimed at specifically disabling autophagy in the lungs have been shown to promote malignant transformation driven by several oncogenes, including mutated B-Raf proto-oncogene, serine/threonine kinase (*BRAF*) [91], epidermal growth factor receptor (*EGFR*) [100], Kirsten rat sarcoma viral oncogene homolog (*KRAS*) [24, 77]. Intriguingly enough, in one of these models, accelerated oncogenesis caused by the lung-specific inactivation of ATG5 was linked to increased tumor-infiltration by immunosuppressive $CD4^{+}CD25^{+}FOXP3^{+}$ regulatory T cells [77]. Moreover, the concomitant bi-allelic inactivation of serine/threonine kinase 11 (STK11, best known as LKB1) and phosphatase and tensin homolog (PTEN), two tumor suppressor genes that inhibit autophagy [34, 87], has been shown to cause the formation of pulmonary squamous cell carcinomas that express high levels of the immunosuppressive molecule CD274 (best known as PD-L1) [102]. These latter observations strongly corroborate the notion that autophagy mediates not only cell-intrinsic, but also cell-extrinsic oncosuppression.

9.1 Autophagy in the Progression of Lung Cancer

The capacity of autophagy to preserve cellular homeostasis is beneficial to healthy cells, but also beneficial to transformed cells. This implies that autophagy often (but not always) promotes tumor progression, i.e., the growth and evolution of a transformed cells into an ever more malignant cancer [62]. Indeed, malignant cells are often exposed to relatively adverse microenvironmental conditions, including a shortage of nutrients and oxygen (especially in poorly vascularized tumor areas), and autophagy is instrumental for these cells (as it is for their non-transformed counterparts) to cope with stress and proliferate. Along similar lines, the ability of autophagy to preserve stemness is beneficial for the host when it preserves normal tissue architecture, but detrimental when it sustains the malignant stem-cell compartment. Finally, autophagy supports the survival of malignant cells in key step of tumor progression, the so-called "epithelial-to-mesenchymal transition" (EMT). In this context, epithelial cancer cells "initially growing *in situ*" physically detach from ECM and become able to colonize surrounding tissues as well as distant organs. The EMT is required for all malignancies to become locally and distantly invasive, and critically relies on proficient autophagic responses [4]. In the presence of autophagic defects or pharmacological inhibitors of autophagy, indeed, malignant cells undergoing the EMT and detaching from the ECM, succumb to a form of regulated cell death often referred to as "anoikis" [73].

Corroborating these observations, the genetic and/or pharmacological inhibition of the autophagic machinery in established tumors has been shown to accelerate disease progression in various models of pulmonary oncogenesis, including (but not limited to) *BRAF*- and *KRAS*-driven tumorigenesis [24, 77, 91].

9.2 Autophagy in the Treatment of Lung Cancer

Autophagy provides malignant cells with an increased resistance to various perturbations of homeostasis, including the lack of nutrient and oxygen that cancer cells normally experience in poorly vascularized tumor areas, as well as the presence of xenobiotics like chemotherapeutic agents and physical stress conditions like irradiation. An abundant amount of literature demonstrates indeed that chemical inhibitors of autophagy as well as genetic interventions that compromise autophagic responses accelerate (rather than inhibit) the demise of malignant cells exposed to a wide panel of chemotherapeutics or to irradiation, both *in vitro* and *in vivo*. These observations provided a strong rationale to the development of combinatorial therapeutic strategies involving chemo- or radiotherapy given in combination with an inhibitor of autophagy [19].

Clinical grade highly specific chemical inhibitors of autophagy, however, have not yet been developed, and currently available molecules that can be used in the clinic, like chloroquine (a widely employed antimalarial agent) often operate as lysosomal inhibitors, i.e., they target several processes other than autophagy [83]. Moreover, concerns have been raised that inhibiting autophagy at the whole-body level may *de facto* favor malignant transformation in healthy tissues, reflecting the prominent oncosuppressive functions of autophagy in physiological conditions [51]. Finally, recent data highlight the differential role of autophagy in cancer therapy in immunocompromised *versus* immunocompetent hosts [44]. In this setting, the response to radiotherapy of human non-small cell lung carcinoma (NSCLC) or murine colorectal carcinoma (CRC) cells xenografted in nude mice was significantly improved when cells were rendered autophagy-deficient by the stable depletion of ATG5 or Beclin 1 [44]. However, when murine CRC cells were implanted in immunocompetent syngeneic mice, the stable knockdown of ATG5 compromised the therapeutic activity of irradiation, a defect that could be restored (at least in part) by the intratumoral administration of a chemical inhibitor of extracellular ATPases [44]. These findings demonstrate that inhibiting autophagy in immunocompetent hosts may prevent the elicitation of a therapeutically relevant immune response against dying cancer cells.

In summary, although autophagy generally (but not always) promote the progression of pulmonary malignancies and increases the resistance of lung cancer cells to chemo- and radiotherapeutic regimens, additional experiments are required to understand whether combinatorial treatments involving autophagy inhibitors constitute a clinically viable approach against pulmonary neoplasms. Similarly, further work is needed to clarify whether biomarkers of autophagy such as the expression levels of Beclin 1 or the lipidation of LC3 have a positive or negative prognostic/predictive value in patients with lung cancer, as preliminary results are rather controversial [40, 110].

10 Conclusions

Abundant evidences indicate that autophagy actively participates in a wide range of cellular responses to both physiologic- and pathologic-related events in the diverse tissues and cell types that constitute the lung system. Nevertheless, much is yet to be learnt about its biological relevance, functional targets, and role in development and disease. As described in this chapter, lungs are the first line of defence against several insults and associated diseases are growing both in number and chronicisation. A clear deregulation of the autophagic machinery has been highlighted in most of the lung diseases, suggesting that this process mainly exerts a defensive role. However, in some pathological contexts, it has been reported that the activation of the autophagic process contributes to damage. As a consequence, a detailed knowledge of the molecular mechanisms at the basis of autophagy in lung pathologies is required for the development of novel diagnostic tools and promising therapeutic strategies.

Acknowledgements VI is supported by FRM (SPF20140129085).

Conflict of Interest The authors report no conflict of interest.

References

1. Abdulrahman BA, Khweek AA, Akhter A et al (2011) Autophagy stimulation by rapamycin suppresses lung inflammation and infection by Burkholderia cenocepacia in a model of cystic fibrosis. Autophagy 7:1359–1370
2. Abdulrahman BA, Khweek AA, Akhter A et al (2013) Depletion of the ubiquitin-binding adaptor molecule SQSTM1/p62 from macrophages harboring cftr ΔF508 mutation improves the delivery of Burkholderia cenocepacia to the autophagic machinery. J Biol Chem 288: 2049–2058
3. Araya J, Hara H, Kuwano K (2013) Autophagy in the pathogenesis of pulmonary disease. Intern Med 52:2295–2303
4. Avivar-Valderas A, Bobrovnikova-Marjon E, Alan Diehl J, Bardeesy N, Debnath J, Aguirre-Ghiso JA (2013) Regulation of autophagy during ECM detachment is linked to a selective inhibition of mTORC1 by PERK. Oncogene 32:4932–4940. doi:10.1038/onc.2012.512
5. Ban GY, Pham DL, Trinh HK et al (2015) Autophagy mechanisms in sputum and peripheral blood cells of patients with severe asthma: a new therapeutic target. Clin Exp Allergy. doi:10.1111/cea.12585
6. Cabrera S, Maciel M, Herrera I et al (2015) Essential role for the ATG4B protease and autophagy in bleomycin-induced pulmonary fibrosis. Autophagy 11:670–684
7. Capellier G, Maupoil V, Boussat S, Laurent E, Neidhardt A (1999) Oxygen toxicity and tolerance. Minerva Anestesiol 65:388–392

8. Chen ZH, Kim HP, Sciurba FC, Lee SJ, Feghali-Bostwick C, Stolz DB, Dhir R, Landreneau RJ, Schuchert MJ, Yousem SA, Nakahira K, Pilewski JM, Lee JS, Zhang Y, Ryter SW, Choi AM (2008) Egr-1 regulates autophagy in cigarette smoke-induced chronic obstructive pulmonary disease. PLoS One 3, e3316. doi:10.1371/journal.pone.0003316
9. Chen ZH, Lam HC, Jin Y, Kim HP, Cao J, Lee SJ, Ifedigbo E, Parameswaran H, Ryter SW, Choi AM (2010) Autophagy protein microtubule-associated protein 1 light chain-3B (LC3B) activates extrinsic apoptosis during cigarette smoke-induced emphysema. Proc Natl Acad Sci U S A 107:18880–18885
10. Cheresh P, Kim SJ, Tulasiram S, Kamp DW (2013) Oxidative stress and pulmonary fibrosis. Biochim Biophys Acta 1832:1028–1040
11. Chitra P, Saiprasad G, Manikandan R, Sudhandiran G (2015) Berberine inhibits Smad and non-Smad signaling cascades and enhancesautophagy against pulmonary fibrosis. J Mol Med (Berl) 93:1015–1031
12. Coussens LM, Zitvogel L, Palucka AK (2013) Neutralizing tumor-promoting chronic inflammation: a magic bullet? Science 339:286–291. doi:10.1126/science.1232227
13. Denton D, Shravage B, Simin R, Mills K, Berry DL, Baehrecke EH, Kumar S (2009) Autophagy, not apoptosis, is essential for midgut cell death in Drosophila. Curr Biol 19:1741–1746
14. Deretic V, Saitoh T, Akira S (2013) Autophagy in infection, inflammation and immunity. Nat Rev Immunol 13:722–737. doi:10.1038/nri3532
15. Dickinson JD, Alevy Y, Malvin NP et al (2015) IL13 activates autophagy to regulate secretion in airway epithelial cells. Autophagy. 2016;12(2):397–409
16. Fujii S, Hara H, Araya J, Takasaka N, Kojima J, Ito S, Minagawa S, Yumino Y, Ishikawa T, Numata T, Kawaishi M, Hirano J, Odaka M, Morikawa T, Nishimura S, Nakayama K, Kuwano K (2012) Insufficient autophagy promotes bronchial epithelial cell senescence in chronic obstructive pulmonary disease. Oncoimmunology 1:630–641
17. Galluzzi L, Bravo-San Pedro JM, Vitale I et al (2015) Essential versus accessory aspects of cell death: recommendations of the NCCD 2015. Cell Death Differ 22:58–73
18. Galluzzi L, Pietrocola F, Levine B, Kroemer G (2014) Metabolic control of autophagy. Cell 159:1263–1276
19. Galluzzi L, Pietrocola F, Bravo-San Pedro JM et al (2015) Autophagy in malignant transformation and cancer progression. EMBO J 34:856–880. doi:10.15252/embj.201490784
20. Grainge C, Thomas PS, Mak JC, Benton MJ, Lim TK, Ko FW (2016) Asthma and chronic obstructive pulmonary disease. Respirology. doi: 10.1111/resp.12771. [Epub ahead of print]
21. Green DR, Levine B (2014) To be or not to be? How selective autophagy and cell death govern cell fate. Cell 157:65–75
22. Green DR, Galluzzi L, Kroemer G (2011) Mitochondria and the autophagy-inflammation-cell death axis in organismal aging. Science 333:1109–1112. doi:10.1126/science.1201940
23. Greim H, Kaden DA, Larson RA, Palermo CM, Rice JM, Ross D, Snyder R (2014) The bone marrow niche, stem cells, and leukemia: impact of drugs, chemicals, and the environment. Ann N Y Acad Sci 1310:7–31. doi:10.1111/nyas.12362
24. Guo JY, Karsli-Uzunbas G, Mathew R et al (2013) Autophagy suppresses progression of K-ras-induced lung tumors to oncocytomas and maintains lipid homeostasis. Genes Dev 27:1447–1461
25. Hanahan D, Weinberg RA (2011) Hallmarks of cancer: the next generation. Cell 144:646–674
26. He C, Klionsky DJ (2009) Regulation mechanisms and signaling pathways of autophagy. Annu Rev Genet 43:67–93
27. Herrmann U, Dockter G, Lammert F (2010) Cystic fibrosis-associated liver disease. Best Pract Res Clin Gastroenterol 24:585–592
28. Hidvegi T, Ewing M, Hale P, Dippold C, Beckett C, Kemp C, Maurice N, Mukherjee A, Goldbach C, Watkins S, Michalopoulos G, Perlmutter DH (2010) An autophagy-enhancing drug promotes degradation of mutant alpha1-antitrypsin Z and reduces hepatic fibrosis. Science 329:229–232
29. Hunt JM, Tuder R (2012) Alpha 1 anti-trypsin: one protein, many functions. Curr Mol Med 12:827–835

30. Im J, Hergert P, Nho RS (2015) Reduced FoxO3a expression causes low autophagy in idiopathic pulmonary fibrosis fibroblasts on collagen matrices. Am J Physiol Lung Cell Mol Physiol 309:L552–L561
31. Inoue D, Kubo H, Taguchi K et al (2011) Inducible disruption of autophagy in the lung causes airway hyper-responsiveness. Biochem Biophys Res Commun 405:13–18
32. Ito S, Araya J, Kurita Y, Kobayashi K, Takasaka N, Yoshida M, Hara H, Minagawa S, Wakui H, Fujii S, Kojima J, Shimizu K, Numata T, Kawaishi M, Odaka M, Morikawa T, Harada T, Nishimura SL, Kaneko Y, Nakayama K, Kuwano K (2015) PARK2-mediated mitophagy is involved in regulation of HBEC senescence in COPD pathogenesis. Autophagy 11:547–559
33. Jara P, Calyeca J, Romero Y et al (2015) Matrix metalloproteinase (MMP)-19-deficient fibroblasts display a profibrotic phenotype. Am J Physiol Lung Cell Mol Physiol 308:L511–L522
34. Ji H, Ramsey MR, Hayes DN et al (2007) LKB1 modulates lung cancer differentiation and metastasis. Nature 448:807–810. doi:10.1038/nature06030
35. Johnson ER, Matthay MA (2010) Acute lung injury: epidemiology, pathogenesis, and treatment. J Aerosol Med Pulm Drug Deliv 23:243–252
36. Jounai N, Takeshita F, Kobiyama K et al (2007) The Atg5 Atg12 conjugate associates with innate antiviral immune responses. Proc Natl Acad Sci U S A 104:14050–14055
37. Junkins RD, Shen A, Rosen K et al (2013) Autophagy enhances bacterial clearance during P. aeruginosa lung infection. PLoS One 8, e72263. doi:10.1371/journal.pone.0072263
38. Jyothula SS, Eissa NT (2013) Autophagy and role in asthma. Curr Opin Pulm Med 19:30–35
39. Kamimoto T, Shoji S, Hidvegi T, Mizushima N, Umebayashi K, Perlmutter DH, Yoshimori T (2006) Intracellular inclusions containing mutant alpha1-antitrypsin Z are propagated in the absence of autophagic activity. J Biol Chem 281:4467–4476
40. Karpathiou G, Sivridis E, Koukourakis MI, Mikroulis D, Bouros D, Froudarakis ME, Giatromanolaki A (2011) Light-chain 3A autophagic activity and prognostic significance in non-small cell lung carcinomas. Chest 140:127–134. doi:10.1378/chest.10-1831
41. Kaushal S, Annamali M, Blomenkamp K, Rudnick D, Halloran D, Brunt EM, Teckman JH (2010) Rapamycin reduces intrahepatic alpha-1-antitrypsin mutant Z protein polymers and liver injury in a mouse model. Exp Biol Med (Maywood) 235:700–709
42. Kepp O, Senovilla L, Vitale I et al (2014) Consensus guidelines for the detection of immunogenic cell death. Oncoimmunology 3, e955691. doi:10.4161/21624011.2014.955691
43. Kerem B, Kerem E (1996) The molecular basis for disease variability in cystic fibrosis. Eur J Hum Genet 4:65–73
44. Ko A, Kanehisa A, Martins I et al (2014) Autophagy inhibition radiosensitizes in vitro, yet reduces radioresponses in vivo due to deficient immunogenic signalling. Cell Death Differ 21:92–99. doi:10.1038/cdd.2013.124
45. Kroemer G, Galluzzi L, Kepp O, Zitvogel L (2013) Immunogenic cell death in cancer therapy. Annu Rev Immunol 31:51–72
46. Kroemer G, Mariño G, Levine B (2010) Autophagy and the integrated stress response. Mol Cell 40:280–293
47. Kuperman DA, Huang X, Koth LL et al (2002) Direct effects of interleukin-13 on epithelial cells cause airway hyperreactivity and mucus overproduction in asthma. Nat Med 8: 885–889
48. Lam HC, Cloonan SM, Bhashyam AR, Haspel JA, Singh A, Sathirapongsasuti JF, Cervo M, Yao H, Chung AL, Mizumura K, An CH, Shan B, Franks JM, Haley KJ, Owen CA, Tesfaigzi Y, Washko GR, Quackenbush J, Silverman EK, Rahman I, Kim HP, Mahmood A, Biswal SS, Ryter SW, Choi AM (2013) Histone deacetylase 6-mediated selective autophagy regulates COPD-associated cilia dysfunction. J Clin Invest 123:5212–5230
49. Lee SJ, Smith A, Guo L, Alastalo TP, Li M, Sawada H et al (2011) Autophagic protein LC3B confers resistance against hypoxia-induced pulmonary hypertension. Am J Respir Crit Care Med 183:649–658. doi:10.1164/rccm.201005-0746OC
50. Li L, Wang X, Wang L et al (2015) Mammalian target of rapamycin overexpression antagonizes chronic hypoxia-triggered pulmonary arterial hypertension via the autophagic pathway. Int J Mol Med 36:316–322. doi:10.3892/ijmm.2015.2224

51. Liang C, Feng P, Ku B, Dotan I, Canaani D, Oh BH, Jung JU (2006) Autophagic and tumour suppressor activity of a novel Beclin1-binding protein UVRAG. Nat Cell Biol 8:688–699
52. Long L, Yang X, Southwood M, Lu J, Marciniak SJ, Dunmore BJ, Morrell NW (2013) Chloroquine prevents progression of experimental pulmonary hypertension via inhibition of autophagy and lysosomal bone morphogenetic protein type II receptor degradation. Circ Res 112:1159–1170
53. López-Campos JL, Tan W, Soriano JB (2015) Global burden of COPD. Respirology. doi:10.1111/resp.12660
54. Luciani A, Villella VR, Esposito S et al (2010) Defective CFTR induces aggresome formation and lung inflammation in cystic fibrosis through ROS-mediated autophagy inhibition. Nat Cell Biol 12:863–875
55. Ma Y, Galluzzi L, Zitvogel L, Kroemer G (2013) Autophagy and cellular immune responses. Immunity 39:211–227
56. Maiuri L, De Stefano D, Raia V, Kroemer G (2015) The holy grail of cystic fibrosis research: pharmacological repair of the F508del-CFTR mutation. Ann Transl Med 3:S24. doi:10.3978/j.issn.2305-5839.2015.02.32
57. Maiuri L, Luciani A, Giardino I et al (2008) Tissue transglutaminase activation modulates inflammation in cystic fibrosis via PPARgamma down-regulation. J Immunol 180:7697–7705
58. Martin LJ, Gupta J, Jyothula SS et al (2012) Functional variant in the autophagy-related 5 gene promotor is associated with childhood asthma. PLoS One 7, e33454. doi:10.1371/journal.pone.0033454
59. Mathew R, Kongara S, Beaudoin B et al (2007) Autophagy suppresses tumor progression by limiting chromosomal instability. Genes Dev 21:1367–1381. doi:10.1101/gad.1545107
60. Menzel DB, Amdur MO (1986) Toxic response of the respiratory system. In: Klaassen K, Amdur MO, Doull J (eds) Casarett and Doull's toxicology: the basic science of poisons, 3rd edn. Macmillan, New York, pp 330–358
61. Mi S, Li Z, Yang HZ et al (2011) Blocking IL-17A promotes the resolution of pulmonary inflammation and fibrosis via TGF-beta1-dependent and –independent mechanisms. J Immunol 187:3003–3014
62. Mihaylova MM, Shaw RJ (2011) The AMPK signalling pathway coordinates cell growth, autophagy and metabolism. Nat Cell Biol 13:1016–1023. doi:10.1038/ncb2329
63. Mizumura K, Cloonan SM, Nakahira K, Bhashyam AR, Cervo M, Kitada T, Glass K, Owen CA, Mahmood A, Washko GR, Hashimoto S, Ryter SW, Choi AM (2014) Mitophagy-dependent necroptosis contributes to the pathogenesis of COPD. J Clin Invest 124:3987–4003
64. Mizushima N, Komatsu M (2011) Autophagy: renovation of cells and tissues. Cell 147: 728–741
65. Monick MM, Powers LS, Walters K, Lovan N, Zhang M, Gerke A, Hansdottir S, Hunninghake GW (2010) Identification of an autophagy defect in smokers' alveolar macrophages. J Immunol 185:5425–5435
66. Morse D, Lin L, Choi AM, Ryter SW (2009) Heme oxygenase-1, a critical arbitrator of cell death pathways in lung injury and disease. Free Radic Biol Med 47:1–12
67. Murai H, Okazaki S, Hayashi H et al (2015) Alternaria extract activates autophagy that induces IL-18 release from airway epithelial cells. Biochem Biophys Res Commun 464: 969–974
68. Murata K, Fujimoto K, Kitaguchi Y et al (2014) Hydrogen peroxide content and pH of expired breath condensate from patients with asthma and COPD. COPD 11:81–87
69. Ojha R, Bhattacharyya S, Singh SK (2015) Autophagy in cancer stem cells: a potential link between chemoresistance, recurrence, and metastasis. Biores Open Access 4:97–108
70. Pampliega O, Orhon I, Patel B, Sridhar S, Díaz-Carretero A, Beau I, Codogno P, Satir BH, Satir P, Cuervo AM (2013) Functional interaction between autophagy and ciliogenesis. Nature 502:194–200
71. Pastore N, Blomenkamp K, Annunziata F et al (2013) Gene transfer of master autophagy regulator TFEB results in clearance of toxic protein and correction of hepatic disease in alpha-1-anti-trypsin deficiency. EMBO Mol Med 5:397–412

72. Patel AS, Lin L, Geyer A et al (2012) Autophagy in idiopathic pulmonary fibrosis. PLoS One 7, e41394. doi:10.1371/journal.pone.0041394
73. Peng YF, Shi YH, Ding ZB et al (2013) Autophagy inhibition suppresses pulmonary metastasis of HCC in mice via impairing anoikis resistance and colonization of HCC cells. Autophagy 9:2056–2068. doi:10.4161/auto.26398
74. Perlmutter DH (2009) Autophagic disposal of the aggregation-prone protein that causes liver inflammation and carcinogenesis in α-1-antitrypsin deficiency. Cell Death Differ 16:39–45
75. Poon AH, Chouiali F, Tse SM et al (2012) Genetic and histologic evidence for autophagy in asthma pathogenesis. J Allergy Clin Immunol 129:569–571
76. Rangarajan S, Kurundkar A, Kurundkar D et al (2015) Novel mechanisms for the anti-fibrotic action of nintedanib. Am J Respir Cell Mol Biol. 2016;54(1):51–9. doi: 10.1165/rcmb.2014-0445OC
77. Rao S, Tortola L, Perlot T et al (2014) A dual role for autophagy in a murine model of lung cancer. Nat Commun 5:3056. doi:10.1038/ncomms4056
78. Ratjen F, Döring G (2003) Cystic fibrosis. Lancet 361:681–689
79. Redington AE (2000) Airway fibrosis in asthma: mechanisms, consequences, and potential for therapeutic intervention. Monaldi Arch Chest Dis 55:317–323
80. Rello-Varona S, Lissa D, Shen S et al (2012) Autophagic removal of micronuclei. Cell Cycle 11:170–176. doi:10.4161/cc.11.1.18564
81. Ricci A, Cherubini E, Scozzi D et al (2013) Decreased expression of autophagic beclin 1 protein in idiopathic pulmonary fibrosisfibroblasts. J Cell Physiol 228:1516–1524
82. Robinson DS, Hamid Q, Ying S et al (1992) Predominant Th2-like bronchoalveolar T-lymphocyte population in atopic asthma. N Engl J Med 326:298–304
83. Rubinsztein DC, Codogno P, Levine B (2012) Autophagy modulation as a potential therapeutic target for diverse diseases. Nat Rev Drug Discov 11:709–730. doi:10.1038/nrd3802
84. Ryter SW, Choi AM (2009) Heme oxygenase-1/carbon monoxide: from metabolism to molecular therapy. Am J Respir Cell Mol Biol 41:251–260
85. Ryter SW, Choi AMK (2010) Autophagy in the lung. Proc Am Thorac Soc 7:13–21
86. Ryter SW, Choi AMK (2015) Autophagy in lung disease pathogenesis and therapeutics. Redox Biol 4:215–225
87. Sansal I, Sellers WR (2004) The biology and clinical relevance of the PTEN tumor suppressor pathway. J Clin Oncol 22:2954–2963. doi:10.1200/JCO.2004.02.141
88. Semenza GL (2011) Oxygen sensing, homeostasis, and disease. N Engl J Med 365:537–547
89. Settembre C, Di Malta C, Polito VA, Garcia Arencibia M et al (2011) TFEB links autophagy to lysosomal biogenesis. Science 332:1429–1433
90. Sica V, Galluzzi L, Bravo-San Pedro JM et al (2015) Organelle-specific initiation of autophagy. Mol Cell 59:522–539
91. Strohecker AM, Guo JY, Karsli-Uzunbas G et al (2013) Autophagy sustains mitochondrial glutamine metabolism and growth of BrafV600E-driven lung tumors. Cancer Discov 3:1272–1285. doi:10.1158/2159-8290.CD-13-0397
92. Takasaka N, Araya J, Hara H, Ito S, Kobayashi K, Kurita Y, Wakui H, Yoshii Y, Yumino Y, Fujii S, Minagawa S, Tsurushige C, Kojima J, Numata T, Shimizu K, Kawaishi M, Kaneko Y, Kamiya N, Hirano J, Odaka M, Morikawa T, Nishimura SL, Nakayama K, Kuwano K (2014) Autophagy induction by SIRT6 through attenuation of insulin-like growth factor signaling is involved in the regulation of human bronchial epithelial cell senescence. J Immunol 192:958–968
93. Tanaka A, Jin Y, Lee SJ, Zhang M, Kim HP, Stolz DB, Ryter SW, Choi AM (2012) Hyperoxia-induced LC3B interacts with the Fas apoptotic pathway in epithelial cell death. Am J Respir Cell Mol Biol 46:507–514
94. Tanaka H, Miyazaki N, Oashi K et al (2001) IL-18 might reflect disease activity in mild and moderate asthma exacerbation. J Allergy Clin Immunol 107:331–336
95. Tang Z, Lin MG, Stowe TR, Chen S, Zhu M, Stearns T, Franco B, Zhong Q (2013) Autophagy promotes primary ciliogenesis by removing OFD1 from centriolar satellites. Nature 502: 254–257

96. Teckman JH, An JK, Loethen S, Perlmutter DH (2002) Fasting in alpha1-antitrypsin deficient liver: constitutive activation of autophagy. Am J Physiol 283:G1156–G1165
97. Teckman JH, Perlmutter DH (2000) Retention of mutant alpha1-antitrypsin Z in endoplasmic reticulum is associated with an autophagic response. Am J Physiol 279:G961–G974
98. Valvano MA (2015) Intracellular survival of Burkholderia cepacia complex in phagocytic cells. Can J Microbiol 61:607–615
99. Wang XM, Kim HP, Nakahira K et al (2009) The heme oxygenase-1/carbon monoxide pathway suppresses TLR4 signaling by regulating the interaction of TLR4 with caveolin-1. J Immunol 182:3809–3818
100. Wei Y, Zou Z, Becker N et al (2013) EGFR-mediated Beclin 1 phosphorylation in autophagy suppression, tumor progression, and tumor chemoresistance. Cell 154:1269–1284
101. Wrighton KH (2013) Cytoskeleton: autophagy and ciliogenesis come together. Nat Rev Mol Cell Biol 14:687. doi:10.1038/nrm3686
102. Xu C, Fillmore CM, Koyama S et al (2014) Loss of Lkb1 and Pten leads to lung squamous cell carcinoma with elevated PD-L1 expression. Cancer Cell 25:590–604
103. Yao H, Rahman I (2011) Current concepts on oxidative/carbonyl stress, inflammation and epigenetics in pathogenesis of chronic obstructive pulmonary disease. Toxicol Appl Pharmacol 254:72–85
104. Yeganeh B, Serebrin A, Mutawe MM et al (2011) Autophagy regulates TGF-Beta1 induced fibrosis in human airway smooth muscle cells. Am J Respir Crit Care Med 183:A2110
105. Yoshida T, Mett I, Bhunia AK, Bowman J, Perez M, Zhang L, Gandjeva A, Zhen L, Chukwueke U, Mao T, Richter A, Brown E, Ashush H, Notkin N, Gelfand A, Thimmulappa RK, Rangasamy T, Sussan T, Cosgrove G, Mouded M, Shapiro SD, Petrache I, Biswal S, Feinstein E, Tuder RM (2010) Rtp801, a suppressor of mTOR signaling, is an essential mediator of cigarette smoke-induced pulmonary injury and emphysema. Nat Med 16:767–773
106. Yu G, Kovkarova-Naumovski E, Jara P et al (2012) Matrix metalloproteinase-19 is a key regulator of lung fibrosis in mice and humans. Am J Respir Crit Care Med 186:752–762
107. Zhang PX, Murray TS, Villella VR et al (2013) Reduced caveolin-1 promotes hyperinflammation due to abnormal heme oxygenase-1 localization in lipopolysaccharide-challenged macrophages with dysfunctional cystic fibrosis transmembrane conductance regulator. J Immunol 190:5196–5206
108. Zhang Y, Sauler M, Shinn AS, Gong H, Haslip M, Shan P, Mannam P, Lee PJ (2014) Endothelial PINK1 mediates the protective effects of NLRP3 deficiency during lethal oxidant injury. J Immunol 192:5296–5304. doi:10.4049/jimmunol.1400653
109. Zhou H, Lu F, Latham C, Zander DS, Visner GA (2004) Heme oxygenase-1 expression in human lungs with cystic fibrosis and cytoprotective effects against Pseudomonas aeruginosa in vitro. Am J Respir Crit Care Med 170:633–640
110. Zhou W, Yue C, Deng J et al (2013) Autophagic protein Beclin 1 serves as an independent positive prognostic biomarker for non-small cell lung cancer. PLoS One 8, e80338. doi:10.1371/journal.pone.0080338
111. Zhu Z, Homer RJ, Wang Z et al (1999) Pulmonary expression of interleukin-13 causes inflammation, mucus hypersecretion, subepithelial fibrosis, physiologic abnormalities, and eotaxin production. J Clin Invest 103:779–788
112. Zitvogel L, Galluzzi L, Smyth MJ, Kroemer G (2013) Mechanism of action of conventional and targeted anticancer therapies: reinstating immunosurveillance. Immunity 39:74–88

Autophagy and Liver Diseases

Emanuela Esposito, Michela Campolo, Marika Cordaro, Giovanna Casili, and Salvatore Cuzzocrea

Abstract Autophagy is a mechanism involved in cellular homeostasis under basal and stressed conditions delivering cytoplasmic content to the lysosomes for degradation to macronutrients. Autophagy provides essential components (amino acids, lipids and carbohydrates) needed to meet the cell's energy necessities, and it also regulates energy supply by controlling the number, quality, and dynamics of the mitochondria. Lastly, autophagy also modulates the levels of enzymes in metabolic pathways. It is generally recognised that autophagy plays a role in the hepatic lipid metabolism. It is not surprising that dysregulation of autophagy has been connected with liver-specific disorders such as fatty liver, non-alcoholic steatohepatitis and hepatocellular carcinoma. However, controversy of the exact role of autophagy in the lipid metabolism exists: some publications report a lipolytic function of autophagy, whereas others claim a lipogenic function. This chapter aims to give an update of the present knowledge (last 5 years) on autophagy in the hepatic lipid metabolism, hepatic insulin resistance, steatohepatitis and hepatic fibrogenesis. An improved understanding of the autophagic process should lead to potentially innovative therapies with direct relevance to surgical diseases.

Abbreviations

A-1ATD	Alpha-1 antitrypsin deficiency
AASLD	American Association for the Study of Liver Diseases
ACC	Acetyl-CoA carboxylase
ADH	Alcohol dehydrogenase
ALD	Alcoholic liver disease
ALT	Alanine aminotransferase
AST	Aspartate aminotransferase
ATZ	z mutation of alpha-1-antitrypsin
CDKs	Cyclin-dependent kinases
CDT	Carbohydrate deficient transferrin

E. Esposito (✉) • M. Campolo • M. Cordaro • G. Casili • S. Cuzzocrea
Department of Chemical, Biological, Pharmaceutical and Environmental Sciences, University of Messina, Messina, Italy
e-mail: eesposito@unime.it

M.C. Maiuri, D. De Stefano (eds.), *Autophagy Networks in Inflammation*, Progress in Inflammation Research, DOI 10.1007/978-3-319-30079-5_17

CKD	Chronic kidney disease
CPT-1	Carnitine palmitoyltransferase 1
CVD	Cardiovascular disease
CYP2E1	Cytochrome P450 2E1
DIC	Disseminated intravascular coagulopathy
EC	Endothelial cells
GGT	Gamma glutamyl transferase
HCC	Hepatocellular carcinoma
HCV	Hepatitis C virus
HMGB1	High mobility group box 1
HSPs	Heat shock proteins
I/R	Ischemia-reperfusion
IL-1	Interleukin-1
IL-12	Interleukin-12
INF-γ	Interferon-γ
KC	Kupffer cells
LCHAD	Long-chain 3-hydroxyl coenzyme A dehydrogenase
LDs	Lipid droplets
MAPK	Mitogen-activated protein kinase
miRNAs	MicroRNAs
MPT	Mitochondrial permeability transition
MUFAs	Monounsaturated fatty acids
NAE	NEDD8-activating enzyme
NAFLD	Nonalcoholic fatty liver disease
NASH	Nonalcoholic steatohepatitis
OSA	Obstructive sleep apnoea
PAS	Phagophore assembly site
PL	Phospholipid
ROS	Reactive oxygen species
SNAREs	NSF attachment protein receptors
Spred protein	Sprouty-related protein with Ena/vasodilator-stimulated phosphoprotein homology-1 domain
TAG	Triacylglycerol
TNF-α	Tumour necrosis factor-α
TPN	Total parenteral nutrition
TTP	Tumor progression
VLDL	Very low-density lipoprotein

1 Introduction

1.1 Steatosis

Steatosis is a cellular pathology due to the accumulation of intracellular triglycerides, which involves a series of damage up to necrosis of the cell. It suggests an impairment of the physiological processes of synthesis and elimination of triglycerides. Additional lipid accumulates in vesicles that dislocate the cytoplasm. When the vesicles are sufficient large to distort the nucleus, the condition is known as macrovesicular steatosis. Macrovesicular steatosis is the most common form of fatty deterioration and may be initiated by overflow of lipids due to obesity, obstructive sleep apnoea (OSA), insulin resistance, or most important to alcoholism. An unbalanced diet may also cause the recruitment of fat from adipocytes and create a local surplus in the liver where the majority of lipid metabolism happens. On the other hand, tetracyclines, Reye's syndrome, and hepatitis C can cause the onset of microvesicular steatosis characterized, in contrast to the macrovesicular steatosis, by small intracytoplasmic fat vacuoles (called liposomes), which accumulate in the cell. Organs most affected are the kidney cortex, striated muscle, myocardial muscle fibres of the heart, intestines and liver. This organ is particularly sensitive to the steatotic processes not only because it is involved in the metabolism of lipids but it is also responsible for the inactivation of numerous toxic substances and, finally, liver has a circulation principally venous (therefore constantly close to a situation of hypoxia).

1.2 Fatty Liver Disease (FLD)

The standard adult human liver may have up to 5 % of its mass as lipid. But if it makes up more than 5–10 % of the organ's weight, it can be considered a fatty liver disease. The total of fatty acid in the liver depends on the balance between the processes of distribution and elimination. In several patients, fatty liver may be attended by hepatic inflammation and liver cell death (steatohepatitis).

Potential pathophysiologic mechanisms for fatty liver include the following:

- Diminished mitochondrial fatty acid beta-oxidation
- Augmented endogenous fatty acid synthesis or improved delivery of fatty acids to the liver
- Lacking incorporation or export of triglycerides as very low-density lipoprotein (VLDL)

This process is modified if too much fat is in the liver [56]. While liver commonly repairs itself by reconstruction new liver cells when the old ones are damaged, when, there's repeated injury to the liver, perpetual scarring takes place, called cirrhosis.

Fatty liver disease is typically asymptomatic. It may experience fatigue or vague abdominal discomfort. Liver may become slightly puffed-up, and doctor can perceive this during a physical exam. Excess fat can cause liver inflammation. If liver becomes inflamed, may have a poor hungriness, weight loss, abdominal pain, weakness, and misunderstanding.

One of the condition most frequently related with fatty liver disease is metabolic syndrome. This includes carrying the analysis of type II diabetes, obesity, or hypertriglyceridemia.

Other factors, such as drugs (e.g., amiodarone, tamoxifen, methotrexate), metabolic aberrations (e.g., galactosemia, glycogen storage diseases, homocystinuria, and tyrosinemia), alimentary status (e.g., overnutrition, severe malnourishment, total parenteral nutrition [TPN], or starvation diet), or additional health problems (e.g., celiac sprue and Wilson disease) may contribute to fatty liver disease.

Blood tests may be used to control if the liver is working properly. A liver biopsy, where a small sample of tissue is removed with a long needle or through a small incision, can be used to check fatty liver. Laboratory irregularities include elevations of the SGOT (serum glutamic-oxaloacetic transaminase) and SGPT (serum glutamic pyruvic transaminase). In many cases the alkaline phosphatase will be expressively elevated due to cholestasis produced by the fatty infiltration. Fatty liver is frequently reversible if recognized and preserved. There may be some long-term predisposition toward other types of liver problems depending on how long and how severe the fatty liver disorder was. If left untreated, there is a high risk of death for both the mother and baby. Severe liver damage that may require a liver transplant can occur in the mother if the condition is not recognized early.

There are two main types of fatty liver disease: alcoholic liver disease (ALD), nonalcoholic fatty liver disease (NAFLD). Moreover, when the fat builds up abundant, it will cause the liver to swell. If the original cause is not from alcohol, it's called nonalcoholic steatohepatitis (NASH). Last, acute fatty liver of pregnancy, a rare, but serious form of fatty liver that starts late in gestation.

1.3 Alcoholic Liver Disease (ALD)

ALD is the most predominant cause of progressive liver disease in Europe, but mortality due to alcoholic cirrhosis distinct out from non-alcoholic cirrhosis is not easy to determine. It represents a spectrum of liver pathology that starts with fatty liver alteration, which is present in almost all heavy alcohol drinkers and is usually asymptomatic. The mechanism of alcohol-induced hepatotoxicity encompasses interactions between the direct toxic effects of alcohol and its metabolites on different cell types in the liver, induction of reactive oxygen species (ROS) as well as up-regulation of the inflammatory cascade and other cell-specific effects in the liver [124]. Clinical assortment of ALD includes steatosis in the presence or absence of inflammation (in this second case it's called steatohepatitis), leading to cirrhosis and an increased risk of hepatocellular carcinoma (HCC) [180]. Twenty percent to 40 % of alcoholics progress into fibrosis, 10–20 % eventually development to cirrhosis, and 1–2 % of cirrhotics are diagnosed with hepatocellular carcinoma every year [134, 156]. Fatty liver develops in approximately 90 % of individuals who drink more than 60 g alcohol/day. This disorder is completely reversible after 4–6 weeks of abstinence, even if fibrosis has already developed.

Recent studies designated that alcohol consumption increases the percentage of reduced nicotinamide adenine dinucleotide/oxidized nicotinamide adenine dinucleotide in hepatocytes, which disturbs mitochondrial β-oxidation of fatty acids [8]. Alcohol can directly (via acetaldehyde) or indirectly (via regulation of multiple factors) up-regulate the expression of SREBP-1c and down-regulate the expression of PPAR-α, which results in the development of alcoholic fatty liver. Alcohol exposure also inhibits AMPK and subsequently increases acetyl-CoA carboxylase

(ACC) activity but diminutions carnitine palmitoyltransferase 1 (CPT-1) activity, leading to an increase in fatty acid synthesis and a decrease in fatty acid β-oxidation. Moreover, alcohol consumption can also modify many factors, including HIF-1, C3, C1qa, PKCε, and iNOS, that subsequently contribute to the development of fatty liver [152].

An initial step in evaluating patients with ALD is the documentation of heavy alcohol ingesting. In its early stages, ALD is a taciturn disease that can only be detected by laboratory analysis including aspartate aminotransferase (AST), alanine aminotransferase (ALT), mean corpuscular volume, carbohydrate deficient transferrin (CDT), and gamma glutamyl transferase (GGT). This last is extremely sensitive and is the most commonly used test to evaluate prolonged alcoholism, but CDT is more specific for detecting daily ethanol ingestion. AST and ALT can be elevated with a typical AST:ALT ratio >1.0, although this is not highly specific for ALD [9, 60]. Moreover, patients with early ALD can show stigmata of alcohol abuse such as bilateral parotid gland hypertrophy, muscle wasting, malnutrition, Dupuytren's sign, and signs of peripheral neuropathy. In patients with cirrhosis, most physical findings are not specific of the etiology. However, some signs such as gynecomastia and extensive spider angiomas may be more frequently seen in those with alcohol as the main cause of liver disease [113]. Liver biopsy remains the key standard diagnostic modality for identifying and staging liver fibrosis; however, it has its own limitations, including expense, sampling errors, and inter-observer variability, which could lead to understating of cirrhosis [11]. The keystone of ALD management at any stage is abstinence from alcohol. Development in fatty liver histology can occur as early as 2 weeks following cessation of alcohol use, while continuous alcohol consumption has been shown to increase portal pressure and worsen complication of portal hypertension, including variceal bleeding [33, 102]. A careful assessment of the nutritional status of these patients is significant and proper nutrition should be emphasized. Addressing vitamins and trace minerals deficiencies (e.g. vitamin A, vitamin D, thiamine, folate, pyridoxine, and zinc) while maintaining a daily intake of 1.2–1.5 g of protein/kg and 35–40 kcal/kg is recommended by the American Association for the Study of Liver Diseases (AASLD) in order to increase nitrogen balance [125]. Pharmacological agents can complement psychosocial treatments for alcoholism. Disulfiram, an inhibitor of acetaldehyde dehydrogenase that leads to an unpleasant reaction when alcohol is consumed, has been traditionally used in alcoholics. However, its use in ALD has been hindered by potential hepatotoxicity. Naltrexone (opioid receptor antagonist), acamprosate (glutamatergic receptor modulator), and topiramate (anticonvulsant) have also demonstrated efficacy in management of alcoholism, but they have not been assessed in patients with end stage liver disease and are consequently not recommended in patients with ALD. Baclofen, a GABA receptor agonist, is a promising agent for cumulative abstinence rates in alcoholic individuals and has demonstrated both efficacy and safety in patients with alcoholic cirrhosis [46]. In add-on to the general ALD therapeutic measures outlined earlier, patients with alcoholic cirrhosis necessitate treatment for cirrhosis-specific complications while evaluating their candidacy for liver transplantation [121].

1.4 Non-alcoholic Fatty Liver Disease (NAFLD)

NAFLD is the most common analysis in subjects with altered aminotransferases in the western world [26]. NAFLD is defined as any excessive fat accumulation in the liver with more than 5 % of hepatocytes containing observable intracellular triglycerides or steatosis affecting at least 5 % of the liver volume or weight in patients consuming less than 30 g (three units) of alcohol per day for men and less than 20 g (two units) of alcohol per day for women, where one unit of alcohol (10 g) is defined as one glass of beer (25 cL), one glass of wine (20 cL) or one glass of whisky (3 cL) [114, 143]. It begins with simple hepatocyte steatosis, and progresses to nonalcoholic steatohepatitis (NASH), fibrosis of the hepatocytes, and liver cirrhosis, which can additional progress to HCC [1]. The prevalence of NAFLD is different among men and women, and it increases with age, occurring in less than 20 % of persons younger than 20 years of age, and in more than 40 % of those over the age of 60 [40]. Obesity, type 2 DM, and hyperlipidemia are simultaneous conditions frequently linked with NAFLD. The reported prevalence of obesity in several series of patients with NASH varied between 30 and 100 %, the prevalence of type 2 DM varied between 10 and 75 %, and prevalence of hyperlipidemia varied between 20 and 92 %. Some kids with NASH have type 1 diabetes [24, 39]. A key focus of the NAFLD-related chronic diseases during the last 10 years has complicated chronic liver disease, cardiovascular disease (CVD) and T2DM; e.g., a recent meta-analysis showed that NAFLD increased overall mortality by 57 % mainly from liver-related and CVD causes, and increased risk of incident T2DM by approximately twofold. Additionally, and even more lately, cumulative attention has also focused on NAFLD-related chronic kidney disease (CKD) and an additional recent meta-analysis reported that NAFLD was connected with an approximate twofold increased risk of CKD. While there is also emerging suggestion that NAFLD is linked to other chronic diseases, such as sleep apnea, colorectal cancers, osteoporosis, psoriasis and various endocrinopathies [15]. The pathogenesis of NAFLD is intricate and while its exact mechanism remains essentially unknown, different genetic factors and/or environmental elements appear to influence it. The "two-hit hypothesis" of NASH, initially explained by Day and James suggests that lipid accumulation in the liver (first hit) is followed by a sequence of oxidative and hepatotoxic progressions (second hit), caused by a mechanism not yet known. Some factors such as genetics, epigenetic mechanisms as well as environmental elements, seems to promote hepatocyte fat statement and insulin resistance, both of which further lead to the secondary pathologic events, such as oxidative stress, lipid peroxidation, improved inflammatory responses, hepatic fibrosis and apoptosis. Other causes such as lipotoxicity, endotoxemia, and adipocytokines or additional inflammatory signals released from fat-infiltrated hepatocytes and adipose tissue, may encourage oxidative stress in the liver and stimulating the progression of NAFLD to NASH [45]. Routine modification to achieve weight loss and promote fitness has conventionally been the cornerstone of management in NASH, with dietetic advice frequently focused on the need for low fat and restricted calorie content. Recent data suggest that lessening in body weight of 7 % or greater is related with reduction in hepatic inflammation and steatosis. Additionally, a growing body of evidence supports the hypothesis that a diet high in macronutrients such as monounsaturated fatty acids (MUFAs) and omega-3 (n-3), and low in carbohydrates such as fructose, can improve NAFLD independent of weight loss. Demarcating the benefits of specific dietary macronutrients and foods is imperative in order to give patients a sense of regulator over their disease and an aptitude to maintain a healthy and stimulating diet that may also increase hepatic and metabolic outcomes. Some of dietary components that have shown a potential benefits for NAFLD/NASH and metabolic disease, are oily fish, coffee, nuts, tea, red wine, avocado and olive oil [55]. There is no proven pharmacologic treatment for NAFLD, it is critically significant to find dietary lines to the prevention, reduction, or reversal of hepatic steatosis, and its progression to steatohepatitis.

1.5 Nonalcoholic Steatohepatitis (NASH)

NASH was first coined by Dr. Ludwig 3 decades ago and represents a part of a wide spectrum of non-alcoholic fatty liver disease (NAFLD), which ranges from simple steatosis and steatohepatitis to progressive fibrosis and cirrhosis [103]. The pathogenetic processes of NASH and its evolution are multifactorial and are inclined by both environmental and genetic factors [78, 137]. Hepatic steatosis, is a common condition characterized by the accumulation of triglyceride-filled vacuoles inside the hepatocyte cytoplasm. Although it is considered a relatively benign situation, the fact that it serves as a necessary precursor lesion to NASH, insinuates that a fatty liver is subject to injury to which a normal liver is not. Insulin resistance is the only metabolic syndrome consistently associated with NASH [89]. Steatosis in NASH is typically macrovesicular and inflammation of steatohepatitis is principally lobular, however intense portal inflammation with interface movement is more potent in other causes like virus, autoimmune or drug-induced hepatitis [31]. Nonspecific constitutional signs of weakness fatigue and disease precede in a third of NASH patients [6]. Despite similarities the patients of NASH are mostly asymptomatic whereas patients of alcoholic hepatitis are always symptomatic [34]. NASH due to drugs nucleoside analogs antimitotic agents or tetracyclines can present intensely with rapid onset of fulminant hepatic failure. Hepatomegaly (75 %) and splenomegaly (25 %) are the most usual signs in NASH patients. Presence of ascites and spider angiomata, indicates improvement of cirrhosis [71]. Pharmacological interventions for NAFLD/NASH patients are aimed to treat underlying MS components (e.g., obesity, diabetes, hypertension, and dyslipidemia) as well as liver dysfunction itself. Many experimental studies have investigated several drugs and supplements, but researchers have not yet recognized a completely safe and effective treatment that can be recommended for NASH management [7, 166].

1.6 Acute Fatty Liver of Pregnancy (AFLP)

AFLP was first explained in 1940 by Sheehan. Is a rare but serious condition with a frequency of approximately 1 per 10,000 deliveries, and it typically occurs during the third trimester between the 30th and 38th week of gestation. It may be connected with "jaundice" (skin and whites of the eyes take on a yellowish color due to an increase of bilirubin in the blood) and liver failure. Left untreated, there is a high risk of death for both the mother and baby [12, 67]. The etiology of this disorder has not yet fully elucidated; however, abnormal fetal mitochondrial β-oxidation of fatty acids has been reported to be involved as a foundation of this condition in the mother [68]. In particular, a fetal defect of long-chain 3-hydroxyl coenzyme A dehydrogenase (LCHAD) due to genetic mutation has been reported to contribute the disease [127]. The diagnosis of AFLP can be challenging because the initial clinical appearance may be nonspecific but general fatigue, vomiting, headache, hypoglycemia, and lactic acidosis can be detected. Although important aminotransferases is estimated, the severity of liver dysfunction is not always replicated by the degree of elevation. Alkaline phosphatase is frequently elevated. Other results such as leukocytosis, thrombocytopenia, disseminated intravascular coagulopathy (DIC), abnormal prothrombin time, partial thromboplastin time, and normal fibrinogen can happen [14, 17, 61]. Ketonuria and proteinuria can exist. Elevated blood urea nitrogen and creatinine indicate renal deficiency. Low serum albumin and hypoglycemia can happen. Uric acid and ammonia levels can be augmented. Hyperuricemia can be an early pointer and develop before hyperbilirubinemia [63, 86]. These symptoms are based on the occurrence of microvesicular fat deposition in organs. In women with AFLP, fat content can range from 13 to 19 %. Liver biopsy is superfluous for the diagnosis and should be evaded in cases with bleeding tendencies; however, in some circumstances, it is helpful if it is the early stage of the disease or the symptoms and laboratory data show mild irregularities. The histology includes microvesicular steatosis, principally in the third zone of the liver and cytoplasmic ballooning [115]. Early analysis and prompt delivery are indispensable in AFLP. Severe therapeutic support is necessary for both maternal and fetal existence and plasmapheresis and liver transplantation should be contemplated in some severe cases. Although AFLP does not have a tendency to recur in subsequent pregnancies in most cases, since the recurrence rate is higher in cases with genetic mutation in LCHAD, close follow up is necessary for the groups [68, 145].

2 Autophagy and Steatosis

As previously described the most accredited hypotheses for the development of steatosis concern a decreased mitochondrial fatty acid beta-oxidation an increased endogenous fatty acid synthesis and an enhanced delivery of fatty acids to the liver and a deficient incorporation or export of triglycerides as VLDL.

Autophagy is a cytoprotective pathway for clearance of damaged proapoptotic cellular components following multiple forms of stress as well as mitochondrial damage and excessive enlargement of lipid droplets (LDs). Recent studies have demonstrated that pharmacological upregulation of autophagy decreases hepatotoxicity and steatosis in a model of fatty liver disease [159]. Moreover it has clearly been demonstrated that the selective autophagic clearance of damaged mitochondria (mitophagy) and excessive LDs (lipophagy) in hepatocytes of chronic ethanol-treated rats may be a prosurvival mechanism for prevention of hepatocytes apoptosis and progression of hepatic steatosis [43]. Understanding of these mechanisms can bring accelerative new therapeutic modalities to recover the disease outcome. While enhancing autophagy via general autophagy-inducing agents, such as rapamycin, has led to the decrease of cellular injury in ethanol intoxication and in other scenarios but this approach may have limitations. Specific improvement of selective autophagy relevant to particular settings may be the decisive choice for a better control of individual disease processes [37, 38]. However, upcoming studies are needed before autophagy activation can be used as a global treatment against this disease, because, for example, upregulation of the autophagic process in hepatic stellate cells has been shown to favour their activation and consequently initiate liver fibrosis [159].

2.1 *Mitophagy*

Mitophagy is a form of discriminatory autophagy specific for degradation of injured mitochondria in the lysosome [91]. Alcohol metabolism occurs mainly in the liver and alcohol is metabolized by both oxidative and non-oxidative pathways. Oxidative pathways are the predominant mechanism for alcohol metabolism. The most usual pathway for oxidative metabolism in the liver is catalysed by alcohol dehydrogenase (ADH), which metabolizes alcohol into acetaldehyde. Alcohol can also be oxidized into acetaldehyde by Cytochrome P450 2E1 (CYP2E1) and catalase [20]. Acetaldehyde is a very reactive metabolite that forms adducts with other macromolecules [178]. Proteins adducted by acetaldehyde have different function, which occasionally results in loss of activity, moreover metabolism of alcohol by CYP2E1 results in production of ROS, both indications of liver injury [21, 144, 152]. Mitophagy protects against alcohol-induced liver injury and steatosis by selectively removing injured mitochondria, because it serves to maintain a healthy population of mitochondria, which prevents cell death by reducing oxidative stress and preserving respiratory chain function as well as mitochondrial bioenergetics for efficient energy production as well as prevent lipid accumulation in the liver by maintaining a healthy population of mitochondria capable of performing β-oxidation [170]. Consequently, targeting removal of damaged mitochondria may be a successful therapeutic option for blocking progression of liver disease.

2.2 *Lipophagy*

Lipophagy was initially described in hepatocytes, which become a major site of excessive lipid growth in obesity and the metabolic syndrome [108, 139]. LDs are encompassed of a core containing primarily triacylglycerol (TAG) and sterol esters surrounded by a phospholipid (PL) monolayer. Lipophagy is a discriminatory form of autophagy, but the mechanism by which LDs are recognized as substrate and how the relative amounts of lipids targeted for degradation are regulated by nutritional status is unknown. Studies to date have begun to reveal a number of disparate functions for lipophagy in cellular physiology and pathophysiology. The most noticeable function for lipophagy is as a regulator of cellular lipid content [98]. In this context, removing stored lipid droplets is most pertinent for protecting host cells from injury, whereas degrading long-lived proteins would be significant for other scenarios, such as nutrient deprivation [37, 38]. Lipophagy is ubiquitous as it functions in other cells that do not store lipids in as large quantities as hepatocytes including fibroblasts, neurons and stellate cells [77, 149]. One of the most common interrogations remains as to how autophagy targets LDs and how lipophagy is selectively regulated in answer to environmental stimuli. Possible candidates to mediate LD targeting are the soluble NSF attachment protein receptors (SNAREs) [10]. Long implicated in LD fusion, SNAREs have recently been designated to mediate autophagosome biogenesis [119, 122]. Another possibility is LC3, which is a protein critical for autophagosome membrane formation [150]. The finding that LC3 associates with LDs in the apparent absence of an autophagosomal membrane suggests that this protein may function in LD recognition [117]. In conclusion the existence of lipophagy suggests that compromised autophagy may be a fundamental mechanism of disorders of lipid metabolism such as obesity and the metabolic syndrome. Individual variation in autophagic role may determine the development or outcome from these human diseases. Lipophagy, as well as mitophagy, has the potential to serve as a significant therapeutic target for the management of these disorders [98].

3 Liver Cancer

Liver cancer is the third leading cause of cancer mortality worldwide, with an annual death toll of approximately 700,000. In contrast to the decreasing mortality rates, liver cancer incidence and overall mortality have significantly increased in the United States over the past 20 years [44]. The liver can be affected by mature primary liver cancer, which arises in the liver (HCC and cholangiocarcinoma), or metastatic cancer to the liver from a distant primary site [73].

The most common type of adult primary liver cancer is HCC, which is usually discovered late in the disease course and generally has poor prognosis, accounting for approximately 75 % of all primary liver cancers [73]. HCC is a cancer formed by liver cells, hepatocytes; another type of cancer formed in the hepatocytes is the hepatoblastoma, specifically formed by immature liver cell [90].

The leading cause of liver cancer is cirrhosis due to hepatitis B, hepatitis C, or alcohol. In 2013, 300,000 deaths from liver cancer were due to hepatitis B, 343,000 to hepatitis C and 92,000 to alcohol ([51] Mortality and Causes of Death Collaborators 2015). Less common causes include hereditary hemochromatosis, autoimmune hepatitis, alpha1-antitrypsin deficiency, and Wilson's disease. There is different evidence showing that the virus of hepatitis B and C can directly elicit oncogenic effects or contribute to enhanced risk of hepatocellular transformation in cooperation with the hyperproliferative response induced by chronic inflammation. Therefore, an inflammatory and proliferative tissue microenvironment could represent an important target of hepatocarcinogenesis [35].

Epidemiologically, around 748,300 new cases of HCC were diagnosed in 2008, and 695,900 patients died of the disease [49]. HCC occurrence starts to rise from the age of 40 years and reaches a peak at around 70 years. In men, it is the fifth most frequently diagnosed cancer worldwide, while in women represents a low percentage positioning HCC as the seventh most commonly diagnosed cancer [73]. The incidence of HCC largely varies according to the geographic area, although it is usually higher in developing countries than in developed countries, such as the United States, as a result of hepatitis C virus (HCV) infection and NASH [2]. East and Southeast Asia, as well as middle and West Africa, have the highest rates of HCC, whereas the rates are low in south-central and western Asia and northern and eastern Europe. The global temporal trend of HCC incidence is not consistent. While the age-adjusted incidence rate of HCC has been decreasing in a few cities in China and Japan, where the main cause of HCC is HBV and HCV, respectively [154, 155], national registries show that the overall incidence of HCC increased, even in United States, between 1992 and 2008 [3].

Traditionally, the curative treatment of liver cancer has involved either surgical resection, liver transplantation, or local ablation, whereas transarterial tumor embolization has been used for palliative treatment. Drug treatment for the more advanced stages of liver cancer has been attempted in the form of conventional chemotherapy in numerous usually uncontrolled clinical trials over the past 50 years [128].

3.1 Molecular Mechanism of Hepatocarcinogenesis

The pathophysiology of HCC is not clearly understood. Hepatocarcinogenesis is a complex process associated with accumulation of genetic and epigenetic changes that occur during initiation, promotion, and progression of the disease. Cellular events are often accompanied by increasing of several factors that influence the survival of cancerous cells by suppressing apoptosis and regulating cell cycle [4].

3.1.1 Wnt/β-Catenin Pathway

The Wnt signaling pathway is a highly conserved pathway involved in homeostasis, cell proliferation, differentiation, motility, and apoptosis [123]. It was shown to be deregulated in a number of cancers, including HCC [57]. In most cases, either the inactivation of the tumor suppressor gene adenomatous polyposis coli or mutation of the proto-oncogene β-catenin and the activation of Wnt signaling was observed. This pathway is involved in HCC arising from HBV/HCV infections and alcoholic liver cirrhosis [157]. Therefore, targeted inactivation of Wnt pathway is a potential therapy for cancer.

3.1.2 p53 Pathway

In about half of all human tumors, the tumor suppressor TP53 gene is inactivated by a single point mutation [52]. In the remaining cancers, p53 is expressed at normal levels but the p53 signaling that leads to cell cycle arrest and subsequent apoptosis is defective [58]. In general, cellular levels of p53 are low; however, in response to intracellular and extracellular stress signals, p53 expression is up-regulated [64].

Several studies have reported that p53 mutations and inactivation play a critical role in liver cancer. For example, in a clinical study of 16 Chinese patients with HCC, 8 had a point mutation at the third base position of codon 249. Moreover, the $G \rightarrow T$ transversion in seven HCC DNA samples and the $G \rightarrow C$ transversion in the other HCC were consistent with mutations caused by AFB1 in mutagenesis experiments, and no mutations were found in exons 5, 6, or 8 or in the remainder of exon 7 [85]. In a case-control study, serum hepatitis B surface antigen and liver AFB1-DNA adducts were found to be significantly elevated in HCC samples compared with controls [66]. Thus, detection of mutant p53 in plasma serves as a potential biomarker for AFB1 exposure and presence of HCC.

3.1.3 pRb Pathway

The tumor suppressor retinoblastoma protein pRb1 is a major cellular barrier to cancer development [42]. It controls cell cycle progression via repression of the E2F transcription factor family of proteins. The activity of cyclin-dependent kinases (CDKs) correlates with the onset of pRb phosphorylation and G1/S cell cycle transition [106]. Up to 16 possible CDK phosphorylation sites exist on pRb, and multiple CDKs can phosphorylate pRb with some site specificity [42].

Several studies have demonstrated that the pRb pathway is harshly disrupted in HCC patients. When pRb expression was examined in 25 patients with HBV-induced HCC using histochemical staining, it was found that pRb expression was altered in eight patients. Another study examined the expression of pRb, cyclin D1, and p16 in 47 HCC specimens and found that 38 of them had been inactivated in either pRb or p16 expression, whereas cyclin D1 was overexpressed in only five samples [132]. This disruption in the pRb pathway in HCC was similar to that observed in various cancers, demonstrating that pRb is a critical player in carcinogenesis.

3.1.4 Mitogen-Activated Protein Kinase Pathway (MAPK)

The intracellular MAPK family has five MAPK subgroups. MAPKs were implicated in diverse cellular processes such as cell survival, differentiation, adhesion, and proliferation [171].

Proteins of HBV, HCV, and hepatitis E virus modulate MAPK signaling by targeting multiple phases of the signaling pathway [179]. In human HCC, the expression levels of Spred protein (Sprouty-related protein with Ena/vasodilator-stimulated phosphoprotein homology-1 domain), an inhibitor of the Ras/Raf-1/ERK pathway, are deregulated [140]. Forced expression of Spred caused inhibition of ERK activation *in vivo* and *in vitro*, resulting in reduced proliferation of cancer cells 2 and 9. This finding suggests that Spred and the correlation between MAPK-ERK could serve as a therapeutic target for human HCC.

3.1.5 Ras Pathway

Human ras proteins H-Ras, N-Ras, K-ras4A, and K-Ras4B are small GTP-binding proteins that function as molecular switches to influence cell growth, differentiation and apoptosis [25]. Activation of Ras and expression of Ras pathway proteins such as p21 were also reported in solid tumors [16] as well as in cell lines [96]. However, in a recent study, it was reported that RASSF1A and NORE1A, members of the RASSF family of Ras inhibitors, are inactivated in human HCC, demonstrating the role for Ras pathway in liver cancer [96]

3.1.6 Stress Response Signaling

Heat shock proteins (HSPs) are important players in cellular stress response. Under stress conditions, they undergo phosphorylation and/or dephosphorylation. A study conducted with 48 clinical specimens, HCC progression was found to be associated with the decrease in serine phosphorylation of HSP27 [104]. In another study, several members of the HSP family were found to be associated with the presence of HCC [177].

3.1.7 Vascular Endothelial Growth Factor and Transforming Growth Factor-β Pathways

Vascular endothelial growth factor (VEGF) and fibroblast growth factor (TGF) can play an important role in liver cancer development [13]. It was reported recently that inflammation is associated with cancer and cytokines are involved in promoting cancer development and progression, especially during infection with HV [141]. In particular, Th2 cytokines are induced and Th1 cytokines decreased in metastases.

Therefore, the modulation of cytokines and the use inflammatory cytokines inhibitors might be critical in alleviating HCC progression; in fact, a recent study showed that the use of inhibitors of VEGF and TGF-β prevented the development of HCC in rat liver [142].

3.1.8 Crucial Role of Autophagy in Liver Cancer

Autophagy refers to a process in which cellular organelles and macromolecules are degraded for recycling of bioenergetics components [116]. The autophagic process includes a series of steps, including initiation, elongation and expansion of the phagophore assembly site (PAS), phagophore, formation and maturation of double-membrane vesicle termed autophagosome, and autophagosomes subsequently fuse with lysosomes to form autolysosomes for degradation [174]. This process is regulated by Atgs through different signaling pathways [47].

With the identification of Atgs, especially the design of various mouse models with Atg deletion liver specific, has been revealed the importance of autophagy in liver physiology and pathology, e.g., clearing misfolded proteins, nutrient and energy-metabolism in hepatocytes lipid and alcohol metabolism, regulating selective organelle degradation, and HV infection [36, 168].

Basically, in the mammalian cells the Atg proteins form several important functional groups in control of autophagosome formation: the ULK1 complex, consisting of the serine/threonine kinase ULK1, Atg13, focal adhesion kinase family interacting protein of 200 kDa (FIP200) and Atg101, controls the induction or initiation of autophagy for the formation of phagophore and is negatively regulated by mechanistic target of rapamycin (mTOR). The Beclin 1-class III PI3K complex controls the nucleation step of autophagosome formation. Subsequently, the two ubiquitin-like conjugation systems (the Atg12–Atg5 system and the LC3 system) mediate the elongation stage, leading to the formation of a complete autophagosome [117, 174].

One key function of autophagy is known to clear intracellular protein aggregates and works together with the ubiquitin–proteasome system to continue intracellular protein homeostasis [80, 109]. Deficiency of Atg7 in mouse liver causes marked accumulation of polyubiquitinated proteins and deformed mitochondria, as well as an increased number of peroxisomes and lipid droplets in hepatocytes [83]. Then, autophagy plays a significant role in protein homeostasis in normal liver [81]. Moreover, the homozygous z mutation of alpha-1-antitrypsin (ATZ) can result in protein misfolding and causes pulmonary emphysema, chronic liver inflammation and HCC [130].

It is well know that liver starvation can cause the largest proportion of protein loss. For example, wild type mice lose 25–40 % of their liver protein in the first 48 h of fasting [120]. Interestingly, the protein levels in the liver tissue of the liver-specific Atg7-deficient mice failed to exhibit similar protein loss, suggesting that autophagy plays a critical role in protein degradation in the liver. As a result, there was a transient increase of amino acid levels in the liver tissue and blood after 24 h of fasting in the wild type mice, but not in Atg7-deficient mice [47].

As above, several studies have indicated the crucial role of autophagy in liver diseases; the deregulation of autophagy is often associated with HV, NAFLD, alcoholic liver disease, fibrosis, cirrhosis, and HCC [87, 136]. Newly, many investigators have proposed that tumor cells rely on autophagy for survival in HCC, although it is still controversial whether autophagy serves as an anti-cancer or pro-cancer mechanism [28, 100].

First, autophagy acts as a tumor-suppression mechanism inhibiting inflammation, to prevent tumor cell necrosis, clearance of the scaffold protein p62/SQSTM1 and promotion of genomic stability. As a result, animals with deletion of Atgs are at high risk of developing tumors [95, 169]. Second, autophagy may act as a pro-survival mechanism to protect cancer cells from various forms of cellular stress. Therefore, inhibition of autophagy may sensitize cancer cells to chemotherapeutic agents in cancer treatment [172]. Recent studies are struggling to reveal the complex paradoxical role of autophagy in cancer development as well as in cancer therapy [90].

4 Autophagy as an Anti-cancer Target

4.1 Beclin 1

The first relation between autophagy and cancer development was established with the finding that Beclin 1 inhibits tumorigenesis [99]. It is found that Beclin 1 is usually monoallelically deleted in many human cancers such as prostate and ovarian cancers [160]. Additional studies showed that homozygous Beclin 1 knockout mice are embryonic lethal while Beclin $1^{+/-}$ mutant mice develop high incidence of spontaneous tumors including HCC [160]. Other studies showed that Beclin $1^{+/-}$ mutant increases the frequency of spontaneous malignancies and accelerates the development of HBV-induced premalignant injury, together with the increasing of cellular proliferation and reduced autophagy *in vivo* [165]. This means that autophagy can exert a mechanistic purpose in tumor suppression.

4.2 Atg5 and Atg7

A mouse model with long-term systemic mosaic deletion of Atg5 demonstrated development of benign liver adenomas, thus clearly suggesting a tumor-suppressive function of autophagy. Temporarily, swollen mitochondria, oxidative stress and genomic damage responses were also detected in the hepatic tumor cells. The similar phenotype, such as enlarged mitochondria and a large number of peroxisomes, had also been observed in liver-specific $Atg7^{-/-}$ mice developing liver tumors [27].

4.3 p62

Mammalian sequestosome 1 (p62/SQSTM1) is a multifunctional ubiquitin-binding scaffolding protein that serves multiple cellular functions in bone metabolism, obesity, caspase activation, inclusion body formation, and tumorigenesis [76, 88]. A recent study showing that p62 is a selective autophagic substrate, which shows the significant role of p62 in the process of autophagy [88]. The low level of p62 accumulation, caused by autophagy degradation, could benefit the liver [82]. Furthermore, human HCC is associated with p62 accumulation in MDBs, and heterozygous mutation of Beclin1 displays p62 accumulation with liver tumorigenesis suggesting that autophagy defects may play an important role in HCC pathogenesis [82, 112]. Tumor size in mice with liver-specific knockouts of Atg7 is clearly reduced in combination with p62 knockout [153]. These results suggested that p62 accumulation caused by autophagy deficiency contributes to tumor progression. Importantly, sustained p62 expression caused by autophagy defects contributed to liver tumorigenesis is via NF-κB regulation and gene expression [112]. In addition, aggregates positive for p62 and KEAP1 are often detected in human cancers including HCC, and induction of NRF2 target genes has also been observed in most of these tumors. In mice, liver-specific Atg7 knockout develop hepatocellular adenoma with excess p62 accumulation followed by NRF2 activation [69]. However, it is intriguing that high expression of Nrf2 promotes tumor cell growth; one option is that activation of Nrf2 could stimulate cell-cycle progression and modulate the G1-S transition by regulation of p21 and pRb phosphorylation [69]. Together, understanding the cellular functions of p62–KEAP1–NRF2 pathways provide a new understanding into the development of liver cancer and support the anti-cancer function of autophagy in HCC.

4.3.1 Autophagy Induction for Liver Cancer Prevention

Cancer cells can commandeer autophagy to limit inflammation, tissue damage and genome instability, which can promote cancer initiation, suggesting that stimulation of autophagy, may be helpful for cancer prevention [153]. In mice, p62 knockout partly rescue the tumor development [112, 153], proposing that p62 may be a valuable target for cancer prevention and treatment. In addition, the activated hepatic autophagy level in Alpha-1 Antitrypsin Deficiency (A-1ATD) patients can suppress liver inflammation and carcinogenesis [131]. Furthermore, the increased autophagy level induced by carbamazepine treatment reduces the Alpha-1 Antitrypsin load and alleviates the associated hepatic fibrosis and hepatic hydroxyproline concentration as well as the cancer risk in mice [59]. A recent in vitro model [62] demonstrate that Sedanolide suppressed J5 cell viability by inducing autophagy, by regulating PI3K, p53 and NF-κB autophagy-associated signaling pathways. This suggests that enhancement of autophagy may be an effective advance for prevention of liver disease including cancer.

5 Autophagy as a Pro-cancer Mechanism

There is important evidences demonstrated that autophagy constitutes an important pro-survival mechanism in response to cellular stress [169]. For example, hypoxia can enhance the chemoresistance of HCC cells *in vitro* via induction of autophagy [151]. In addition, in hypoxic tumor regions, autophagosomes are most prominent and Beclin 1 deletion results in tumor cell death specifically in these hypoxic regions [32]. Supplementary, inhibition of autophagy which repairs hepatoma cell sensitivity to chemotherapy, suggests that autophagy plays a pro-survival role in chemotherapeutic agent-induced cell death [151]. HCC cells treated with AS-6 (a synthetic derivative of ascochlorin, -O-carboxymethylascochlorin) induce ER stress and activate autophagic response characterized by increased expression of Beclin 1, Atg5, and LC3-II as well as autophagosome formation [75].

5.1 High Basal Autophagy Is Required for Liver Cancer Development

The basal autophagy in normal cells is induced by stress in the process of tumorigenesis. In many cancer cells, such as pancreatic cancer, the elevated basal autophagy is required for continued cell growth [175]. Similarly, Ras-driven tumorigenesis risen in elevated basal autophagy level; In their study, autophagy is found to be required to maintain the pool of functional mitochondria required to support growth of Ras-driven tumors, based on the observations that deletion of Atg5 or Atg7 can suppress Ras-driven tumorigenesis *in vivo* [54]. Moreover, these observations may not be limited to Ras. The role of autophagy in tumorigenesis is also supported by a recent study using FIP200 (an Atg17 homolog of yeast) knockout mice in which mammary cancer growth was impaired by inactivation of FIP200 [167]. Notably, deletion of Atg5 or Atg7 in the mice only causes hepatoma formation without progression to HCC [153], which suggests that autophagy could be required to support the growth of aggressive cancers.

6 Autophagy Like Liver Cancer Therapy

As autophagy acts a dual role in the initiation and development of liver cancer, many researches have evaluated its mechanisms and applications to HCC treatment. Increasing evidence supports the fact that autophagy also contributes to tumor cell responses to therapies and changing environmental stimuli.

7 Autophagy Inducer

7.1 *Rapamycin and Its Analogues*

One central cascade in autophagy as well as in liver cancer, including HCC, is the PI3K/Akt/mTOR pathway that regulates cell growth, angiogenesis, proliferation, and apoptosis [138]. This pathway is activated in 15–41 % of HCCs, and mTOR inhibitors showed anti-tumor activity in HCC [148]. Rapamycin and its derivatives such as everolimus (RAD001, 40-O-(2-hydroxyethyl)) showed anti-tumor activity in preclinical studies of HCC [148]. Rapamycin (sirolimus), an mTOR kinase inhibitor which had been widely used as an autophagy inducer showed anti-proliferative and anti-angiogenesis activities [65]. Temsirolimus (CCI-779) a derivative of Rapamycin also showed therapeutic effects on Recurrent Glioblastoma Multiforme. Radiographic improvement was observed in 36 % of Temsirolimus-treated patients, and was associated with significantly longer time to tumor progression (TTP) (Median 2.3 months) [126]. So, Rapamycin and its analogues have been used as cancer therapeutic agents and induction of autophagy is believed to be part of the underlying mechanisms for its therapeutic effects [18]

7.2 *Tyrosine Kinase Inhibitors*

Tyrosine kinases play significant roles in tumor progression and the inhibitors of tyrosine kinases have been developed for cancer therapy [5]. Sorafenib, a multi-tyrosine-kinase inhibitor in combination with a HDAC inhibitor SAHA showed improvement of cancer cell death by induction of autophagy in liver cancer and pancreatic one [110]. However, doxorubicin-induced autophagic cell death was suppressed by sorafenib which can facilitate cell cycle progression, increased survival, and reduced autophagy in HCC cells. Anyway, the possible antagonistic effects in the combination of sorafenib and DOX which enhances anti-cancer efficacy need further consideration [107].

7.3 *Others*

NPC-16, a novel naphthalimide-polyamine conjugate can stimulate autophagy and apoptosis in liver cancer cells and demonstrate that mTOR signal pathway was involved in NPC-16-mediated autophagy [173]. Also berberine, a quaternary ammonium salt from the protoberberine group of isoquinolinealkaloids, is derived from Coptidisrhizoma, can induce autophagic cell death in HCC cells through inhibition of the mTOR-signaling pathway by suppressing the activity of Akt and up-regulating P38MAPK signaling [164]. Furthermore, MLN4924, a potent and selective small molecule NEDD8-activating enzyme (NAE) inhibitor, can suppress the outgrowth of liver cancer cells *in vitro* and *in vivo* by induction of autophagy which was attributed to the inhibition of mTOR activity due to Deptor (a mTOR binding protein) accumulation [105].

A recent study showed that cannabinoids (Δ9-THC) and its agonist (JWH-015) reduced the growth of HCC subcutaneous xenografts via autophagy induction by inhibition of the Akt–mTORC1 axis and AMPK stimulation [161]. This result also demonstrated that Δ9-THC and the agonist JWH-015 promote HCC death via autophagy stimulation [162, 163]. As a novel anti-tumor agent, fangchinoline can induce autophagy, and the transcriptional activity of p53 is required for the initiation of autophagy. Senstrin2 but not DRAM is involved in fangchinoline-induced autophagy in hepatoma cells suggesting that fangchinoline induces autophagic cell death via p53/sestrin2/AMPK signaling in human HCC cells [162, 163].

8 Autophagy Inhibitors

8.1 Chloroquine (CQ) and Hydroxychloroquine (HCQ)

CQ and HCQ, which belong to the 4-aminoquinoline class, are originally used in the treatment of malaria and in autoimmune disorders [74]. These agents have been commonly used as autophagy inhibitors via suppression of the lysosomal catalytic function through neutralization of the lysosomal pH [133]. There are many reports showing the sensitizing effect of CQ or HCQ on cell death induced by various cancer therapeutic agents both in vivo and in vitro. For example inhibition of autophagy by CQ further enhanced the oxaliplatin-induced apoptotic cell death and the sensitivity to chemotherapy in HCC cell, suggesting that autophagy may play an important role in releasing the oxaliplatin resistance in liver cancer cells [41]. Similarly, Ding et al. demonstrated that suppression of autophagy using CQ enhanced cell death is induced by oxaliplatin in Huh7 and SMMC-7721 cell lines Hence the combination of oxaliplatin with CQ can lead to more pronounced tumor suppression in liver cancer xenografts [37, 38].

Moreover, it has been show that the combination of CQ with sorafenib produced more tumor suppression in HCC both in cell line and mice [146]. Further, co-administration of sorafenib and CQ significantly suppressed tumor growth compared to sorafenib alone in liver xenograft tumors in mice. At present, there are 25 ongoing clinical trials using CQ/HCQ alone or in combination with other drugs in cancer via targeting autophagy. However, so far no clinical trials targeting liver cancer and autophagy have been recruited and further work is needed to utilize CQ/HCQ as a therapeutic strategy in treatment of liver cancer [28].

8.2 siRNA

siRNA is a natural process through which a targeted gene is silenced with high specificity and selectivity [135]. The role of siRNA in autophagy inhibition by knockdown of specific Atg genes as well as in cancer treatment has been described.

For instance, Chen et al. showed that autophagy level could be inhibited by knocking down of Beclin 1, which results in an enhanced cell death in HCC cell line [23]. In addition, siRNA silencing of Atg5 and Beclin 1 partially blocked the autophagy that improved cell death response upon MLN4924 treatment in HepG2 cells [105]. Moreover, inhibition of autophagy by siRNA silencing of Beclin 1 also enhanced melatonin-induced cell death in H22 cell; these data suggest that melatonin may activate a protective autophagic reaction to protect H22 cells from death, and inhibition of autophagy by siRNA may enhance the anti-tumor effect of melatonin [97]. Taken together, autophagy inhibition by siRNA of specific Atg genes enhanced the chemotherapy agent-induced cell death, although it remains to be cleared whether autophagy inhibition using the siRNA approach truly affects the efficacy of treatment of liver cancer.

8.3 MicroRNAs

MicroRNAs (miRNAs) are small, non-coding endogenous RNAs ~ 22 nucleotides (nt) in length that may play the critical role for regulation of cell death, such as apoptosis and autophagy by targeting multiple genes and pathways [50]. Moreover, accumulated evidence also demonstrates that administration of miRNAs (miR-26a) results in inhibition of cancer cell proliferation, induction of tumor-specific apoptosis in a mouse model of HCC [84]. Furthermore, many miRNAs, such as, miR-101, miR-30a, miR-34a, miR-204, and miR-375, identified as the inhibitors of autophagy are down regulated or lost in cancer [28]. Therefore, inhibiting autophagy by miRNAs has been an emerging focus via limiting therapeutic resistance for liver cancer therapy. miR-375, for instance, effectively inhibited hypoxia-induced autophagy by direct targeting of ATG7 in HCC cells [22], suggesting the possibility of utilizing miR-375 for autophagy modulation in HCC therapy.

9 Hepatic Ischemia-Reperfusion Injury and Autophagy

Ischemia-reperfusion (I/R) injury is an important cause of liver damage during surgical procedures such as hepatic resection and liver transplantation. I/R injury is a biphasic phenomenon whereby cellular damage due to hypoxia and lack of biomechanical stimulus is accentuated upon restoration of oxygen delivery and shear stress [129]. I/R injury is a complex phenomenon involving intracellular injury processes and also an injurious inflammatory response. The signaling events contributing to local hepatocellular damage are diverse and complex and involve the interaction between hepatocytes, sinusoidal endothelial cells, Kupffer cells (KC), as well as infiltrating neutrophils, macrophages and platelets [118]. During the ischemic period, the lack of energetic substrate interferes with active transmembrane transport, producing edema in KC and endothelial cells (EC). Several relevant factors and mediators such as nitric oxide (NO) are involved in the ischemic injury of the liver the loss of the delicate equilibrium between NO and endothelin induces vasoconstriction and narrowing of the sinusoidal lumen, compromising leukocyte flow and inducing accumulation, adherence, and extravasation of leukocytes in both hepatic sinusoids and postsinusoidal venules. The increase in contact between leukocytes and EC promotes leukotaxis, and the trapped leukocytes interfere with the flow of blood through the sinusoidal capillaries [53]. Also, platelets play an important role in hepatic I/R injury, synthesizing and release several factors that intervene in liver transplant and hepatic regeneration; in fact, platelets activate neutrophils for ROS generation, further contributing to the amplification of the neutrophil response [93]. On reperfusion of the ischemic liver, the collapse of the microcirculation maintains areas of ischemic liver parenchyma and, in addition to the microcirculatory failure, the activation of KC and neutrophils leads to the synthesis of inflammatory cytokines, further aggravating the severity of the ischemic injury. Concomitantly, KC suffer from a profound activation process that is promoted by neighbor hepatic cells-released damage-associated molecular patterns and KC activation significantly increase their release of ROS and pro-inflammatory cytokines, including tumour necrosis factor-α (TNF-α), interleukin-1 (IL-1), interferon-γ (INF-γ) and interleukin-12 (IL-12) [92]. When the liver is subjected to an ischemic insult, the alterations induced by oxidative stress can exceed the compensatory capacity of the liver, producing cell death. The exact mechanism of cell death in hepatic I/R injury remains unclear; apoptosis of hepatic cells is one of the possible pathway [158], but, on the other hand, other groups oppose the view that the majority of cells undergo apoptosis in response to either warm or cold I/R injury, believing that necrosis is the main form of cell death [111]. As a consequence of the incipient amount of uncovered molecular mechanisms responsible for hepatic I/R injury, a variety of new therapeutic strategies have been developed [70, 72]. An important new pathway is autophagy, because was demonstrated the induction of autophagy is associated with attenuation of I/R injury [48]. Autophagy is involved in various physiological processes, such as liver diseases, but also I/R injury [136]. Autophagy is an intracellular self-digesting pathway responsible for removal of long-lived proteins, damaged organelles, and malformed proteins during biosynthesis by lysosomes. Autophagy is found in normal and diseased liver. Although depending on the type of ischemia, warm and/or cold, the dynamic process of liver I/R results mainly in adenosine triphosphate depletion and in production of ROS, leads to both, a local ischemic insult and an acute inflammatory-mediated reperfusion injury, and results finally in cell death. This process can induce liver dysfunction and can increase patient morbidity and mortality after liver surgery and hemorrhagic shock [29, 30]. In liver ischemia reperfusion injury, autophagy mainly has a prosurvival activity allowing the cell for coping with nutrient starvation and anoxia. As the first known role of autophagy is its action during nutrient starvation, studies on autophagy and liver diseases have rapidly focused on liver ischemia/reperfusion. The physiological role of autophagy in nutrient and energy metabolism in hepatocytes has been reviewed elsewhere [176]; it is now clear that macroautophagy in the liver is important for the balance of energy and nutrients for basic cell functions, the removal of misfolded proteins resulting from genetic mutations or pathophysiological stimulations, and the turnover of major subcellular organelles such as mitochondria, endoplasmic reticulum, and peroxisomes under both normal and pathophysiological conditions. Also, stimulation of autophagy through nutrient depletion and overexpression of Beclin-1 inhibits mitochondrial permeability transition (MPT)-dependent hepatocyte necrosis and apoptosis and enhances ATP recovery after reoxygenation [79]. So, increasing autophagy might ameliorate liver damage and restore mitochondrial function after I/R [162, 163]. The nuclear protein high mobility group box 1 (HMGB1) is an important inflammatory mediator involved in the pathogenesis of liver ischemia/reperfusion (I/R) injury. Strategies aimed at preventing its release from stressed or damaged cells may be beneficial in preventing inflammation after I/R. The mechanism of action of new drugs appears to involve its ability to sequester HMGB1 inside the nucleus of redox-stressed hepatocytes and to modulate liver I/R-induced autophagy [19]. Many studies showed that PPARγ activation is associated with autophagy in the liver, so stimulation of PPARγ in adult mice resulted in increased autophagy; this suggest that hepatoprotective effects of PPARγ may be related to induction of autophagy [147]. Despite very similar protocols, results of studies performed in mice, assessing the impact of autophagy on liver I/R injury, are highly controversial. The mechanism of hepatic I/R has not been clarified due to its complexity, but increasing evidence shows that the production of ROS and inflammatory cytokines are key factors in inducing liver damage; it was demonstrated that there is a relationship between attenuation of hepatic ischemia reperfusion-induced apoptosis and autophagy via the ROS/MAPK pathway in mice and the reduction of inflammatory cytokines [94]. In other studies, increase of hepatocellular autophagy results as a stress stimulus to hepatocytes. Moreover, the stress response of hepatocytes may be involved in their degeneration process [101]. Whether autophagy protects from or promotes liver injury following warm and/or cold I-R remains to be elucidated.

Conflict of Interest The authors report no conflict of interest.

References

1. Adams LA, Lymp JF, Sauver JS, Sanderson SO, Lindor KD, Feldstein A, Angulo P (2005) The natural history of nonalcoholic fatty liver disease: a population-based cohort study. Gastroenterology 129:113–121
2. Ahmed F, Perz JF, Kwong S, Jamison PM, Friedman C, Bell BP (2008) National trends and disparities in the incidence of hepatocellular carcinoma, 1998-2003. Prev Chronic Dis 5(3):A74
3. Altekruse SF, McGlynn KA, Dickie LA, Kleiner DE (2012) Hepatocellular carcinoma confirmation, treatment, and survival in surveillance, epidemiology, and end results registries, 1992-2008. Hepatology 55:476–482
4. Aravalli RN, Steer CJ, Cressman EN (2008) Molecular mechanisms of hepatocellular carcinoma. Hepatology 48:2047–2063

5. Arora A, Scholar EM (2005) Role of tyrosine kinase inhibitors in cancer therapy. J Pharmacol Exp Ther 315:971–979
6. Bacon BR, Farahvash MJ, Janney CG, Neuschwander-Tetri BA (1994) Nonalcoholic steatohepatitis: an expanded clinical entity. Gastroenterology 107:1103–1109
7. Baran B, Akyuz F (2014) Non-alcoholic fatty liver disease: what has changed in the treatment since the beginning? World J Gastroenterol 20:14219–14229
8. Baraona E, Lieber CS (1979) Effects of ethanol on lipid metabolism. J Lipid Res 20: 289–315
9. Bell H, Tallaksen CM, Try K, Haug E (1994) Carbohydrate-deficient transferrin and other markers of high alcohol consumption: a study of 502 patients admitted consecutively to a medical department. Alcohol Clin Exp Res 18:1103–1108
10. Bostrom P, Andersson L, Rutberg M, Perman J, Lidberg U, Johansson BR, Fernandez-Rodriguez J, Ericson J, Nilsson T, Boren J, Olofsson SO (2007) SNARE proteins mediate fusion between cytosolic lipid droplets and are implicated in insulin sensitivity. Nat Cell Biol 9:1286–1293
11. Bravo AA, Sheth SG, Chopra S (2001) Liver biopsy. N Engl J Med 344:495–500
12. Browning MF, Levy HL, Wilkins-Haug LE, Larson C, Shih VE (2006) Fetal fatty acid oxidation defects and maternal liver disease in pregnancy. Obstet Gynecol 107:115–120
13. Budhu A, Forgues M, Ye QH, Jia HL, He P, Zanetti KA, Kammula US, Chen Y, Qin LX, Tang ZY, Wang XW (2006) Prediction of venous metastases, recurrence, and prognosis in hepatocellular carcinoma based on a unique immune response signature of the liver microenvironment. Cancer Cell 10:99–111
14. Burroughs AK, Seong NH, Dojcinov DM, Scheuer PJ, Sherlock SV (1982) Idiopathic acute fatty liver of pregnancy in 12 patients. Q J Med 51:481–497
15. Byrne CD, Targher G (2015) NAFLD: a multisystem disease. J Hepatol 62:S47–S64
16. Calvisi DF, Ladu S, Gorden A, Farina M, Conner EA, Lee JS, Factor VM, Thorgeirsson SS (2006) Ubiquitous activation of Ras and Jak/Stat pathways in human HCC. Gastroenterology 130:1117–1128
17. Cano RI, Delman MR, Pitchumoni CS, Lev R, Rosenthal WS (1975) Acute fatty liver of pregnancy. Complication by disseminated intravascular coagulation. JAMA 231:159–161
18. Cao C, Subhawong T, Albert JM, Kim KW, Geng L, Sekhar KR, Gi YJ, Lu B (2006) Inhibition of mammalian target of rapamycin or apoptotic pathway induces autophagy and radiosensitizes PTEN null prostate cancer cells. Cancer Res 66:10040–10047
19. Cardinal J, Pan P, Dhupar R, Ross M, Nakao A, Lotze M, Billiar T, Geller D, Tsung A (2009) Cisplatin prevents high mobility group box 1 release and is protective in a murine model of hepatic ischemia/reperfusion injury. Hepatology 50:565–574
20. Cederbaum AI (2012) Alcohol metabolism. Clin Liver Dis 16:667–685
21. Cederbaum AI, Lu Y, Wu D (2009) Role of oxidative stress in alcohol-induced liver injury. Arch Toxicol 83:519–548
22. Chang Y, Yan W, He X, Zhang L, Li C, Huang H, Nace G, Geller DA, Lin J, Tsung A (2012) miR-375 inhibits autophagy and reduces viability of hepatocellular carcinoma cells under hypoxic conditions. Gastroenterology 143:177–187
23. Chen LH, Loong CC, Su TL, Lee YJ, Chu PM, Tsai ML, Tsai PH, Tu PH, Chi CW, Lee HC, Chiou SH (2011) Autophagy inhibition enhances apoptosis triggered by BO-1051, an N-mustard derivative, and involves the ATM signaling pathway. Biochem Pharmacol 81:594–605
24. Chitturi S, Abeygunasekera S, Farrell GC, Holmes-Walker J, Hui JM, Fung C, Karim R, Lin R, Samarasinghe D, Liddle C, Weltman M, George J (2002) NASH and insulin resistance: Insulin hypersecretion and specific association with the insulin resistance syndrome. Hepatology 35:373–379
25. Clark GJ, Quilliam LA, Hisaka MM, Der CJ (1993) Differential antagonism of Ras biological activity by catalytic and Src homology domains of Ras GTPase activation protein. Proc Natl Acad Sci U S A 90:4887–4891

26. Clark SJ, Shojaee-Moradie F, Croos P, Seed PT, Umpleby AM, Wendon JA, Miell J (2001) Temporal changes in insulin sensitivity following the development of acute liver failure secondary to acetaminophen. Hepatology 34:109–115
27. Cotler SJ, Hay N, Xie H, Chen ML, Xu PZ, Layden TJ, Guzman G (2008) Immunohistochemical expression of components of the Akt-mTORC1 pathway is associated with hepatocellular carcinoma in patients with chronic liver disease. Dig Dis Sci 53:844–849
28. Cui J, Gong Z, Shen HM (2013) The role of autophagy in liver cancer: molecular mechanisms and potential therapeutic targets. Biochim Biophys Acta 1836:15–26
29. Cursio R, Colosetti P, Codogno P, Cuervo AM, Shen HM (2015) The role of autophagy in liver diseases: mechanisms and potential therapeutic targets. Biomed Res Int 2015:480508. doi:10.1155/2015/480508
30. Cursio R, Colosetti P, Gugenheim J (2015) Autophagy and liver ischemia-reperfusion injury. Biomed Res Int 2015:417590. doi:10.1155/2015/417590
31. Das K, Kar P (2005) Non-alcoholic steatohepatitis. J Assoc Physicians India 53:195–199
32. Degenhardt K, Mathew R, Beaudoin B et al (2006) Autophagy promotes tumor cell survival and restricts necrosis, inflammation, and tumorigenesis. Cancer Cell 10:51–64
33. Diehl AM (1997) Alcoholic liver disease: natural history. Liver Transpl Surg 3:206–211
34. Diehl AM (1999) Nonalcoholic steatohepatitis. Semin Liver Dis 19:221–229
35. Ding J, Wang H (2014) Multiple interactive factors in hepatocarcinogenesis. Cancer Lett 346:17–23
36. Ding WX (2010) Role of autophagy in liver physiology and pathophysiology. World J Biol Chem 1:3–12
37. Ding WX, Li M, Yin XM (2011) Selective taste of ethanol-induced autophagy for mitochondria and lipid droplets. Autophagy 7:248–249
38. Ding ZB, Hui B, Shi YH, Zhou J, Peng YF et al (2011) Autophagy activation in hepatocellular carcinoma contributes to the tolerance of oxaliplatin via reactive oxygen species modulation. Clin Cancer Res 17:6229–6238
39. Dixon JB, Bhathal PS, O'Brien PE (2001) Nonalcoholic fatty liver disease: predictors of nonalcoholic steatohepatitis and liver fibrosis in the severely obese. Gastroenterology 121:91–100
40. Dowman JK, Tomlinson JW, Newsome PN (2011) Systematic review: the diagnosis and staging of non-alcoholic fatty liver disease and non-alcoholic steatohepatitis. Aliment Pharmacol Ther 33:525–540
41. Du H, Yang W, Chen L, Shi M, Seewoo V, Wang J, Lin A, Liu Z, Qiu W (2012) Role of autophagy in resistance to oxaliplatin in hepatocellular carcinoma cells. Oncol Rep 27:143–150
42. Dynlacht BD, Flores O, Lees JA, Harlow E (1994) Differential regulation of E2F transactivation by cyclin/cdk2 complexes. Genes Dev 8:1772–1786
43. Eid N, Ito Y, Otsuki Y (2015) Ethanol-induced hepatic autophagy: friend or foe? World J Hepatol 7:1154–1156
44. El-Serag HB (2011) Hepatocellular carcinoma. N Engl J Med 365:1118–1127
45. Eslamparast T, Eghtesad S, Poustchi H, Hekmatdoost A (2015) Recent advances in dietary supplementation, in treating non-alcoholic fatty liver disease. World J Hepatol 7:204–212
46. European Association for the Study of Liver (2012) EASL clinical practical guidelines: management of alcoholic liver disease. J Hepatol 57:399–420
47. Ezaki J, Matsumoto N, Takeda-Ezaki M, Komatsu M et al (2011) Liver autophagy contributes to the maintenance of blood glucose and amino acid levels. Autophagy 7:727–736
48. Fang H, Liu A, Dahmen U, Dirsch O (2013) Dual role of chloroquine in liver ischemia reperfusion injury: reduction of liver damage in early phase, but aggravation in late phase. Cell Death Dis 4, e694. doi:10.1038/cddis.2013.225
49. Ferlay J, Shin HR, Bray F, Forman D, Mathers C, Parkin DM (2010) Estimates of worldwide burden of cancer in 2008: GLOBOCAN 2008. Int J Cancer 127:2893–2917
50. Fu LL, Wen X, Bao JK, Liu B (2012) MicroRNA-modulated autophagic signaling networks in cancer. Int J Biochem Cell Biol 44:733–736

51. GBD (2013) Mortality and Causes of Death Collaborators (2015) Global, regional, and national age-sex specific all-cause and cause-specific mortality for 240 causes of death, 1990-2013: a systematic analysis for the Global Burden of Disease Study 2013. Lancet 385: 117–171
52. Giaccia AJ, Kastan MB (1998) The complexity of p53 modulation: emerging patterns from divergent signals. Genes Dev 12:2973–2983
53. Gracia-Sancho J, Villarreal G Jr, Zhang Y, Yu JX, Liu Y, Tullius SG, Garcia-Cardena G (2010) Flow cessation triggers endothelial dysfunction during organ cold storage conditions: strategies for pharmacologic intervention. Transplantation 90:142–149
54. Guo JY, Chen HY, Mathew R, Fan J, Strohecker AM, Karsli-Uzunbas G et al (2011) Activated Ras requires autophagy to maintain oxidative metabolism and tumorigenesis. Genes Dev 25:460–470
55. Gupta V, Mah XJ, Garcia MC, Antonypillai C, van der Poorten D (2015) Oily fish, coffee and walnuts: dietary treatment for nonalcoholic fatty liver disease. World J Gastroenterol 21: 10621–10635
56. Guy CD, Suzuki A, Zdanowicz M, Abdelmalek MF, Burchette J, Unalp A, Diehl AM, Nash CRN (2012) Hedgehog pathway activation parallels histologic severity of injury and fibrosis in human nonalcoholic fatty liver disease. Hepatology 55:1711–1721
57. Herbst A, Kolligs FT (2007) Wnt signaling as a therapeutic target for cancer. Methods Mol Biol 361:63–91
58. Hickman ES, Moroni MC, Helin K (2002) The role of p53 and pRB in apoptosis and cancer. Curr Opin Genet Dev 12:60–66
59. Hidvegi T, Ewing M, Hale P et al (2010) An autophagy-enhancing drug promotes degradation of mutant alpha1-antitrypsin Z and reduces hepatic fibrosis. Science 329:229–232
60. Hock B, Schwarz M, Domke I, Grunert VP et al (2005) Validity of carbohydrate-deficient transferrin (%CDT), gamma-glutamyltransferase (gamma-GT) and mean corpuscular erythrocyte volume (MCV) as biomarkers for chronic alcohol abuse: a study in patients with alcohol dependence and liver disorders of non-alcoholic and alcoholic origin. Addiction 100:1477–1486
61. Holzbach RT (1974) Acute fatty liver of pregnancy with disseminated intravascular coagulation. Obstet Gynecol 43:740–744
62. Hsieh SL, Chen CT, Wang JJ, Kuo YH, Li CC, Hsieh LC, Wu CC (2015) Sedanolide induces autophagy through the PI3K, p53 and NF-kappaB signaling pathways in human liver cancer cells. Int J Oncol 47:2240–2246
63. Hsiung R, Hasselmann M, Lutun P, Gordji M (1988) Acute fatty liver of pregnancy. Diagnostic value of hyperuricemia in the pre-jaundice stage. J Gynecol Obstet Biol Reprod (Paris) 17:901–905
64. Hsu IC, Metcalf RA, Sun T, Welsh JA, Wang NJ, Harris CC (1991) Mutational hotspot in the p53 gene in human hepatocellular carcinomas. Nature 350:427–428
65. Huang S, Houghton PJ (2002) Inhibitors of mammalian target of rapamycin as novel antitumor agents: from bench to clinic. Curr Opin Investig Drugs 3:295–304
66. Hussain SP, Raja K, Amstad PA, Sawyer M et al (2000) Increased p53 mutation load in non-tumorous human liver of Wilson disease and hemochromatosis: oxyradical overload diseases. Proc Natl Acad Sci U S A 97:12770–12775
67. Ibdah JA (2006) Acute fatty liver of pregnancy: an update on pathogenesis and clinical implications. World J Gastroenterol 12:7397–7404
68. Ibdah JA, Bennett MJ, Rinaldo P, Zhao Y, Gibson B, Sims HF, Strauss AW (1999) A fetal fatty-acid oxidation disorder as a cause of liver disease in pregnant women. N Engl J Med 340:1723–1731
69. Inami Y, Waguri S, Sakamoto A, Kouno T, Nakada K, Hino O et al (2011) Persistent activation of Nrf2 through p62 in hepatocellular carcinoma cells. J Cell Biol 193:275–284
70. Iniguez M, Dotor J, Feijoo E, Goni S, Prieto J, Berasain C, Avila MA (2008) Novel pharmacologic strategies to protect the liver from ischemia-reperfusion injury. Recent Pat Cardiovasc Drug Discov 3:9–18

71. Itoh S, Yougel T, Kawagoe K (1987) Comparison between nonalcoholic steatohepatitis and alcoholic hepatitis. Am J Gastroenterol 82:650–654
72. Jaeschke H, Woolbright BL (2012) Current strategies to minimize hepatic ischemia-reperfusion injury by targeting reactive oxygen species. Transplant Rev (Orlando) 26:103–114
73. Jemal A, Bray F, Center MM, Ferlay J, Ward E, Forman D (2011) Global cancer statistics. CA Cancer J Clin 61:69–90
74. Kametaka S, Matsuura A, Wada Y, Ohsumi Y (1996) Structural and functional analyses of APG5, a gene involved in autophagy in yeast. Gene 178:139–143
75. Kang JH, Chang YC, Maurizi MR (2012) 4-O-carboxymethyl ascochlorin causes ER stress and induced autophagy in human hepatocellular carcinoma cells. J Biol Chem 287:15661–15671
76. Katoh H, Ojima H, Kokubu A, Saito S, Kondo T, Kosuge T, Hosoda F, Imoto I, Inazawa J, Hirohashi S, Shibata T (2007) Genetically distinct and clinically relevant classification of hepatocellular carcinoma: putative therapeutic targets. Gastroenterology 133:1475–1486
77. Kaushik S, Rodriguez-Navarro JA, Arias E, Kiffin R, Sahu S, Schwartz GJ, Cuervo AM, Singh R (2011) Autophagy in hypothalamic AgRP neurons regulates food intake and energy balance. Cell Metab 14:173–183
78. Kim CH, Younossi ZM (2008) Nonalcoholic fatty liver disease: a manifestation of the metabolic syndrome. Cleve Clin J Med 75:721–728
79. Kim JS, Nitta T, Mohuczy D, O'Malley KA, Moldawer LL, Dunn WA Jr, Behrns KE (2008) Impaired autophagy: a mechanism of mitochondrial dysfunction in anoxic rat hepatocytes. Hepatology 47:1725–1736
80. Knaevelsrud H, Simonsen A (2010) Fighting disease by selective autophagy of aggregate-prone proteins. FEBS Lett 584:2635–2645
81. Komatsu M (2012) Liver autophagy: physiology and pathology. J Biochem 152:5–15
82. Komatsu M, Waguri S, Koike M, Sou YS et al (2007) Homeostatic levels of p62 control cytoplasmic inclusion body formation in autophagy-deficient mice. Cell 131:1149–1163
83. Komatsu M, Waguri S, Ueno T, Iwata J, Murata S, Tanida I, Ezaki J, Mizushima N et al (2005) Impairment of starvation-induced and constitutive autophagy in Atg7-deficient mice. J Cell Biol 169:425–434
84. Kota J, Chivukula RR, O'Donnell KA, Wentzel EA, Montgomery CL et al (2009) Therapeutic microRNA delivery suppresses tumorigenesis in a murine liver cancer model. Cell 137:1005–1017
85. Kruse KB, Dear A, Kaltenbrun ER, Crum BE, George PM, Brennan SO, McCracken AA (2006) Mutant fibrinogen cleared from the endoplasmic reticulum via endoplasmic reticulum-associated protein degradation and autophagy: an explanation for liver disease. Am J Pathol 168:1299–1308
86. Kueh YK, Wang TL, Yeo TC, Guan R (1991) A non-fatal case report of acute fatty liver of pregnancy. Ann Acad Med Singapore 20:789–791
87. Kwanten WJ, Martinet W, Michielsen PP, Francque SM (2014) Role of autophagy in the pathophysiology of nonalcoholic fatty liver disease: a controversial issue. World J Gastroenterol 20:7325–7338
88. Lau A, Villeneuve NF, Sun Z, Wong PK, Zhang DD (2008) Dual roles of Nrf2 in cancer. Pharmacol Res 58:262–270
89. Lee RG (1989) Nonalcoholic steatohepatitis: a study of 49 patients. Hum Pathol 20:594–598
90. Lee YJ, Jang BK (2015) The role of autophagy in hepatocellular carcinoma. Int J Mol Sci 16:26629–26643
91. Lemasters JJ (2014) Variants of mitochondrial autophagy: types 1 and 2 mitophagy and micromitophagy (type 3). Redox Biol 2:749–754
92. Lentsch AB, Kato A, Yoshidome H, McMasters KM, Edwards MJ (2000) Inflammatory mechanisms and therapeutic strategies for warm hepatic ischemia/reperfusion injury. Hepatology 32:169–173
93. Lesurtel M, Graf R, Aleil B, Walther DJ, Tian Y, Jochum W, Gachet C, Bader M, Clavien PA (2006) Platelet-derived serotonin mediates liver regeneration. Science 312:104–107

94. Li AM, Li JZ, Chen ZQ, Wu YH, Wu LD, Lu GJ, Wang CH, Zhang G (2015) Growth and spectral properties of Yb3+/Ho3+ co-doped NaGd(MoO4)(2) crystal. Mater Express 5:527–533
95. Liang C, Jung JU (2010) Autophagy genes as tumor suppressors. Curr Opin Cell Biol 22: 226–233
96. Liao Y, Tang ZY, Liu KD, Ye SL, Huang Z (1997) Apoptosis of human BEL-7402 hepatocellular carcinoma cells released by antisense H-ras DNA – in vitro and in vivo studies. J Cancer Res Clin Oncol 123:25–33
97. Liu C, Jia Z, Zhang X, Hou J, Wang L, Hao S, Ruan X, Yu Z, Zheng Y (2012) Involvement of melatonin in autophagy-mediated mouse hepatoma H22 cell survival. Int Immunopharmacol 12:394–401
98. Liu K, Czaja MJ (2013) Regulation of lipid stores and metabolism by lipophagy. Cell Death Differ 20:3–11
99. Llovet JM, Chen Y, Wurmbach E et al (2006) A molecular signature to discriminate dysplastic nodules from early hepatocellular carcinoma in HCV cirrhosis. Gastroenterology 131:1758–1767
100. Lozy F, Karantza V (2012) Autophagy and cancer cell metabolism. Semin Cell Dev Biol 23:395–401
101. Lu Z, Dono K, Gotoh K et al (2005) Participation of autophagy in the degeneration process of rat hepatocytes after transplantation following prolonged cold preservation. Arch Histol Cytol 68:71–80
102. Luca A, Garcia-Pagan JC, Bosch J, Feu F, Caballeria J, Groszmann RJ, Rodes J (1997) Effects of ethanol consumption on hepatic hemodynamics in patients with alcoholic cirrhosis. Gastroenterology 112:1284–1289
103. Ludwig J, Viggiano TR, McGill DB, Oh BJ (1980) Nonalcoholic steatohepatitis: Mayo Clinic experiences with a hitherto unnamed disease. Mayo Clin Proc 55:434–438
104. Luk JM, Lam CT, Siu AF, Lam BY, Ng IO, Hu MY, Che CM, Fan ST (2006) Proteomic profiling of hepatocellular carcinoma in Chinese cohort reveals heat-shock proteins (Hsp27, Hsp70, GRP78) up-regulation and their associated prognostic values. Proteomics 6:1049–1057
105. Luo Z, Yu G, Lee HW et al (2012) The Nedd8-activating enzyme inhibitor MLN4924 induces autophagy and apoptosis to suppress liver cancer cell growth. Cancer Res 72:3360–3371
106. Malumbres M, Barbacid M (2001) To cycle or not to cycle: a critical decision in cancer. Nat Rev Cancer 1:222–231
107. Manov I, Pollak Y, Broneshter R, Iancu TC (2011) Inhibition of doxorubicin-induced autophagy in hepatocellular carcinoma Hep3B cells by sorafenib – the role of extracellular signal-regulated kinase counteraction. FEBS J 278:3494–3507
108. Marchesini G, Brizi M, Bianchi G et al (2001) Nonalcoholic fatty liver disease: a feature of the metabolic syndrome. Diabetes 50:1844–1850
109. Marfany G, Farras R, Salido E, Xirodimas DP, Rodriguez MS (2008) Much to know about proteolysis: intricate proteolytic machineries compromise essential cellular functions. Biochem Soc Trans 36:781–785
110. Martin AP, Park MA, Mitchell C, Walker T, Rahmani M, Thorburn A, Haussinger D, Reinehr R, Grant S, Dent P (2009) BCL-2 family inhibitors enhance histone deacetylase inhibitor and sorafenib lethality via autophagy and overcome blockade of the extrinsic pathway to facilitate killing. Mol Pharmacol 76:327–341
111. Massip-Salcedo M, Rosello-Catafau J, Prieto J, Avila MA, Peralta C (2007) The response of the hepatocyte to ischemia. Liver Int 27:6–16
112. Mathew R, Karp CM, Beaudoin B et al (2009) Autophagy suppresses tumorigenesis through elimination of p62. Cell 137:1062–1075
113. Mathurin P, Bataller R (2015) Trends in the management and burden of alcoholic liver disease. J Hepatol 62:S38–S46
114. McCullough AJ (2006) Pathophysiology of nonalcoholic steatohepatitis. J Clin Gastroenterol 40:S17–S29

115. Minakami H, Takahashi T, Tamada T (1991) Should routine liver biopsy be done for the definite diagnosis of acute fatty liver of pregnancy? Am J Obstet Gynecol 164:1690–1691
116. Mizushima N (2007) Autophagy: process and function. Genes Dev 21:2861–2873
117. Mizushima N, Yoshimori T, Ohsumi Y (2011) The role of Atg proteins in autophagosome formation. Annu Rev Cell Dev Biol 27:107–132
118. Montalvo-Jave EE, Escalante-Tattersfield T, Ortega-Salgado JA, Pina E, Geller DA (2008) Factors in the pathophysiology of the liver ischemia-reperfusion injury. J Surg Res 147: 153–159
119. Moreau K, Ravikumar B, Renna M, Puri C, Rubinsztein DC (2011) Autophagosome precursor maturation requires homotypic fusion. Cell 146:303–317
120. Mortimore GE, Poso AR (1987) Intracellular protein catabolism and its control during nutrient deprivation and supply. Annu Rev Nutr 7:539–564
121. N'Kontchou G, Paries J, Htar MT, Ganne-Carrie N, Costentin L, Grando-Lemaire V, Trinchet JC, Beaugrand M (2006) Risk factors for hepatocellular carcinoma in patients with alcoholic or viral C cirrhosis. Clin Gastroenterol Hepatol 4:1062–1068
122. Nair U, Jotwani A, Geng J, Gammoh N et al (2011) SNARE proteins are required for macroautophagy. Cell 146:290–302
123. Nelson WJ, Nusse R (2004) Convergence of Wnt, beta-catenin, and cadherin pathways. Science 303:1483–1487
124. Neuman MG (2003) Cytokines – central factors in alcoholic liver disease. Alcohol Res Health 27:307–316
125. O'Shea RS, Dasarathy S, McCullough AJ, Diseases Practice Guideline Committee of the American Association for the Study of Liver, Gastroenterology Practice Parameters Committee of the American College of Gastroenterology (2010) Alcoholic liver disease. Hepatology 51:307–328
126. Ohta T, Iijima K, Miyamoto M et al (2008) Loss of Keap1 function activates Nrf2 and provides advantages for lung cancer cell growth. Cancer Res 68:1303–1309
127. Ong SS, Baker PN, Mayhew TM, Dunn WR (2002) No difference in structure between omental small arteries isolated from women with preeclampsia, intrauterine growth restriction, and normal pregnancies. Am J Obstet Gynecol 187:606–610
128. Peck-Radosavljevic M (2014) Drug therapy for advanced-stage liver cancer. Liver Cancer 3:125–131
129. Peralta C, Jimenez-Castro MB, Gracia-Sancho J (2013) Hepatic ischemia and reperfusion injury: effects on the liver sinusoidal milieu. J Hepatol 59:1094–1106
130. Perlmutter DH (2006) The role of autophagy in alpha-1-antitrypsin deficiency: a specific cellular response in genetic diseases associated with aggregation-prone proteins. Autophagy 2:258–263
131. Perlmutter DH (2009) Autophagic disposal of the aggregation-prone protein that causes liver inflammation and carcinogenesis in alpha-1-antitrypsin deficiency. Cell Death Differ 16:39–45
132. Pimienta G, Pascual J (2007) Canonical and alternative MAPK signaling. Cell Cycle 6: 2628–2632
133. Poole B, Ohkuma S (1981) Effect of weak bases on the intralysosomal pH in mouse peritoneal macrophages. J Cell Biol 90:665–669
134. Poynard T, Mathurin P, Lai CL et al (2003) A comparison of fibrosis progression in chronic liver diseases. J Hepatol 38:257–265
135. Rao DD, Vorhies JS, Senzer N, Nemunaitis J (2009) siRNA vs. shRNA: similarities and differences. Adv Drug Deliv Rev 61:746–759
136. Rautou PE, Mansouri A, Lebrec D, Durand F, Valla D, Moreau R (2010) Autophagy in liver diseases. J Hepatol 53:1123–1134
137. Rector RS, Thyfault JP, Wei Y, Ibdah JA (2008) Non-alcoholic fatty liver disease and the metabolic syndrome: an update. World J Gastroenterol 14:185–192
138. Sabatini DM (2006) mTOR and cancer: insights into a complex relationship. Nat Rev Cancer 6:729–734

139. Sanyal AJ, Campbell-Sargent C, Mirshahi F et al (2001) Nonalcoholic steatohepatitis: association of insulin resistance and mitochondrial abnormalities. Gastroenterology 120: 1183–1192
140. Satoh T, Kaziro Y (1992) Ras in signal transduction. Semin Cancer Biol 3:169–177
141. Schiffer E, Housset C, Cacheux W et al (2005) Gefitinib, an EGFR inhibitor, prevents hepatocellular carcinoma development in the rat liver with cirrhosis. Hepatology 41:307–314
142. Schulze-Bergkamen H, Fleischer B, Schuchmann M, Weber A, Weinmann A, Krammer PH, Galle PR (2006) Suppression of Mcl-1 via RNA interference sensitizes human hepatocellular carcinoma cells towards apoptosis induction. BMC Cancer 6:232
143. Serfaty L, Lemoine M (2008) Definition and natural history of metabolic steatosis: clinical aspects of NAFLD, NASH and cirrhosis. Diabetes Metab 34:634–637
144. Setshedi M, Wands JR, Monte SM (2010) Acetaldehyde adducts in alcoholic liver disease. Oxid Med Cell Longev 3:178–185
145. Seyyed Majidi MR, Vafaeimanesh J (2013) Plasmapheresis in acute fatty liver of pregnancy: an effective treatment. Case Rep Obstet Gynecol 2013:615975. doi:10.1155/2013/615975
146. Shi YH, Ding ZB, Zhou J, Hui B, Shi GM, Ke AW, Wang XY, Dai Z, Peng YF, Gu CY, Qiu SJ, Fan J (2011) Targeting autophagy enhances sorafenib lethality for hepatocellular carcinoma via ER stress-related apoptosis. Autophagy 7:1159–1172
147. Shin T, Kuboki S, Huber N, Eismann T et al (2008) Activation of peroxisome proliferator-activated receptor-gamma during hepatic ischemia is age-dependent. J Surg Res 147:200–205
148. Sieghart W, Fuereder T, Schmid K et al (2007) Mammalian target of rapamycin pathway activity in hepatocellular carcinomas of patients undergoing liver transplantation. Transplantation 83:425–432
149. Singh R, Kaushik S, Wang Y, Xiang Y, Novak I, Komatsu M, Tanaka K, Cuervo AM, Czaja MJ (2009) Autophagy regulates lipid metabolism. Nature 458:1131–1135
150. Singh R, Xiang Y, Wang Y, Baikati K et al (2009) Autophagy regulates adipose mass and differentiation in mice. J Clin Invest 119:3329–3339
151. Song J, Qu Z, Guo X, Zhao Q, Zhao X, Gao L, Sun K, Shen F, Wu M, Wei L (2009) Hypoxia-induced autophagy contributes to the chemoresistance of hepatocellular carcinoma cells. Autophagy 5:1131–1144
152. Sun AY, Ingelman-Sundberg M, Neve E et al (2001) Ethanol and oxidative stress. Alcohol Clin Exp Res 25:237S–243S
153. Takamura A, Komatsu M, Hara T, Sakamoto A et al (2011) Autophagy-deficient mice develop multiple liver tumors. Genes Dev 25:795–800
154. Tanaka H, Imai Y, Hiramatsu N, Ito Y, Imanaka K, Oshita M et al (2008) Declining incidence of hepatocellular carcinoma in Osaka, Japan, from 1990 to 2003. Ann Intern Med 148:820–826
155. Tanaka M, Katayama F, Kato H, Tanaka H, Wang J, Qiao YL, Inoue M (2011) Hepatitis B and C virus infection and hepatocellular carcinoma in China: a review of epidemiology and control measures. J Epidemiol 21:401–416
156. Teli MR, Day CP, Burt AD, Bennett MK, James OF (1995) Determinants of progression to cirrhosis or fibrosis in pure alcoholic fatty liver. Lancet 346:987–990
157. Terris B, Pineau P, Bregeaud L, Valla D, Belghiti J, Tiollais P, Degott C, Dejean A (1999) Close correlation between beta-catenin gene alterations and nuclear accumulation of the protein in human hepatocellular carcinomas. Oncogene 18:6583–6588
158. Theruvath TP, Czerny C, Ramshesh VK, Zhong Z, Chavin KD, Lemasters JJ (2008) C-Jun N-terminal kinase 2 promotes graft injury via the mitochondrial permeability transition after mouse liver transplantation. Am J Transplant 8:1819–1828
159. Thoen LF, Guimaraes EL, Dolle L, Mannaerts I, Najimi M, Sokal E, van Grunsven LA (2011) A role for autophagy during hepatic stellate cell activation. J Hepatol 55:1353–1360
160. Tsai CC, Huang KW, Chen HF et al (2006) Gene expression analysis of human hepatocellular carcinoma by using full-length cDNA library. J Biomed Sci 13:241–249
161. Vara D, Salazar M, Olea-Herrero N, Guzman M, Velasco G, Diaz-Laviada I (2011) Antitumoral action of cannabinoids on hepatocellular carcinoma: role of AMPK-dependent activation of autophagy. Cell Death Differ 18:1099–1111

162. Wang JH, Ahn IS, Fischer TD et al (2011) Autophagy suppresses age-dependent ischemia and reperfusion injury in livers of mice. Gastroenterology 141:2188–2199
163. Wang N, Pan W, Zhu M, Zhang M, Hao X, Liang G, Feng Y (2011) Fangchinoline induces autophagic cell death via p53/sestrin2/AMPK signalling in human hepatocellular carcinoma cells. Br J Pharmacol 164:731–742
164. Wang N, Feng Y, Zhu M, Tsang CM, Man K, Tong Y, Tsao SW (2010) Berberine induces autophagic cell death and mitochondrial apoptosis in liver cancer cells: the cellular mechanism. J Cell Biochem 111:1426–1436
165. Wang SM, Ooi LL, Hui KM (2007) Identification and validation of a novel gene signature associated with the recurrence of human hepatocellular carcinoma. Clin Cancer Res 13:6275–6283
166. Watanabe S, Hashimoto E, Ikejima K et al (2015) Evidence-based clinical practice guidelines for nonalcoholic fatty liver disease/nonalcoholic steatohepatitis. J Gastroenterol 50:364–377
167. Wei H, Wei S, Gan B, Peng X, Zou W, Guan JL (2011) Suppression of autophagy by FIP200 deletion inhibits mammary tumorigenesis. Genes Dev 25:1510–1527
168. Weidberg H, Shvets E, Elazar Z (2011) Biogenesis and cargo selectivity of autophagosomes. Annu Rev Biochem 80:125–156
169. White E (2012) Deconvoluting the context-dependent role for autophagy in cancer. Nat Rev Cancer 12:401–410
170. Williams JA, Ding WX (2015) A mechanistic review of mitophagy and its role in protection against alcoholic liver disease. Biomolecules 5:2619–2642
171. Wong CH, Cheng CY (2005) Mitogen-activated protein kinases, adherens junction dynamics, and spermatogenesis: a review of recent data. Dev Biol 286:1–15
172. Wu WK, Coffelt SB, Cho CH, Wang XJ, Lee CW, Chan FK, Yu J, Sung JJ (2012) The autophagic paradox in cancer therapy. Oncogene 31:939–953
173. Xie SQ, Li Q, Zhang YH, Wang JH, Mei ZH, Zhao J, Wang CJ (2011) NPC-16, a novel naphthalimide-polyamine conjugate, induced apoptosis and autophagy in human hepatoma HepG2 cells and Bel-7402 cells. Apoptosis 16:27–34
174. Yang Z, Klionsky DJ (2009) An overview of the molecular mechanism of autophagy. Curr Top Microbiol Immunol 335:1–32
175. Yang S, Wang X, Contino G, Liesa M et al (2011) Pancreatic cancers require autophagy for tumor growth. Genes Dev 25:717–729
176. Yin XM, Ding WX, Gao W (2008) Autophagy in the liver. Hepatology 47:1773–1785
177. Yoshiji H, Noguchi R, Kuriyama S, Yoshii J, Ikenaka Y, Yanase K, Namisaki T, Kitade M, Yamazaki M, Uemura M, Fukui H (2005) Different cascades in the signaling pathway of two vascular endothelial growth factor (VEGF) receptors for the VEGF-mediated murine hepatocellular carcinoma development. Oncol Rep 13:853–857
178. Zakhari S (2006) Overview: how is alcohol metabolized by the body? Alcohol Res Health 29:245–254
179. Zhao LJ, Wang L, Ren H, Cao J, Li L, Ke JS, Qi ZT (2005) Hepatitis C virus E2 protein promotes human hepatoma cell proliferation through the MAPK/ERK signaling pathway via cellular receptors. Exp Cell Res 305:23–32
180. Zimmerman HJ (1955) The evolution of alcoholic cirrhosis; clinical, biochemical and histologic correlations. Med Clin North Am 12:241–259

Index

M.C. Maiuri, D. De Stefano (eds.), *Autophagy Networks in Inflammation*,
Progress in Inflammation Research, DOI 10.1007/978-3-319-30079-5

MIX
Papier aus verantwortungsvollen Quellen
Paper from responsible sources
FSC® C105338

If you have any concerns about our products,
you can contact us on
ProductSafety@springernature.com

In case Publisher is established outside the EU,
the EU authorized representative is:
Springer Nature Customer Service Center GmbH
Europaplatz 3, 69115 Heidelberg, Germany

Printed by Libri Plureos GmbH
in Hamburg, Germany